W9-BKI-146

FINANCIAL ECONOMICS, RISK AND INFORMATION

An Introduction to Methods and Models

Marcelo Bianconi
Tufts University, USA

World Scientific
New Jersey • London • Singapore • Hong Kong

Published by

World Scientific Publishing Co. Pte. Ltd.

5 Toh Tuck Link, Singapore 596224

USA office: Suite 202, 1060 Main Street, River Edge, NJ 07661

UK office: 57 Shelton Street, Covent Garden, London WC2H 9HE

British Library Cataloguing-in-Publication Data
A catalogue record for this book is available from the British Library.

FINANCIAL ECONOMICS, RISK AND INFORMATION
An Introduction to Methods and Models

ISBN 981-238-501-0
ISBN 981-238-502-9 (pbk)

Printed by FuIsland Offset Printing (S) Pte Ltd, Singapore

To Celia and Giampaolo

FOREWORD

A few years ago, I started a new course for economics quantitative undergraduate and graduate students with the intent of teaching how economists and financial economists apply statistical and stochastic methods to economic decision-making. My interest grew out of the apparent lack of a unified course that focuses on risk and information applied to economics and finance. This book is the result of teaching and doing research in those areas in the last few years.

In working on this enterprise, I have benefited from several colleagues who shared their thoughts and research with me over the years and several students who have taken the courses I teach. In particular, I thank Stephen Turnovsky for the fruitful collaborations, discussions and guidance over the last several years; Yannis Ioannides for thoughtful comments on the early stages of this project; and all my colleagues at Tufts University who keep providing a lively and productive intellectual environment. I thank Jun-Ho Bae, Paul Nmeke, Benjamin Tarlow, John Tilney and all students from Tufts University and other schools in the Boston area who made comments; and Dali Jing and Jamie Maldonado who helped in typing some parts of the manuscript at early stages. At World Scientific Publishing Co., I am most grateful for the incentive and comments of the editor Yubing Zhai and the referees she has provided; and the superb editorial execution and assistance of Juliet Lee Ley Chin.

M. Bianconi
April 2003
Tufts University
Medford, MA

CONTENTS

Introduction 1

1. Basic Mathematical Tools

1.1 Introduction 6
1.2 Integration 6
1.3 Basic Statistics 10
1.4 Basic Linear Algebra 17
1.5 Static Optimization 20
1.6 Notes on Stochastic Difference Equations 29
1.7 Dynamic Optimization in Discrete Time: Heuristics of Dynamic Programming in the Certainty Case 33
1.8 Stochastic Dynamic Optimization in Discrete Time 41
1.9 Notes on Stochastic Differential Equations 48
1.10 Stochastic Dynamic Optimization in Continuous Time 49
1.11 Summary 56
Problems 57
Notes on the Literature 61
References 64

2. Mean-Variance Approach to Financial Decision-Making

2.1 Introduction 67
2.2 Portfolio Mean Return and Variance 68
2.3 The Efficient Frontier 76
2.4 Two-Fund Theorem, the Risk-Free Asset and One-Fund Theorem 82

2.5 The Pricing of Assets in the Mean-Variance Framework and the No-Arbitrage Theorem 92
2.6 Summary 100
Problems 101
Notes on the Literature 104
References 105

3. Expected Utility Approach to Financial Decision-Making

3.1 Introduction 107
3.2 The Von-Neumann-Morgenstern (VNM) Framework: Probability Distributions over Outcomes 108
3.3 Measurement of Risk Aversion 114
3.4 The VNM Framework: State Dependent Utility 118
3.5 Portfolio Choice and Comparative Statics with VNM State Independent Expected Utility 121
3.6 The Quadratic Utility Function 126
3.7 Diversification, Risk Aversion and Non-Systematic Risk 128
3.8 Summary 130
Problems 132
Notes on the Literature 134
References 136
Appendix to Chapter 3
A3.1 Introduction to Mean-Variance Analysis with Expected Utility based on Normal Distribution of Payoffs 138
A3.2 The Payoffs 138
A3.3 The Specific Functional Form for the Utility Function 139
A3.4 The Individual Budget Constraint 143
A3.5 The Equilibrium Allocation 146
A3.6 Comparative Statics 149
A3.7 Closed Form Solutions 151
A3.8 Summary I 153
A3.9 Introduction to Non-additive Probabilities 153
A3.10 The Cost of Knightian Uncertainty with Uncertainty Aversion 157
A3.11 Risk Averse Bayesian Behavior 161
A3.12 Summary II 164
Notes on the Literature 166
References 167

4. Introduction to Systems of Financial Markets

4.1 Introduction 169
4.2 Pricing Securities in a Linear Fashion 169
4.3 Optimal Portfolio Choice Problems 171
4.4 Pricing the States of Nature 176
4.5 Market Regimes: Complete versus Incomplete 182
4.6 Optimal Portfolio Choice and the Price of Elementary Securities (State Prices) 185
4.7 Individual Optimal Allocation under Complete Markets Regime: An Example 189
4.8 General Equilibrium under Complete Markets Regime: Full Risk Sharing 193
4.9 General Equilibrium under Incomplete Markets Regime 196
4.10 A Simple Geometrical Illustration of Market Regimes 199
4.11 Application to the Neoclassical Theory of the Firm 202
4.12 Summary 206
Questions and Problems 207
Notes on the Literature 208
References 210

5. Contracts, Contract Design, and Static Agency Relationships

5.1 Introduction to Bilateral Relationships and Contracts 212
5.2 Theories of the Firm: Agency, Transactions Costs and Property Rights 212
5.3 Summary I 215
5.4 Introduction to Adverse Selection 215
5.5 The Principal-Agent Relationship 216
5.6 A Simple Example in Insurance Markets 217
5.7 Mechanism Design: A Problem of Price Discrimination 219
5.8 Adverse Selection in Credit Markets and the Possibility of Credit Rationing 234
5.9 Signaling 241
5.10 Summary II 244
5.11 Introduction to Moral Hazard 244
5.12 Finite Number of Actions and Outcomes, Principal and Agent both Risk Neutral 246

5.13 Variations with Finite Number of Actions, Principal is Risk Neutral, Agent is Risk Averse 251
5.14 Infinite Number of Actions and First Order Approach, Principal is Risk Neutral, Agent is Risk Averse 263
5.15 Summary III 269
5.16 Asset Returns and Moral Hazard 270
5.17 Basic Model 271
5.18 Equilibrium Asset Returns 275
5.19 Summary IV 278
Problems and Questions 279
Notes on the Literature 280
References 282

6. **Non-convexities and Lotteries in General Equilibrium**

6.1 Introduction 284
6.2 A Static Decentralized Competitive Framework 285
6.3 Competitive Equilibrium 286
6.4 Trade in Lotteries 289
6.5 Implications for the Elasticity of Labor Supply 292
6.6 Summary I 293
6.7 General Equilibrium Approach to Asymmetric Information 294
6.8 Basic Structure, Pareto Optimality and Decentralized Competitive Equilibrium 294
6.9 An Insurance Problem with Adverse Selection 300
6.10 Summary II 308
6.11 Unemployment Insurance, Asset Returns and Adverse Selection 308
6.12 Basic Structure 310
6.13 Heterogeneity, Efficiency, and Market Completeness 312
6.14 Consequences for Asset Allocation 319
6.15 Summary III 324
Problems 326
Notes on the Literature 327
References 330

7. **Dynamics I: Discrete Time**

7.1 Time and Markets 332
7.2 Introduction to Financial Contracts 333

7.3 Summary I 344
7.4 General Equilibrium and Asset Pricing under Uncertainty with Complete Markets 344
7.5 General Equilibrium under Uncertainty: Two Equivalent Approaches 345
7.6 Pricing Contingent Claims in the Two-Period Economy with Complete Markets 349
7.7 Introduction to the Multi-Period Economy 358
7.8 Conditional and Transitional Probabilities, Markov Processes, and Conditional Moments 362
7.9 The Multi-Period Economy Again 366
7.10 Asset Prices in an Infinite Horizon Exchange Economy 369
7.11 Excess Returns 375
7.12 Summary II 382
7.13 Stochastic Monetary Theory 382
7.14 Fisher Equation and Risk 386
7.15 Summary III 388
7.16 The Financial Problem of the Firm in General Equilibrium 388
7.17 Summary IV 398
7.18 Private Information, Stochastic Growth and Asset Prices 398
7.19 Recursive Contracts, General Equilibrium and Asset Prices 400
7.20 Growth and Asset Prices with Alternative Arrangements 407
7.21 Summary V 419
7.22 Risk Aversion, Intertemporal Substitution and Asset Returns 421
7.23 Summary VI 435
Problems 437
Notes on the Literature 439
References 445

8. **Dynamics II: Continuous Time**

8.1 Asset Price Dynamics, Options and the Black-Scholes Model 452
8.2 Discrete Time Random Walks 452
8.3 A Multiplicative Model in Discrete Time and a Preview of the Lognormal Random Variable 454
8.4 Introduction to Random Walk Models of Asset Prices in Continuous Time 457
8.5 A Multiplicative Model of Asset Prices in Continuous Time 462
8.6 Introduction to Ito's Lemma and the Lognormal Distribution Again 464
8.7 Ito's Formula: The General Case 469
8.8 Asset Price Dynamics and Risk 470
8.9 Options 471

8.10 The Black-Scholes Partial Differential Equation 473
8.11 The Black-Scholes Formula for a European Call Option 475
8.12 Summary I 479
8.13 Introduction to Equilibrium Stochastic Models 479
8.14 Consumption Growth and Portfolio Choice with Logarithmic Utility 480
8.15 Consumption Growth and Portfolio Choice with *CRRA* Utility 486
8.16 Capital Accumulation and Asset Returns 491
8.17 Risk Aversion and Intertemporal Substitution 499
8.18 Summary II 504
Problems 505
Notes on the Literature 507
References 510

Index 514

FINANCIAL ECONOMICS, RISK AND INFORMATION

An Introduction to Methods and Models

INTRODUCTION

This book grew out of my intellectual interest in two areas of the general economics discipline. One field has been generally referred to as the economics of uncertainty and information. It started with the seminal contributions of Kenneth Arrow, Gerard Debreu, and Frank Hahn among others, in the area of general equilibrium theory under uncertainty. The next generation of models under uncertainty evolved roughly into two distinct branches. One has been the economics of uncertainty and information with seminal contributions by James Mirrlees, Oliver Hart, Sanford Grossman, and Joseph Stiglitz, among others. The other has been the extension of the general equilibrium under uncertainty approach to the pricing of financial assets where the seminal contributors were Kenneth Arrow, Robert Lucas, and Robert Merton among others. This latter trend also has evolved parallel to the more systematic distinction between the disciplines of pure economics and pure finance advanced by Eugene Fama, Merton Miller, and Franco Modigliani among others. After a fruitful evolution over the last 30 years, the fields of economics of uncertainty and information and modern finance theory as a discipline per se are well established at the present time.

However, even though their potential interrelations and interactions are obvious, there has been little effort to present a concise introduction to the methods and techniques applied to the most important problems in these two areas. This book is an effort in this direction. One main objective is to call the attention of young well-tooled and talented undergraduates and graduate students to potential new avenues of research that explore the economics of uncertainty and information and modern finance theory together.

At a general level, the material focuses on the role of risk in trading environments that can be classified in two categories. At one level, trade can occur in atomistic competitive environments under conditions of symmetric information so that an individual's action is independent, taking all prices as given; for example anonymous trading relationships.

At another level, trade can occur in a bilateral environment under conditions of asymmetric information where the action of one individual can have a significant impact on the surplus of another individual; for example the trading relationship involving a specific contractual arrangement to reveal private information. We examine the role of risk in those alternative environments when the time horizon is short, or the static case, and when the time horizon is long, or the dynamic case.

In particular, this book presents an introduction to the use of stochastic methods in economics and finance. The topics range from financial risk and asset returns, portfolio choice, asset pricing, individual behavior towards risk, general equilibrium under uncertainty, indivisibilities and non-convexities in a general equilibrium context, international portfolio diversification, contract theory and mechanism design, principal-agent relationships, credit markets, and option pricing.

We start the book with a review of the basic mathematical tools needed to follow the material. Chapter 1 presents mathematical background in integral calculus, probability and statistics, linear algebra, optimization and dynamic analysis. Chapters 2 to 6 are devoted to static, or short horizon equilibrium analysis. In Chapter 2 we examine the simple and well-known mean-variance approach to financial decision-making in a general structure of full information. We present a derivation of the efficient frontier, the Two-Fund and One-Fund theorems, and the static capital asset pricing model (SCAPM) where the excess return of an asset is driven by market risk only.

Chapter 3 is dedicated to the expected utility approach to financial decision-making. We discuss behavior towards risk and portfolio choice from an expected utility perspective with alternative utility functions. We start with the traditional Von-Neumann-Morgenstern (VNM) expected utility approach and discuss its implications for behavior towards risk and portfolio choice. In an Appendix to Chapter 3, we present mean-variance analysis with expected utility and some new developments in choice under uncertainty based on alternative approaches to decision and uncertainty. The Appendix to Chapter 3 is the only part in this book where we discuss decision-making under Knight's concept of uncertainty. The rest of the book uses risk and uncertainty interchangeably to denote traditional statistical analysis with known distribution of the states of nature.

Chapter 4 presents the static general equilibrium approach to portfolio choice and asset pricing under full information. We introduce the consumption-based capital asset pricing model (CCAPM) and engage in a thorough, but simple discussion of complete versus incomplete financial market regimes with special attention to the pricing of the states of nature and risk sharing. We end the chapter with an application of the material to the basic neoclassical theory of the firm.

Chapter 5 preserves the qualification of static, short horizon equilibrium framework, but moves from the full information paradigm to the case of asymmetric information among agents in the economy, focusing on bilateral relationships such as contracts and agency. We first present an introduction to contracts and bilateral relationships. Then, we focus on the basic models that characterize principal-agent relationships and contract design in a partial equilibrium framework, the well-known adverse selection, and signaling and moral hazard problems. We then present a discussion of the effects of moral hazard on asset returns with a comparison to the case of full information.

Chapter 6 focuses on non-convexities and lotteries. We examine a general framework of asymmetric information as an extension to traditional static general equilibrium analysis under uncertainty. We first discuss problems of labor supply decisions under an exogenous indivisibility constraint. Then, we characterize and examine efficiency in a general equilibrium framework under asymmetric information, with an application to insurance under adverse selection. Finally, we use the basic framework to analyze labor supply decisions and adverse selection effects on asset returns and on international portfolio diversification under heterogeneity.

Chapters 7 and 8 advance on the time dimension extending the time horizon. We present dynamic stochastic models that include behavior of random variables over time in both full information and asymmetric information settings. Chapter 7 studies dynamic problems in discrete time. We introduce futures and forward asset markets and simple hedging problems. Then we move to dynamic stochastic general equilibrium analysis in discrete time with particular attention to issues relating to the characterization of equilibrium under uncertainty, portfolio choice and asset pricing with the consumption-based capital asset pricing model (CCAPM), excess returns, monetary theory, and the financial problem of the firm. We present an extension to asymmetric information in a dynamic framework where we focus on dynamic contract design for adverse selection problems and its effect on economic growth and asset prices. Finally, we analyze the implications of alternative preferences that separate risk aversion from intertemporal substitution.

Chapter 8 presents dynamic stochastic problems in continuous time. We start with simple models of asset prices driven by Wiener or Brownian motion processes. Then we apply the stochastic calculus techniques to financial derivatives presenting the Black-Scholes model. Finally, we introduce general equilibrium under uncertainty in continuous time focusing on stochastic economic growth, asset pricing, welfare analysis, and separation of risk aversion and intertemporal substitution.

It is obvious that this book does not cover every material in the uncertainty and information and modern finance theory fields. I have used my own areas of research interest, which include macroeconomic and international economic theory and applications from an equilibrium perspective, as a benchmark to choose the topics covered in the book. Each chapter aims to present the material and techniques as clearly and straightforward as possible. At the end of each chapter, we have Notes on the Literature which provide references and discussions of the specific material presented in the chapter. Some practice problems and questions are provided as well. Answers to problems and questions, additional problems and questions and any corrections will be posted on my web site at: *http://www.tufts.edu/~mbiancon.*

I also point out that the book does not cover computational issues. Pure computational methods are beyond the scope of this introduction to methods and models. This is intended to be a pencil-and-paper book. Students and readers can follow the material and learn the techniques and solutions with the use of pencil-and-paper only. In some chapters, I present some simple numerical examples that are meant to be illustrative of the results obtained; those can be calculated with pencil-and-paper or a simple calculator. In the Notes on the Literature, I provide some references to the computational approach.

I hope this is useful reading and guidance for your intellectual, academic and professional development.

Notes: Sections 6.11-6.15 draw on my paper entitled "Heterogeneity, Efficiency and Asset Allocation with Endogenous Labor Supply: The Static Case," which appeared in *The Manchester School,* 69 (3), pages 253-268, June 2001. Sections 7.18-7.21 draw on my paper entitled "Private Information, Growth and Asset Prices with Stochastic Disturbances," which appeared in the *International Review of Economics and Finance*, 12, pages 1-24, Spring 2003.

1. Basic Mathematical Tools

1.1 Introduction

We offer in this chapter an introduction to some basic mathematical tools used throughout this book. The selection of topics reflects the usage in the chapters ahead. We assume some basic knowledge of differential calculus and matrix algebra. We adopt some simple basic notation. Let $\mathbb{R}^1$ be the set of all real numbers on the line. Then a function is a rule that assigns a number in $\mathbb{R}^1$ to each number in $\mathbb{R}^1$. The notation $f{:}D \rightarrow \mathbb{R}^1$ pertains to a real valued function $y=f(x)$ whose domain is $D \subset \mathbb{R}^1$. The notation $(a,b) \equiv \{x \in \mathbb{R}^1 : a < x < b\}$ pertains to the open interval between a and b, and $[a,b] \equiv \{x \in \mathbb{R}^1 : a \leq x \leq b\}$ pertains to the closed interval including the endpoints a and b. All logarithms are natural logarithms on base $exp=2.71828...$, denoted log.

We start with basic integral calculus and statistics. Then we present some rudiments of linear algebra and move on to static optimization. Finally, we present dynamics in discrete and continuous time. The material is simple and self-contained. However, if the reader senses extreme difficulty in following this chapter, it is useful to supplement the mathematics background with the references provided in the Notes on the Literature.

1.2 Integration

A useful first approach to study integration is the concept of area under a curve. Consider a well defined function $f{:}\ \mathbb{R}^1 \rightarrow \mathbb{R}^1$. Suppose we find another arbitrary function F, continuous on the closed interval $[a,b]$, so that

$$dF(x)/dx = f(x) \text{ for all } x \in (a,b),$$

then F is called the anti-derivative of f. In general, continuous functions are integrable. Moreover, the area under the curve f, denoted $\mathcal{A}$, is

$$\mathcal{A} = F(b)-F(a)$$

as shown in Figure 1.1. The area under the curve is like a summation of infinitesimally "small" intervals. With this interpretation, integration can be studied in unbounded domains as well.

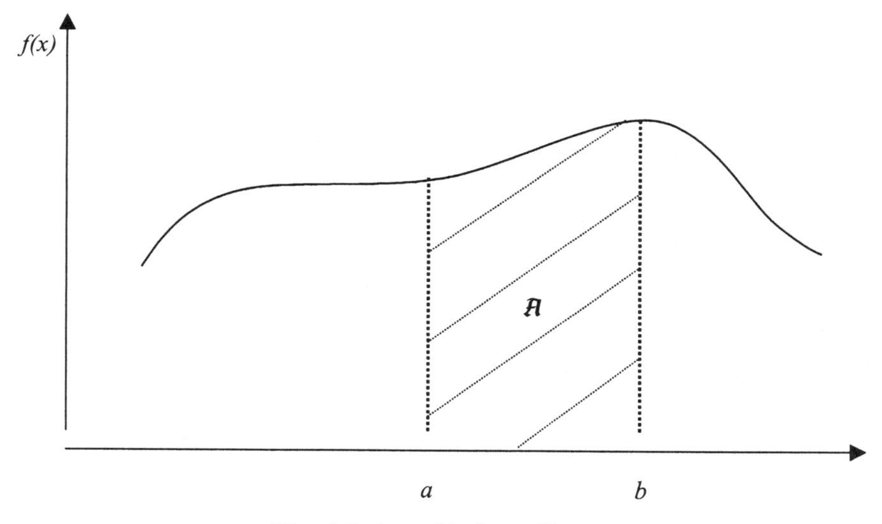

Fig. 1.1 Area Under a Curve

(a) Indefinite Integrals

Whenever $dF(x)/dx = f(x)$, we call F the indefinite integral of f

$$\int f(x)\, dx = F(x) + c \tag{1.1}$$

for c an arbitrary constant of integration.

For example, let $f(x)=x^2$, then

$$\int x^2\, dx = (1/3)\, x^3 + c$$

since

$$d[(1/3)\, x^3 + c]/dx = x^2.$$

Properties and rules of integration are:

- for a a real number,

$$\int a\, f(x)\, dx = a \int f(x)\, dx; \tag{1.2}$$

- for the function $g: \mathbb{R}^1 \rightarrow \mathbb{R}^1$,

$$\int [f(x) + g(x)]\, dx = \int f(x)\, dx + \int f(x)\, dx; \tag{1.3}$$

- for constant $a \neq -1$,

$$\int x^a\, dx = [1/(a+1)]\, x^{a+1} + c; \tag{1.4}$$

- for $a = -1,\ x > 0$,

$$\int x^{-1}\, dx = |\log x| + c; \tag{1.5}$$

- for exp the exponential operator, constant $a \neq 0$,

$$\int exp(ax)\, dx = [(1/a) exp(ax)] + c; \tag{1.6}$$

- for $a > 0$ and $a \neq 1$,

$$\int a^x\, dx = (a^x / \log a) + c; \tag{1.7}$$

An initial value problem determines the arbitrary constant of integration in the following manner. Let

$$F(x^*) = \nu \tag{1.8}$$

where ν is a real number obtained by evaluating F at x^*. Rewrite (1.1) as

$$F(x) = \int f(x)\, dx - c$$

so that using (1.8)

$$c = \int f(x^*)\, dx - \nu$$

where f is evaluated at x^*.

(b) Definite Integrals

The Fundamental Theorem of Calculus determines that for a well defined function $f: \mathbb{R}^1 \rightarrow \mathbb{R}^1$, and F continuous on the closed interval $[a,b]$, so that we have

$$dF(x)/dx = f(x) \quad for\ all\ x \in (a,b),$$

then $F(b)-F(a)$ is the definite integral over $[a,b]$, denoted

$$\int_a^b f(x)\, dx = F(b)-F(a). \tag{1.9}$$

For example, let $f(x)=x^2$, and $a=0$, $b=1$ we have

$$\int_a^b x^2\, dx = [(1/3)\, x^3|_0^{\,1}\,] = F(1)-F(0) = 1/3.$$

Properties of definite integrals are:

$$\int_a^b f(x)\, dx = -\int_b^a f(x)\, dx; \tag{1.10}$$

$$\int_a^a f(x)\, dx = 0\,; \tag{1.11}$$

- for α a real number,

$$\int_a^b \alpha f(x)\, dx = \alpha \int_a^b f(x)\, dx\,; \tag{1.12}$$

- for $a \le c \le b$ real numbers,

$$\int_a^b f(x)\, dx = \int_a^c f(x)\, dx + \int_c^b f(x)\, dx; \tag{1.13}$$

- for t a real number,

$$d[\int_a^t f(x)\, dx]/dt = dF(t)/dt = f(t); \tag{1.14a}$$

$$d[\int_t^a f(x)\, dx]/dt = -dF(t)/dt = -f(t); \tag{1.14b}$$

- for a,b continuous functions, by (1.14a,b),

$$d[\int_{a(t)}^{b(t)} f(x)\, dx]/dt = f(b(t))\,[db/dt] - f(a(t))\,[da/dt] \tag{1.15}$$

a special case of Liebniz's formula, used in Chapter 5.

In the possible case that the integral is improper, say $b\to\infty$, the improper integral

$$\int_a^\infty f(x)\, dx = lim_{b\to\infty} \int_a^b f(x)\, dx$$

is said to converge if the limit exits and is finite.

(c) Integration by Parts

The method of integration by parts applies for both indefinite and definite integrals studied above. For continuously differentiable functions f and g, the product rule of differentiation yields

$$d[f(x)g(x)]/dx = [df(x)/dx]\, g(x) + f(x)\, [dg(x)/dx].$$

Integrating both sides gives the formula for integration by parts

$$f(x)g(x) = \int [df(x)/dx]\, g(x)\, dx + \int f(x)\, [dg(x)/dx]\, dx. \tag{1.16}$$

For example, we evaluate the indefinite integral $\int x\ exp(x)\ dx$ by noting that we may write the integrand as $f(x)\ [dg(x)/dx]$ for the functions $f(x)=x$ and $g(x)=exp(x)$. Then, $df(x)/dx=1$ and $dg(x)/dx=exp(x)$. We can apply the integration by parts formula in (1.16) to obtain

$$x\ exp(x) = \int exp(x)\ dx + \int x\ exp(x)\ dx = exp(x) + c + \int x\ exp(x)\ dx$$

$$\Rightarrow \int x\ exp(x)\ dx = (x-1)exp(x) - c$$

for c the arbitrary constant of integration.

(d) Integration by Substitution

Again, this method applies for both indefinite and definite integrals studied above. For well behaved functions f and g, with $u=g(x) \Rightarrow du=[dg(x)/dx]dx$, the substitution method is of the general form

$$\int f(g(x))\ [dg(x)/dx]\ dx = \int f(u)\ (du/dx)\ dx = \int f(u)\ du. \qquad (1.17)$$

The key to apply integration by substitution is to be able to write the integrand as $f(u)(du/dx)dx$ where $u=g(x)$. For example, consider the indefinite integral $\int 2x\ (x^2+1)^3\ dx$. If we let $u=g(x)=(x^2+1)$, then $du/dx=2x \Rightarrow du= 2x\ dx$, and the original integral can be easily solved as $\int u^3\ du = (1/4)\ u^4 + c$, for c a constant of integration.

1.3 **Basic Statistics**

Suppose you flip a coin. The two possible outcomes are head or tail. Suppose you roll a die, the six possible outcomes are $\{1,2,3,4,5,6\}$. In those two experiments, the set of all possible or basic outcomes is

$$S = \{heads, tails\} \quad or \quad S= \{1,2,3,4,5,6\}$$

and we call S the <u>sample space</u> of the experiment. Define an <u>event</u> as a subset of the sample space, denoted $\mathcal{E} \subset S$. Two events $\mathcal{E}_i$ and $\mathcal{E}_j$ are said to be <u>mutually exclusive</u> if $\mathcal{E}_i \cap \mathcal{E}_j = \emptyset$, or there are no common outcomes in $\mathcal{E}_i$ and $\mathcal{E}_j$. Events $\mathcal{E}_i$ and $\mathcal{E}_j$ are said to be <u>collectively exhaustive</u> if $\mathcal{E}_i \cup \mathcal{E}_j = S$, or the elements combined give the sample space. To each event $\mathcal{E}$, $\mathcal{E}^c$ denotes the <u>complement</u> of $\mathcal{E}$, so that $\mathcal{E}$ and $\mathcal{E}^c$ are both mutually exclusive, $\mathcal{E} \cap \mathcal{E}^c=\emptyset$, and collectively exhaustive, $\mathcal{E} \cup \mathcal{E}^c=S$. Also, to each event $\mathcal{E}$, $Prob(\mathcal{E})$ denotes the probability event $\mathcal{E}$ occurs in the sample space S. Probabilities satisfy the properties

$$0 \le Prob(\mathcal{E}) \le 1 \tag{1.18a}$$

$$Prob(S) = 1. \tag{1.18b}$$

For a collection of mutually exclusive events $\mathcal{E}_1, \mathcal{E}_2, \mathcal{E}_3 \ldots \mathcal{E}_m$,

$$Prob(\mathcal{E}_1 \cup \mathcal{E}_2 \cup \mathcal{E}_3 \cup \ldots \cup \mathcal{E}_m) = Prob(\mathcal{E}_1) + Prob(\mathcal{E}_2) + \ldots + Prob(\mathcal{E}_m). \tag{1.19}$$

Also, for $\mathcal{E}^c$ the complement of $\mathcal{E}$,

$$Prob(\mathcal{E}^c) = 1 - Prob(\mathcal{E}). \tag{1.20}$$

For two events $\mathcal{E}_i$ and $\mathcal{E}_j$ in S, the probability of $\mathcal{E}_i$ given that $\mathcal{E}_j$ has occurred is called the conditional probability of $\mathcal{E}_i$ given $\mathcal{E}_j$, given by the formula

$$Prob(\mathcal{E}_i|\mathcal{E}_j) = Prob(\mathcal{E}_i \cap \mathcal{E}_j)/Prob(\mathcal{E}_j). \tag{1.21}$$

Two events $\mathcal{E}_i$ and $\mathcal{E}_j$ in S, are said to be statistically independent if

$$Prob(\mathcal{E}_i|\mathcal{E}_j) = Prob(\mathcal{E}_i) \Rightarrow Prob(\mathcal{E}_i \cap \mathcal{E}_j) = Prob(\mathcal{E}_i)Prob(\mathcal{E}_j). \tag{1.22}$$

When a sample space is composed of n equally likely outcomes, the probability of an outcome is $1/n$. For example, if you toss a fair die, the sample space is $S=\{1,2,3,4,5,6\}$, the probability of observing one of the outcomes is $1/6$. Suppose we define the event 'odd numbers' or $\mathcal{E}_o=\{1,3,5\}$, then $Prob(\mathcal{E}_o)=1/2$. Similarly, we define the event 'even numbers' or $\mathcal{E}_e=\{2,4,6\}$, then $Prob(\mathcal{E}_e)=1/2$. Note that $\mathcal{E}_o$, $\mathcal{E}_e$ are mutually exclusive and collectively exhaustive. Are $\mathcal{E}_o$, $\mathcal{E}_e$, statistically independent? Let's check formula (1.22). $Prob(\mathcal{E}_o \cap \mathcal{E}_e)=Prob(\mathcal{E}_o|\mathcal{E}_e)=0$ since mutually exclusive, but $Prob(\mathcal{E}_o)=1/2$. Hence, $Prob(\mathcal{E}_o|\mathcal{E}_e) \neq Prob(\mathcal{E}_o)$ and $\mathcal{E}_o$, $\mathcal{E}_e$, are not statistically independent. In fact, if $\mathcal{E}_e$ happens, then $\mathcal{E}_o$ cannot happen, thus they are not independent.

The two basic laws of probability are:

(i) Multiplicative Law, for $\mathcal{E}_i$, $\mathcal{E}_j$ in S,

$$Prob(\mathcal{E}_i \cap \mathcal{E}_j) = Prob(\mathcal{E}_i)Prob(\mathcal{E}_j|\mathcal{E}_i); \tag{1.23}$$

(ii) Additive Law, for $\mathcal{E}_i$, $\mathcal{E}_j$ in S,

$$Prob(\mathcal{E}_i \cup \mathcal{E}_j) = Prob(\mathcal{E}_i) + Prob(\mathcal{E}_j) - Prob(\mathcal{E}_i \cap \mathcal{E}_j). \tag{1.24}$$

A mechanism to modify probability assessments when additional information becomes available is provided by Bayes Theorem. For $\mathcal{E}_i$, $\mathcal{E}_j$ in S, Bayes theorem states

$$Prob(\mathcal{E}_j \mid \mathcal{E}_i) = Prob(\mathcal{E}_j)\ Prob(\mathcal{E}_i \mid \mathcal{E}_j) / Prob(\mathcal{E}_i). \qquad (1.25a)$$

where $Prob(\mathcal{E}_i \mid \mathcal{E}_j) / Prob(\mathcal{E}_i)$ provides an updating mechanism for the assessment of the probability of $\mathcal{E}_j$. An alternative statement of Bayes Theorem is available when there are $\mathcal{E}_1$, $\mathcal{E}_2$, $\mathcal{E}_3$, ...$\mathcal{E}_m$ mutually exclusive and collectively exhaustive events in S and some other event $\mathcal{E}$. Then the conditional probability of $\mathcal{E}_i$, $i=1,2,..,m$ given $\mathcal{E}$ can be expressed as

$$Prob(\mathcal{E}_i \mid \mathcal{E}) = Prob(\mathcal{E}_i)\ Prob(\mathcal{E} \mid \mathcal{E}_i) / \Sigma_{j=1}^{m} Prob(\mathcal{E}_j)\ Prob(\mathcal{E} \mid \mathcal{E}_j). \qquad (1.25b)$$

(a) Discrete Random Variables and Their Probability Distribution

By definition, a random variable is a real valued function defined over the sample space,

$$\{y: y=Y(s),\ s \in S\}$$

where Y is a random variable with possible basic outcome, or realization $y=Y$. For example, in the fair die example, Y is a possible number when the die is rolled with $2=y=Y$, an outcome in the sample space.

A random variable is discrete if it can take only a countable number of values. For a discrete random variable, Y on S, the probability distribution provides probabilities, $P(Y=y)$, for each and every value of $y \in S$, that is $Prob(y)=P(Y=y)$ is the probability distribution of Y. It thus satisfies:

$$Prob(y) \geq 0, \quad all\ y \in S; \qquad (1.26a)$$

$$\Sigma_{y \in S} Prob(y) = 1. \qquad (1.26b)$$

The probability of an arbitrary outcome, $y^* \in S$, is then

$$P(Y \in y^*) = \Sigma_{y \in y^*} Prob(y)$$

or the sum of probabilities of all sample points that are assigned y^*. Commonly, the sample space S is called the support of the probability distribution of Y, because for $Y \not\subset S$, $Prob(y)=P(Y=y)=0$.

For example, let the sample space be $S=\{1,2,3,4\}$, with each outcome equally likely. Then, the probability distribution of Y is

$$Prob(y)=P(Y=y)=1/4 \qquad all\ y\in S,$$

that satisfies (1.26a,b). Now, for the same sample space consider a different experiment that yields a probability distribution

$$Prob(y) = P(Y=y) = (2y-1)/16 \qquad all\ y\in S.$$

Then the probability of each outcome is

$$Prob(1) = P(Y=1) = 1/16$$

$$Prob(2) = P(Y=2) = 3/16$$

$$Prob(3) = P(Y=3) = 5/16$$

$$Prob(4) = P(Y=4) = 7/16$$

that again satisfies (1.26a,b). This last example is useful to illustrate the computation of probabilities of events, in particular

$$P(Y\leq 2) = Prob(1) + Prob(2) = 1/4$$

$$P(Y\leq 3) = Prob(1) + Prob(2) + Prob(3) = 9/16$$

$$P(Y\leq 4) = Prob(1) + Prob(2) + Prob(3) + Prob(4) = 1.$$

In general, for Y on S, this function,

$$F(y) = P(Y\leq y) = \Sigma_{Y\leq y}\, Prob(y) \qquad (1.27)$$

is called the Cumulative Distribution Function of Y with properties

$$F(y) \in [0,1] \quad all\ y\in S, \qquad (1.28a)$$

$$if\ \{y,y^*\}\in S,\ and\ y < y^*,\ then\ F(y) \leq F(y^*). \qquad (1.28b)$$

The mathematical expectation of a discrete random variable is defined as follows. Let $Prob(y)$ be a well defined probability distribution of a discrete random variable Y. Let $u(Y)$ be a real valued function of Y. The mathematical expectation or expected value of u is defined as

$$E[\,u(Y)\,] = \Sigma_{y\in S}\ u(y)\, Prob(y) \qquad (1.29)$$

i.e. the average of u. Hence, if $u(Y)=Y$, then applying (1.29) yields

$$E[Y] = \Sigma_{y\in S}\ y\,Prob(y) \equiv \mu \qquad (1.30)$$

i.e. the average or mean of the distribution of Y.

For example, Y on $S=\{-1,0,1\}$ has a well defined probability distribution $Prob(y) = 1/3$. The expected value of Y is, using (1.30)

$$\mu = E[Y] = \Sigma_{y\in S}\ y\,Prob(y) = -1\times(1/3) + 0\times(1/3) + 1\times(1/3) = 0$$

i.e. the mean of Y is zero. If we let $u(Y) = Y^2$, using (1.29), the expected value is

$$E[u(Y)] = \Sigma_{y\in S}\ y^2\,Prob(y) =$$

$$(-1)^2\times(1/3) + (0)^2\times(1/3) + (1)^2\times(1/3) = 2/3.$$

Note that since $(-1)^2=1$, the support could be as well $S'=\{0,1\}$ and similarly

$$E[u(Y)] = \Sigma_{y\in S'}\ y^2\,Prob(y) = (0)^2\times(1/3) + (1)^2\times(2/3) = 2/3.$$

In general, if we let $u(Y)= (Y - E[Y])^2$, the square of the discrepancy from the mean, then

$$E[u(Y)] = E[(Y - E[Y])^2] = \Sigma_{y\in S}\ (y-\mu)^2\,Prob(y) \equiv \sigma^2 \qquad (1.31a)$$

is defined as the variance of Y or $var(Y)= \sigma^2$. The positive square root of the variance of Y is the standard deviation of Y, or

$$\sigma = (var(Y))^{1/2}. \qquad (1.31b)$$

We note that given that expectation is a linear operator,

$$\begin{aligned}\sigma^2 = E[(Y - E[Y])^2] &= E[Y^2 - 2Y E[Y] + (E[Y])^2] \\ &= E[Y^2] - 2E[Y]E[Y] + (E[Y])^2 \\ &= E[Y^2] - 2\mu^2 + \mu^2 \\ &= E[Y^2] - \mu^2\end{aligned}$$

an alternative formula for the variance of Y.

(b) Continuous Random Variables and Their Probability Distribution

A random variable is continuous if it is not composed of a countable number of values, i.e. has uncountable infinite possible values. For a continuous random variable, Y on S, we cannot obtain the probability of a specific value and weak and strong inequalities do not matter. In the case of a discrete random variable, the probability mass in the distribution falls in particular points on the real line. In the case of continuous random variables, the probability mass falls in some interval of the real line.

Suppose we let the function f be continuously differentiable. We call $f(y)$ the "probability distribution" of a continuous random variable Y on S. Then,

$$f(y)>0, \quad all\ y \in S; \qquad (1.32a)$$

$$\int_{y \in S} f(y)\, dy = 1 \qquad (1.32b)$$

and

$$P(a<Y<b) = \int_a^b f(y)\, dy, \qquad for\ (a,b) \subset S \qquad (1.33)$$

i.e. the probability of the event $Y \subset (a,b)$ is the area under the function f. The function f satisfying (1.32a,b) is the <u>probability density function</u> (p.d.f.) of the continuous random variable Y.

For example, let $S=\{s: s \in [0,\infty)\}$, and the continuous random variable Y on S has p.d.f.

$$f(y) = (1/4)\ exp(-y/4), \quad all\ y \in S.$$

This p.d.f. satisfies (1.32a), and we check (1.32b) using the integration techniques of section **1.2**:

$$\int_{y \in S} f(y)\, dy = (1/4) \int_0^\infty exp(-y/4)\, dy = (1/4)\ [\ -4\ exp(-y/4)|_0^\infty] = 1.$$

The probability that this random variable falls in the interval $(0,1)$ is, using (1.33)

$$P(0<Y<1) = \int_0^1 f(y)\, dy = (1/4) \int_0^1 exp(-y/4)\, dy =$$

$$(1/4)\ [\ -4\ exp(-y/4)|_0^1] = -exp(-1/4) + 1 = 0.2212.$$

In general, in the support $S=\{ s: s \in (-\infty,\infty)\}$, the <u>cumulative distribution function</u> of the continuous random variable Y on S is defined as

$$F(y) = P(Y \leq y) = \int_{-\infty}^{y} f(\tau)\, d\tau \qquad (1.34)$$

where τ is the index of integration and F is continuously differentiable. Consequently, the relationship between the p.d.f. and the cumulative distribution function of a continuous random variable is (recall section 1.2)

$$dF(y)/dy = f(y), \qquad all\ y \in S, \qquad (1.35)$$

and we can rewrite expression (1.34) as

$$F(y) = P(Y \leq y) = \int_{-\infty}^{y} dF(\tau). \qquad (1.34')$$

The mathematical expectation of a continuous random variable Y on S with p.d.f. $f(y)$ and $u(Y)$ a real valued function of Y is

$$E[\, u(Y)\,] = \int_{y \in S} u(y) f(y)\, dy = \int_{y \in S} u(y)\, dF(y) \qquad (1.36)$$

i.e. the average of u. Hence, if $u(Y)=Y$, then applying (1.36) yields

$$E[\, Y\,] = \int_{y \in S} y f(y)\, dy \equiv \mu \qquad (1.37)$$

i.e. the average or mean of the distribution of Y. Similarly, the variance is

$$E[(Y - E[Y])^2] = \int_{y \in S} (y - \mu)^2 f(y)\, dy \equiv \sigma^2. \qquad (1.38)$$

A widely used continuous distribution is the Normal or Gaussian distribution. A continuous random variable is said to follow a normal distribution on the support $S=\{ s: s \in (-\infty, \infty)\}$ when it has a p.d.f.

$$f(y) = \sigma^{-1} (2\pi)^{-1/2} \exp(-(y-\mu)^2/2\sigma^2) \quad all\ y \in S.$$

where $\pi=3.141592...$ is a physical constant, and $\{\mu, \sigma\}$ are the mean and standard deviation of the normally distributed random variable, i.e. a two-parameter distribution. The normal random variable is symmetric about the mean (which is equal to the median) and bell-shaped.

(c) Topics

Let Y and X be well defined random variables, continuous or discrete, in their respective supports. The covariance between Y and X is defined as

$$cov(Y,X) = E[\, (Y - E[Y\,])\, (X - E[X])]. \qquad (1.39)$$

The covariance between two random variables is a relative measure of linear association among them, with dependence on the scale of measurement of each. Given that expectation is a linear operator,

$$\begin{aligned} cov(Y,X) &= E\,[(\, Y - E[Y])\, (X - E[X])] \\ &= E\,[\, Y X - Y E[X] - X E[Y] + E[Y]\, E[X]\,] \\ &= E\,[\, Y X\,] - 2\, E[Y]\, E[X] + E[Y]E[X] \\ &= E\,[\, Y X\,] - E[Y]\, E[X] \\ &= E\,[\, Y X\,] - \mu_Y\, \mu_X \end{aligned}$$

where μ_Y, μ_X are the means of Y and X respectively, an alternative formula for the covariance of $\{Y,X\}$, widely used in this book. Covariance and statistical independence are subtly related. The covariance between two random variables may be nonexistent, but they may or may not be statistically independent. Covariance is a measure of linear association whereas statistical independence is a more general measure of association of events. Hence, when two random variables are statistically independent, their covariance must be zero, however the reverse is not true since there may exist a nonlinear relation between the random variables not captured by the covariance.

An absolute measure of linear dependence that filters out units of account is the correlation between Y and X defined as

$$corr(Y,X) = cov(Y,X) / \sigma_Y \sigma_X \tag{1.40a}$$

satisfying

$$-1 \leq corr(Y,X) \leq 1 \tag{1.40b}$$

where $\{\sigma_Y, \sigma_X\}$ are the standard deviations of Y and X respectively.

An important result useful for the analysis of risk and uncertainty in economics and finance is Jensen's Inequality. Let a continuous random variable Y on S with p.d.f. $f(y)$ and $u(Y)$ a real valued function of Y. Jensen's Inequality states that if u is a convex function of Y, then

$$E[u(Y)] > u(E[Y]). \tag{1.41}$$

This result is particularly useful to understand comparative statics results of changes in the variance of random variables when the mean remains constant, a so-called mean-preserving spread of the distribution about the mean. For a convex function of a random variable Y, an increase in the variance of Y increases $E[u(Y)]$ by Jensen's inequality.

1.4 Basic Linear Algebra

We give a basic introduction to linear algebra and examine the issues of linear independence and spanning.

A geometric representation of the set of all real numbers is the real line, or $\mathbb{R}^1$. A geometric representation of pairs of ordered numbers is the

Cartesian plane, $\mathbb{R}^2$. A space consisting of ordered n-tuples of numbers is $\mathbb{R}^n$. The Cartesian product of two spaces, $\mathbb{R}^n \times \mathbb{R}^n$, is a collection of ordered pairs of dimension $n\times n$, or $\mathbb{R}^{n\times n}$. A vector in a well defined space is the locus of points describing the movement from one point to another point in the space. A collection, $(v_1, v_2, v_3, \ldots, v_n) \equiv \boldsymbol{\upsilon}$ denotes a vector from the origin in $\mathbb{R}^n$. The length of this vector is

$$\|\boldsymbol{\upsilon}\| = (v_1^{\,2} + v_2^2 + v_3^2 + \ldots + v_n^{\,2})^{1/2} \tag{1.42}$$

a direct application of Pithagoras theorem. Thus, for r a scalar in $\mathbb{R}^1$, we have $\| r\,\boldsymbol{\upsilon}\| = |r|\,\|\boldsymbol{\upsilon}\|$.

Let $\boldsymbol{\upsilon}$ and $\boldsymbol{u} \equiv (u_1, u_2, u_3, \ldots, u_n)$ be well defined vectors from the origin in $\mathbb{R}^n$. The dot product, or inner or scalar product of the two vectors is

$$\boldsymbol{\upsilon} \bullet \boldsymbol{u} = \Sigma_{i=1}^{\;n}\, v_i\; u_i \tag{1.43}$$

resulting in a scalar in $\mathbb{R}^1$. Note from (1.42)-(1.43) that

$$\|\boldsymbol{\upsilon}\| = (\boldsymbol{\upsilon} \bullet \boldsymbol{\upsilon})^{1/2}. \tag{1.44}$$

The triangle inequality states that any side of a triangle is less than or equal to the sum of the lengths of the two other sides, i.e.

$$\|\boldsymbol{\upsilon} + \boldsymbol{u}\| \le \|\boldsymbol{\upsilon}\| + \|\boldsymbol{u}\|. \tag{1.45}$$

(a) Linear Independence and Spanning

We define the set

$$\mathcal{L}[\boldsymbol{\upsilon}] \equiv \{ r\,\boldsymbol{\upsilon} : \boldsymbol{\upsilon} = (v_1, v_2, v_3, \ldots, v_n),\ r \in \mathbb{R}^1\} \tag{1.46}$$

as the set of all multiples of the vector $\boldsymbol{\upsilon}$ from the origin; that is the line generated or spanned by $\boldsymbol{\upsilon}$. For example, in $\mathbb{R}^2$ if $\boldsymbol{\upsilon} = (1,1)$ then $\mathcal{L}$ is the $45°$ line. Now let $\boldsymbol{\upsilon}$ and $\boldsymbol{u} \equiv (u_1, u_2, u_3, \ldots, u_n)$ be well defined vectors from the origin in $\mathbb{R}^n$, and consider the set of all linear combinations of $\boldsymbol{\upsilon}$ and $\boldsymbol{u}$ defined as

$\mathcal{L}[\boldsymbol{\upsilon}, \boldsymbol{u}] \equiv$

$$\{ r\,\boldsymbol{\upsilon} + s\,\boldsymbol{u} : \boldsymbol{\upsilon} = (v_1, v_2, v_3, \ldots, v_n),\ \boldsymbol{u} = (u_1, u_2, u_3, \ldots, u_n),\ (r, s) \in \mathbb{R}^1\} \tag{1.47}$$

i.e. the set spanned by $\boldsymbol{\upsilon}$ and $\boldsymbol{u}$.

Henceforth, $\boldsymbol{\upsilon}$ and $\boldsymbol{u}$ are said to be linear independent if and only if

$$r\,\boldsymbol{\upsilon} + s\,\boldsymbol{u} = 0 \;\Rightarrow\; r = s = 0. \tag{1.48}$$

Alternatively, $\boldsymbol{\upsilon}$ and $\boldsymbol{u}$ are said to be linear dependent if and only if at least r or s are not all zero and

$$r\,\boldsymbol{\upsilon} + s\,\boldsymbol{u} = 0. \tag{1.49}$$

It is simple to check linear independence using matrix algebra. Let $\mathbb{A}$ be a matrix consisting of n rows and two columns, that is an $n \times 2$ matrix, whose columns are the vectors $\boldsymbol{\upsilon}$ and $\boldsymbol{u}$. Then, $\boldsymbol{\upsilon}$ and $\boldsymbol{u}$ are said to be linear independent if and only if the linear system

$$\mathbb{A}\ |r\ \ s|^T = 0 \tag{1.50}$$

has a nonzero solution $\{r,s\}$, where superscript T denotes transpose. Alternatively, if $n=2$ and the matrix $\mathbb{A}$ is square, $\boldsymbol{\upsilon}$ and $\boldsymbol{u}$ are said to be linear independent if and only if

$$det\,\mathbb{A} = |\,\mathbb{A}\,| \neq 0 \tag{1.51}$$

where $det\,\mathbb{A} = |\,\mathbb{A}\,|$ is the determinant of the matrix $\mathbb{A}$, i.e. $\mathbb{A}$ is nonsingular.

Of course, in the material above we simplified to the case of two vectors but all is valid for any arbitrary number of vectors. We continue our discussion for the case of two vectors.

Recall the definition of a spanned set in (1.47) and let $V \subset \mathbb{R}^2$, or V is a subset of $\mathbb{R}^2$. Consider the following inquiry. Is there a set of vectors, say $\boldsymbol{\upsilon}$ and $\boldsymbol{u}$ in $\mathbb{R}^2$, so that every vector in V can be written as a linear combination of $\boldsymbol{\upsilon}$ and $\boldsymbol{u}$? Specifically, are there $\boldsymbol{\upsilon}$ and $\boldsymbol{u}$ so that $V = \mathcal{L}[\boldsymbol{\upsilon}, \boldsymbol{u}]$? If the answer is yes, we say that the set of vectors $(\boldsymbol{\upsilon}, \boldsymbol{u})$ span the set V.

The matrix criteria to determine whether the set of vectors $(\boldsymbol{\upsilon}, \boldsymbol{u})$ span the set V is as follows. Let $\mathbb{A}$ be a matrix consisting of two rows and two columns, that is a 2×2 square matrix, whose columns are the vectors $\boldsymbol{\upsilon}$ and $\boldsymbol{u}$, and let $\boldsymbol{b}$ be a vector in $\mathbb{R}^2$. The first theorem is that $\boldsymbol{b} \in \mathcal{L}[\boldsymbol{\upsilon}, \boldsymbol{u}]$ if and only if the linear system $\mathbb{A}\ \mathbb{C} = \boldsymbol{b}$ has a solution $\mathbb{C}$. Then, another theorem states that $\boldsymbol{\upsilon}$ and $\boldsymbol{u}$ in $\mathbb{R}^2$ span the set $\mathbb{R}^2$ if and only if $\mathbb{A}\ \mathbb{C} = \boldsymbol{b}$ has a solution $\mathbb{C}$ for every vector $\boldsymbol{b}$ in $\mathbb{R}^2$. Of course, the spanning sets extend to $\mathbb{R}^n$ as long as the set of vectors that span $\mathbb{R}^n$ contain at least n vectors. We use the concepts above in Chapter 4 when we study systems of financial markets.

1.5 Static Optimization

In this section we present some basic mathematics of static optimization used throughout this book. The material is primarily presented in the deterministic case. Given that the expectation operator is linear, taking expected values accommodates the stochastic case. For example, in an appendix to Chapter 3, we examine in detail this procedure for objective functions that depend on the mean and variance only. In Chapter 4, we examine this procedure for more general utility functions, the so-called expected utility approach. We start with the case of unconstrained optimization and extend to the constrained case with both equality and inequality constraints. We provide sufficient conditions for determining the qualitative aspects of extrema only, those interested in necessary (second order) conditions may use sources in the Notes on the Literature.

(a) Unconstrained Optimization

Consider a well defined twice continuously differentiable function $f: \mathbb{R}^1 \to \mathbb{R}^1$. The point x^* is an extremum or a critical point of this function when the derivative with respect to x evaluated at x^* is zero, i.e. $df(x^*)/dx=0$. Now, let $f: A \to \mathbb{R}^1$, where $A \subset \mathbb{R}^n$. If $x^* = (x_1^*, x_2^*, x_3^*, \ldots, x_n^*)$ is a local extremum of f and it is an interior point, then a necessary condition is

$$\partial f(x^*)/\partial x_i = 0, \text{ for all } i=1,2,\ldots,n. \qquad (1.52)$$

Expression (1.52) forms a system of n equations in n unknowns, $x^* = (x_1^*, x_2^*, x_3^*, \ldots, x_n^*)$, all candidates for local extrema of f.

For example, let $f: A \to \mathbb{R}^1$, $A \subset \mathbb{R}^2$, $f(x_1, x_2) = x_1^2 + x_2^3 - x_1 x_2 - 1$. Candidates for local extrema satisfy the first order necessary conditions

$$\partial f(x^*)/\partial x_1 = 2x_1^* - x_2^* = 0$$

$$\partial f(x^*)/\partial x_2 = 3x_2^{*2} - x_1^* = 0$$

yielding candidates $(x_1^*, x_2^*) = (0,0), (1/12, 1/6)$.

It is clear that, with (1.52) alone, we cannot guarantee the qualitative nature of the extrema in terms of maximum, minimum, or none. We must use conditions of the second derivatives of the function $f: A \to \mathbb{R}^1$, $A \subset \mathbb{R}^n$

to determine the type of extrema, the so-called second order conditions. The Hessian $n \times n$ square matrix of f evaluated at x^* is

$$D^2 f(x^*) = \begin{bmatrix} \partial^2 f(x^*)/\partial x_1^2 & \partial^2 f(x^*)/\partial x_1 \partial x_2 & \dots & \dots & \partial^2 f(x^*)/\partial x_1 \partial x_n \\ . & \partial^2 f(x^*)/\partial x_2^2 & & & . \\ . & . & & . & . \\ . & . & & . & . \\ \partial^2 f(x^*)/\partial x_n \partial x_1 & \partial^2 f(x^*)/\partial x_n \partial x_2 & \dots & \dots & \partial^2 f(x^*)/\partial x_n^2 \end{bmatrix}$$

where by Young's theorem, the cross partials are equal and the Hessian is symmetric.

(i) A second order sufficient condition for x^* to be a strict local maximum of f is that the n leading principal minors of $D^2f(x^*)$ alternate in sign at x^* starting from negative

$$\left|\partial^2 f(x^*)/\partial x_1^2\right| < 0, \quad \begin{vmatrix} \partial^2 f(x^*)/\partial x_1^2 & \partial^2 f(x^*)/\partial x_1 x_2 \\ \partial^2 f(x^*)/\partial x_2 x_1 & \partial^2 f(x^*)/\partial x_2^2 \end{vmatrix} > 0, \dots$$

Equivalently, if $D^2f(x^*)$ is a negative definite symmetric square matrix, then x^* is a strict local maximum of f at x^*. Under these conditions, the function f is strictly concave;

(ii) A second order sufficient condition for x^* to be a strict local minimum of f is that the n leading principal minors of $D^2f(x^*)$ are all positive at x^*

$$\left|\partial^2 f(x^*)/\partial x_1^2\right| > 0, \quad \begin{vmatrix} \partial^2 f(x^*)/\partial x_1^2 & \partial^2 f(x^*)/\partial x_1 x_2 \\ \partial^2 f(x^*)/\partial x_2 x_1 & \partial^2 f(x^*)/\partial x_2^2 \end{vmatrix} > 0, \dots$$

Equivalently, if $D^2f(x^*)$ is a positive definite symmetric square matrix, then x^* is a strict local minimum of f at x^*. Under these conditions, the function f is strictly convex;

(iii) If the sequences above are violated, then x^* is neither a maximum nor a minimum; it is a saddle point of f at x^*, and f is neither concave nor convex.

In our example above, $f: A \to \mathbb{R}^1$, $A \subset \mathbb{R}^2$, $f(x_1, x_2) = x_1^2 + x_2^3 - x_1 x_2 - 1$; the Hessian is

$$D^2 f(x^*) = \begin{bmatrix} 2 & -1 \\ -1 & 6x_2 \end{bmatrix}.$$

The first leading principal minor is $|\partial^2 f(x^*)/\partial x_1^2| = 2 > 0$. The second leading principal minor is the Hessian itself that gives $D^2 f(x^*) = 12\, x_2^* - 1$. Evaluating $D^2 f(x^*) = 12\, x_2^* - 1$ at the first candidate $(x_1^*, x_2^*) = (0,0)$, yields $D^2 f(x^*) = -1 < 0$ and the sequences (i) and (ii) are violated. Hence, $(x_1^*, x_2^*) = (0,0)$ is a saddle point of f. Evaluating $D^2 f(x^*) = 12\, x_2^* - 1$ at the second candidate $(x_1^*, x_2^*) = (1/12, 1/6)$, yields $D^2 f(x^*) = 1 > 0$ and the sequence (ii) is satisfied. Hence, $(x_1^*, x_2^*) = (1/12, 1/6)$ is a local minimum of f.

(b) Constrained Optimization with Equality Constraints

Consider two well defined twice continuously differentiable functions $f: \mathbb{R}^n \to \mathbb{R}^1$ and $g: \mathbb{R}^n \to \mathbb{R}^1$. Now, consider a constant c in $\mathbb{R}^1$ and the convex set $\{x: g(x) = c\}$. Typically, a constrained optimization problem with one equality constraint involves finding the *n-tuple*,

$$x^* = (x_1^*, x_2^*, x_3^*, \ldots, x_n^*),$$

that is an extremum of f subject to $g(x) = c$. A convenient way to solve those problems is to write the Lagrangean function, $\ell(x, \lambda; c)$; or $\ell: \mathbb{R}^n \times \mathbb{R}^1 \to \mathbb{R}^1$ as

$$\ell = f(x) + \lambda\,(c - g(x)) \qquad (1.53)$$

where λ is a real number, the so-called Lagrange multiplier of the function ℓ. Then, the first order necessary conditions for a constrained extremum of f subject to $g(x) = c$ can be obtained by finding the critical points of the function ℓ with respect to $\{x, \lambda\}$ or

$$\partial \ell (x^*, \lambda^*)/\partial x_i = 0, \ \text{for all } i = 1, 2, \ldots, n. \qquad (1.54a)$$

$$\partial \ell (x^*, \lambda^*)/\partial \lambda = 0 \qquad (1.54b)$$

Expressions (1.54a,b) form a system of $n+1$ equations in $n+1$ unknowns, $\{x^*, \lambda^*\}$ all candidates for critical points of ℓ, that deliver extremum of f subject to $g(x)=c$.

As before, with (1.54a,b) alone, we cannot guarantee the qualitative nature of the extrema in terms of maximum, minimum, or none. We must use conditions of the second derivatives of the function ℓ to determine the type of extrema. The so-called bordered Hessian $n+1 \times n+1$ square matrix of ℓ evaluated at $\{x^*, \lambda^*\}$ is

$$H\,\ell(x^*,\lambda^*) = \begin{bmatrix} \partial^2\ell(x^*,\lambda^*)/\partial x_1{}^2 & \partial^2\ell(x^*,\lambda^*)/\partial x_1\partial x_2 & \dots & \dots & \partial^2\ell(x^*,\lambda^*)/\partial x_1\partial\lambda \\ . & \partial^2\ell(x^*,\lambda^*)/\partial x_2{}^2 & & & . \\ . & . & & . & . \\ . & . & & . & . \\ \partial^2\ell(x^*,\lambda^*)/\partial\lambda\partial x_1 & \partial^2\ell(x^*,\lambda^*)/\partial\lambda\partial x_2 & \dots & \dots & \partial^2\ell(x^*,\lambda^*)/\partial\lambda^2 \end{bmatrix}.$$

(i) A second order <u>sufficient</u> condition for x^* to be a strict local constrained maximum of f subject to $g(x)=c$ is that the leading principal minors of $H\,\ell(x^*,\lambda^*)$ alternate in sign at $\{x^*,\lambda^*\}$ starting from positive, or

$$\begin{vmatrix} \partial^2\ell(x^*,\lambda^*)/\partial x_1{}^2 & \partial^2\ell(x^*,\lambda^*)/\partial x_1\partial x_2 & \partial^2\ell(x^*,\lambda^*)/\partial x_1\partial\lambda \\ \partial^2\ell(x^*,\lambda^*)/\partial x_2\partial x_1 & \partial^2\ell(x^*,\lambda^*)/\partial x_2{}^2 & \partial^2\ell(x^*,\lambda^*)/\partial x_2\partial\lambda \\ \partial^2\ell(x^*,\lambda^*)/\partial\lambda\partial x_1 & \partial^2\ell(x^*,\lambda^*)/\partial\lambda\partial x_2 & \partial^2\ell(x^*,\lambda^*)/\partial\lambda^2 \end{vmatrix} > 0,\ |.| < 0, \dots$$

in this case, the Lagrangean function $\ell(x^*,\lambda^*)$ is concave, similarly, f is concave and g is convex;

(ii) A second order <u>sufficient</u> condition for x^* to be a strict local constrained minimum of f subject to $g(x)=c$ is that all leading principal minors of $H\,\ell(x^*,\lambda^*)$ are strictly negative at $\{x^*,\lambda^*\}$ or

$$\begin{vmatrix} \partial^2\ell(x^*,\lambda^*,\lambda^*)/\partial x_1{}^2 & \partial^2\ell(x^*,\lambda^*)/\partial x_1\partial x_2 & \partial^2\ell(x^*,\lambda^*)/\partial x_1\partial\lambda \\ \partial^2\ell(x^*,\lambda^*)/\partial x_2\partial x_1 & \partial^2\ell(x^*,\lambda^*)/\partial x_2{}^2 & \partial^2\ell(x^*,\lambda^*)/\partial x_2\partial\lambda \\ \partial^2\ell(x^*,\lambda^*)/\partial\lambda\partial x_1 & \partial^2\ell(x^*,\lambda^*)/\partial\lambda\partial x_2 & \partial^2\ell(x^*,\lambda^*)/\partial\lambda^2 \end{vmatrix} < 0,\ |.| < 0, \dots$$

in this case, the Lagrangean function $\ell(x^*,\lambda^*)$ is convex, similarly, f is convex and g is concave;

(iii) If the sequences above are violated, then x^* is neither a maximum nor a minimum; it is a <u>saddle point</u> of f subject to $g(x)=c$ at x^*.

For example, let $f(x_1, x_2)= (1/3)\, x_1^{\,3} - (3/2)\, x_2^{\,2} + 2\, x_1$, $g(x_1, x_2)= x_1 - x_2$, and $c=0$. The Lagrangean function is then $\ell(x\, ,\, \lambda\, ;\, c)$; $\ell\colon \mathbb{R}^2\times\mathbb{R}^1\to\mathbb{R}^1$,

$$\ell = (1/3)\, x_1^{\,3} - (3/2)\, x_2^{\,2} + 2\, x_1 + \lambda\, (\, x_2 - x_1\,);$$

the first order necessary conditions are

$$\partial\ell\,(x^*,\lambda^*)/\partial x_1 = x_1^{\,2} + 2 - \lambda = 0$$

$$\partial\ell\,(x^*,\lambda^*)/\partial x_2 = -\, 3\, x_2 + \lambda = 0$$

$$\partial\ell\,(x^*,\lambda^*)/\partial\lambda = x_2 - x_1 = 0.$$

The system of equations yields candidates for local extrema $\{x_1^*,\ x_2^*,\ \lambda^*\}=\{(1,1,3),(2,2,6)\}$. The bordered Hessian is

$$H\,\ell(x^*,\lambda^*) = \begin{vmatrix} 2x_1 & 0 & -1 \\ 0 & -3 & 1 \\ -1 & 1 & 0 \end{vmatrix}$$

which upon evaluation at the candidates yields

$$H\,\ell(1,1,3)=1>0, \quad H\,\ell(2,2,6)= -1<0.$$

Hence, $\{x_1^*,\ x_2^*\ ,\ \lambda^*\}=\{1,1,3\}$ is a strict local maximum and $\{x_1^*,\ x_2^*,\ \lambda^*\}=\{2,2,6\}$ is a strict local minimum of the constrained optimization.

Of course, the problems studied above can be appropriately extended to the case of several equality constraints by expanding the set of Lagrange multipliers attached to each constraint.

(c) Constrained Optimization with Inequality Constraints: Kuhn-Tucker Theory

Consider again two well defined twice continuously differentiable functions $f{:}\mathbb{R}^n \to \mathbb{R}^1$, $g{:}\mathbb{R}^n \to \mathbb{R}^1$, a constant c in $\mathbb{R}^1$ and the convex set $\{x: g(x) \leq c\}$. Typically, a constrained optimization problem with one inequality constraint involves finding the *n-tuple*, $x^* = (x_1^*, x_2^*, x_3^*, \ldots, x_n^*) \geq 0$, that is an extremum of f subject to $g(x) \leq c$. Kuhn and Tucker proposed a Lagrangean function that does not include the n non-negativity constraint, $x = (x_1, x_2, x_3, \ldots, x_n) \geq 0$, but satisfies it implicitly. The function is $\ell(x, \lambda; c)$; $\ell{:}\ \mathbb{R}^n \times \mathbb{R}^1 \to \mathbb{R}^1$ as

$$\ell = f(x) - \lambda\,(g(x) - c) \qquad (1.55)$$

where λ is the Lagrange multiplier of ℓ. The first order necessary conditions are written

$$\partial \ell(x^*,\lambda^*)/\partial x_i \leq 0, \quad \text{if} < 0 \text{ then } x_i^* = 0, \quad \text{for all } i=1,2,\ldots,n \qquad (1.56a)$$

$$\partial \ell(x^*,\lambda^*)/\partial \lambda \leq 0, \quad \text{if} < 0 \text{ then } \lambda^* = 0 \qquad (1.56b)$$

or equivalently

$$x_i\, \partial \ell(x^*,\lambda^*)/\partial x_i = 0, \quad \text{for all } i=1,2,\ldots,n \qquad (1.57a)$$

$$\lambda\ \partial \ell(x^*,\lambda^*)/\partial \lambda = 0 \qquad (1.57b)$$

the so-called complementary slackness conditions. Expressions (1.56a,b), or (157a,b) form a system of $n+1$ equations in $n+1$ unknowns, $\{x^*, \lambda^*\} \geq 0$, all candidates for critical points of ℓ, that deliver extremum of f subject to $g(x) \leq c$.

For example, let $n=2$ and g be linear and given by $g(x) = p_1 x_1 + p_2 x_2$, for given positive real numbers $\{p_1, p_2\} > 0$. The Lagrangean function in this case is

$$\ell = f(x_1, x_2) - \lambda\,(p_1 x_1 + p_2 x_2 - c)$$

with first order necessary conditions given by

$$\partial \ell(x^*,\lambda^*)/\partial x_i = [\partial f(x_1, x_2)/\partial x_i] - \lambda p_i \leq 0, \quad \text{if} < 0 \text{ then } x_i^* = 0, \quad \text{for } i=1,2$$

$$\partial \ell(x^*,\lambda^*)/\partial \lambda = -\,(p_1 x_1 + p_2 x_2 - c) \leq 0, \quad \text{if} < 0 \text{ then } \lambda^* = 0$$

where Figure 1.2 provides a geometric description of alternative cases in the convex constraint set.

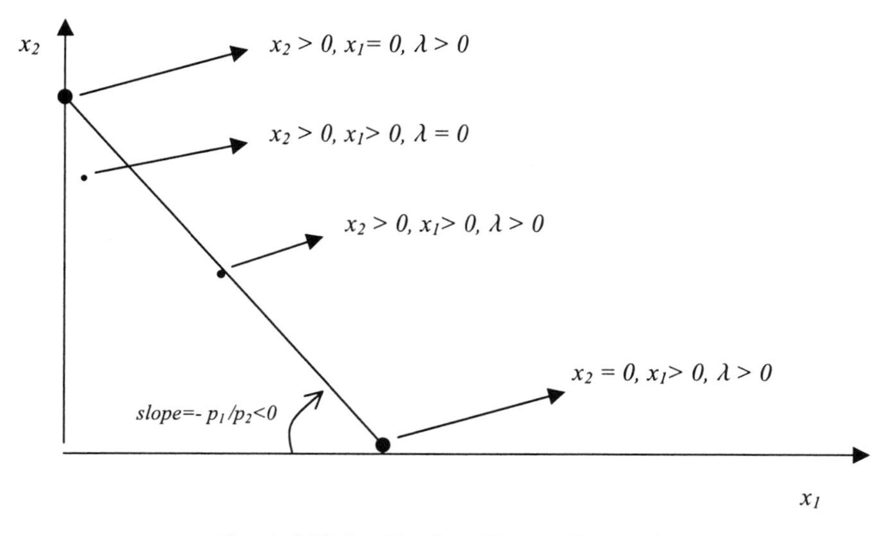

Fig. 1.2 Kuhn-Tucker Theory Example

Sufficient second order conditions are obtained as in the case with equality constraints above, and the problems studied can be appropriately extended to the case of several inequality constraints by expanding the set of Lagrange multipliers attached to each constraint.

(d) Interpretation of the Lagrange Multiplier and the Envelope Theorem

Typically, an economic problem takes the form of finding an extremum of a function subject to constraints. As in case (b) above, consider two well defined twice continuously differentiable functions $f:\mathbb{R}^n \to \mathbb{R}^1$, and $g:\mathbb{R}^n \to \mathbb{R}^1$, a constant real number c in $\mathbb{R}^1$ and the convex set $\{x: g(x) \leq c\}$. Let the *n-tuple*, $x^* = (x_1^*, x_2^*, x_3^*, \ldots, x_n^*)$, be an extremum of f subject to $g(x)=c$. Then, the solution is of the general form, $x^* = x^*(c)$, $\lambda^* = \lambda^*(c)$, functions of the constant parameter c. Substituting the optimal values $x^* = x^*(c)$, $\lambda^* = \lambda^*(c)$ into the objective function f yields the value function or the indirect objective function $f(x^*(c))$. It can be easily shown that

$$\lambda^* = df(x^*(c)) / dc \qquad (1.58)$$

or the Lagrange multiplier is the derivative of the value function with respect to small changes in the parameter governing the constraint. This result is generalized for several equality or inequality constraints.

The result in (1.58) has an interesting implication for general static and dynamic constrained optimization problems. Suppose we consider the Lagrangean function in (1.53) evaluated at the optimum,

$$\ell^* = f(x^*(c)) - \lambda^*(c)\,(g(x^*(c)) - c).$$

If we take the differential of the Lagrangean function with respect to the parameter c, we obtain, using the chain rule of differentiation,

$$\begin{aligned}
\partial \ell^*/\partial c = {} & \\
& [\partial f(x^*)/\partial x^*]\,(\partial x^*/\partial c) - (\partial\lambda^*/\partial c)\,(g(x^*(c)) - c) - \lambda^* \times \\
& \qquad \{[\partial g(x^*)/\partial x^*]\,(\partial x^*/\partial c) - 1\} \\
= {} & (\partial x^*/\partial c)\,\{ [\partial f(x^*)/\partial x^*] - \lambda^*\,[\partial g(x^*)/\partial x^*] \} + \lambda^* \\
= {} & \lambda^*
\end{aligned}$$

where the second line is obtained using (1.54b), or $(g(x^*(c)) - c)=0$, and the third line follows from (1.54a), or $[\partial f(x^*)/\partial x^*]-\lambda^*[\partial g(x^*)/\partial x^*]=0$. The result in (1.59) is important because it shows that differentiating the Lagrangean function at the optimum allows us to ignore the dependence of the optimized values on the parameter c. Thus, using (1.58) and (1.59) yields

$$\lambda^* = d\,f\,(x^*(c))\,/\,dc = \partial \ell^*/\partial c \tag{1.60}$$

the so-called Envelope Theorem. The intuition for the Envelope Theorem is clear in the simple case of unconstrained optimization, say find the extremum of the well behaved function $f(x;c)$. The value function is $f(x^*(c);c)$ and computing $d\,f(x^*(c);c)/dc$ yields

$$\begin{aligned}
d\,f(x^*(c);c)/dc &= \Sigma_{i=1}^{n}\,[\partial f\,(x^*;c)/\partial x_i^*]\,(\partial x_i^*/\partial c) + \partial f(.;c)/\partial c \\
&= \partial f(.;c)/\partial c \qquad\qquad (1.61)
\end{aligned}$$

by the necessary first order conditions (1.52), obtaining the Envelope Theorem. Hence, we can ignore the dependence of the optimized solution on the parameter because, at the optimum, that effect vanishes.

(e) Systems of Implicit Functions

In the extrema problems studied in (a)-(b)-(c) above, we ultimately end up with a system of necessary first order conditions of the form

$$h_1(x_1, x_2, c) = 0$$

$$h_2(x_1, x_2, c) = 0$$

where $\{x_1, x_2\}$ are endogenous variables, c is a constant real number and h_1, h_2 are twice continuously differentiable functions. In fact, this is a system of implicit functions that admits an explicit unique solution for the endogenous variables

$$x_1^* = x_1(c)$$

$$x_2^* = x_2(c)$$

under certain conditions. We can differentiate the system of implicit functions to obtain

$$\begin{pmatrix} \partial h_1 / \partial x_1 & \partial h_1 / \partial x_2 \\ \partial h_2 / \partial x_1 & \partial h_2 / \partial x_2 \end{pmatrix} \begin{pmatrix} \partial x_1 * / \partial c \\ \partial x_2 * / \partial c \end{pmatrix} = \begin{pmatrix} -\partial h_1 / \partial c \\ -\partial h_2 / \partial c \end{pmatrix}$$

and, as long as the Jacobian determinant of this system is nonsingular,

$$J \equiv \begin{vmatrix} \partial h_1 / \partial x_1 & \partial h_1 / \partial x_2 \\ \partial h_2 / \partial x_1 & \partial h_2 / \partial x_2 \end{vmatrix} \neq 0$$

there will be a unique solution for $\partial x_1^*/\partial c$, and $\partial x_2^*/\partial c$. Of course, the condition on the Jacobian determinant $J \neq 0$, is implied by the sufficient second order conditions for an extrema studied in (a)-(b)-(c) above.

1.6 Notes on Stochastic Difference Equations

In this section, we discuss a simple example of a stochastic linear difference equation highlighting solution issues. Let $t=1,2,..$ be a discrete time index and let ϵ_t be a discrete random variable with well defined probability distribution, zero expected value, finite variance, and uncorrelated across periods, or

$$E[\epsilon_t] = 0$$

$$var(\epsilon_t) = \sigma^2 < \infty \qquad (1.62)$$

$$cov(\epsilon_i, \epsilon_j) = 0 \;\; for \;\; i \neq j,$$

we usually say this in an independent and identically distributed (i.i.d.) random variable. Suppose the variable y, a scalar, evolves over time according to the first order linear process

$$y_t = a_0 + a_1 y_{t-1} + \epsilon_t \qquad (1.63)$$

where $\{a_0, a_1\}$ are constant real numbers. This is a simple discrete time-discrete state process, also known as a Markov process because the current value only depends on the immediate past value, plus constant and shock (see Chapter 7 for a detailed discussion of Markov processes).

A simple way to understand how this equation can be solved for is to iterate on it recursively. Suppose there is an initial condition known to be y_0. Then starting from $t=0$ and substituting at each step to an arbitrary t, we have

$$y_1 = a_0 + a_1 y_0 + \epsilon_1$$

$$y_2 = a_0(1 + a_1) + a_1^2 y_0 + a_1 \epsilon_1 + \epsilon_2$$

$$y_3 = a_0(1 + a_1 + a_1^2) + a_1^3 y_0 + a_1^2 \epsilon_1 + a_1 \epsilon_2 + \epsilon_3$$

....

$$y_t = a_0 [\, \Sigma_{i=0}^{t-1} (a_1)^i \,] + a_1^t y_0 + [\, \Sigma_{i=0}^{t-1} (a_1)^i \epsilon_{t-i} \,]. \qquad (1.64)$$

Expression (1.64) is one possible solution for (1.63) when the initial condition is known. This solution is backwards because the current value of y is determined by a cumulated sum of the past errors,

$$\Sigma_{i=0}^{t-1} (a_1)^i \epsilon_{t-i},$$

plus terms involving the constant a_0 and initial condition y_0. It is easy to examine some even more specialized cases with (1.63)-(1.64). Suppose $a_1=1$, then the process in (1.63) becomes the so-called random walk with drift with solution, e.g. Chapter 7, from known y_0

$$y_t = a_0\, t + y_0 + \Sigma_{i=1}^{t} \epsilon_i \,. \tag{1.65}$$

A random walk without drift is when $a_0=0$ and $a_1=1$ yielding, from known y_0

$$y_t = y_0 + \Sigma_{i=1}^{t} \epsilon_i \,. \tag{1.66}$$

Of course, the case where $a_0=0$ and $a_1 \neq 1$ yields, from known y_0

$$y_t = a_1^{\,t} y_0 + [\, \Sigma_{i=0}^{t-1} (a_1)^i \epsilon_{t-i} \,]. \tag{1.67}$$

Referring back to the original equation (1.63), it may not be the case that y_0 is known at $t=0$. In this case, we have to go further backwards from y_0 to an arbitrary m, or

$$y_0 = a_0 + a_1 y_{-1} + \epsilon_0$$

$$y_{-1} = a_0 + a_1 y_{-2} + \epsilon_{-1}$$

$$\dots \quad \dots \quad \dots$$

$$y_{-m} = a_0 + a_1 y_{-m-1} + \epsilon_{-m} \,. \tag{1.68}$$

Substituting the recursion in (1.68) into (1.64), we obtain

$$y_t = a_0 [\, \Sigma_{i=0}^{t+m} (a_1)^i \,] + a_1^{\,t+m+1} y_{-m-1} + [\, \Sigma_{i=0}^{t+m} (a_1)^i \epsilon_{t-i} \,] \tag{1.69}$$

for an arbitrary m. The main issue now is the behavior of (1.69) when m becomes large. Consider the first term $a_0 [\, \Sigma_{i=0}^{t+m} (a_1)^i \,]$ and compute the limit

$$\lim_{m\to\infty} [\, \Sigma_{i=0}^{t+m} (a_1)^i \,]. \tag{1.70a}$$

It is clear that if $|\, a_1| < 1$, then the limit converges to

$$\lim_{m\to\infty} [\, \Sigma_{i=0}^{t+m} (a_1)^i \,] = 1/(1 - a_1). \tag{1.70b}$$

However, if $|\, a_1| > 1$, then the limit does not converge, or

$$\lim_{m\to\infty} [\, \Sigma_{i=0}^{t+m} (a_1)^i \,] = \infty, \tag{1.70c}$$

and the same is true for the terms $a_1^{\,t+m+1} y_{-m-1}$ and $[\Sigma_{i=0}^{t+m} (a_1)^i \epsilon_{t-i}]$.

Thus, for the case when a_1 is inside the unit circle, or $|\, a_1| < 1$, the backward solution in (1.69) for $m \to \infty$ is

$$y_t = [a_0/(1-a_1)] + [\Sigma_{i=0}^{\infty}(a_1)^i \epsilon_{t-i}] \qquad (1.71)$$

and a constant plus the cumulated infinite sum of the past errors determine the current value of y.

However, because y_0 is not known, (1.71) is not the unique backward solution of (1.63). For any arbitrary real number, or constant c,

$$y_t = c\,(a_1)^t + [a_0/(1-a_1)] + [\Sigma_{i=0}^{\infty}(a_1)^i \epsilon_{t-i}] \qquad (1.72)$$

is also a solution to (1.63) when y_0 is not known. To see this, equate (1.72) to the original equation (1.63) substituting (1.72) and note that both sides become identical after some manipulation. Of course, when y_0 is known, then the arbitrary constant c can be appropriately determined in (1.72) evaluated at $t=0$ and substituted back to yield the solution obtained in (1.64).

In all cases examined above, we obtained solutions that ultimately depend on the cumulated sum of the past disturbances, i.e. backward solution, and we noticed that those solutions work in the special case when $|a_1|<1$. What happens in the case when $|a_1|>1$? It turns out that given the linearity of the process we can also seek a forward solution for the original equation in (1.63). Suppose we update (1.63) one period forward and obtain

$$y_{t+1} = a_0 + a_1 y_t + \epsilon_{t+1}$$

which may be appropriately written as

$$y_t = -[(a_0 + \epsilon_{t+1})/a_1] + (y_{t+1}/a_1). \qquad (1.73)$$

Expression (1.73) may be iterated n arbitrary periods forward to obtain upon recursive substitution

$$y_t = -[(a_0 + \epsilon_{t+1})/a_1] + (y_{t+1}/a_1)$$

$$y_{t+1} = -[(a_0 + \epsilon_{t+2})/a_1] + (y_{t+2}/a_1)$$

$$\dots \quad \dots \quad \dots$$

$$y_t = -a_0[\Sigma_{i=1}^{n}(1/a_1)^i] - [\Sigma_{i=1}^{n}(1/a_1)^i \epsilon_{t+i}] + y_{t+n}(1/a_1)^n \qquad (1.74)$$

where the current value of y is now determined in part by the cumulated sum of the future disturbances. Again, the issue is the behavior of (1.74) when n becomes large. Consider the first term $-a_0[\Sigma_{i=1}^{n}(1/a_1)^i]$ and compute the limit

$$lim_{n\to\infty}[\Sigma_{i=1}^{n}(1/a_1)^i]. \qquad (1.75a)$$

It is clear that if $|a_1|<1$, then the limit diverges to

$$\lim_{n\to\infty} [\Sigma_{i=1}^{n}(1/a_1)^i] = \infty. \qquad (1.75b)$$

However, if $|a_1|>1$, then the limit converges, or

$$\lim_{n\to\infty} [\Sigma_{i=1}^{n}(1/a_1)^i] = (1/a_1)\{1/[1-(1/a_1)]\}. \qquad (1.75c)$$

The same convergence properties hold for the terms $[\Sigma_{i=1}^{n}(1/a_1)^i \epsilon_{t+i}]$ and $y_{t+n}(1/a_1)^n$. Thus, for the case when $|a_1|>1$, the forward solution in (1.74) for $n\to\infty$ is, after simplification,

$$y_t = [a_0/(1-a_1)] - [\Sigma_{i=1}^{\infty}(1/a_1)^i \epsilon_{t+i}] \qquad (1.76)$$

and the current value of y is determined by a constant minus the cumulated infinite sum of the future errors.

Since the original process in (1.63) is linear, the general solution involves any linear combination of the backward part (1.72) and the forward part (1.76). The solution is of the general form

$$y_t = [a_0/(1-a_1)] + g(t) + \textit{forward error component} + \textit{backward error component} \qquad (1.77)$$

where $g(t)$ takes the form $g(t) = c\,(a_1)^t$ for an arbitrary constant c, possibly determined by the initial condition, and the forward and backward error components are respectively:

$$-[\Sigma_{i=1}^{\infty}(1/a_1)^i \epsilon_{t+i}]; \quad [\Sigma_{i=0}^{\infty}(a_1)^i \epsilon_{t-i}].$$

Choosing a solution for the process depends on the particular economic application. For example, when the application determines that $|a_1|<1$, then we use the backward solution whereas if the application determines that $|a_1|>1$, then we use the forward solution. Note, however, that in the backward solution the arbitrary constant c can be tied to the initial condition whereas in the forward solution the arbitrary constant can provide multiple paths to y. To avoid the multiplicity of paths in the forward solution, a terminal or boundary condition must be imposed. In many economic applications, choice of the unique convergent path is achieved by setting arbitrarily c to zero, or $c=0$.

(a) Lag Operators

The lag or backshift operator, L, is such that when applied to a variable it lags it, or

$$L^{j} x_{t} \equiv x_{t-j}, \quad j=...-2,-1,0,1,2... \tag{1.78}$$

This operator provides simplified notation for dynamic processes in discrete time. For example, the original process (1.63) can be equivalently expressed as

$$\begin{aligned} y_t &= a_0 + a_1 L y_t + \epsilon_t \\ &= (a_0 + \epsilon_t) / (1 - a_1 L). \end{aligned} \tag{1.79}$$

In particular, for $|a_1|<1$,

$$a_0 / (1 - a_1 L) = a_0 \; [\Sigma_{i=0}^{\infty} (a_1 L)^i] = a_0 \; [\Sigma_{i=0}^{\infty} (a_1)^i]$$

$$\epsilon_t / (1 - a_1 L) = [\Sigma_{i=0}^{\infty} (a_1 L)^i \epsilon_t] = [\Sigma_{i=0}^{\infty} (a_1)^i \epsilon_{t-i}]$$

and for $|a_1|>1$,

$$a_0 / (1 - a_1 L) = - a_0 [1/(a_1 L)] \; [\Sigma_{i=0}^{\infty} (1/a_1 L)^i] = - a_0 \; [\Sigma_{i=1}^{\infty} (1/a_1)^i]$$

$$\epsilon_t / (1 - a_1 L) = - [1/(a_1 L)] [\Sigma_{i=0}^{\infty} (1/a_1 L)^i \epsilon_t] = - [\Sigma_{i=1}^{\infty} (1/a_1)^i \epsilon_{t+i}].$$

Finally, discussion of higher order processes is beyond the scope here, but the interested reader is referred to the Notes on the Literature.

1.7 **Dynamic Optimization in Discrete Time: Heuristics of Dynamic Programming in the Certainty Case**

This is an heuristic presentation of Bellman's dynamic programming in discrete time under certainty. The horizon of the decision maker is long, $t, t+1, t+2, t+3,...T$, with $T \rightarrow \infty$ possibly. Consider an individual whose current economic status (at time t) is described by a state variable, x_t, assumed to be a scalar for simplicity (it could be a vector of state variables as well) as a sequence $(x_t, x_{t+1}, x_{t+2}, ...)$ taken as given at t. Denote the choices available for this individual, in period t, by u_t, where u_t can be thought of as a control variable used to achieve some objective, a sequence $(u_t, u_{t+1}, u_{t+2}, ...)$. Let the function $r_t(x_t, u_t)$ denote the individual immediate reward function; individuals are assumed to discount future periods using the constant discount factor $\beta \in [0,1)$, so that the dependence of the reward function on t is only through the exponential discount function β^t, and

$$r_t(x_t, u_t) = \beta^t r(x_t, u_t).$$

The general problem involves choosing a sequence of controls $(u_t, u_{t+1}, u_{t+2}, \ldots)$ to achieve an extremum of the present discounted value of rewards, $\Sigma_{t=0}^{T} \beta^t r(x_t, u_t)$, $T \rightarrow \infty$ possibly, given a sequence of states $(x_t, x_{t+1}, x_{t+2}, \ldots)$. Richard Bellman provided a key insight for the solution of these kinds of problems. Suppose the sequence of controls is chosen optimally $(u^*_t, u^*_{t+1}, u^*_{t+2}, \ldots)$, then the extremum value of the objective, or the value function is given by

$$r(x_t, u^*_t) + \beta\, r(x_{t+1}, u^*_{t+1}) + \beta^2 r(x_{t+2}, u^*_{t+2}) + \beta^3 r(x_{t+3}, u^*_{t+3}) + \ldots$$

and we note that this is simply

$$r(x_t, u^*_t) + \beta [\, r(x_{t+1}, u^*_{t+1}) + \beta\, r(x_{t+2}, u^*_{t+2}) + \beta^2 r(x_{t+3}, u^*_{t+3}) + \ldots].$$

Thus, the sequential decision process can be split into two parts: the immediate period and the future beyond that. In the current date t, with current state x_t, the individual choice of u_t yields immediate reward denoted by the function $r(x_t, u^*_t)$. Now imagine the following: from period $t+1$ onwards the individual makes all decisions optimally. Denote by V_{t+1} the present discounted value of all the rewards when from period $t+1$ onwards the individual makes all decisions optimally. Hence V_{t+1} is called the continuation value of the sequential problem. Then, generally in period t, the sum of the current immediate value plus the continuation value is expressed as

$$r(x_t, u_t) + \beta\, V_{t+1}.$$

It is simple to understand the current value function V_t now. For a maximum problem, V_t and V_{t+1} are related by

$$V_t = \underset{\{u_t\}}{Max} [\, r(x_t, u_t) + \beta\, V_{t+1}\,]$$

where the maximization in the right-hand side is with respect to the control u_t, given the current state x_t, and we note that V_t is a function of the current state x_t, $V_t(x_t)$. And the same will be the case for V_{t+1}, $V_{t+1}(x_{t+1})$. Hence, the sequential problem can be written as

$$V_t(x_t) = \underset{\{u_t\}}{Max} \{ r(x_t, u_t) + \beta [V_{t+1}(x_{t+1})]\}, \qquad x_t \text{ given.}$$

This expression is called the Bellman functional equation of the sequential optimization problem. It is an equation in the function V_t (that is, a solution to this equation is achieved by a solution to V_t that satisfies this equation).

Examining Bellman's equation closely, we note that a complication arises because of the dependence of the function V upon the time index t. This is particularly troublesome for the infinite horizon case because when applying Bellman's equation for period T, we obtain

$$V_T(x_T) = \underset{\{u_T\}}{Max} \{ r(x_T, u_T) + \beta [V_{T+1}(x_{T+1})]\}$$

and it is not exactly clear how to solve for the function V_{T+1} when $T \to \infty$.

The important question here is whether iterating on this equation, from some initial value, converges to a time invariant function as $T \to \infty$, or $lim\ V_{T\to\infty} = V$. It turns out that the answer to this question is affirmative. To understand this, we have to grasp some mathematical concepts, ideas and theorems.

(i) The first concept to examine is distance. In the real line, the distance between points a and b is easy to understand. In general dimensions, distance is well defined within a metric space, e.g. section 1.4.

(ii) The second concept is the operator. For example, in statistics we all understand the expectation operator. In the context of analysis, an operator is a function that maps a metric space into itself. Given a metric space and an operator (note that this simply gives a well defined meaning to distance), we consider the concept of a contraction mapping by example. Consider an operator T applied to a variable (or function) X from $[0,1] \to [0,1]$ (that is mapped from $[0,1]$ into itself). Then T is a contraction mapping of modulus β if for all $x, y \in X$:

$$|Tx - Ty| \,/\, |x - y| \leq \beta < 1,$$

i.e. the slope of the function is uniformly less than one in absolute value (note that $|Tx - Ty|$, $|x - y|$ are "metric distances"). Figure 1.3 illustrates the idea of a contraction mapping: since the function in the relevant domain has slope uniformly less than one in absolute value, the ratio $|Tx - Ty| / |x - y|$ is less than one for all the choices of $\{x,y\}$ in the domain.

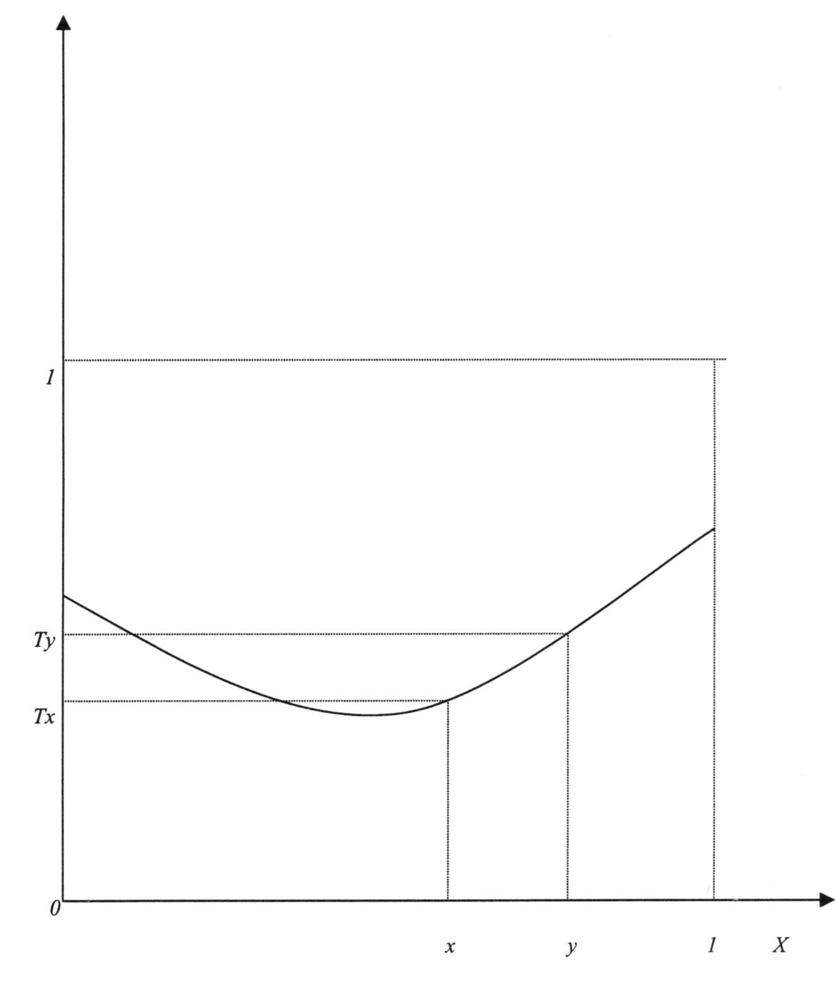

Fig. 1.3 Example of a Contraction Mapping

(iii) The next tool to use is the contraction mapping theorem. The contraction mapping theorem is as follows. Suppose T is a contraction mapping of modulus $\beta < 1$ as in Figure 1.3. Then, there exists a fixed point $x^* \in [0,1]$ so that $Tx^* = x^*$. In particular, let $x_0 \in [0,1]$, and apply the operator T inductively:

$$x_1 = T x_0$$

$$x_2 = Tx_1 = T^2 x_0$$

$$x_3 = Tx_2 = T^3 x_0$$

$$x_4 = Tx_3 = T^4 x_0$$

$$x_5 = Tx_4 = T^5 x_0$$

$$\dots \quad \dots \quad \dots$$

$$x_{n+1} = Tx_n = T^{n+1} x_0;$$

as $n \to \infty$, x_{n+1} converges uniformly to the fixed point $x^* = Tx^*$, i.e. as $n \to \infty$ the distance between Tx and Ty, $\{x,y\} \in X$, converges to zero. Figure 1.4 illustrates the contraction mapping theorem by example. Point A is where the contraction mapping crosses the 45° line (the fixed point); starting from x_0 and iterating we reach $Tx_0 = x_1$ back in the 45° line; continuing the iteration we ultimately converge to point A where $x^* = Tx^*$, i.e. x converges uniformly to x^*.

(iv) Finally, we make use of Blackwell's sufficient conditions for a contraction mapping to check whether our function V satisfies the contraction mapping theorem. Blackwell's sufficient conditions in the context of our example are:

- Monotonicity: for all $x,y \in [0,1]$, $x \geq y \Rightarrow Tx \geq Ty$;
- Discounting: for any real number $c \geq 0$ and every $x \in [0,1]$, there exists some $\beta \in [0,1)$ so that

$$T(x+c) \leq Tx + \beta c;$$

then T is a contraction mapping of modulus β and uniformly converges.

We can now consider Bellman's equation

$$V_t(x_t) = \underset{\{u_t\}}{Max} \{ r(x_t, u_t) + \beta [V_{t+1}(x_{t+1})] \}, \qquad x_t \text{ given,}$$

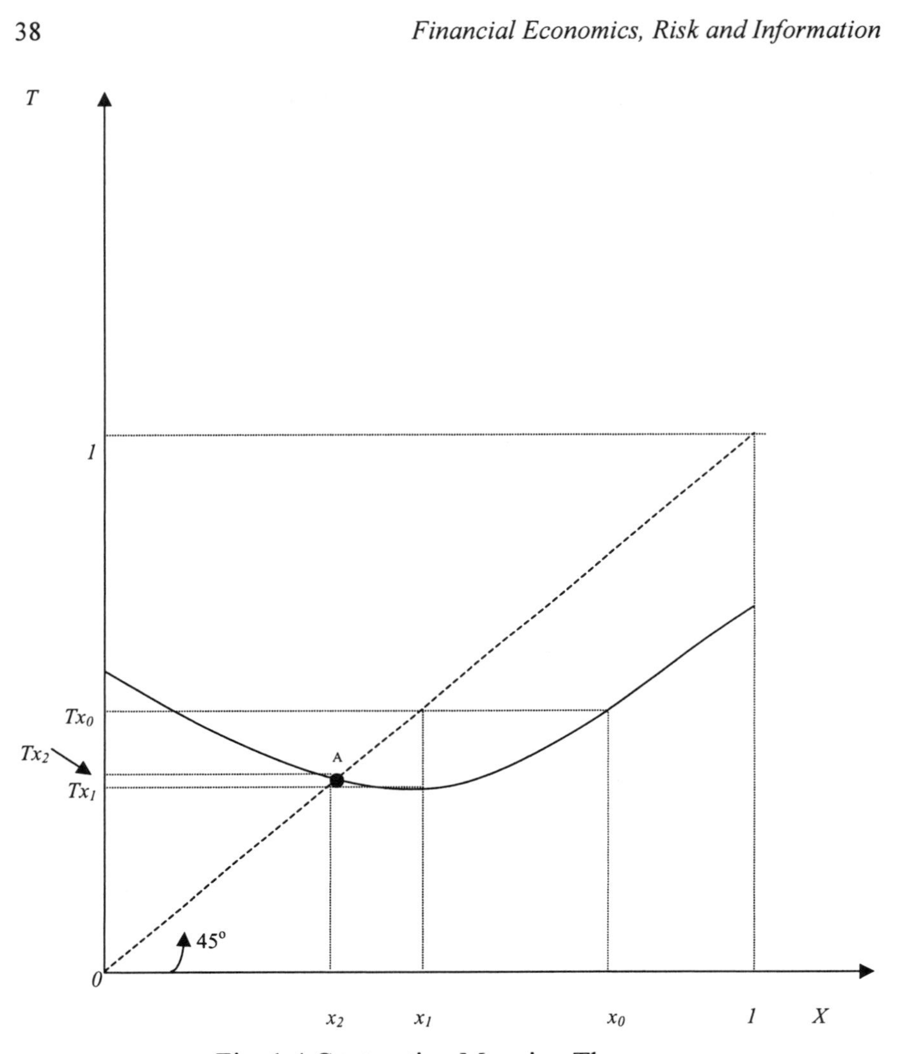

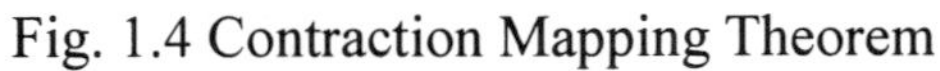
Fig. 1.4 Contraction Mapping Theorem

in the function V, apply the operator T to this equation and check Blackwell's sufficient conditions. Thus, applying T we obtain

$$TV(x_t) = \underset{\{u_t\}}{Max} \{ r(x_t, u_t) + \beta [V(x_{t+1})] \},$$

and verify Blackwell's conditions:

• Monotonicity: consider V_1 and V_2 so that $V_1 \geq V_2$ then

$$TV_1 = \underset{\{u_t\}}{Max}\, \{ r(x_t, u_t) + \beta V_1 \},$$

$$TV_1 \geq \underset{\{u_t\}}{Max}\, \{ r(x_t, u_t) + \beta V_2 \},\ since\ V_1 \geq V_2$$

$$TV_1 \geq TV_2$$

so T is monotonic when applied to V;

- Discounting: let $c \geq 0$ be a constant real number

$$T(V+c) = \underset{\{u_t\}}{Max}\, \{ r(x_t, u_t) + \beta (V + c) \},$$

$$T(V+c) = \underset{\{u_t\}}{Max}\, \{ r(x_t, u_t) + \beta V + \beta c \},$$

$$T(V+c) = \underset{\{u_t\}}{Max}\, \{ r(x_t, u_t) + \beta V \} + \beta c,$$

$$T(V+c) = TV + \beta c,$$

and T discounts.

Hence, T applied to V satisfies Blackwell's sufficient conditions for a contraction mapping and iterating on the value function converges uniformly. Bellman's equation is the solution to the time invariant function V that satisfies

$$V(x_t) = \underset{\{u_t\}}{Max}\, \{ r(x_t, u_t) + \beta [V(x_{t+1})] \}, \qquad x_t\ given,$$

for $\beta \in [0, 1)$.

Instead of trying to solve for the function V directly, we provide a procedure that leads ultimately to the solution for V, but that provides insights in the intermediate steps. This procedure is more useful in economic applications and shows sequences of state and control variables that solve a dynamic optimization problem with dynamic programming techniques.

First, we specifically model the evolution of the state variable as a deterministic difference equation, or

$$x_{t+1} = g(x_t, u_t), \qquad x_0\ given \qquad (1.80)$$

where the set $\{x_{t+1}, x_t: x_{t+1} \leq g(x_t, u_t)\}$ is a convex set. The infinite horizon problem facing an individual is to maximize discounted utility by choice of a sequence of controls subject to the evolution of the state in (1.80), or

$$\underset{\{u_t\}_{t=0}^{\infty}}{Max} \sum_{t=0}^{\infty} \beta^t r(x_t, u_t)$$

$$subject\ to \quad x_{t+1} = g(x_t, u_t), \quad x_0\ given.$$

The solution to the recursive problem involves finding a contingency plan, or policy function that maps the state into the control, or

$$u_t = h(x_t)$$

where h is a well defined time invariant function.

The Bellman equation for problem (1.81) is

$$V(x_t) = \underset{\{u_t\}}{Max} \{ r(x_t, u_t) + \beta[V(x_{t+1})]\}$$

$$= \underset{\{u_t\}}{Max} \{ r(x_t, u_t) + \beta\ [V(g(x_t, u_t)]\}. \tag{1.83}$$

The first order necessary condition for the right-hand side of (1.83) is

$$[\partial r(x_t, u_t)/\partial u_t] + \beta[\{\partial V(x_{t+1})/\partial x_{t+1}\}\{\partial g(x_t, u_t)/\partial u_t\}] = 0. \tag{1.84}$$

The value function (1.83) can be appropriately differentiated yielding a version of the Benveniste and Scheinkman formula:

$$\partial V(x_t)/\partial x_t = [\partial r(x_t, u_t)/\partial x_t] + \beta[\{\partial V(x_{t+1})/\partial x_{t+1}\}\{\partial g(x_t, u_t)/\partial x_t\}]. \tag{1.85}$$

Expression (1.85) can be updated and substituted into (1.84) to obtain the so-called Euler equation for the problem:

$$[\partial r(x_t, u_t)/\partial u_t] + \beta[(\{\partial r(x_{t+1}, u_{t+1})/\partial x_{t+1}\} + \beta\ [\{\partial V(x_{t+2})/\partial x_{t+2}\} \times \{\partial g(x_{t+1}, u_{t+1})/\partial x_{t+1}\}])\ \{\partial g(x_t, u_t)/\partial u_t\}] = 0. \tag{1.86}$$

Fortunately, in many economic applications we can arrange state and controls such that we make the evolution of the state independent of x_t (that is, the choice of state and controls is not unique), or

$$\partial g(x_t, u_t)/\partial x_t = 0, \quad all\ t \tag{1.87}$$

and the Euler equation becomes

$$[\partial r\,(x_t\,,\,u_t)/\partial u_t] + \beta E_t[\{\partial r\,(x_{t+1}\,,\,u_{t+1})/\partial x_{t+1}\}\ \{\partial g(x_t\,,\,u_t)/\partial u_t\}\,] = 0. \tag{1.88}$$

Equation (1.88) is very useful because it provides, together with the evolution of the state, (1.80), the needed mapping (1.82) that solves for the control problem. Ultimately, the value function can be solved by working with the Euler equation (1.88) as well.

One method used to find the invariant function h that solves the Euler equation is a guess-and-verify method. For example, substituting (1.82) into (1.88) we obtain,

$$[\partial r\,(x_t\,,\,h(x_t))/\partial u_t] + \beta[\{\partial r\,(x_{t+1},\,h(x_{t+1}))/\partial x_{t+1}\} \times \{\partial g(x_t\,,\,h(x_t))/\partial u_t\}] = 0 \tag{1.89}$$

which can be appropriately solved for h . In this case, the verify part can be checked using the transversality condition

$$lim_{t\to\infty}\, \beta^{t+1}\,(\,x_{t+1}\,[\partial V\,(x_{t+1}\,)/\,\partial x_{t+1}\,]\,) = 0$$

which asserts that the discounted value of the state, priced by its marginal value, $\partial V(x_t\,)/\partial x_t$, at infinity is null. In the next section, we present Bellman's method in the case of dynamic stochastic problems in discrete time, while the last section extends to the continuous time case.

1.8 Stochastic Dynamic Optimization in Discrete Time

In many economic problems, we want to extend the horizon of the decision maker to include a sequence of time periods, possibly to infinity. We describe here dynamic programming methods under conditions of risk. In the discrete time case of section 1.7, consider an individual whose current economic status is described by a state variable, x_t , assumed to be a scalar for simplicity. In the current period t, x_t , is fully observed but its future values are random as of time t, or the sequence $(x_{t+1},\ x_{t+2},\ ...)$ are random variables. Denote the choices available for this individual in period t by u_t , where u_t can be thought of as a control variable used to achieve some objective. Taking u_t as given, the process describing the evolution of the state variable, x_t, is a assumed to be a Markov process with well defined support, and we denote by $\Phi_t(x_{t+1} \mid x_t, u_t)$ the cumulative probability distribution function of the state conditional on information available at time t, or the pair $\{x_t, u_t\}$, i.e.

$$Prob_t(x_{t+1} \leq x \mid x_t, u_t) = \Phi_t(x_{t+1} \mid x_t, u_t).$$

Individuals are assumed to discount future periods using the constant discount factor $\beta \in [0, 1)$.

In the finite horizon case, the decision process ends in some arbitrary period T, with terminal payoff depending on the state at T, x_T. However, in the infinite horizon case, the final payoff is null and the decision process becomes recursive. We shall see that to be able to analyze the sequential problem in a recursive manner is a useful simplification.

Let us first consider the finite horizon case. We look at the sequential decision process by splitting it in two parts: the immediate period and the future beyond that. In the current date t, with current state x_t, the individual choice of u_t yields immediate reward denoted by the function $r_t(x_t, u_t)$. Now imagine the following: from period t onwards the individual makes all decisions optimally. Denote by $V_t(x_t)$ the expected present discounted value of all the rewards when from period t onwards the individual makes all decisions optimally. Similarly, $E_t[V_{t+1}(x_{t+1})]$ is the expected present discounted value, at period $t+1$, of all the rewards when from period $t+1$ onwards the individual makes all decisions optimally. Hence $V_{t+1}(x_{t+1})$ is called the continuation value of the sequential problem. It is clear that in period t, the sum of the immediate value plus the continuation value are expressed as

$$r_t(x_t, u_t) + \beta E_t[V_{t+1}(x_{t+1})] \tag{1.90}$$

where $\beta E_t[V_{t+1}(x_{t+1})]$ denotes the expected discounted (to time t) value of V_{t+1} at the current period t. Given the definition of V_t and expression (1.90), it is easy to conclude that V_t can be written as

$$V_t(x_t) = \underset{\{u_t\}}{Max} \{ r_t(x_t, u_t) + \beta E_t[V_{t+1}(x_{t+1})] \} \tag{1.91}$$

where the maximization in the right-hand side is with respect to the control u_t and

$$E_t[V_{t+1}(x_{t+1})] = \int [V_{t+1}(x_{t+1})] \, d\Phi_t(x_{t+1} \mid x_t, u_t).$$

Expression (1.91) is called the Bellman equation of the sequential optimization problem, i.e. dynamic programming. Examining (1.91) closely, we note that a complication arises because of the dependence of the functions r, V and Φ upon the time index t. However, under the assumed conditions examined in section 1.7, this dependence disappears in the infinite horizon case and the problem becomes recursive. For an infinite horizon, the Bellman equation can be appropriately written, for any t, as

$$V(x_t) = \underset{\{u_t\}}{Max} \{ r(x_t, u_t) + \beta E_t [V(x_{t+1})]\} \qquad (1.92)$$

where the maximization in the right-hand side is with respect to the control u_t and

$$E_t [V(x_{t+1})] = \int [V(x_{t+1})] \, d\Phi(x_{t+1} \mid x_t, u_t).$$

Bellman equation (1.92) is a functional equation in the function V and can be solved appropriately for the unknown function V, yielding a solution for the dynamic programming problem.

(a) The Stochastic Control Problem

Instead of trying to solve for the function V directly in (1.92), we provide a procedure that leads ultimately to the solution for V, but that provides insights in the intermediate steps. This procedure is more useful in economic applications and shows sequences of state and control variables that solve the stochastic dynamic optimization problem with dynamic programming techniques.

First, we specifically model the evolution of the state variable as a stochastic difference equation with the Markov property, or

$$x_{t+1} = g(x_t, u_t, \epsilon_{t+1}) \qquad (1.93)$$

where the set $\{x_{t+1}, x_t: x_{t+1} \leq g(x_t, u_t, \epsilon_{t+1})\}$ is a convex set. In the difference equation (1.93), x_0 is known with certainty and ϵ_t is a sequence of identically and independently distributed (i.i.d.) random variables with cumulative probability distribution function

$$Prob(\epsilon_t \leq e) = F(e), \quad all\ t$$

for an infinite state space. The realization of ϵ_{t+1} is after the current decision u_t. The infinite horizon problem facing an individual is to maximize expected discounted utility by choice of a sequence of controls subject to the evolution of the state in (1.93), or

$$\underset{\{u_t\}_{t=0}^{\infty}}{Max}\ E_0\,[\Sigma_{t=0}^{\infty}\ \beta^t\ r\,(x_t\,,\,u_t)] \tag{1.94}$$

$$subject\ to \qquad x_{t+1} = g\,(x_t\,,\,u_t,\,\epsilon_{t+1}),$$

$$\{x_0\}\ and\ probability\ distribution\ of\ \{\epsilon\}\ given.$$

The solution to the recursive problem involves finding a contingency plan, or policy function that maps the state into the control, or

$$u_t = h\,(x_t)$$

where h is a well defined time invariant function.

Given (1.92), the Bellman equation for problem (1.94) is

$$V\,(x_t) = \underset{\{u_t\}}{Max}\,\{\,r\,(x_t\,,\,u_t) + \beta\,E_t\,[V(x_{t+1})]\}$$

$$= \underset{\{u_t\}}{Max}\,\{\,r\,(x_t\,,\,u_t) + \beta\,E_t\,[V(g(x_t\,,\,u_t,\,\epsilon_{t+1})]\} \tag{1.96}$$

where

$$E_t\,[V\,(g(x_t\,,\,u_t,\,\epsilon_{t+1}))]= \int [V(g(x_t\,,\,u_t,\,\epsilon_{t+1}))]\ dF(e).$$

The first order necessary condition for the right-hand side of (1.96) is

$$[\partial r\,(x_t\,,\,u_t)/\,\partial u_t] + \beta\,E_t\,[\{\partial V(x_{t+1})/\,\partial x_{t+1}\}\ \{\partial g(x_t\,,\,u_t\,,\,\epsilon_{t+1})/\partial u_t\,\}\,] = 0. \tag{1.97}$$

The value function (1.96) can be appropriately differentiated yielding a version of the Benveniste and Scheinkman formula:

$$\partial V\,(x_t\,)/\,\partial x_t = [\partial r\,(x_t\,,\,u_t)/\,\partial x_t] + \beta\,E_t\,[\{\partial V(x_{t+1})/\,\partial x_{t+1}\} \times \{\partial g(x_t,\,u_t\,,\epsilon_{t+1})/\partial x_t\,\}\,]. \tag{1.98}$$

Expression (1.98) can be updated and substituted into (1.97) to obtain the so-called stochastic Euler equation for the problem:

$$[\partial r(x_t, u_t)/\partial u_t] + \beta E_t[(\{\partial r(x_{t+1}, u_{t+1})/\partial x_{t+1}\} +$$
$$\beta E_{t+1}[\{\partial V(x_{t+2})/\partial x_{t+2}\}\{\partial g(x_{t+1}, u_{t+1}, \epsilon_{t+2})/\partial x_{t+1}\}]) \times$$
$$\{\partial g(x_t, u_t, \epsilon_{t+1})/\partial u_t\}] = 0. \qquad (1.99)$$

Fortunately, in many economic applications we can arrange state and controls such that we make the evolution of the state independent of x_t, or

$$\partial g(x_t, u_t, \epsilon_{t+1})/\partial x_t = 0, \quad all\ t \qquad (1.100)$$

and the stochastic Euler equation becomes

$$[\partial r(x_t, u_t)/\partial u_t] + \beta E_t[\{\partial r(x_{t+1}, u_{t+1})/\partial x_{t+1}\} \times$$
$$\{\partial g(x_t, u_t, \epsilon_{t+1})/\partial u_t\}] = 0. \qquad (1.101)$$

Equation (1.101) is very useful because it provides, together with the evolution of the state, (1.93), the needed mapping (1.95) that solves for the stochastic control problem. Substituting (1.95) into (1.101) we obtain,

$$[\partial r(x_t, h(x_t))/\partial u_t] + \beta E_t[\{\partial r(x_{t+1}, h(x_{t+1}))/\partial x_{t+1}\} \times$$
$$\{\partial g(x_t, h(x_t), \epsilon_{t+1})/\partial u_t\}] = 0 \qquad (1.102)$$

which can be appropriately solved as $u_t = h(x_t)$. Ultimately, the value function can be solved by working with the stochastic Euler equation as well.

(b) Example: The Stochastic Ramsey Problem

Consider an economic planner choosing sequences of consumption and investment, with *100%* depreciation of physical capital, $\{c_t, k_{t+1}\}_{t=0}^{\infty}$, to maximize expected discounted utility subject to a resources constraint, or

$$\underset{\{c_t, k_{t+1}\}_{t=0}^{\infty}}{Max} E_0[\Sigma_{t=0}^{\infty} \beta^t u(c_t)]$$

$$subject\ to \quad k_{t+1} = \theta_t f(k_t) - c_t, \qquad k_0\ given$$

where θ_t is a random variable that follows an i.i.d. lognormal distribution (see Chapter 8 for a discussion of lognormal random variables), $\log \theta_t \sim N(0,\sigma^2)$, and the functions u and f are twice continuously differentiable functions and strictly increasing and concave, $u' > 0$, $u'' < 0$, $f' > 0$, $f'' < 0$. For this application, we can appropriately choose the state and control as

$$x_t \equiv (k_t\,,\,\theta_t)$$

$$u_t \equiv k_{t+1}.$$

Rewriting the resources constraint as

$$c_t = \theta_t f(k_t) - k_{t+1}$$

and substituting into the function u, the Bellman equation can be written as

$$V(k_t\,,\,\theta_t) = \underset{\{k_{t+1}\}}{Max}\,\{\,u\,(\theta_t f(k_t) - k_{t+1}) + \beta\, E_t\,[V(k_{t+1}\,,\,\theta_{t+1})]\}$$

and the first order necessary condition is

$$[-\,u'\,(\theta_t f(k_t) - k_{t+1})\,] + \beta\, E_t\,[\{\partial\, V(k_{t+1},\theta_{t+1})/\,\partial\, k_{t+1}\}\,] = 0.$$

The Benveniste and Scheinkman equation is

$$\partial\, V\,(k_t,\,\theta_t)/\,\partial\, k_t = u'\,(\theta_t f(k_t) - k_{t+1}\,)\;\theta_t f\,'\,(k_t).$$

The last equation can be updated and substituted into the first order condition to yield the stochastic Euler equation

$$[-\,u'\,(\theta_t f(k_t) - k_{t+1})\,] + \beta\, E_t\,[u'\,(\theta_{t+1} f(k_{t+1}) - k_{t+2}\,)\;\theta_{t+1} f\,'\,(k_{t+1})\,] = 0$$

as a second order stochastic difference equation in physical capital, k_t.

When we assume special cases for the functions u and f, the stochastic Euler equation delivers the policy function directly with pencil and paper only. For example, let utility be logarithmic and production be Cobb-Douglas, or

$$u(c) = log\, c\;;$$

$$f(k) = k^{\alpha},\ \alpha\in(0,1).$$

In this case, the stochastic Euler equation becomes

$$[1/(\theta_t\, k_t^{\alpha} - k_{t+1})\,] = \beta\, E_t[\,(\theta_{t+1}\,\alpha\;\; k_{t+1}^{\alpha-1})/\,(\theta_{t+1}\, k_{t+1}^{\alpha} - k_{t+2})]$$

i.e. a second order stochastic difference equation in k_t. We use the guess-and-verify method to find the policy function from the last expression.

We guess that the policy function,

$$u_t = k_{t+1} = h\,(x_t) = h\,(k_t\,,\theta_t)$$

takes the form

$$k_{t+1} = h\,(k_t\,,\theta_t) = X\;\;\theta_t\;\; k_t^{\alpha}$$

where X is an undetermined coefficient to be determined. Substituting the policy function into the stochastic Euler equation, we obtain

$$[1/(\theta_t k_t^{\alpha} - X\ \theta_t\ k_t^{\alpha})] = \beta E_t[\ (\theta_{t+1}\ \alpha\ k_{t+1}^{\alpha-1})/(\theta_{t+1} k_{t+1}^{\alpha} - X\ \theta_{t+1}\ k_{t+1}^{\alpha})]$$

$$= \beta E_t[\ \alpha / \{X \theta_t k_t^{\alpha} (1-X)\}]$$

which may be appropriately simplified to yield

$$X = \alpha\beta$$

so that the policy function becomes

$$k_{t+1} = h(k_t, \theta_t) = \alpha\beta\ \theta_t\ k_t^{\alpha}$$

where k_0 is given and $\log \theta_t \sim N(0, \sigma^2)$.

The verify part can be checked by substituting the solution into the transversality condition

$$lim_{t\to\infty} \beta^{t+1} E_t[\ \{\partial V(k_{t+1}, \theta_{t+1}) / \partial k_{t+1}\}\ k_{t+1}] = 0$$

which is satisfied for $k_{t+1} = h(k_t, \theta_t) = \alpha\beta\ \theta_t\ k_t^{\alpha}$ as long as $\beta \in [0,1)$, thus the solution is correct. Using the resources constraint, $k_{t+1} = \theta_t f(k_t) - c_t$ and the policy function, the evolution of consumption becomes

$$c_t = (1-\alpha\beta)\ \theta_t\ k_t^{\alpha}$$

and consumption and investment follow the recursive linear system in logarithms

$$\log\ k_{t+1} = (\log \alpha\beta) + (\alpha \log k_t) + \log \theta_t, \qquad k_0 \text{ given}, \ \log \theta_t \sim N(0, \sigma^2)$$

$$\log c_t = [\log(1-\alpha\beta)] + \alpha \log k_t + \log \theta_t$$

that may be appropriately analyzed with the methods learned in section 1.6.

In this example, we can also find the value function $V(k_t, \theta_t)$ using the stochastic Euler equation, the policy function and Bellman equation. The value function has the form

$$V(k_t, \theta_t) = X_1 + X_2 (\log\ k_t) + X_3 \log \theta_t$$

where $\{X_1, X_2, X_3\}$ are undetermined coefficients to be determined. Using the stochastic Euler equation we obtain

$$X_2 = \alpha / (1-\alpha\beta)$$

and then using the Bellman equation, we obtain

$$X_3 = 1/(1-\alpha\beta)$$

$$X_1 = [1/(1-\beta)] \{ [\log (1-\alpha\beta)] + [\alpha\beta/(1-\alpha\beta)] (\log \alpha\beta) \}$$

which solves for the undetermined coefficients in the value function. We use the methods presented in this section extensively in Chapter 7.

1.9 Notes on Stochastic Differential Equations

In Chapter 8, we discuss in detail the continuous time version of a discrete random walk process like the one discussed in section 1.6. Thus, for a dynamic variable $x \geq 0$, its infinitesimal change is denoted dx and a random walk with drift in continuous time is expressed as

$$dx(t) = \mu x(t)\, dt + \sigma x(t)\, dZ \tag{1.103}$$

where $\{\mu, \sigma\}$ are constants and dZ denotes a series of increments of Z, the Brownian motion process, with $x(0)$ given. The increments are usually written as $dZ = \varepsilon_t (dt)^{1/2}$, which is expression (8.16) in Chapter 8, where $\varepsilon_t \sim N(0,1)$, i.e. ε_t follows a standard normal distribution with mean 0 and variance 1. Thus, $E[dZ] = 0$, and $var\ (dZ) = dt$ and it is assumed that $Z(0)=0$ with probability one. Expression (1.103) is a stochastic differential equation. Here, our interest is in describing a solution for the stochastic differential equation.

Recalling Chapter 8, section 8.7, we apply Ito's Lemma to a transformation of x,

$$y = \log x$$

to obtain

$$dy = (1/x)\, dx(t) + (1/2)(-1/x^2)\, [dx(t)]^2. \tag{1.104}$$

We then substitute (1.103) into (1.104) to obtain

$$\begin{aligned} dy &= (1/x)\, [\mu x(t)\, dt + \sigma x(t)\, dZ] - (1/2x^2)\, [\mu x(t)\, dt + \sigma x(t)\, dZ]^2 \\ &= (1/x)\, [\mu x(t)\, dt + \sigma x(t)\, dZ] - (1/2x^2)\, [\mu^2 x(t)^2 (dt)^2 + \sigma^2 x(t)^2 (dZ)^2 \\ &\qquad + 2\mu x(t)^2\, dt\, \sigma\, dZ] \\ &\approx (1/x)\, [\mu x(t)\, dt + \sigma x(t)\, dZ] - (1/2x^2)\, [\sigma^2 x(t)^2\, dt] \end{aligned}$$

where the last step is the approximation that ignores the terms in dt of order higher than one, or $(dt)^2$ and $dtdZ=(dt)^{3/2}$. Thus, we have that

$$dy = [\mu - (1/2)\, \sigma^2]\, dt + \sigma\, dZ. \tag{1.105}$$

We can now integrate both sides from 0 to t:

$$\int_0^t dy = \int_0^t [\mu - (1/2)\, \sigma^2]\, dt + \int_0^t \sigma\, dZ$$

yielding

$$\begin{aligned} y(t) - y(0) &= \int_0^t [\mu - (1/2)\, \sigma^2]\, dt + \int_0^t \sigma\, dZ \\ &= [\mu - (1/2)\, \sigma^2]\, t + \sigma\, Z(t) \end{aligned}$$

since $Z(0)$ is zero with probability one. Hence, we obtain

$$y(t) = y(0) + [\mu - (1/2)\, \sigma^2]\, t + \sigma\, Z(t). \tag{1.106}$$

Recalling that

$$y = \log x \;\Rightarrow\; x = exp(y), \quad x(0) = exp(y(0))$$

we note that

$$\begin{aligned} x(t) &= exp\,(y(t)) \\ &= exp\,(y(0))\, [\, exp([\mu - (1/2)\, \sigma^2]\, t + \sigma\, Z(t))\,] \\ &= x(0)\, [\, exp(\, [\mu - (1/2)\, \sigma^2]\, t + \sigma\, Z(t)\,)] \end{aligned} \tag{1.107}$$

yields a solution for the stochastic differential equation (1.103). It is clear that the intuition behind the technique to solve stochastic differential equations lies in the specific transformation of the right-hand side of (1.103).

The transformation used above allows us to use integration directly to solve for the stochastic differential equation. Notice also that taking expected value of (1.107) yields

$$\begin{aligned} E[x(t)] &= x(0)\, [\, exp([\mu - (1/2)\, \sigma^2]\, t + (1/2)\, \sigma^2\, t)\,] \\ &= x(0)\, [\, exp\, (\, \mu\, t\,)\,]. \end{aligned}$$

1.10 **Stochastic Dynamic Optimization in Continuous Time**

We follow analogously to section 1.7, the continuous time version of stochastic dynamic optimization. In general, continuous time stochastic dynamic optimization, applied to economic problems, imply the maximization over the infinite horizon of an objective function subject to a state evolving according to a stochastic differential equation. In particular, we let the state be denoted $x(t)$, the control be denoted $u(t)$, and the problem takes the form

$$Max\ E_o \int_0^\infty R\,(\,x(t),\ u(t),\ t\,)\ dt \qquad (1.108)$$

by choice of $u(t)$ subject to the evolution of the state according to the stochastic process

$$dx(t) = f(x(t),u(t),t)\ dt + dw \qquad (1.109)$$

where $x(0)$ is given. The functions R and f are twice continuously differentiable and dw is a proportional Brownian motion, or $dw \equiv \sigma\, x\ dZ$, where σ is the standard deviation of the growth of x, and dZ is a standard Brownian motion. Here R is the reward function and f is the "technology" of the evolution of the state.

According to Bellman's principle of optimality, the continuous time analogue of the discrete time expression (1.92) is, for an arbitrary interval Δt,

$$V(\,x(t),\ t\,) = \underset{\{u(t)\}}{Max}\ \{\ R(\,x(t),\ u(t),\ t\,)\ \Delta t + E[\ V(\,x(t{+}\Delta t),\ t{+}\Delta t\,)\ ]\ \} \qquad (1.110)$$

where $V(x(t),t)$ is the maximum value of the objective from time t onwards, i.e. an indirect "return" function, or the value function. First, we have to make expression (1.110) operational. We note that the expectation term in the inner maximum can be written as

$$E[\ V(\,x(t{+}\Delta t),\ t{+}\Delta t\,)\ ] = V(\,x(t),\ t) + E[\ dV(\,x(t),\ t)\ ]$$

which involves the change in the value function dV. However, for a continuously differentiable function V and the underlying evolution of the state (1.109), Ito's Lemma, e.g. Chapter 8, can be applied yielding

$$dV = [\ (\partial V/\partial t) + (\partial V/\partial x)\,f + (1/2)\ (\partial^2 V/\partial x^2)\sigma_w^2\ ]\ dt + (\partial V/\partial x)dw$$

where $\sigma_w^2 \equiv \sigma^2 x^2$. Taking expected value of the last expression yields

$$E[dV]/dt = [\ (\partial V/\partial t) + (\partial V/\partial x)\,f + (1/2)\ (\partial^2 V/\partial x^2)\sigma_w^2\ ]. \qquad (1.111)$$

Sometimes, it is useful in applications to define $E[dV]/dt \equiv L(V)$ as the Differential Generator of the value function. We can substitute (1.111) into the Bellman equation (1.110), divide by Δt and take the appropriate limit to obtain a more operational version of the Bellman equation

$$0 = \underset{\{u(t)\}}{Max} \{ R(x(t), u(t), t) + L(V) \}$$

$$= \underset{\{u(t)\}}{Max} \{ R(x(t), u(t), t) + (\partial V/\partial t) + (\partial V/\partial x) f + (1/2) (\partial^2 V/\partial x^2)\sigma_w^2 \}. \qquad (1.112)$$

The inner maximum and the Bellman equation must be satisfied for a solution of the maximization problem (1.108)-(1.109). However, an important simplification in economic applications is that the return function and consequently the value function take the time separable form:

$$R(x(t),u(t),t) = r(x(t),u(t)) \exp(-\beta t) \qquad (1.113a)$$

$$V(x(t),t) = J(x(t)) \exp(-\beta t) \qquad (1.113b)$$

where $\beta > 0$ is a constant discount rate and the functions r and J are twice continuously differentiable. In this case, the Bellman equation becomes

$$0 = \underset{\{u(t)\}}{Max} \{ r(x(t), u(t)) - \beta J(x(t)) + (\partial J/\partial x) f + (1/2) (\partial^2 J/\partial x^2)\sigma_w^2 \}. \qquad (1.114)$$

with a necessary first order condition for the inner maximum given by

$$(\partial r/\partial u) + (\partial J/\partial x) (\partial f/\partial u) = 0. \qquad (1.115)$$

Then, the solution of the maximization problem (1.108)-(1.109) involves finding the function J and control u that satisfy the Bellman equation and the first order condition for the inner maximum, given the evolution of the state dx. A feasible solution must also satisfy the transversality condition

$$lim_{t\to\infty} E [V(x(t),t)] = lim_{t\to\infty} E [J(x(t)) \exp(-\beta t)] = 0. \quad (1.116)$$

It is useful to obtain the continuous time analogue to the discrete time stochastic Euler equation in (1.89)-(1.91). We can apply Ito's Lemma to the function $(\partial J/\partial x)$ to obtain

$$d (\partial J/\partial x) = (\partial^2 J/\partial x^2)\, dx + (1/2) (\partial^3 J/\partial x^3)\sigma_w^2\, dt. \quad (1.117a)$$

Next, we differentiate Bellman equation (1.104) recognizing that the control and state are related by the first order condition (1.105) to obtain:

$$(\partial r/\partial u)(\partial u/\partial x) - \beta (\partial J/\partial x) + (\partial^2 J/\partial x^2) f + (\partial J/\partial x) (\partial f/\partial x) +$$

$$(\partial J/\partial x) (\partial f/\partial u) (\partial u/\partial x) + (1/2) (\partial^3 J/\partial x^3)\sigma_w^2 = 0$$

which upon substitution of the first order condition (1.115), rearranging and multiplying through by dt simplifies to

$$- (\partial J/\partial x)\, (\partial f/\partial x)\, dt + \beta\, (\partial J/\partial x)\, dt - (\partial^2 J/\partial x^2)\, f\, dt \;=\; (1/2)\, (\partial^3 J/\partial x^3) \sigma_w^{\,2}\, dt.$$

This last expression can be substituted into (1.117a) to obtain

$$d\, (\partial J/\partial x) = (\partial^2 J/\partial x^2)\, dx - (\partial J/\partial x)\, (\partial f/\partial x)\, dt + \beta\, (\partial J/\partial x)\, dt - (\partial^2 J/\partial x^2)\, f\, dt. \qquad (1.117b)$$

Substituting for dx from (1.109) and simplifying we obtain

$$d\, (\partial J/\partial x)/\, (\partial J/\partial x) = [\beta - (\partial f/\partial x)]\, dt + \; [(\partial^2 J/\partial x^2)/\, (\partial J/\partial x)]\, dw \qquad (1.118)$$

which describes the evolution of the change in the value function, i.e. the stochastic Euler equation in continuous time analogous to the discrete time version (1.99)-(1.101). The mean and variance of the change in the value function are given by

$$E[d\, (\partial J/\partial x)/\, (\partial J/\partial x)]/dt = \beta - (\partial f/\partial x) \qquad (1.118a)$$

$$E[\{d\, (\partial J/\partial x)/\, (\partial J/\partial x)\}^2]/dt = [(\partial^2 J/\partial x^2)/\, (\partial J/\partial x)]^{\,2} \sigma^{\,2}. \qquad (1.118b)$$

(a) Example: Stochastic Wealth Accumulation

Consider a simple example of a consumer who wants to maximize expected discounted utility over the infinite horizon subject to a stochastic wealth accumulation equation. The instantaneous utility, u, over consumption stream, c, is denoted $u(c)$, where as usual $u' > 0$ $(du/dc > 0)$, $u'' < 0$ $(d^2 u/dc^2 < 0)$. The problem can be stated as

$$\underset{\{c\}}{Max}\; E_o \int_0^\infty u\,(c)\, exp(-\beta t)\, dt$$

subject to the evolution of wealth according to the process

$$dW = [\,f(W) - c]\, dt + dw$$

where $W(0)$ is given, the function f satisfies $f' > 0$ $(df/dW > 0)$, $f'' < 0$ $(d^2 f/dW^2 < 0)$ and $dw \equiv W\sigma\, dZ$, for dZ a standard Brownian motion with $E[dw] = 0$, $E[dw^2] = W^2 \sigma^2 dt$, and σ is the standard deviation of the growth of wealth.

In this example, wealth is the state $(W \equiv x)$ and consumption is the control $(c \equiv u)$. The Bellman equation in (1.114) can be written as

$$0 = \underset{\{c(t)\}}{Max} \{u(c(t)) - \beta J(W(t)) + (\partial J/\partial W)[f(W) - c] + (1/2)(\partial^2 J/\partial W^2) W^2 \sigma^2\}.$$

The solution involves finding the consumption stream that satisfies the necessary first order condition for the inner maximum

$$u'(c) - (\partial J/\partial W) = 0.$$

and the function J that satisfies the Bellman equation

$$u(c(t)) - \beta J(W(t)) + (\partial J/\partial W)[f(W) - c] + (1/2)(\partial^2 J/\partial W^2) W^2 \sigma^2 = 0 \qquad (1.119a)$$

and transversality condition

$$lim_{t \to \infty} E[J(W(t)) exp(-\beta t)] = 0$$

given the evolution of the state

$$dW = [f(W) - c]\, dt + dw, \qquad W(0) \text{ given.}$$

In this case, the stochastic Euler equation can be found exactly as in (1.117)-(1.118) above. Applying Ito's Lemma to $J(W)$ we obtain

$$d(\partial J/\partial W) = (\partial^2 J/\partial W^2)\, dW + (1/2)(\partial^3 J/\partial W^3) W^2 \sigma^2\, dt.$$

Differentiating Bellman's equation and recognizing the necessary first order condition for the inner maximum yields

$$u'(\partial c/\partial W) - \beta(\partial J/\partial W) + (\partial^2 J/\partial x^2)(f - c) +$$

$$(\partial J/\partial W)[f' - (\partial c/\partial W)] + (1/2)(\partial^3 J/\partial W^3) W^2 \sigma^2 = 0.$$

Then, substituting for $u'(c) = (\partial J/\partial W)$, multiplying both sides by dt and rearranging yields

$$(\beta - f')(\partial J/\partial W)\, dt - (\partial^2 J/\partial W^2)(f - c)\, dt = (1/2)(\partial^3 J/\partial W^3) W^2 \sigma^2 dt$$

which may be appropriately substituted into the evolution of $(\partial J/\partial W)$ to obtain

$$d(\partial J/\partial W) = (\partial^2 J/\partial W^2)\, dW + (\beta - f')(\partial J/\partial W)\, dt - (\partial^2 J/\partial W^2)(f - c)\, dt.$$

Substituting for the evolution of the state dW and simplifying yields

$$d(\partial J/\partial W) = (\beta - f')(\partial J/\partial W)\, dt + (\partial^2 J/\partial W^2)\, dw.$$

We recall the first order condition

$$u'(c) - (\partial J/\partial W) = 0 \quad \Leftrightarrow \quad u''(c)\,(\partial c/\partial W) = (\partial^2 J/\partial W^2)$$

and substitute into the last expression to obtain the stochastic Euler equation as

$$d\,u'(c) \,/\, u'(c) = (\beta - f'\,)\,dt + [u''(c)\,(\partial c/\partial W)/u'(c)]\,dw$$

analogous to (1.108). Its expected value is simply

$$E[\,d\,u'(c) \,/\, u'(c)]/dt = \beta - f'$$

as in a model without risk; and the variance is

$$E[\{d\,u'(c) \,/\, u'(c)\}^2]/dt = [u''(c)\,(\partial c/\partial W)/u'(c)]^2\, W^2\sigma^2.$$

In many cases, it is not easy to find closed form solutions for the value function J that solves for the stochastic dynamic optimization problem. However, for well defined specialized cases of the functions u and f in this example, we can find closed form solutions. Let u take the power form and f take the linear form, or

$$u(c) = c^{\gamma}/\gamma, \qquad \gamma < 1$$

$$f(W) = \alpha\,W, \qquad \alpha > 0.$$

The specific form of u is studied in great detail in Chapter 4. For these functional forms, the Bellman equation in (1.119) becomes

$$c^{\gamma}/\gamma - \beta J(W(t)) + (\partial J/\partial W)\,[\alpha\,W - c] + (1/2)\,(\partial^2 J/\partial W^2)\,W^2\sigma^2 = 0$$

which can be rearranged and expressed in terms of the ratio (c/W) as

$$(W^{\gamma})\,[(c/W)^{\gamma}/\gamma] - \beta J(W(t)) + (\partial J/\partial W)\,[\alpha - (c/W)]\,W +$$
$$(1/2)\,(\partial^2 J/\partial W^2)\,W^2\sigma^2 = 0\,. \qquad (1.119b)$$

We conjecture that c/W is constant and treat the last equation as an ordinary differential equation in the function J. We apply a guess-and-verify method to solve the differential equation. Let

$$J(W) = W^{\gamma} \;\Rightarrow\; \partial J/\partial W = \gamma\,W^{\gamma-1} \;\Rightarrow\; \partial^2 J/\partial W^2 = \gamma\,(\gamma-1)W^{\gamma-2}$$

and substitute into the Bellman equation (1.119b) to obtain

$$[(c/W)^{\gamma}/\gamma] - \beta + \gamma\,[\alpha - (c/W)] + (1/2)\,\gamma\,(\gamma-1)\,\sigma^2 = 0. \qquad (1.119c)$$

Recall that for the specific function u and the guess J, the first order condition becomes

$$(c/W)^{\gamma} = \gamma\,(c/W)$$

which can be substituted into the Bellman equation (1.119c) to obtain a closed form solution for c/W:

$$c/W = [1/(1-\gamma)][\beta - \gamma\alpha - (1/2)\gamma(\gamma-1)\sigma^2]$$

confirming that the marginal propensity to consume out of wealth is constant. The stochastic Euler equation in this case can be easily computed. For the special functions and definitions above,

$$u' = c^{\gamma-1}$$

$$u'' = (\gamma-1)c^{\gamma-2}$$

$$\partial c/\partial W = [1/(1-\gamma)][\beta - \gamma\alpha - (1/2)\gamma(\gamma-1)\sigma^2]$$

$$f' = \alpha$$

$$dw \equiv W\sigma\, dZ,$$

and the stochastic Euler equation becomes

$$d\,u'(c)\,/\,u'(c) = (\beta - \alpha)\,dt + (\gamma-1)\,\sigma\, dZ$$

with

$$E[\,d\,u'(c)\,/\,u'(c)]/dt = (\beta - \alpha)$$

as in a model without risk; and variance

$$E[(\,d\,u'(c)\,/\,u'(c))^2\,]/dt = (\gamma-1)^2\,\sigma^2.$$

The evolution of the state in this case is

$$dW\,/\,W = [\alpha - (c/W)]\,dt + \sigma\, dZ$$

with

$$E[dW\,/\,W]/dt = [\alpha - (c/W)]$$

$$E[(dW\,/\,W)^2]/dt = \sigma^2$$

so that in equilibrium, c and W grow at the same average rate $\alpha - (c/W)$.

Finally, we must verify whether our guess is indeed correct by checking the transversality condition, which in this case takes the form

$$lim_{t\to\infty} E\,[\,W^{\gamma} exp(-\beta\, t)] = 0.$$

In order to find $W(t)$ along the equilibrium path we must solve the stochastic differential equation

$$dW\,/\,W = [\alpha - (c/W)]\,dt + \sigma\, dZ\,.$$

We have learned how to solve equations of this type in section 1.8. From expression (1.103),

$$\mu \equiv \alpha - (c/W) \quad and \quad \sigma \equiv \sigma,$$

thus the solution from (1.107) is

$$W(t) = W(0)\, exp(\{[\alpha - (c/W)] - (1/2)\, \sigma^2\}\, t + \sigma\, Z(t)\,)$$

so that

$$W(t)^{\gamma}\, exp(-\beta\, t) = W(0)^{\gamma}\, exp(\{\gamma\, [\alpha - (c/W)] - (1/2)\, \gamma\, \sigma^2 - \beta\}\, t + \gamma\, \sigma\, Z(t)).$$

We compute the expected value of the last expression as

$$E\, [W(t)^{\gamma}\, exp(-\beta\, t)] =$$

$$W(0)^{\gamma}\, exp(\, \{\gamma\, [\alpha - (c/W)] - (1/2)\, \gamma\, \sigma^2 - \beta\}\, t + (1/2) \gamma^2 \sigma^2\, t\,).$$

$$= W(0)^{\gamma}\, exp(\, \{\gamma\, [\alpha - (c/W)] - (1/2)\, \gamma\, \sigma^2 - \beta + (1/2) \gamma^2 \sigma^2\, \}\, t\,).$$

$$= W(0)^{\gamma}\, exp(\, \{\gamma\, [\alpha - (c/W)] - (1/2)\, \gamma\, (1 - \gamma) \sigma^2 - \beta\}\, t\,).$$

Substituting into the transversality condition and computing requires that

$$\gamma\, [\alpha - (c/W)] - (1/2)\, \gamma\, (1 - \gamma) \sigma^2 - \beta < 0$$

for the condition to be satisfied. But, recalling the solution for c/W,

$$c/W = [1/(1-\gamma)][\beta \, - \gamma\, \alpha \, - (1/2)\, \gamma \, (\gamma - 1)\, \sigma^2]$$

it is easy to show that the transversality condition is satisfied and the guess is correct when

$$c/W > 0$$

which we assume to be the case, and thus the solution is complete. It turns out that this condition applies to several extensions of this simple model. We use the methods described in this section extensively in Chapter 8.

1.11 Summary

In this chapter, we provided some of the basic tools that are going to be used in this book. The reader is invited to consult back and forth this initial summary to recall some of the techniques used in the models and applications presented. The Notes on the Literature and references provide further reading if necessary.

Problems (Note: Many problems present material complementary to this chapter)

1. Find the local extrema of the function in $\mathbb{R}^1$:

$$F(x) = x^4 - 4x^3 + 4x^2 + 4.$$

2. Find the local extrema of the function in $\mathbb{R}^1$:

$$F(x) = x^3 - x^2 + 2x$$

3. Let $f(x) = a^x$, for $a>0$ a real number. Compute $f'(x)$.

4. Compute the indefinite integral: $\int (1/x)\, dx$.

5. Consider the function $F(x,y)=kx^a y^b$ for $\{k,a,b\}$ strictly positive constants. Compute the Hessian of F.

6. Consider the system of implicit functions:

$$F(x, y; z) = a$$

$$G(x, y; z) = b$$

where x,y are endogenous variables, z is an exogenous variable, and a,b are constants. Compute dx/dz and dy/dz.

7. Find the local extrema of the function in $\mathbb{R}^2$:

$$F(x,y) = x^2 + y^3 - xy - 1.$$

8. Compute a maximum for the function $f(x,y)$ subject to the constraint (convex) set $\{g(x,y)=c\}$, for a constant c.

9. Let the continuous random variable Y have a probability density function

$$f(y) = (1/40)\, e^{-y/40}$$

on the support $0 \leq y < \infty$. Show that $\int_0^\infty f(y)\, dy = 1$.

10. Let the matrix $\boldsymbol{A}$ be defined as

$$\begin{vmatrix} 1 & 3 \\ 0 & -5 \end{vmatrix}$$

Compute the eigenvalues of $\boldsymbol{A}$ (see Chapter 6).

11. Let the value of the assets held by an individual after t periods be given by

$$V(t) = A \;\; exp(t^{\alpha})$$

where $A>0$ and $\alpha>0$ are constants and exp is the exponential operator.
(i) If the market interest rate is $r>0$, what is the present discounted value of the assets?
(ii) What is the holding time of assets that maximizes the present discounted value?

12. Let a linear system of equations be denoted by

$$y = X\beta + u$$

where $\boldsymbol{y}$ is a $(N\times 1)$ column vector, $\boldsymbol{X}$ is a $(N\times k)$ matrix, $\boldsymbol{\beta}$ is a $(k\times 1)$ column vector and $\boldsymbol{u}$ is a $(N\times 1)$ column vector.
(i) Show that $\boldsymbol{u}^T\boldsymbol{u}$ is a quadratic form.
(ii) Give the solution for the first order necessary conditions for

$$\underset{\{\beta\}}{Min}\; \boldsymbol{u}^T\boldsymbol{u}\,.$$

for the case of $N=k=2$.

13. Let $z = f(x_1,x_2)$, with f: $\mathbb{R}^2_+ \rightarrow \mathbb{R}^1$ and f is C^2.
(i) Give the first order necessary condition for an extremum of z.
(ii) Derive the second order sufficient condition for a maximum and a minimum of z.

14. Let there be $n=1,2,\ldots N$ observations of the variables: y_n , x_{in}, $i=1,2,\ldots k$. Consider the matrix system of these observations embedded in the linear model:

$$y = X\beta + u$$

where $\boldsymbol{y}$ is a $(N\times 1)$ column vector, $\boldsymbol{X}$ is a $(N\times k)$ matrix, $\boldsymbol{\beta}$ is a $(k\times 1)$ column vector and $\boldsymbol{u}$ is a $(N\times 1)$ column vector. Let $E[\boldsymbol{u}]=0$ and $Var[\boldsymbol{u}]=\sigma^2\boldsymbol{I}$. Consider the matrix

$$A = [\,I - X(X^T X)^{-1} X^T\,].$$

Show that:

$$E\,[\hat{\boldsymbol{u}}^T \hat{\boldsymbol{u}}\,] = E\,[\boldsymbol{u}^T \boldsymbol{A}\,\boldsymbol{u}\,] = \sigma^2\,(N-k).$$

15. Let $y = f\,(x_1,x_2\,)$ be a production function where $(x_1,x_2\,) \geq 0$ are inputs and $y \geq 0$ is output and f is given by

$$f\,(x_1,x_2\,) = [\,\alpha\, x_1{}^r + (1-\alpha)\, x_2{}^r\,]^{1/r}$$

for $r \leq 1$. Show, using Taylor's approximation methods, that

$$1 + r\log f\,(x_1,x_2\,) \approx 1 + r\log x_1{}^{\alpha}\, x_2{}^{1-\alpha}.$$

[Hint: Recall that $x_1{}^r = exp\,(\,r\log x_1\,)$]

16. Solve a utility maximization problem by the Lagrangean method where utility is

$$u = u\,(x_1,x_2\,),$$

with $u\colon \mathbb{R}^2_+ \to \mathbb{R}^1$, u is C^2, and the budget constraint is

$$M = p_1\, x_1 + p_2\, x_2.$$

Compute comparative statics with respect to $(\,M, p_1, p_2\,)$.

17. Let $U(c,n) = (\,\log c) - \alpha\, n$, with $\alpha \in (0,1)$, where c is consumption and n is hours of work, and

$$c = a\, f(n) + Y_0$$

where $a \in (0,\infty)$, $Y_0 \in [0,\infty)$, and $f(.)$ is strictly increasing and strictly concave. Maximize utility subject to the technology constraint by choice of (c,n).

18. In a two-period problem, if

$$U(x_t, x_{t+1}) = \log x_t + [1/(1+\theta)] \log x_{t+1};$$

Solve for x_t and x_{t+1}, by maximizing U subject to the relevant intertemporal budget constraint (please, check the sufficient second order condition!).

19. Let X be a continuous random variable on the real line with probability density function:

$$g_X(x) = (2\pi)^{-1/2} \exp(-x^2/2).$$

Show that the expected value of X is zero.

20. A firm produces two outputs denoted y_1 and y_2, with the use of two inputs z_1 and z_2. Let the constants a_{ij} denote the amount of input i required to produce one unit of output j, organized in the input-output matrix of coefficients:

$$A = \begin{vmatrix} 3 & 1 \\ 2 & 5 \end{vmatrix}$$

and inputs are related to outputs according to

$$\begin{vmatrix} z_1 \\ z_2 \end{vmatrix} = A \begin{vmatrix} y_1 \\ y_2 \end{vmatrix} \quad or \quad z = A\,y.$$

Suppose that you are given the level of inputs, $z_1=10$ and $z_2=20$. Find the levels of output generated by the inputs and interpret the results.

21. Let a producer have $n=1,2,\ldots,N$ outputs, $x_1, x_2, \ldots, x_N$ at given prices $p_1, p_2, \ldots, p_N$ facing $C(x_1, x_2, \ldots, x_N)$ as operating costs and $D(k)$ the cost of maintaining output capacity k for each output $x_1, x_2, \ldots, x_N$, i.e. $x_1 \leq k, x_2 \leq k, \ldots, x_N \leq k$.
Find the nonnegative optimal demands and capacity that maximize the producer's profit by the Kuhn-Tucker method.

Notes on the Literature

General references on mathematics for economists include: Allen (1976); Binmore (1982), which is a simple text that presents some useful topics (it is good for a quick consultation); Chiang (1984) is widely known, but outdated; De la Fuente (2000) is a medium to upper level treatment of mathematical methods with focus useful for those interested in macroeconomics; Hoy et al (2001) is a good low to medium level book in the mathematical methods with a variety of applications suitable for both undergraduates and graduate students; Klein (1998) is a low to medium level text which is useful for its good applications; Simon and Blume (1994) is one of the most clear and concise modern texts, it is rigorous in its mathematics and presents a variety of examples and applications; Sydseater and Hammond (1995) is a book that covers most of the topics that are of interest to economists at a basic low to medium level; and Takayama (1985, 1993) is more advanced and rigorous. Varian (1996) and Huang and Crooke (1997) present computational applications with Mathematica.

Section 1.2 on integration follows Sydseater and Hammond (1995) and a more interested reader may consult Bugrov and Nikolsky (1982).

Section 1.3 on statistics is very rudimentary. Useful references in this subject are: Chung (1979); Gut (1995); Hogg and Tanis (1993); Hogg and Craig (1978) which is a classic for an introduction to probability theory at a more advanced level; Mendenhall et al (1990) is a very competent and widely used medium level mathematical statistics graduate textbook; and Pratt, Raiffa and Schlaifer (1995) is a classic on the economic applications.

Section 1.4 on linear algebra follows Simon and Blume (1994).

Section 1.5 on static optimization includes several references: Dixit (1990) is a very well written monograph which gives a useful introduction to the relevant economic applications; Intriligator (1971) is a classic for Ph.D. students in economics and alike, the material is medium to advanced and some of the topics such as Nonlinear Programming are very useful for consultation; Lambert (1985); Leonard and Long (1992); Mas-Colell et al (1995) is the most modern synthesis of basic economic theory from a rigorous perspective and it has some of the main economic applications; Novshek (1993) is a good book for basic microeconomic applications and exercises are mostly solved in the book; Silberberg (1990) presents well written explanations of the mathematical methods applied to economic problems at a medium level; Simon and Blume (1994); Sudaram (1996) is a very good introduction to the mathematics of (mostly static) optimization used in economics; Sydseater and Hammond (1995); Takayama (1985, 1993), and Gollier (2001).

Section 1.6 on stochastic difference equations follows Enders (1995), Sargent (1976) and Ljungqvist and Sargent (2000). Gandolfo (1980) presents straightforward mathematical methods in deterministic difference and differential equations.

Section 1.7 on optimization in discrete time is a basic introduction to Dynamic Programming in the certainty case and the interested reader can consult Bellman (1957) and Stokey and Lucas (1989), for a more in depth treatment.

Section 1.8 on stochastic optimization in discrete time has several references: Chow (1997) examines several existing applications in dynamic economics and finance using the Lagrangean technique as opposed to Dynamic Programming; Dixit and Pindyck (1994) and Dixit (1990) are very useful and present very well written explanations of the methodology of dynamic programming; Ferguson and Lim (1998) is a low level introduction to some of the dynamic models used in economics in general with some attention to macroeconomic theory; Harris (1987) is a monograph that gives a rigorous treatment of topics in capital theory and the dynamics of capital accumulation; Stokey and Lucas (1989) give an in depth presentation of the recursive approach to economics; Sargent (1987) and Ljungqvist and Sargent (2000) are classic presentations of dynamic programming techniques in discrete time; Ross (1996) presents the underlying stochastic processes.

The Benveniste and Scheinkman formula is in Benveniste and Scheinkman (1979) and Ekeland and Scheinkman (1986) discuss the transversality condition.

Section 1.9 on stochastic differential equations follows Oksendal (1985). Chow (1997); Dixit and Pindyck (1994); Dixit (1990); Ferguson and Lim (1998); Kamien and Schwartz (1991); Karatzas and Shevre (1991, 1996); Malliaris and Brock (1982); Musiela and Rutkowsky (1998); Neftci (1996); Turnovsky (2000) and Varian (1996) all present theory and applied material. Finally, section 1.10 on stochastic optimization in continuous time follows Turnovsky (2000), Merton (1990), Dixit and Pindyck (1994), and Dixit (1990). Several presentations of theory and applications include: Ferguson and Lim (1998); Kamien and Schwartz (1991); Karatzas and Shevre (1991, 1996); Malliaris and Brock (1982); Musiela and Rutkowsky (1998); Neftci (1996); and Varian (1996).

References

Allen, Richard G. D. (1967) *Mathematical Analysis for Economists*. St. Martin's Press, New York, NY.

Bellman, Richard (1957) *Dynamic Programming*. Princeton University Press, Princeton, NJ.

Benveniste, Lawrence M. and Jose A. Scheinkman (1979) "On the Differentiability of the Value Function in Dynamic Models of Economics." *Econometrica*, 47, 727-732.

Binmore, Kenneth G. (1982) *Mathematical Analysis: A Straightforward Approach*, Second Edition. Cambridge University Press, Cambridge, UK.

Bugrov, Y. S. and S. M. Nikolsky (1982) *Differential and Integral Calculus*. Mir Publishers, Moscow.

Chiang, Alpha C. (1984) *Fundamental Methods of Mathematical Economics*, Third Edition. McGraw Hill Book Co., New York, NY.

Chow, Gregory (1997) *Dynamic Economics: Optimization by the Lagrangean Method*. Oxford University Press, Oxford, UK.

Chung, Kai L. (1979) Elementary Probability Theory and Stochastic Processes. Springer-Verlag, New York, NY.

De la Fuente, Angel (2000) *Mathematical Methods and Models for Economists*. Cambridge University Press, Cambridge, UK.

Dixit, Avinash K. and Robert S. Pindyck (1994) *Investment Under Uncertainty*. Princeton University Press, Princeton, NJ.

Dixit, Avinash K. (1990) *Optimization in Economics*, Second Edition. Oxford University Press, New York, NY.

Ekeland, Ivar and Jose A. Scheinkman (1986) "Transversality Conditions for Some Infinite Horizon Discrete Time Optimization Problems." *Mathematics of Operations Research*, 11, 216-229.

Enders, Walter (1995) *Applied Econometric Time Series*. John Wiley&Sons, New York, NY.

Ferguson, Brian S. and G. C. Lim (1998) *Introduction to Dynamic Economic Models*. Manchester University Press, Manchester, U.K.

Gandolfo, Giancarlo (1980) *Economic Dynamics: Methods and Models*. North Holland, Amsterdam.

Gollier, Christian (2001) *The Economics of Risk and Time.* The MIT Press, Cambridge, MA.

Gut, Alan (1995) *An Intermediate Course in Probability Theory.* Springer-Verlag, New York, NY.

Harris, Milton (1987) *Dynamic Economic Analysis.* Oxford University Press, New York, NY.

Hogg, Robert V. and Elliot A. Tanis (1993) *Probability and Statistical Inference.* MacMillan Press, New York, NY.

Hogg, Robert V. and Allen T. Craig (1978) *Introduction to Mathematical Statistics.* Macmillan Publishing Co., New York, NY.

Hoy, Michael, John Livernois, Chris McKenna, Ray Rees, and Thanasis Stengos (2001) *Mathematics for Economics*, Second Edition. The MIT Press, Cambridge, MA.

Huang, Cliff J. and Phillip S. Crooke (1997) *Mathematics and Mathematica for Economists.* Basil Blackwell Publishers, New York, NY.

Intriligator, Michael D. (1971) *Mathematical Optimization and Economic Theory.* Prentice Hall, Englewood Cliffs, NJ.

Kamien, Morton I. and Nancy L. Schwartz (1991) *Dynamic Optimization: The Calculus of Variations and Optimal Control in Economics and Management*, Second Edition. North-Holland, Amsterdam.

Karatzas, Ioannis and Stephen Shevre (1991) *Brownian Motion and Stochastic Calculus.* Springer-Verlag, New York, NY.

Karatzas, Ioannis and Stephen Shreve (1998) *Methods of Mathematical Finance.* Springer-Verlag, New York, NY.

Klein, Michael (1998) *Mathematical Methods for Economics.* Addison-Wesley, Reading, MA.

Lambert, Peter J. (1985) *Advanced Mathematics for Economists: Static and Dynamic Optimization.* Basil Blackwell, Cambridge, MA.

Leonard, David and Ngo V. Long (1992) *Optimal Control Theory and Static Optimization in Economics.* Cambridge University Press, Cambridge, UK.

Ljungqvist, Lars and Thomas J. Sargent (2000) *Recursive Macroeconomic Theory.* The MIT Press, Cambridge, MA.

Malliaris, A. G. and William A. Brock (1982) *Stochastic Methods in Economics and Finance.* North-Holland, Amsterdam.

Mas-Colell, Andreu, Michael Whinston and Jerry Green (1995) *Microeconomic Theory.* Oxford University Press, Cambridge, MA.

Mendenhall, William, Dennis D. Wackerly and Richard L. Scheaffer (1990) *Mathematical Statistics with Applications*, Any edition. PWS-Kent, Boston, MA.

Merton, Robert C. (1990) *Continuous Time Finance.* Basil Blackwell, New York, NY.

Musiela, Marek and Marek Rutkowsky (1998) *Martingale Methods in Financial Modelling*, Springer-Verlag, New York, NY.

Neftci, Salih N. (1996) *Introduction to the Mathematics of Financial Derivatives.* Academic Press, San Diego, CA.

Novshek, William (1993) *Mathematics for Economists.* Academic Press, San Diego, CA.

Oksendal, Bernt (1985) *Stochastic Differential Equations.* Springer-Verlag, New York, NY.

Pratt, John W., Howard Raiffa, and Robert Schlaifer (1995) *Statistical Decision Theory.* The MIT Press, Cambridge, MA.

Ross, Sheldon (1996) *Stochastic Processes*, Second Edition. John Wiley&Sons, New York, NY.

Sargent, Thomas J. (1976) *Macroeconomic Theory.* Academic Press, San Diego, CA.

Sargent, Thomas J. (1987) *Dynamic Macroeconomic Theory.* Harvard University Press, Cambridge, MA.

Silberberg, Eugene (1990) *The Structure of Economics: A Mathematical Analysis,* Second Edition. McGraw Hill Book Co., New York, NY.

Simon, Carl P. and Lawrence Blume (1994) *Mathematics for Economists.* W.W. Norton and Co., New York, NY.

Stokey, Nancy L., Robert E. Lucas, with Edward C. Prescott (1989) *Recursive Methods in Economic Dynamics.* Harvard University Press, Cambridge, MA.

Sudaram, Rangarajan K. (1996) *A First Course in Optimization Theory.* Cambridge University Press, Cambridge, UK.

Sydseater, Knut and Peter J. Hammond (1995) *Mathematics for Economic Analysis.* Prentice Hall, New York, NY.

Takayama, Akira (1985) *Mathematical Economics*, Second Edition. Cambridge University Press, Cambridge, UK.

Takayama, Akira (1993) *Analytical Methods in Economics.* The University of Michigan Press, Ann Harbor, MI.

Turnovsky, Stephen J. (2000) *Methods of Macroeconomic Dynamics*, Second Edition. MIT Press, Cambridge, MA.

Varian, Hal R., Editor (1996) *Computational Economics and Finance: Modeling and Analysis with Mathematica.* Springer-Verlag, New York, NY.

2. Mean-Variance Approach to Financial Decision-Making

2.1 Introduction

We present analysis of a static, one-period investment problem where agents have full information about the first two moments of the probability distribution of the random returns. This is usually referred to as the mean-variance framework.

Suppose an individual invests an amount X_0 at the beginning of the period and receives, at the end-of-period, the amount X_1, possibly a random variable. The gross end-of-period return on the investment is $R=X_1/X_0$ and the rate of return is $r=(X_1/X_0)-1$, both possibly random. An investment may work in both directions:

(i) a long position refers to acquiring the asset for X_0 and selling it for X_1 at the end-of-period, a profit occurs when $X_1 - X_0>0$;

(ii) a short position refers to selling an asset, that you may not own but may borrow from a third party, for X_0 and acquiring it back at the end-of-period for X_1 and possibly returning it to the original lender, then a profit occurs whenever $X_1 - X_0<0$.

A trademark of the mean-variance approach is that the individual decision-making process is driven by information about the mean and variance (or standard deviation) of the asset return only. In the next chapter, we shall discuss the case where individuals have well defined utility functions over random wealth, such that the investment decision-making process is fully modeled.

2.2 Portfolio Mean Return and Variance

Suppose there are n assets available in the market place indexed by $i=1,2,...n$. The amount to be invested initially is X_0. We can form a portfolio, satisfying the budget constraint:

$$X_0 = \Sigma_{i=1}^{n} X_{0i}, \tag{2.1}$$

where X_{0i} is the amount invested in each asset i. The portfolio becomes itself another asset. If one is allowed to go short on the ith asset, then $X_{0i}<0$, whereas if short sales are not allowed we impose the additional constraint $X_{0i} \geq 0$.

Let w_i be defined as the fraction of the ith asset in the portfolio, or

$$X_{0i} = w_i X_0 \quad with \quad \Sigma_{i=1}^{n} w_i=2.$$

Similarly, if short sales of the ith asset are allowed then $w_i<0$, whereas if short sales are not allowed we impose the additional constraint $w_i \geq 0$.

Letting R_i be the gross return on the ith asset and recalling that $R=X_1/X_0$, implies that the portfolio gross return is

$$R = \Sigma_{i=1}^{n} R_i w_i X_0 / X_0 = \Sigma_{i=1}^{n} R_i w_i \tag{2.2}$$

that is, a weighted average of the individual gross returns. Similarly, recalling that the rate of return is $r=(X_1/X_0)-1$, the rate of return on the portfolio is

$$r = \Sigma_{i=1}^{n} r_i w_i \tag{2.3}$$

a weighted average of the individual rates of return. Hence, the portfolio is a new asset with gross return, possibly random, determined by the linear combination in (2.2), and rate of return, possibly random, determined by the linear combination in (2.3).

The next step is to introduce randomness per se. Let each asset have a random rate of return with (unconditional) expected value $E[r_i] = r_i^m$ and (unconditional) variance $E[(r_i - r_i^m)^2]=\sigma_{ii}$, with $\sigma_{ii} \in [0,\infty)$. Thus, forming a portfolio with all risky assets and fixed weights w_i yields a portfolio expected rate of return

$$r^m = \Sigma_{i=1}^{n} r_i^m w_i \tag{2.4}$$

and associated portfolio variance

$$\sigma^2 = E[(\Sigma_{i=1}^{n} r_i w_i - \Sigma_{i=1}^{n} r_i^m w_i)^2]. \tag{2.5}$$

A general formula for the portfolio variance may be obtained by induction starting with the simplest case of two assets, that is $n=2$:

$$\sigma^2|_{n=2} = (w_1^2 E[r_1^2] - w_1^2 r_1^{m\,2}) + (w_2^2 E[r_2^2] - w_2^2 r_2^{m\,2}) + 2 w_1 w_2 (E[r_1 r_2] - r_1^m r_2^m)$$

$$= w_1^2 \sigma_{11} + w_2^2 \sigma_{22} + 2 w_1 w_2 \sigma_{12} \qquad (2.6)$$

where $\sigma_{12} = \sigma_{21} = E[(r_1 - r_1^m)(r_2 - r_2^m)]$ is the (unconditional) covariance between the rates of return of the two assets. Formula (2.6) is a simple positive definite quadratic form, $\boldsymbol{w}^T \Sigma\, \boldsymbol{w}$, where in this case Σ is a (2×2) variance-covariance symmetric matrix and $\boldsymbol{w}$ is a (2×1) vector column of fixed weights. An important remark about (2.6) is that when the sign of the covariance between the rates of return is negative, we can naturally adjust the portfolio weights to reduce the variance of the portfolio, eventually to zero. Thus, the sign of the covariance between the returns is crucial to determine the risk embodied in the portfolio. We explore this characteristic below and in later chapters as well.

Therefore, for a general portfolio with n assets indexed by $i=1,2...n$ or $j=1,2...n$, the variance of the associated portfolio is given by

$$\sigma^2 = \Sigma_{i=1}^n \Sigma_{j=1}^n w_i w_j \sigma_{ij} \qquad (2.7)$$

a linear function of the variances and covariances of the assets in the portfolio, or a quadratic form $\boldsymbol{w}^T \Sigma\, \boldsymbol{w}$, where Σ is a $(n\times n)$ variance-covariance symmetric matrix and $\boldsymbol{w}$ is a $(n\times 1)$ vector column of fixed weights.

(a) Alternative Portfolios in Mean-Variance Space

First, consider the simplest case of $n=2$, or two risky assets. Let $w_1=(1-\alpha)$ and $w_2=\alpha$ for $\alpha\in[0,1]$. Therefore, no short sales are being considered momentarily. Assume the following ordering of means and variances:

$$r_2^m > r_1^m \quad and \quad \sigma_{22} > \sigma_{11} \qquad (2.8)$$

where σ_{ij}, $i=j$, is the variance of the ith asset. Inequality (8) says that asset labeled 2 has a higher mean rate of return but a higher variance relative to asset labeled 1. The mean rate of return of the portfolio is a function of α:

$$r^m(\alpha, \ldots) = (1-\alpha)\, r_1^m + \alpha\, r_2^m \qquad (2.9)$$

and the standard deviation of the portfolio is also a function of α:

$$\sigma(\alpha, ...) = \{(1-\alpha)^2\sigma_{11} + \alpha^2\sigma_{22} + 2\alpha(1-\alpha)\sigma_{12}\}^{1/2}. \quad (2.10)$$

In most cases, considering the standard deviation is equivalent to considering the variance because both move in the same direction. Recall that the coefficient of correlation between the rates of return of the two assets is given by

$$\rho = \sigma_{12}/\sigma_1\sigma_2$$

for $\rho \in [-1,1]$, where σ_i is the standard deviation of the mean rate of return of the ith asset. Therefore, the standard deviation of the portfolio may be rewritten as a function of α and ρ as

$$\sigma(\alpha, \rho, ...) = \{(1-\alpha)^2\sigma_{11} + \alpha^2\sigma_{22} + 2\alpha(1-\alpha)\rho\,\sigma_1\sigma_2\}^{1/2}. \quad (2.10')$$

The mean returns, variances and covariances are given parameters, but we can isolate α and ρ so that for each value of ρ, we examine how the return $r^m(\alpha)$ in (2.9) and $\sigma(\alpha, \rho)$ in (2.10') vary as α varies. Let us consider different values of ρ:

(i) $\rho = 1$ is an upper bound for $\sigma(\alpha, \rho)$ where the two rates of return are perfectly positively correlated. From equation (2.10') we obtain

$$\sigma(\alpha, \rho=1) = (1-\alpha)\sigma_1 + \alpha\sigma_2 \quad (2.11)$$

and (2.9) gives $r^m(\alpha)$. From equations (2.9) and (2.11), we solve for α as

$$\alpha = [\sigma(., \rho=1) - \sigma_1]/(\sigma_2 - \sigma_1) \quad (2.12)$$

so that α varies linearly with $\sigma(., \rho=1)$, or

$$\partial\alpha/\partial\sigma(., \rho=1) = 1/(\sigma_2 - \sigma_1) > 0 \quad (2.13)$$

where the last inequality follows from the assumed ordering (2.8). As the standard deviation of the portfolio increases the share of asset *2* in the portfolio increases. This is because asset *2* is the riskier asset and the correlation of the rates of return is perfect and positive, equal one. Substituting (2.12) into (2.9) we obtain

$$r^m(., \rho=1) = \{1-[\sigma(., \rho=1) - \sigma_1]/(\sigma_2 - \sigma_1)\}\, r_1^m + \{[\sigma(., \rho=1) - \sigma_1]/(\sigma_2 - \sigma_1)\}\, r_2^m \quad (2.14)$$

so that in mean-standard deviation space we have:

$$\partial r^m(.)/\partial\sigma(.)\,|_{\rho=1} = (r_2^m - r_1^m)/(\sigma_2 - \sigma_1) > 0 \quad (2.15)$$

where the last inequality follows from (2.8). Figure 2.1 illustrates the linear relationship: as σ increases, α increases by (2.13) and r^m increases by (2.15). In between the bounds of $\alpha \in [0,1]$, we have the possible alternative portfolios consistent with $\rho = 1$. In this case, the returns are perfectly correlated and there is no scope for diversification in terms of choices of α that could increase mean return and reduce standard deviation. However, the case below where returns are negatively correlated presents opportunity for diversification.

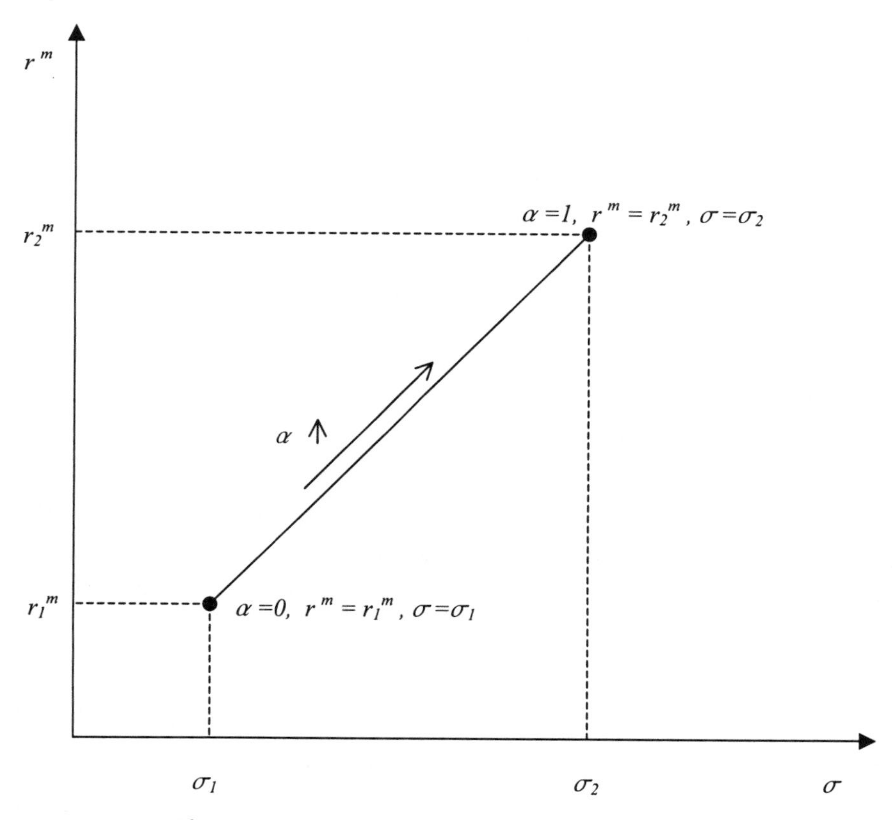

Fig. 2.1 Mean-Standard Deviation Space, $\rho = 1$

(ii) $\rho = -1$ is a lower bound for $\sigma(\alpha, \rho)$ where the two rates of return are perfectly negatively correlated, and from equation (10) we obtain

$$\sigma(\alpha, \rho = -1) = \{ [(1-\alpha)\sigma_1 - \alpha\sigma_2]^2 \}^{1/2}$$

$$= | (1-\alpha)\sigma_1 - \alpha\sigma_2 | \qquad (2.16)$$

where the absolute value operator is applied because the standard deviation is the positive square root of the variance. Hence, in the domain of $\alpha \in [0,1]$, given (σ_1, σ_2), the sign of the expression in (2.16) switches and so does the relationship with $r^m(\alpha)$. First, note that the term in (2.16) switches sign at some critical level of α where the relationship in (2.16) is equal to zero, or

$$\sigma(\alpha, \rho = -1) = (1-\alpha)\sigma_1 - \alpha\sigma_2 = 0 \implies$$

$$\alpha |_{\sigma(\alpha, \rho=-1)=0} = \sigma_1 / (\sigma_1 + \sigma_2). \qquad (2.17)$$

Recalling equation (2.8), for α in the lower domain

$$\alpha \in [0, \sigma_1 / [\sigma_1 + \sigma_2]),$$

the sign of (2.16) is

$$(1-\alpha)\sigma_1 - \alpha\sigma_2 > 0$$

and

$$\alpha |_{\alpha \in [0, \sigma 1/[\sigma 1 + \sigma 2])} = [\sigma_1 - \sigma(., \rho=-1)] / (\sigma_1 + \sigma_2) \qquad (2.18)$$

so that α varies linearly with $\sigma(., \rho=-1)$, or

$$\partial\alpha / \partial\sigma(., \rho=-1) |_{\alpha \in [0, \sigma 1/[\sigma 1 + \sigma 2])} = -1 / (\sigma_1 + \sigma_2) < 0. \qquad (2.19)$$

In this case, as the standard deviation of the portfolio increases, the share of asset *2* in the portfolio decreases because the share of the riskier asset is initially low and the correlation of the rates of return is perfectly negative, equal *-1*. Substituting (2.18) into (2.9) we obtain

$$r^m(\alpha \in [0, \sigma_1/[\sigma_1+\sigma_2]), \rho = -1) =$$

$$\{1 - ([\sigma_1 - \sigma()] / [\sigma_1 + \sigma_2])\} r_1^m + \{[\sigma_1 - \sigma()] / (\sigma_1+\sigma_2)\} r_2^m \qquad (2.20)$$

so that in mean-standard deviation space we have:

$$\partial r^m() / \partial\sigma() |_{\rho=-1, \alpha \in [0, \sigma 1/[\sigma 1 + \sigma 2])} = [r_1^m - r_2^m] / (\sigma_1 + \sigma_2) < 0 \qquad (2.21)$$

where the last inequality follows from (2.8). As σ increases, α decreases by (2.19) and r^m decreases by (2.21).

Recalling (2.8) again, for α in the upper domain

$$\alpha \in (\sigma_1 / (\sigma_1 + \sigma_2), 1],$$

the sign of (2.16) is

$$(1-\alpha)\sigma_1 - \alpha\sigma_2 < 0$$

and by the absolute value rule

$$\alpha \,|_{\alpha \in (\sigma 1/[\sigma 1 + \sigma 2], 1]} = [\sigma(., \rho=-1) + \sigma_1] / (\sigma_1 + \sigma_2) \qquad (2.22)$$

so that α varies linearly with $\sigma(., \rho=-1)$, or

$$\partial\alpha / \partial\sigma(., \rho=-1) \,|_{\alpha \in (\sigma 1/[\sigma 1 + \sigma 2], 1]} = 1 / (\sigma_1 + \sigma_2) > 0. \qquad (2.23)$$

As the standard deviation of the portfolio increases the share of asset *2* in the portfolio increases since the share of the riskier asset is initially high and the correlation of the rates of return is *-1*. Substituting (2.22) into (2.9) we obtain

$$r^m(\alpha \in (\sigma_1 / [\sigma_1 + \sigma_2], 1], \rho = -1) =$$

$$\{1 - ([\sigma(.) + \sigma_1] / [\sigma_1 + \sigma_2])\} r_1^m + \{[\sigma(.) + \sigma_1] / (\sigma_1 + \sigma_2)\} r_2^m \qquad (2.24)$$

so that in mean-standard deviation space we have:

$$\partial r^m(.) / \partial\sigma(.) \,|_{\rho=-1, \alpha \in (\sigma 1/[\sigma 1 + \sigma 2], 1]} = (r_2^m - r_1^m)/(\sigma_1 + \sigma_2) > 0 \qquad (2.25)$$

where the last inequality follows from (2.8). As σ increases, α increases by (2.23) and r^m increases by (2.25).

Figure 2.2 illustrates the relationship when $\rho = -1$: as α increases from below in the lower domain $\alpha \in [0, \sigma_1 / [\sigma_1 + \sigma_2])$ then σ decreases by (2.19) and r^m increases by (2.21); at the critical level, point A, where

$$\alpha \,|_{\sigma(\alpha, \rho=-1) = 0} = \sigma_1 / (\sigma_1 + \sigma_2),$$

using (2.9) the mean rate of return of the portfolio is

$$r^m = (\sigma_2 r_1^m + \sigma_1 r_2^m) / (\sigma_1 + \sigma_2)$$

and

$$\sigma = 0$$

so that risk can be perfectly diversified. It is important to observe that point A is a risk-free point; which means that when the rates of returns are perfectly negatively correlated, risk can be perfectly diversified. In this case, whenever the return on one asset is increasing, the return on the other asset is decreasing. Setting up a portfolio with $\alpha = \sigma_1 / (\sigma_1 + \sigma_2)$,

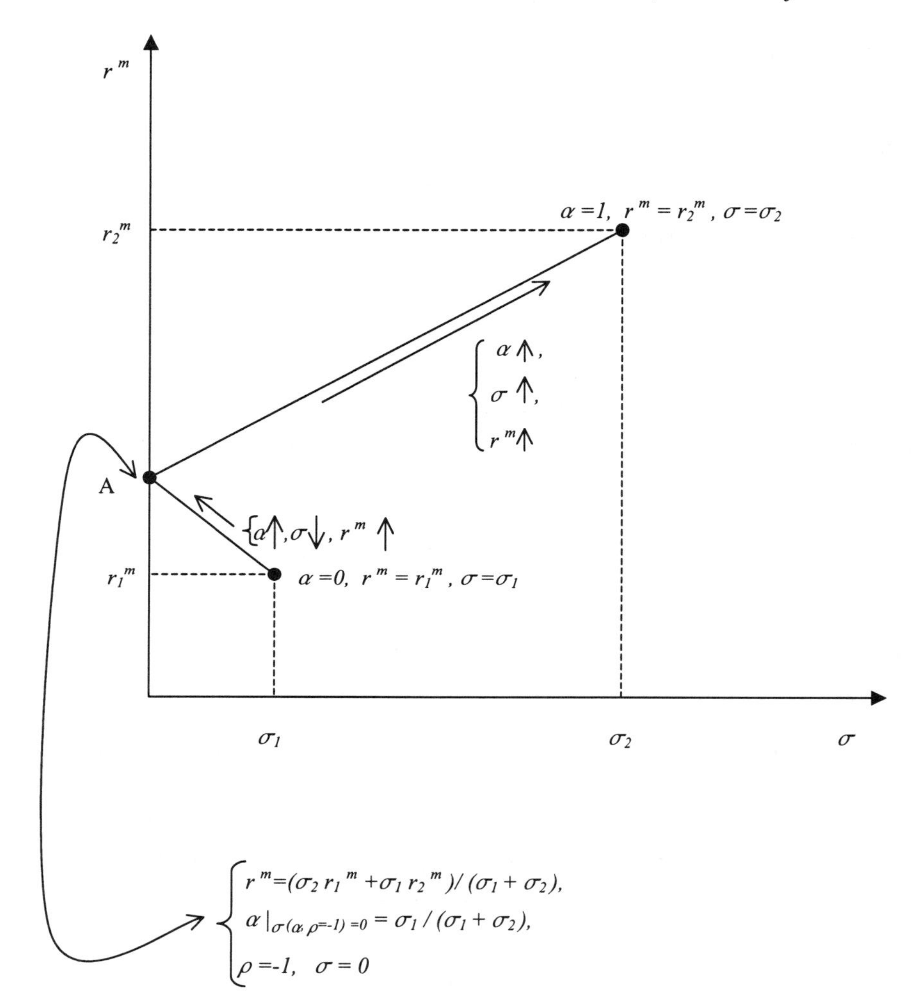

Fig. 2.2 Mean-Standard Deviation Space, $\rho=-1$

creates a risk-free fund. Finally, as α increases in the upper domain $\alpha \in (\sigma_1/(\sigma_1+\sigma_2), 1]$ then, using (2.23) σ increases, and r^m increases by (2.25).

Combining case (i) and (ii), we obtain resulting portfolios in the domain $\rho \in [-1,1]$. Figure 2.3 illustrates the alternative cases in mean-standard deviation space, the so-called cone of diversification. As ρ

decreases from the upper bound of $+1$, the alternative portfolios are described by hyperbolas convex towards the critical mean return

$$r^m=(\sigma_2 r_1^m+\sigma_1 r_2^m)/(\sigma_1+\sigma_2)$$

where $\sigma=0$, a cone of diversification. For each given ρ, choosing $\alpha\in[0,1]$ yields alternative combinations of mean and standard deviation. As α increases from its lower bound, diversification gives a higher mean rate of return eventually at the cost of a higher standard deviation. In cases where the rates of return of the assets are negatively correlated, risks can be diversified so that a higher mean return is consistent with a low standard deviation. Figure 2.3 also illustrates the direction in which the cone of diversification would expand were short sales allowed, that is the cases where $\alpha<0$, and $\alpha>1$.

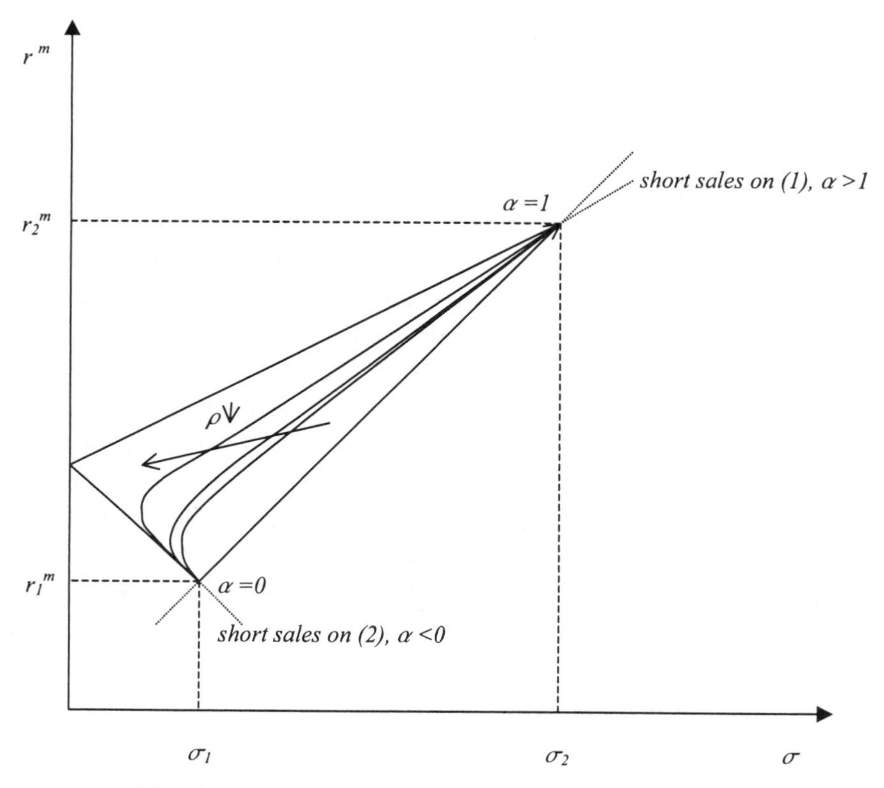

Fig. 2.3 Mean-Standard Deviation Space, $\rho\in[-1,1]$

2.3 The Efficient Frontier

In order to derive the efficient frontier for the general case of several assets, we start with some heuristics relating to the simplest case of two risky assets discussed above. Figure 2.4 illustrates the arguments. We choose an arbitrary ρ and consider the alternative feasible portfolios along the convex hyperbola DE. First note that there will be a unique choice of α that minimizes the standard deviation of the portfolio and yields an implied mean rate of return; this is illustrated by point A where the portfolio α^* yields a minimum standard deviation σ^* and an implied mean rate of return r^{m*}.

Agents supposedly prefer lower standard deviation under the assumption that there is some degree of "aversion to risk" as measured by the standard deviation of the portfolio. Alternatively, consider an arbitrary standard deviation, σ', with the associated portfolio and implied mean rate of return that this arbitrary standard deviation yields.

It is easy to note that, given the convexity of the portfolio hyperbola, there are two portfolios that are consistent with the standard deviation denoted σ'. One is the portfolio at point B with implies mean rate of return $r^{m\prime}$ and the other is at point C with implies rate of return $r^{m\prime\prime} < r^{m\prime}$. Plausibly, assuming that agents have no satiation in mean rate of return, the portfolio at point B, denoted α', is strictly preferred to point C, since at B the standard deviation is identical but the mean rate of return is higher. Therefore, the efficient portion of the portfolio hyperbola is always the set of portfolios, i.e. for the alternative α's, equal or above the minimum variance portfolio, that is the portion AE of the hyperbola. Efficiency in this context refers to a tradeoff between mean rate of return and standard deviation: an agent is willing to take more risk if awarded a higher expected rate of return on the portfolio and vice-versa.

It is important to note in Figure 2.4 that the whole convex hyperbola DE is represented by a simple convex combination of the two assets in the portfolio. This will prove useful below when we derive an efficient frontier in the presence of several assets.

Consider now the case of several risky assets, $n \geq 3$, and retain the assumption of no short sales momentarily. We know from above that the

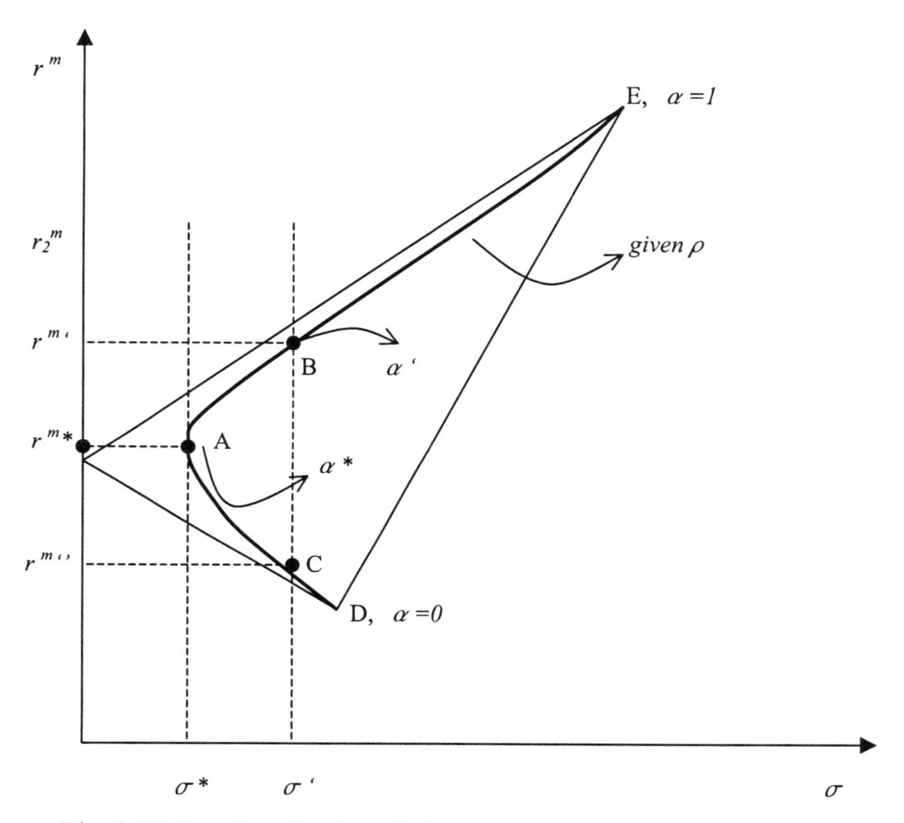

Fig. 2.4 Mean-Standard Deviation Space, Arbitrary $\rho \in [-1,1]$, $n=2$

portfolios formed with the several assets will have expected rates of return and variances that will depend on the alternative weights of each asset in the portfolio and the variances and covariances of the returns of these assets. Therefore, in the two dimensional space of mean and standard deviation, each asset is a point in that space and alternative weights, variances and covariances yield several possible connections between these points, i.e. alternative portfolios. When the number of assets is large, the connections become almost indistinguishable or a shaded area indicating a set of feasible portfolios. The set of feasible portfolios has three important properties:

(i) For each expected rate of return, given the alternative variances and covariances of the returns of the assets, there will be a portfolio with respective weights w_i, $i=1,2,...n$, that yield the smallest variance among the set of feasible portfolios;

(ii) Connecting all the points of smallest variance for the alternative expected rates of return yields a minimum-variance set of portfolios (a subset of the set of feasible portfolios) which is strictly convex towards the mean rate of return axis;

(iii) Within the minimum-variance set of portfolios, there is a unique portfolio, with respective weights w_i^*, $i=1,2,...n$, which yields the smallest variance across all expected rates of return; the portion of the minimum-variance set that is above this point is called the efficient frontier.

Figure 2.5 illustrates the main properties of the set of feasible portfolios in mean-standard deviation space. In Figure 2.5(a), the shaded area represents the set of feasible portfolios with alternative weights, variances and covariances of asset returns. Property (i) says that for each expected rate of return there will be a portfolio that yields the smallest standard deviation as illustrated by points E, A, D, C, and B. Property (ii) is described by the envelope of all the alternative portfolios that give the smallest standard deviation, the line EB, which is strictly convex towards the mean rate of return axis. Recall that in the case of two risky assets studied in section 2.2, the cone of diversification has a risk-free point where the standard deviation is zero for $\rho=-1$. The cone of diversification for several assets cannot be consistent with zero standard deviation portfolios because when $n\geq 3$ it is not possible that the correlations be perfectly negative for all assets. Figure 2.5(b) illustrates property (iii). At point A, the portfolio w_i^* yields the smallest variance, σ^*, with implied expected rate of return r^{m*}. Using the efficiency criteria discussed above regarding the tradeoff between expected rate of return and standard deviation, the efficient frontier is the portion of the minimum-variance set above point A, portion denoted AB. Relaxing the no short sales constraint would simply expand the cone of diversification in all directions but would not change the main properties of the set of feasible portfolios.

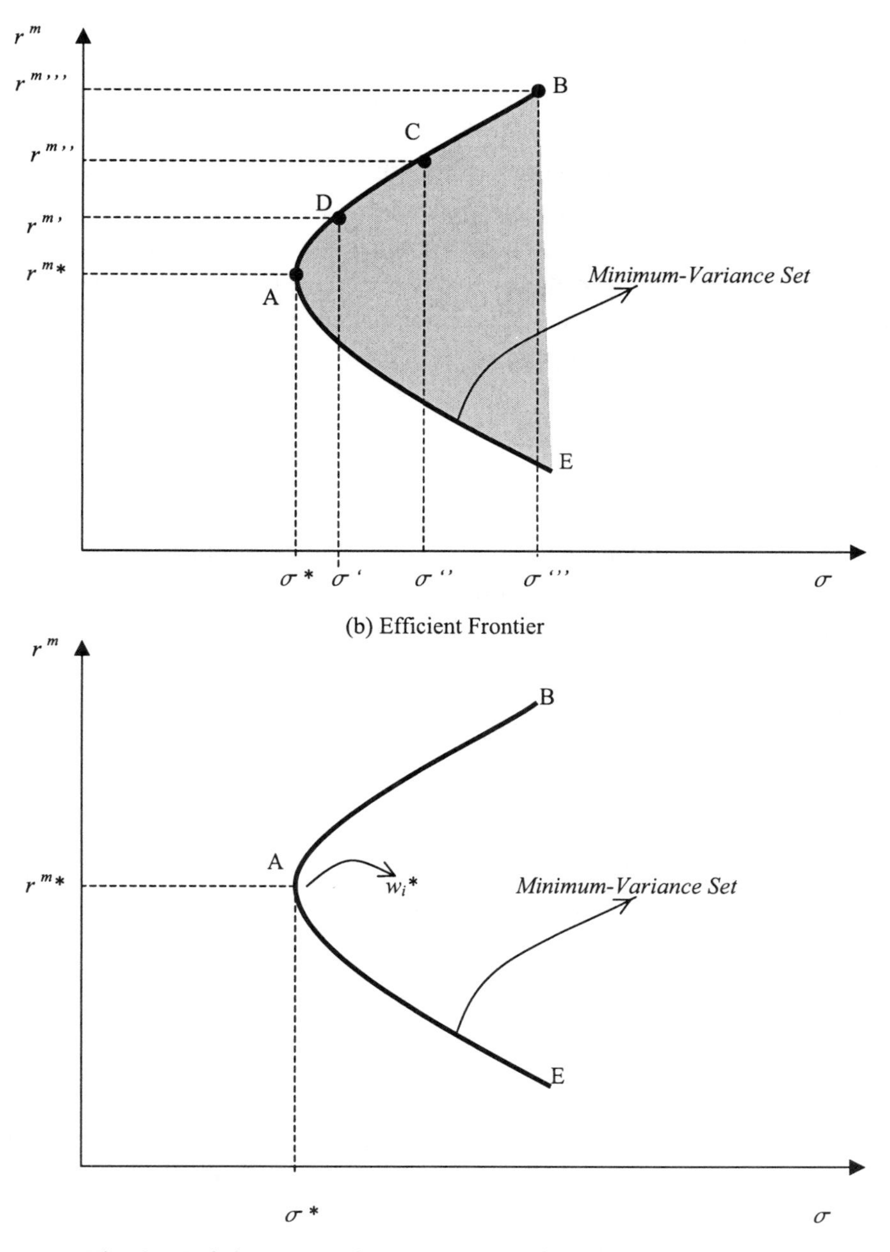

Fig. 2.5 Minimum-Variance Set and Efficient Frontier, $n \geq 3$

(a) Computing the Efficient Frontier

We are indeed able to compute the minimum-variance set and the efficient frontier. We start with the minimum-variance set. We use the Markowitz formulation, which is straightforward. There are $i=1,2,\ldots n$ risky assets, short sales are allowed, and a rational agent chooses the portfolio weights, w_i, $i=1,2,\ldots n$, to minimize the variance of the portfolio in (2.7) subject to the budget constraint (2.1), given expected rates of return as in (2.4), and variances and covariances. Formally,

$$\underset{\{w_i\}_{i=1}^{n}}{Min} \quad [(1/2)\, \Sigma_{i=1}^{n} \Sigma_{j=1}^{n} w_i w_j \sigma_{ij}] \tag{2.26}$$

$$\text{subject to} \qquad r^m = \Sigma_{i=1}^{n} r_i^m w_i$$

$$1 = \Sigma_{i=1}^{n} w_i .$$

Therefore, a solution to this problem yields a portfolio that has minimum variance amongst portfolios that have the same expected rate of return. The objective function is quadratic, formally it is a quadratic form, thus strictly convex, with a constant $(1/2)$ for analytical ease, and the constraints are linear. The Lagrangean function for this problem is

$$\mathcal{L} = (1/2)\, \Sigma_{i=1}^{n} \Sigma_{j=1}^{n} w_i w_j \sigma_{ij} + \lambda\, [r^m - \Sigma_{i=1}^{n} r_i^m w_i] + \mu\, [1 - \Sigma_{i=1}^{n} w_i] \tag{2.27}$$

where $\lambda \geq 0$ is the Lagrange multiplier attached to the expected rate of return of the portfolio and $\mu \geq 0$ is the Lagrange multiplier attached to the budget constraint. The first order necessary conditions for an interior solution are given by

$$\partial\mathcal{L}/\partial w_i = \Sigma_{j=1}^{n} w_j \sigma_{ij} - \lambda\, r_i^m - \mu = 0 \qquad \text{for all } i=1,2,\ldots n \tag{2.28}$$

$$\partial\mathcal{L}/\partial \lambda = r^m - \Sigma_{i=1}^{n} r_i^m w_i = 0 \tag{2.29}$$

$$\partial\mathcal{L}/\partial \mu = 1 - \Sigma_{i=1}^{n} w_i = 0 \tag{2.30}$$

and the convexity of the objective function and linearity of the constraints satisfy the second order sufficient conditions for a minimum. Expressions (2.29)-(2.30) make a total of $n+2$ linear equations in $n+2$ unknowns, with solutions

$$w_i^* = w_i^*(r^m, \sigma_{ij}) \qquad \text{for all } i=1,2,\ldots n \tag{2.31}$$

$$\lambda^* = \lambda^*(r^m, \sigma_{ij}) \qquad (2.32)$$

$$\mu^* = \mu^*(r^m, \sigma_{ij}). \qquad (2.33)$$

In particular, $w_i^*(r^m, \sigma_{ij})$ yields the minimum-variance set of portfolios, for each r^m, where some $w_i^*<0$, i.e. short sales are allowed. If short selling is not allowed, one additional inequality constraint, $w_i^* \geq 0$, must be added to problem (2.26) and some w_i^* may be zero whenever that constraint binds.

An illustration for the simple case $n=2$, two assets, is instructive. The solution is recursive in this case: expressions (2.29) and (2.30) alone solve for the portfolio w_i^* as

$$w_1^* = (r^m - r_2^m)/(r_1^m - r_2^m) \quad \text{and} \quad w_2^* = 1 - w_1^* \qquad (2.34)$$

and equations (2.28) solve for λ^* and μ^*. In effect, the given expected return and budget constraint would be sufficient to yield a solution in this case. The interesting insight is that a valid solution could also be obtained by choosing the Lagrange multiplier on the portfolio mean rate of return, λ arbitrarily so that equations (2.28) and (2.30) solve for w_i and μ^* and equation (2.29) yields the implied expected rate of return r^m.

This insight is useful in understanding the computation of a specific point along the minimum-variance set for the general case of many assets, in particular the point of smallest variance. This global minimum-variance point can be obtained by considering the value function, or indirect objective function of problem (2.26) and recalling that, at the optimum, the Envelope theorem yields

$$\partial \mathcal{L}/\partial r^m = \partial[(1/2)\Sigma_{i=1}^n \Sigma_{j=1}^n w_i^* w_j^* \sigma_{ij}]/\partial r^m = \lambda^* \qquad (2.35)$$

that is the marginal cost in terms of portfolio variance of a small change in the expected rate of return of the portfolio. We have just noted above that an arbitrary choice of λ yields a valid solution, in particular, $\lambda^*=0$ in (2.35) is exactly the point where

$$\partial[(1/2)\Sigma_{i=1}^n \Sigma_{j=1}^n w_i^* w_j^* \sigma_{ij}]/\partial r^m = 0,$$

or the choice of r^m consistent with the global minimum variance point in the minimum-variance set. Basically, from the Lagrangean function (2.27), $\lambda=0$ relaxes the constraint of the exogenously given expected rate of return, $r^m = \Sigma_{i=1}^n r_i^m w_i$, adjusting it for the one that yields the minimum variance point. Therefore, the solution for the portfolio with the smallest variance in the minimum-variance set is obtained by solving, from (2.28)-(2.30) with $\lambda=0$, the set of equations

$$\Sigma_{j=1}^{n} w_j \sigma_{ij} - \mu = 0 \quad \text{for all } i=1,2,\ldots n \qquad (2.36)$$

$$1 - \Sigma_{i=1}^{n} w_i = 0 \qquad (2.37)$$

or the $n+1$ linear equations in $n+1$ unknowns, with solutions

$$w_i^*|_{Global\ Minimum\ Variance(GMV)} = w_i^*|_{GMV}(r^m, \sigma_{ij}) \quad \text{for all } i=1,2,\ldots n \qquad (2.38)$$

$$\mu^*|_{Global\ Minimum\ Variance(GMV)} = \mu^*|_{GMV}(r^m, \sigma_{ij}). \qquad (2.39)$$

The solution for the GMV portfolio in the simplest case of $n=2$ is from (2.36)-(2.37)

$$w_1^*|_{GMV} = (\sigma_{22} - \sigma_{12}) / (\sigma_{11} - 2\sigma_{12} + \sigma_{22}) \quad \text{and} \quad w_2^*|_{GMV} = 1 - w_1^*|_{GMV}. \qquad (2.40)$$

In general, other points along the minimum-variance set can be similarly found by arbitrarily choosing values for λ, and all points above the GMV point supply the efficient frontier. We should note that this is one possible way to compute the frontier, there are other ways that we suggest below as well. Figure 2.6 illustrates the efficient frontier as the portion to the northeast of the GMV point.

2.4 **Two-Fund Theorem, The Risk-Free Asset and One-Fund Theorem**

The mean-variance framework outlined above yields some sophisticated understanding of asset allocation strategy. We shall first examine the Two-Fund theorem in this context. Suppose an agent chooses two arbitrary expected rates of return for given variances and covariances in a world of several risky assets. Denote them r^{m1} and r^{m2} respectively. For each of those arbitrary expected rates of return, there will be solutions of (2.28)-(2.30) yielding minimum-variance portfolios that belong to the minimum-variance set, denoted

$$w_i^{*1} = w_i^{*1}(r^{m1}, \sigma_{ij}), \quad \text{and} \quad w_i^{*2} = w_i^{*2}(r^{m2}, \sigma_{ij})$$

for all assets $i=1,2,\ldots n$. Next, consider a third portfolio consisting of a linear, convex, combination of the two portfolios: $\alpha w_i^{*1} + (1-\alpha) w_i^{*2}$ for some real number α. We ask whether this new portfolio belongs to the minimum-variance set. This can be answered by checking whether or not this new portfolio satisfies the first order necessary conditions in (2.28)-(2.30). Equations (2.28) are satisfied by each $w_i{*}^1$ and $w_i{*}^2$ individually, therefore a linear convex combination of the two portfolios satisfies them as well:

$$\alpha\, \partial\mathcal{L}/\partial w_i{*}^1 + (1-\alpha)\, \partial\mathcal{L}/\partial w_i{*}^2 = 0 \qquad \textit{for all } \alpha.$$

Equation (2.29) is satisfied for each arbitrary expected rate of return r^{m1} and r^{m2} as well. Therefore, the implied expected rate of return of the new portfolio, $\alpha\, r^{m1} + (1-\alpha)\, r^{m2}$, is the corresponding given expected rate of return, and (2.29) is satisfied for the new portfolio and for all α. Equation (2.30) is satisfied since the weights of the new portfolio add up to one, $\alpha+(1-\alpha)=1$. Hence, the linear convex combination belongs to the minimum-variance set for all α. In particular, when we vary α in the range $(-\infty, \infty)$, it spans the whole minimum-variance set. This result is called the Two-Fund Theorem.

Two-Fund Theorem: For two portfolios, elements of the minimum-variance set, any linear convex combination of the portfolios gives a new portfolio that belongs to the minimum-variance set as well, hence, the two portfolios completely span the minimum-variance set.

Intuitively, this result can be traced to the simplest case of two risky assets by considering each portfolio with several assets a new asset. We have seen above that for two risky assets, alternative combinations of the two assets yield a complete characterization of the mean-variance tradeoff, given the covariance structure of the assets. The Two-Fund theorem is an analogous characterization when the two assets are portfolios. This is not a trivial result and has important practical implications. For example, choosing the portfolio at the GMV point and another one, such as $\lambda^*=1$ in Figure 2.6, completely characterizes the minimum-variance set with alternative choices of α. As a practical matter, an investor needs only two mutual funds to span the minimum-variance set with alternative diversification strategies in these funds, which is quite a remarkable result.

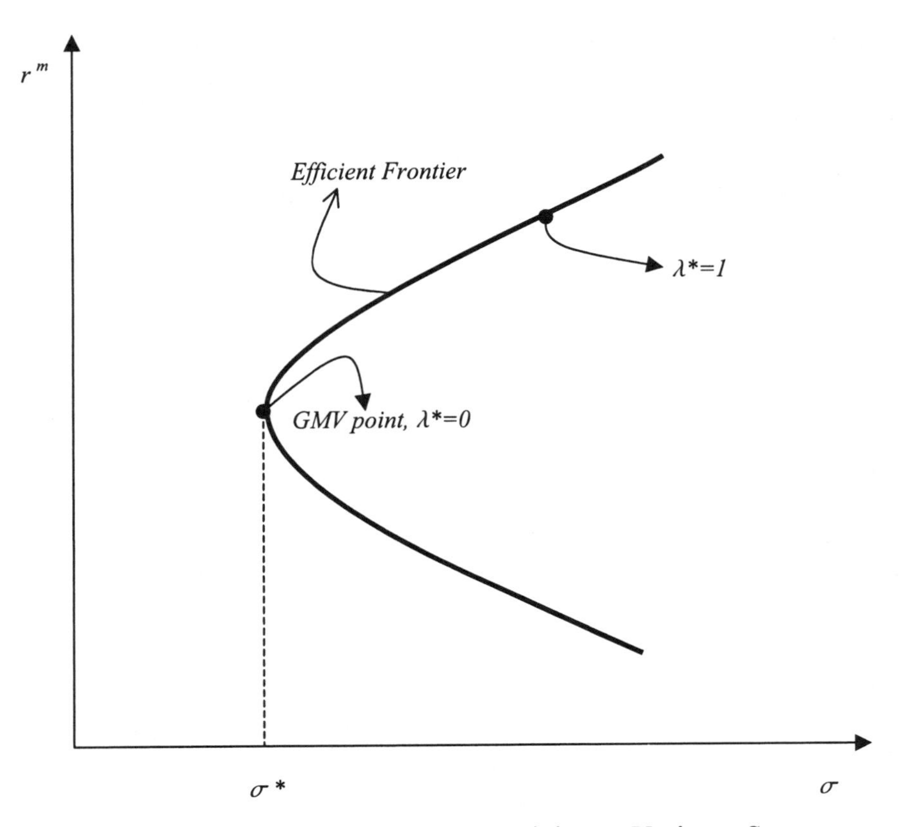

Fig. 2.6 Specific Points along the Minimum-Variance Set

The other important result in this context is the One-Fund theorem. In order to arrive at the main result, we need to formally introduce one asset that has been discussed only briefly so far, the risk-free asset. We define a risk-free asset, in context, as the asset, or set of assets or portfolio, that yields a strictly positive rate of return, $r_f > 0$, without risk or with null variance, $\sigma_{ff}=0$:

$$\underline{\text{Risk-free asset}}: \quad r_f>0, \quad \sigma_{ff}=0. \tag{2.41}$$

The risk-free asset has deterministic rate of return and no risk attached. In the case examined in section 2.2, a risk-free portfolio emerged as a function of the perfectly negative correlation between the rates of return. The difference here is that we introduce exogenously a risk-free asset in the menu of available assets for an agent to invest. This indeed changes the portfolio problem discussed so far in a meaningful way. A risk-free asset corresponds essentially to a borrowing-lending contract and there will be no restrictions regarding short sales. The reason is simple, an investor that goes long on the risk-free asset, $w_i>0$ for the ith risk-free asset, is essentially lending funds to another individual who is willing to pay a sure rate of return. Alternatively, an investor that goes short on the risk-free asset, $w_i<0$ for the ith risk-free asset, is essentially borrowing funds from another individual who is willing to lend for a sure rate of return. In particular, a borrower can go short on the risk-free asset in order to go long on other risky assets expanding the possibilities to trade risk and return.

Consider first a portfolio with two assets, one risky and the other risk-free. The risky asset is characterized by expected rate of return r_1^m and variance $\sigma_{11}>0$ whereas the risk-free asset is characterized in (2.41). Because $(r_f - E[r_f])=0$, the covariance between the two assets is also zero. If we denote by $\alpha \leq 1$, the share of the risk-free asset on the portfolio, where $\alpha < 0$ implies short sales on the risk-free asset, the expected rate of return and standard deviation of the portfolio are respectively

$$r^m(\alpha) = \alpha\, r_f + (1-\alpha)\, r_1^m \tag{2.42}$$

$$\sigma(\alpha) = (1-\alpha)\,\sigma_1. \tag{2.43}$$

From equation (2.43),

$$\alpha = [\sigma_1 - \sigma(.)\,] / \sigma_1 \tag{2.44}$$

so that α varies linearly with $\sigma(\,.\,)$, or

$$\partial\alpha / \partial\sigma(.) = -1 / \sigma_1 < 0. \tag{2.45}$$

As the standard deviation of the portfolio increases, the share of the risk-free asset in the portfolio decreases. Substituting (2.44) into (2.42) we obtain

$$r^m(\alpha) = r_f + [\sigma(.) / \sigma_1]\ (r_1^m - r_f). \tag{2.46}$$

which is linear in $\sigma(.)$. This relationship says that the expected rate of return of the portfolio has two components: (i) The risk-free rate of return, r_f, plus; (ii) A risk premium, $[\sigma(.)/\sigma_1](r_1^m - r_f)$, which is proportional to the discrepancy between the expected return on the risky asset and the risk-free rate of return, i.e. the excess return on the risky asset. Then, in mean-standard deviation space, we have

$$\partial r^{m}(.) / \partial \sigma(.) = (1/\sigma_1)\ (r_1^{m} - r_f) \lesseqgtr 0 \qquad (2.47)$$

and the sign of $\partial r^{m}(.) / \partial \sigma(.)$ ultimately depends on the sign of the excess return $(r_1^m - r_f)$. If the excess return is positive, $(r_1^m - r_f)>0$, as σ increases, α decreases by (2.45) and r^m increases by (2.47). If the excess return is negative, $(r_1^m - r_f)<0$, as σ increases α decreases by (2.45) and r^m decreases by (2.47).

Figure 2.7 illustrates the one risky-one risk-free asset portfolios in mean-standard deviation space. The portfolio with zero standard deviation is a point on the vertical axis with rate of return r_f and $\alpha = 1$. The portion above the risk-free return is where $(r_1^m - r_f)>0$ and the portion below the risk-free return is where $(r_1^m - r_f)<0$. By the same efficiency criteria discussed above in terms of mean-standard deviation tradeoff, the lower portion where $(r_1^m - r_f)<0$ is not relevant since a rational individual would not hold a risky asset that yields a lower expected rate of return than the risk-free rate. Therefore, the upper portion is the efficient frontier for the one risky-one risk-free portfolio, with points to the right of $\alpha = 0$ indicating that short sales on the risk-free asset, used to go long on the risky asset, will give higher expected rate of return with higher risk. Now that we have introduced the risk-free asset, we can discuss how it changes the minimum-variance set obtained in section 2.5.

Consider the more general case where there are several $i=1,2,\ldots n$ risky assets and one risk-free asset. First, the minimum-variance set derived in (2.28)-(2.30) above and the associated efficient frontier are a collection of portfolios with risky assets only. Second, the efficient frontier in the one risky-one risk-free asset derived in (2.47) above has a portion to the right of point $\alpha = 0$, see Figure 2.7, where only risky assets are held. Heuristically, if we combine several risky assets with one risk-free asset we will have a superimposition of the efficient frontier with several risky assets as in Figures 2.5(b) or 2.6 and the efficient frontier with one risky-one risk-free asset in Figure 2.7. This is illustrated in Figure 2.8. The point where the two efficient frontiers touch, point M, is the tangency point where $\alpha = 0$ in the one risky-one risk-free asset efficient frontier. More importantly, at standard deviation σ ', i.e. the minimum variance point A of the minimum-variance set, the expected rate of return of the portfolio is r^{m}'. However, for that standard deviation σ ', the efficient frontier with one risky-one risk-free asset supplies a portfolio of the risky and risk-free assets at point B that yields a higher expected rate of return, r^{m}'' > r^{m}'. Henceforth, adding the risk-free asset to the menu available to the investor expands the minimum-variance set to the left, towards the efficient frontier of the risk-free and risky assets. We show next that the One-Fund theorem is indeed a direct consequence of the expansion of the efficient frontier.

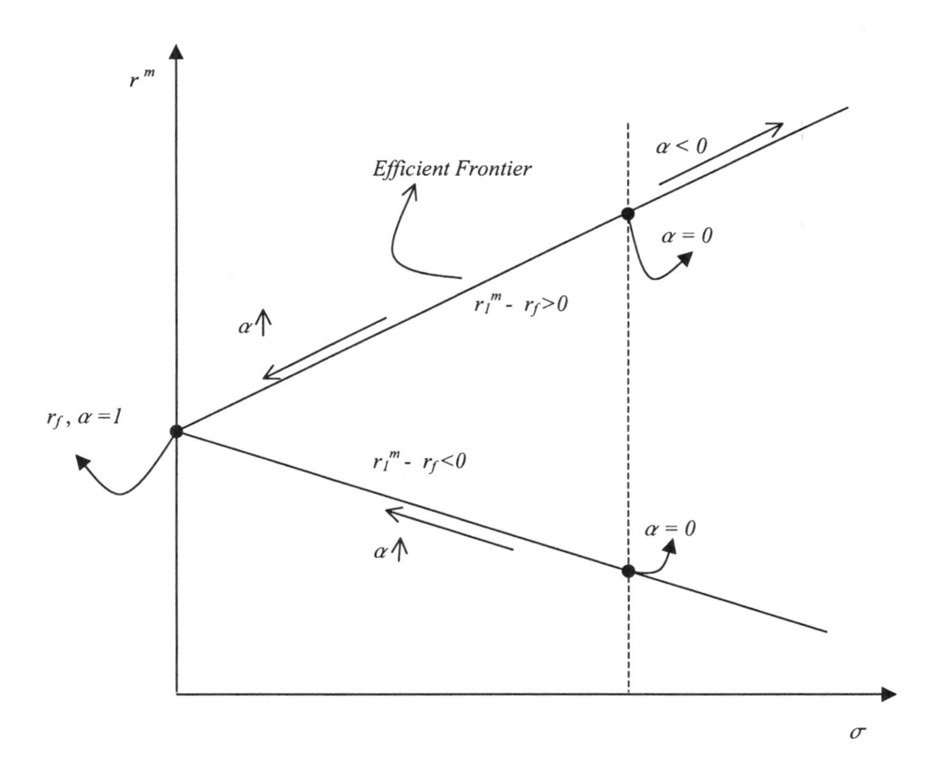

Fig. 2.7 One Risky-One Risk-Free Efficient Frontier

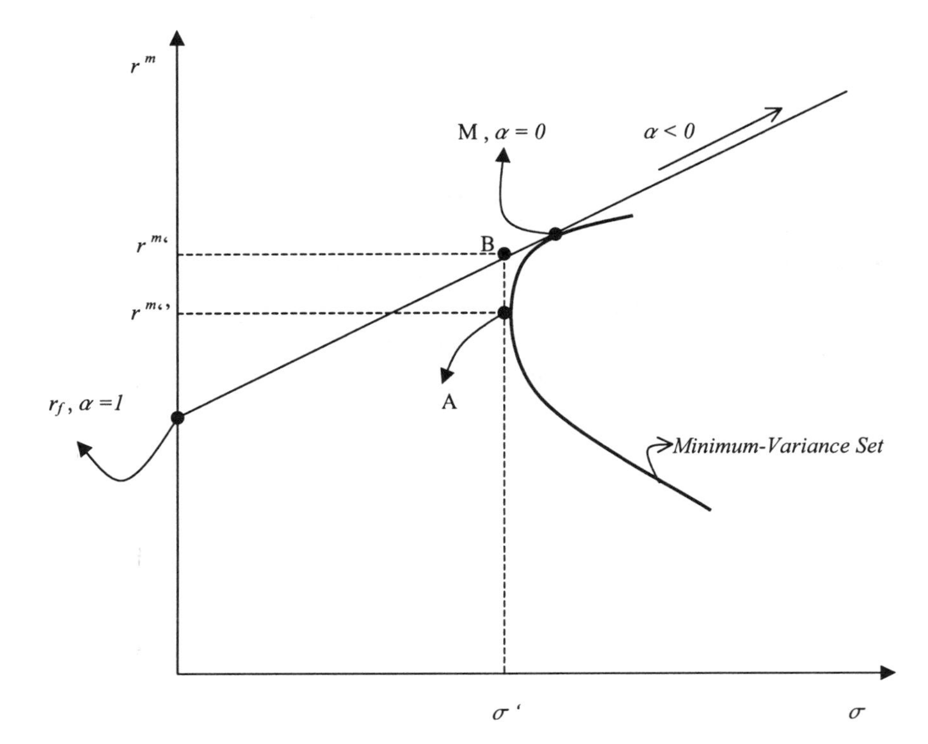

Fig. 2.8 The One-Fund Theorem

We have learned above by the Two-Fund theorem that, with only two funds (assets), we are able to span the minimum-variance set. Thus, we choose the fund of risky assets at point M in Figure 2.8, the tangency point between the two efficient frontiers, as one of the funds, and the risk-free asset is chosen as the other. Then, a linear convex combination of those two completely spans the efficient frontier from points C to M and beyond, i.e. the Two-Fund theorem. The One-Fund theorem basically states that to span the efficient frontier with a risk-free asset, only one fund of risky assets is needed, the fund at the point M of tangency of the two frontiers.

One-Fund Theorem: There is one fund, at the tangency point M, whose linear convex combinations with the risk-free asset completely span the efficient frontier with several risky assets and risk-free assets; in effect everyone should hold some proportion of the one fund at point M.

The portfolio of risky assets at point M represents an important benchmark in the minimum-variance set because it is the one that everyone should hold to span the efficient frontier in the presence of risk-free assets.

How can we find this portfolio? One method to compute this portfolio is to consider the angle θ whose tangent is given by

$$tan\ \theta = (r^{m\prime} - r_f) / \sigma' \tag{2.48}$$

as illustrated in Figure 2.9, where the tangent is in fact the slope of the line r_f–M'. It is clear that at point M, on the minimum-variance set, the tangent of the angle θ is maximized and that at other points along the minimum-variance set the tangent is smaller. Hence, the portfolio at

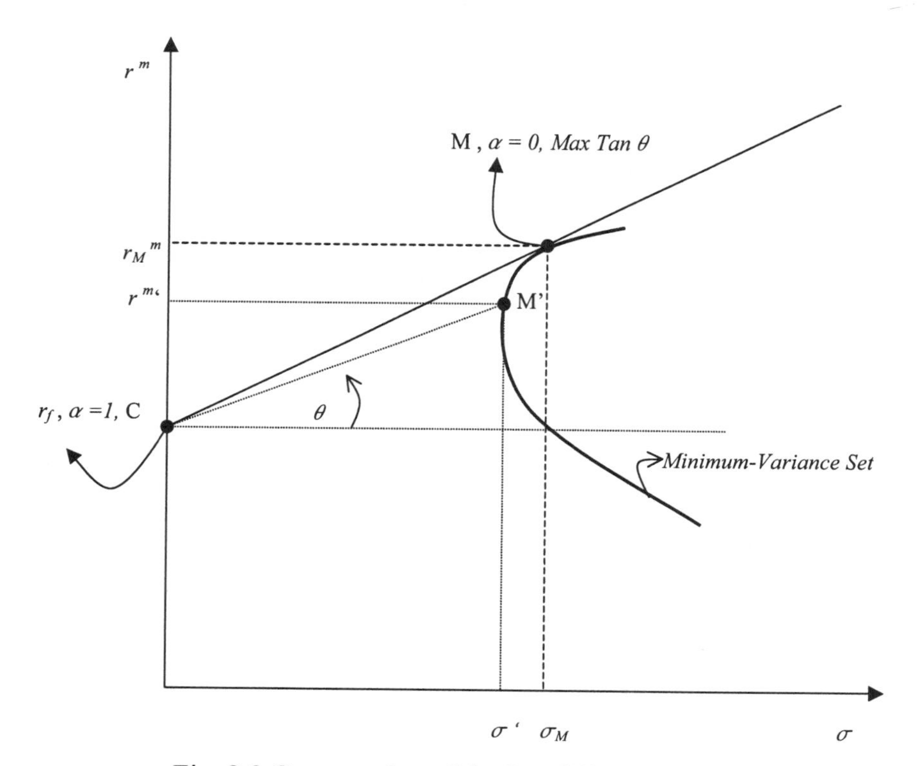

Fig. 2.9 Computation of the Portfolio at Point M

point M is the choice of weights, w_i, elements of the minimum-variance set, that maximize the tangent of the angle θ. You may note here that another plausible way to compute the minimum-variance set itself would be to choose alternative values of r^m in the vertical axis, and maximize the tangent of the angle θ for each r^m. The respective expected rates of returns, $r^{m'}$ and r_f and variance $(\sigma')^2$ are

$$r^{m'} = \Sigma_{i=1}^n w_i r_i^{m'}$$

$$(\sigma')^2 = \Sigma_{i=1}^n \Sigma_{j=1}^n w_i w_j \sigma_{ij}'$$

$$r_f = \Sigma_{i=1}^n w_i r_f$$

$$\sigma_{ff} = 0$$

where the third expression is an identity used for ease of computation. Hence, substituting the relationships above into (2.48) yields

$$tan\, \theta = \Sigma_{i=1}^n w_i (r_i^{m'} - r_f) / [\Sigma_{i=1}^n \Sigma_{j=1}^n w_i w_j \sigma_{ij}']^{1/2}. \quad (2.48')$$

The portfolio choice problem is

$$\underset{\{w_k\}_{k=1}^n}{Max} \Sigma_{i=1}^n w_i (r_i^{m'} - r_f) / [\Sigma_{i=1}^n \Sigma_{j=1}^n w_i w_j \sigma_{ij}']^{1/2} \quad (2.49)$$

and the first order necessary conditions are

$$(r_k^{m'} - r_f)\, \sigma' - \Sigma_{i=1}^n w_i (r_i^{m'} - r_f)(\sigma')^{-1} \Sigma_{j=1}^n w_j \sigma_{kj}' = 0$$

$$for\ all\ k=1,2,\ldots n \quad (2.50)$$

with the second order sufficient conditions satisfied by the convexity of the minimum-variance set. Using the relationships above for $r^{m'}$, σ' and r_f and rearranging, we can rewrite (2.50) as

$$(r_k^{m'} - r_f) = \Sigma_{j=1}^n \{w_j [\Sigma_{i=1}^n w_i (r_i^{m'} - r_f) / \Sigma_{i=1}^n \Sigma_{j=1}^n w_i w_j \sigma_{ij}']\} \sigma_{kj}$$

$$for\ all\ k=1,2,\ldots n \quad (2.51)$$

where $\Sigma_{i=1}^n \Sigma_{j=1}^n w_i w_j \sigma_{ij}'$ is the variance of the portfolio at the maximum. We note in (2.51) that the term in keys

$$\{w_j [\Sigma_{i=1}^n w_i (r_i^{m'} - r_f) / \Sigma_{i=1}^n \Sigma_{j=1}^n w_i w_j \sigma_{ij}']\}$$

is not a function of *k* so that we may simplify (2.51) as

$$(r_k^{m'} - r_f) = \Sigma_{j=1}^n v_j \sigma_{kj} \quad for\ all\ k=1,2,\ldots n \quad (2.52)$$

where

$$v_j \equiv \{w_j [\Sigma_{i=1}^n w_i (r_i^{m'} - r_f) / \Sigma_{i=1}^n \Sigma_{j=1}^n w_i w_j \sigma_{ij}']\}.$$

Equations (2.52) form a set of k linear equations in v_j unknowns that can be solved as

$$v_j * = v_j * (r_k^{m'} - r_f, \sigma_{kj}) \quad \text{for all } j=1,2,\ldots n; \ k=1,2,\ldots n. \quad (2.53)$$

Then, to obtain the portfolio at point M, we normalize the v_j * and obtain the portfolio weights

$$w_j *^M = v_j * / \Sigma_{z=1}^n v_z * \qquad \text{for all } j=1,2,\ldots n \qquad (2.54)$$

which characterize the portfolio at point M.

As an example, consider the simplest case of two risky assets with zero covariance, $\sigma_{12}=0$. Solving the two equations in (2.52) yields

$$v_1 * = (r_1^m - r_f) / \sigma_{11}$$

$$v_2 * = (r_2^m - r_f) / \sigma_{22}$$

and using (2.54) the portfolio at point M is characterized by

$$w_1 *^M = \sigma_{22} \ (r_1^m - r_f) / [r_1^m \sigma_{22} + r_2^m \sigma_{11} - r_f(\sigma_{11} + \sigma_{22})]$$

$$w_2 *^M = \sigma_{11} \ (r_2^m - r_f) / [r_1^m \sigma_{22} + r_2^m \sigma_{11} - r_f(\sigma_{11} + \sigma_{22})]$$

with weights that add up to unity.

Finally, as a preview to our next section, we can identify the portfolio of risky assets at point M with the market portfolio of risky assets since it is the one that every individual should hold together with the risk-free asset. At point M, we have

$$(r_k^{m'} - r_f) = \Sigma_{j=1}^n \{ w_j *^M [\Sigma_{i=1}^n w_i *^M (r_i^{m'} - r_f) / \sigma_{MM}]\} \sigma_{kj}$$

$$\text{for all } k=1,2,\ldots n \qquad (2.55)$$

where

$$\sigma_{MM} \equiv \Sigma_{i=1}^n \Sigma_{j=1}^n w_i *^M w_j *^M \sigma_{ij}'$$

is the variance of the portfolio at point M. Noting that at point M, the rate of return and expected rate of return of the portfolio of risky assets are respectively

$$r_M = \Sigma_{i=1}^n w_i *^M r_i \quad \text{and} \quad r_M^m = \Sigma_{i=1}^n w_i *^M r_i^m$$

we obtain the covariance of an arbitrary rate of return r_k and expected rate of return r_k^m and the market rate of return as

$$cov(r_k, r_M) = \sigma_{kM} = E[(r_k - r_k^m)(\Sigma_{i=1}^n w_i *^M r_i - \Sigma_{i=1}^n w_i *^M r_i^m)]$$

$$= E[(r_k - r_k^m) \ \Sigma_{i=1}^n w_i *^M (r_i - r_i^m)]$$

$$= E\,[\Sigma_{i=1}{}^{n}\; w_i{*}^{M}\,(r_i - r_i{}^{m})\,(r_k - r_k{}^{m})\,]$$
$$= \Sigma_{i=1}{}^{n}\; w_i{*}^{M}\,E\,[(r_i - r_i{}^{m})\,(r_k - r_k{}^{m})\,]$$
$$= \Sigma_{i=1}{}^{n}\; w_i{*}^{M}\,\sigma_{kj}$$

and thus (2.55) may be rewritten as

$$(r_k{}^{m} - r_f) = \sigma_{kM}\;[\Sigma_{i=1}{}^{n}\; w_i{*}^{M}\,(r_i{}^{m} - r_f)\,/\,\sigma_{MM}\,] \qquad \textit{for all } k=1,2,\ldots n. \qquad (2.56)$$

Now, for the portfolio at point M, e.g. the market portfolio,

$$\Sigma_{i=1}{}^{n}\; w_i{*}^{M}\,(r_i{}^{m} - r_f) = r_M{}^{m} - r_f$$

the expected excess return of the market over the risk-free return; substitution into (2.56) yields

$$(r_k{}^{m} - r_f) = (\,\sigma_{kM}/\,\sigma_{MM}\,)\,(r_M{}^{m} - r_f) \qquad \textit{for all } k=1,2,\ldots n. \quad (2.57)$$

Expression (2.57) states that the expected excess return of an arbitrary risky asset k over the risk-free rate is proportional to the expected excess return of the market portfolio over the risk-free rate. The coefficient of proportionality $(\sigma_{kM}/\sigma_{MM})$ is seen to be directly related to the covariance of the risky asset k and the market portfolio. This is generally referred to as the Static Capital Asset Pricing Model (SCAPM) as we discuss it in more detail below.

2.5 The Pricing of Assets in the Mean-Variance Framework and the No-Arbitrage Theorem

As we discussed in the last section, the portfolio at point M can be identified with the market portfolio. How can this be? The answer is that the portfolio at point M contains weights of every risky asset in the market that are proportional to the asset's capital value relative to the total market value. M is the one risky fund that every individual should hold to diversify with the risk-free asset, hence it is the market fund. Therefore, the efficient frontier characterized by the points CM and beyond (short sales) in Figures 2.8 and 2.9 is also denoted the capital market line.

Capital Market Line: the linear relationship between expected rate of return and standard deviation that spans from the risk-free rate, r_f, to the point of tangency of the minimum-variance set and the one risky-one risk-free frontier. The equation for the capital market line is, in general mean-standard deviation space, given by

$$r^m = r_f + [(r_M{}^m - r_f)/\sigma_M]\,\sigma \qquad (2.58)$$

and its slope is

$$\partial r^m / \partial \sigma\,|_{\text{Capital Market Line(CML)}} = [(r_M{}^m - r_f)/\sigma_M] \qquad (2.59)$$

which yields the change in expected rate of return given a unit change in standard deviation, sometimes called the market price of risk.

(a) The Static Capital Asset Pricing Model (SCAPM)

In essence, the SCAPM is an equilibrium model that relates expected rates of return on arbitrary assets to the price of risk. There are several ways to derive the SCAPM formula. We have presented one way to derive it in what led to formula (2.57). Here, we present an alternative way.

First, recall the Two-Fund theorem and span the minimum-variance set with a portfolio ($\alpha \leq 1$) consisting of the risky fund at point M and another risky asset denoted i. The expected rate of return and standard deviation of this portfolio are

$$r^m(\alpha, \ldots) = \alpha\, r^m{}_i + (1-\alpha)\, r_M{}^m \qquad (2.60)$$

$$\sigma(\alpha, \ldots) = \{\alpha^2 \sigma_{ii} + (1-\alpha)^2 \sigma_{MM} + 2\,\alpha\,(1-\alpha)\sigma_{iM}\}^{1/2} \qquad (2.61)$$

so that at $\alpha = 0$ it corresponds to the portfolio at point M, and at $\alpha = 1$ it corresponds to the portfolio with risky asset i. For $\alpha < 0$, it corresponds to short sales on the portfolio at point M.

Second, add a risk-free asset to the menu of available assets so that a capital market line can be drawn from r_f tangent to point M at the minimum-variance set, see e.g. Figures 2.8 or 2.9. The tangency implies that, at point M, the slope of the capital market line is identical to the slope of the minimum-variance set. We explore this relationship as follows: from equations (2.60)-(2.61) above, the slope of the minimum-variance set can be found by computing

$$[\partial r^m(\alpha,\ldots)/\partial\alpha]/[\partial\sigma(\alpha,\ldots)/\partial\alpha]|_{MVS} =$$

$$(r^m{}_i - r_M{}^m)/\{[\alpha\,\sigma_{ii} - (1-\alpha)\,\sigma_{MM} + (1-2\alpha)\sigma_{iM}]/\sigma(.,\ldots)\}. \quad (2.62)$$

Evaluating the expression in (2.62) at $\alpha = 0$, it corresponds to the slope of the minimum-variance set at point M, or

$$[\partial r^m(\alpha,\ldots)/\partial\alpha]/[\partial\sigma(\alpha,\ldots)/\partial\alpha]|_{MVS,M} =$$

$$(r^m{}_i - r_M{}^m)/[(\sigma_{iM} - \sigma_{MM})/\sigma_M]. \quad (2.63)$$

Equating (2.63) with the slope of the capital market line in (2.59) and rearranging yields

$$r^m{}_i - r_f = (\sigma_{iM}/\sigma_{MM})\,(r_M{}^m - r_f) \qquad \text{for all } i=1,2,\ldots n \quad (2.64)$$

the SCAPM formula exactly as in (2.57). Note also from (2.59) and (2.64), that the SCAPM formula is a linear relationship between the excess rate of return of the *i*th asset and the price of risk. It is common practice to define the proportion $(\sigma_{iM}/\sigma_{MM})$ as the *beta* of the asset,

$$\beta_i \equiv \sigma_{iM}/\sigma_{MM} \qquad \text{for all } i=1,2,\ldots n \quad (2.65)$$

or the ratio of the covariance of the *i*th risky asset with the market, σ_{iM}, to the variance of the market, σ_{MM}. Using (2.65), the SCAPM formula may be expressed as

$$r^m{}_i - r_f = \beta_i\,(r_M{}^m - r_f) \qquad \text{for all } i=1,2,\ldots n \quad (2.66)$$

which says that the expected excess rate of return of the *i*th asset relative to the risk-free, the left-hand side, is proportional to the expected excess rate of return of the market asset relative to the risk-free, the right-hand side, where the coefficient of proportionality is the asset β_i related to the covariance of the *i*th asset with the market and the variance of the market. Thus, we may consider the following cases:

(i) $\beta_i = 1 \Longrightarrow \sigma_{iM} = \sigma_{MM}$: the expected excess rate of return of the *i*th asset has variation identical to the expected excess return of the market;

(ii) $\beta_i > 1 \Longrightarrow \sigma_{iM} > \sigma_{MM}$: the expected excess rate of return of the *i*th asset is greater than the expected excess return of the market, the asset is said to be “aggressive;”

(iii) $\beta_i < 1 \Longrightarrow \sigma_{iM} < \sigma_{MM}$: the expected excess rate of return of the *i*th asset is smaller than the expected excess return of the market, the asset is said to be “defensive;”

(iv) $\beta_i = 0 \Longrightarrow \sigma_{iM} = 0 \Longrightarrow r^m{}_i = r_f$: the ith asset is uncorrelated with the market, even if it has a high variance, i.e. σ_{ii} large, the risk can be diversified away since it is uncorrelated with the market;
(v) $\beta_i < 0 \Longrightarrow \sigma_{iM} < 0 \Longrightarrow r^m{}_i < r_f$: the ith asset has negative excess return because it covaries negatively with the market; but one can still hold this asset as it provides a hedge against variations in the market.

The usefulness of the SCAPM formula in (2.66) is that it can be plausibly applied to stock market data using simple statistical techniques such as two-variable linear regression analysis. For example, using ready available stock market data one can construct the variables on the left- and right-hand side of (2.66) and estimate the asset beta, β. The coefficient has an appealing statistical intuition as the proportion of the variation of the excess return of the asset explained by the variation of the excess return of the market. In the case of a portfolio with n risky assets, $i=1,2,...n$, and weights w_i, the β of the portfolio is a weighted average of the individual β's, or

$$\beta = \Sigma_{i=1}{}^n \, w_i \, \beta_i.$$

(b) Systematic versus Non-systematic, Idiosyncratic Risk

In discussing the alternative cases of an asset β, case (iv) is the one where the asset is uncorrelated with the market. This is an interesting special case of a more general property of risky assets. Consider an asset on the capital market line (2.58) and denote it the ith asset. This asset satisfies (2.64) so that, by (2.66), its expected return and standard deviation are respectively

$$r^m{}_i = r_f + \beta_i \, (r_M{}^m - r_f) \tag{2.67}$$

$$\sigma_i = \beta_i \, \sigma_M. \tag{2.68}$$

From equation (2.68), as long as $\beta_i \neq 0$, the standard deviation of the ith asset reflects risk derived from the risk of the market. Therefore, any other asset with $\beta_j \neq 0$ does contain this risk, and it is impossible for the holder of the ith asset to diversify this risk using other assets in the market portfolio. When the source of risk in the asset is completely derived from the market, this is nondiversifiable risk and is usually called systematic risk.

The most intuitive way to think about this is to consider macroeconomic risk that affects all sectors of a closed economy simultaneously or even worldwide macro risk that affects the whole world economy in the international case, e.g. a "meteor shower." In this case, there would be no uncorrelated assets in "Mars" to diversify this kind of worldwide risk. However, it is possible that the *i*th asset has another source of risk which is uncorrelated with the market as a whole. Then, the standard deviation of the asset in (2.68) increases by a component, denoted σ_N, yielding, in terms of variances

$$\sigma_{ii} = \beta_i^2 \ \sigma_{MM} + \sigma_{NN} . \qquad (2.69)$$

The component σ_{NN} is called the Non-systematic, idiosyncratic risk. This risk is uncorrelated with the market so it can be fully diversified away with other assets in the risky portfolio. In this case, the risk is specific to the asset, so you can think of other assets in the market that have uncorrelated returns and can be used to diversify the idiosyncratic risk.

For example, if one holds shares of oil company X and hardware company Y in their portfolio, there are risks specific to the oil industry versus the technology sector such that a combination of the assets in the portfolio can diversify those risks. Suppose there is a technology disturbance that affects company Y in the technology sector but is uncorrelated with the oil sector. A balanced portfolio between X and Y can diversify this risk. However, both of those companies are subject to macroeconomic risk that cannot be diversified away, the systematic component. One useful way to think about equation (2.69) is as a linear regression with the component σ_{NN} denoting an identically and independently distributed (i.i.d.) random error term. In particular, Figure 2.10 illustrates the case of the *i*th asset with standard deviation governed by (2.69). The distance from point *i* on the capital market line, to A is a measure of the Non-systematic risk, σ_N. Since this idiosyncratic component can always be diversified away, an investor will choose a diversified portfolio that will undo σ_N and move back to the capital market line at point *i* along the straight line.

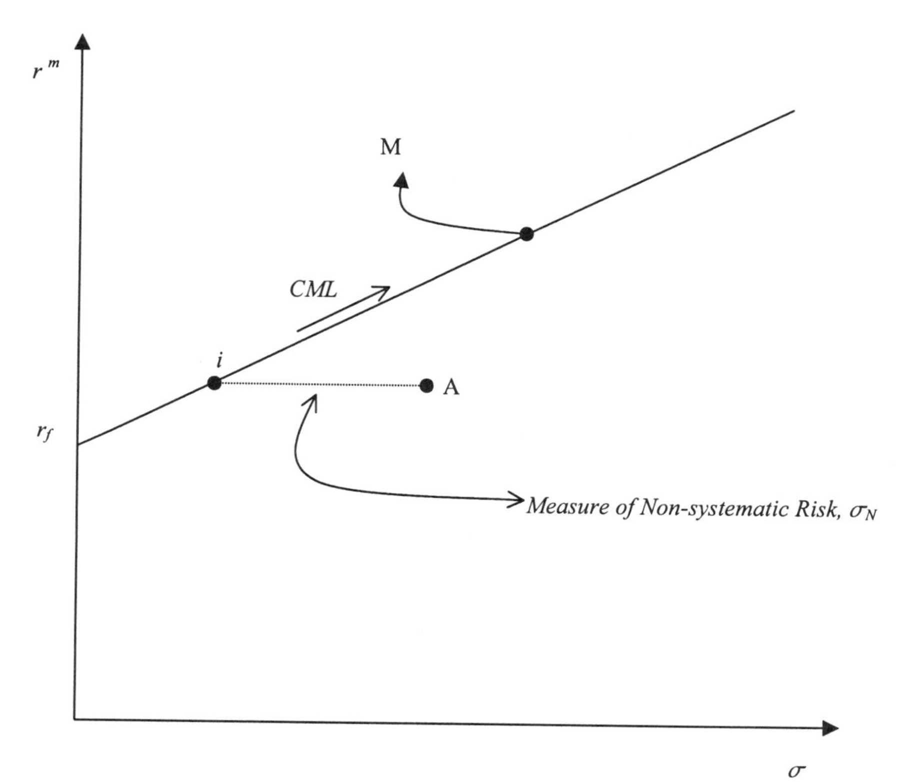

Fig. 2.10 Systematic versus Non-systematic (Idiosyncratic) Risk

(c) The No-Arbitrage Theorem

In context, the SCAPM may be easily reinterpreted as an asset pricing model. An investment of an amount X_0 at the beginning of the period with a payoff of X_1 at the end-of-period yields a gross rate of return of $R=1+r=(X_1/X_0)$, possibly random. Thus, the discounted value of the investment in the beginning of the period is

$$X_0 = X_1/(1+r). \tag{2.70}$$

Now consider the ith risky asset with several other risky assets and the risk-free asset as well. The random rate of return on this asset is

$$r_i = (X_{1i}/X_{0i}) - 1 \tag{2.71}$$

so that the expected rate of return is

$$r_i^{\,m} = (X_{1i}^{\,m} / X_{0i}) - 1 \qquad (2.72)$$

where $X_{1i}^{\,m}$ is the expected end-of-period payoff of the ith asset. Using the SCAPM formula in (2.67), we can rewrite (2.72) as

$$r_f + \beta_i\ (r_M^{\,m} - r_f) = (X_{1i}^{\,m} / X_{0i}) - 1 \qquad (2.73)$$

and solve for X_{0i} as

$$X_{0i} = X_{1i}^{\,m} / [\,1 + r_f + \beta_i\ (r_M^{\,m} - r_f)\,]. \qquad (2.74)$$

Formula (2.74) denotes the price of one unit of the ith risky asset, which equals the discounted value of its expected payoff $X_{1i}^{\,m}$ where the discount factor is $[1 + r_f + \beta_i\ (r_M^{\,m} - r_f)]$. If we compare the discount factor in (2.70) with the discount factor in (2.74), we note that (2.74) is adjusted by the term

$$\beta_i\ (r_M^{\,m} - r_f) = r^{\,m}_{\ i} - r_f$$

i.e. the excess return of the ith asset over the risk-free return. The pricing formula (2.74) has an important linearity property that is consistent with the absence of arbitrage opportunities in the market. We define an arbitrage opportunity as an investment strategy that yields a gain over and above the risk-free rate of return without any risk. The absence of arbitrage opportunities is particularly convenient from an economic point of view where we expect rational agents under full information to take full instantaneous advantage of all profit opportunities available in the market place. The consequence is that any gain without risk is not feasible. To see this, we consider from (2.65) and (2.71)-(2.72) that the β for the ith asset is given by

$$\beta_i \equiv \sigma_{iM} / \sigma_{MM} = cov[(X_{1i}/X_{0i}) - 1,\ r_M] / \sigma_{MM}. \qquad (2.75)$$

But, note that

$$cov\,[(X_{1i}/X_{0i}) - 1,\ r_M] = E[\{(X_{1i}/X_{0i}) - 1 - (X_{1i}^{\,m}/X_{0i}) + 1\}(r_M - r_M^{\,m})]$$

$$= (1/X_{0i})\ cov(X_{1i},\ r_M)$$

which when substituted into (2.75) above yields

$$\beta_i = cov[X_{1i}, r_M] / X_{0i}\ \sigma_{MM}. \qquad (2.76)$$

Substituting (2.76) into the pricing formula (2.74), dividing by X_{0i} and manipulating yields

$$X_{0i} = [1/(1+r_f)]\ \{X_{1i}^{\,m} - (cov[X_{1i}, r_M]\ (r_M^{\,m} - r_f) / \sigma_{MM})\}. \qquad (2.77)$$

Relationship (2.77) is important because when we compare it with (2.70), we notice that the term in keys is the total payoff, for this risky asset, that makes the individual treat it as a risk-free asset, i.e. discounts it by the risk-free discount factor $1/(1+r_f)$. Hence, the term in keys has an important property: it is an adjusted payoff that makes the individual indifferent to the risk embodied in the *i*th asset relative to the risk-free asset. We call this term the certainty equivalent of X_{1i},

$$CE_{X1i} = X_{1i}^m - (cov[X_{1i}, r_M] (r_M^m - r_f) / \sigma_{MM}) \qquad (2.78)$$

The certainty equivalent is the random expected payoff, X_{1i}^m, plus a premium,

$$-cov[X_{1i}, r_M] (r_M^m - r_f) / \sigma_{MM}$$

that depends upon the covariance of the payoff with the market and an adjusted price of risk, see (2.59). In particular, if the asset's payoff covaries negatively with the market rate of return, the premium must be positive to offset the opportunity cost of holding other risky assets that covary positively with the market. On the other hand, if the asset's payoff covaries positively with the market rate of return, the premium is negative since the risk embodied in the asset is the risk of the market. If the covariance is zero, the asset does not covary with the market and its risk can be fully diversified so that it is just as the risk-free asset, see (2.67)-(2.69). Therefore, the certainty equivalent of an asset is the amount of payoff that makes the individual indifferent to the risk embodied in the asset.

The extraordinary feature of the certainty equivalent in this context is that it is linear in X_{1i}^m. If we have two risky assets, *i* and *j*, with unit prices and payoffs, X_{0i}, X_{0j}, X_{1i}^m, X_{1j}^m respectively, the sum of their prices separately is, using (2.77)-(2.78), given by

$$X_{0i} + X_{0j} = [1/(1+r_f)] \{CE_{X1i} + CE_{X1j}\}. \qquad (2.79)$$

What is the price of this new asset? To answer, examine the right-hand side of (2.79) and note that linearity of the certainty equivalent in X_{1i}^m makes the price of the sum of the two assets identical to the sum of the prices of each asset separately. This property yields the no-arbitrage theorem in this mean-variance full information context:

No-Arbitrage Theorem: The linearity of asset prices in terms of initial prices and expected payoffs guarantees that no profit opportunities without risk remain available in the market.

Intuitively, if the sum of the prices diverges from the price of the sum, i.e. no linear pricing measure, one would be able to make a gain by, for instance, buying (selling) the two assets separately and selling (buying) a package of the two for whichever direction a profit entails. In essence, this would be an arbitrage opportunity without risk. The linear pricing measure rules out any such opportunities.

2.6 **Summary**

We have examined a problem of allocation of wealth when individuals only focus on the mean and the variance of the returns of the assets available in the market place. This framework ignores explicit utility over wealth (or bundles of consumption), so that a fully thought decision-making process for investment (or saving-investment) is not specified. However, as seen from above, this framework is worth studying because many insights can be drawn at a relatively simple analytical level.

Problems

1. Suppose there are only two risky assets denoted A and B. Monthly price data (in US$) for 12 periods for units of the two assets are given below.

Asset Prices

Month	*Asset A*	*Asset B*
0	*25*	*45*
1	*24.12*	*44.85*
2	*23.37*	*46.88*
3	*24.75*	*45.25*
4	*26.62*	*50.87*
5	*26.50*	*58.50*
6	*28*	*57.25*
7	*28.88*	*62.75*
8	*29.75*	*65.50*
9	*31.38*	*74.38*
10	*36.25*	*78.5*
11	*37.13*	*78*
12	*36.88*	*78.12*

Month zero denotes the initial price of the asset.

Use some spreadsheet to do the following computations:

i. Construct a table where you compute the monthly rate of return for each asset. Use a continuous compounding method such that the monthly rate of return is:

$$r_{it} = \log\left(P_{it} / P_{it-1}\right)$$

for i=A,B and P_t the price at month t. Is there any difference between the continuous compounding method and the discrete compounding method

$$r_{it} = \left(P_{it} - P_{it-1}\right) / P_{it-1} \ ?$$

Why or why not?

ii. Now make a heroic assumption:

A1: The twelve month rate of return data represent the known distribution of the returns in the coming months.

This allows us to use it as a proxy for the expected monthly return from each asset. Hence, compute the following moments from your rate of return table: mean, variance, standard deviation, and covariance.

iii. Now suppose you form a portfolio with the two assets: let the proportion $\alpha \in (0,1)$ denote the share of asset A in the portfolio. Set $\alpha = 0.5$ and compute the mean rate of return and standard deviation of your portfolio (note the mean of the portfolio is the average of the mean return of the two assets, but the variance is not because of the covariance!).

iv. Using your procedure in iii. above, construct a table for a grid of α, say α starting at zero and up to one in intervals of *0.05 (i.e. 0, 0.05, 0.1, 0.15...)*, with the respective mean and standard deviation.

v. Plot a graph in mean/standard deviation space from your results in vi. above. What does your graph look like? Is this a minimum-variance set? Why or why not?

vi. Compute, with pencil and paper, the minimum variance portfolio and its associated rate of return.

2. Show exactly how you obtain expression (2.6).

3. Expression (2.8): By assuming this specific ordering, is there any loss of generality in mean-variance space? Why or why not?

4. Explain why the mean-variance set of Figure 2.5 does not touch the vertical axis.

5. In Mean-Variance space, how do you derive an efficient frontier?

6. In Mean-Variance space, what is the Two-Fund Theorem? What is the One-Fund Theorem?

7. Derive the formula characterizing the Static Capital Asset Pricing Model (SCAPM).

8. In the context of the SCAPM, what is the difference between systematic and non-systematic risk?

9. In the context of the SCAPM, let X_1 be a final random payoff for an initial investment of X_0. What is the certainty equivalent of X_1?

10. In the context of the SCAPM, let X_1 be a final random payoff for an initial investment of X_0. Derive and explain the No-Arbitrage theorem.

Notes on the Literature

One of the first proponents of the mean-variance approach was Markowitz (1952, 1987a, 1987b). The One-Fund argument was introduced by Tobin (1978). Other developments in this area were presented by Fama (1976), Sharpe (1967), and Merton (1972). The SCAPM model was developed by Sharpe (1964), Lintner (1965) and Mossin (1966). Black (1972) has extended it to the case of absence of a riskless asset. Mutual Fund cases were developed by Sharpe (1966), Jensen (1969) and more recently Ross (1978), Chamberlain (1983) and Chamberlain and Rothschild (1983).

Several books present some of the material in this chapter at different levels of complexity. For example, Dixit (1990) and Luenberger (1998) are very accessible, Ingersoll (1987), Huang and Litzenberger (1987), and Cochrane (2001) are more advanced. Berndt (1991), Chapter 2, presents a simple and clear regression analysis application of the SCAPM of section 2.5. A recent useful discussion and evaluation is in Jagannathan and McGrattan (1995). Simple computational examples and exercises are available in Benninga (1997) and, more recently, Myerson (2002) presents a useful introduction to the computational methods. Bernstein (1992, 1996) provides vivid historical developments, at a non-technical level, of some of the theories studied in this chapter.

Campbell and Viceira (2000), Chapter 2, provide an excellent and concise review of the mean-variance framework from the point of view of short term investment strategies.

References

Benninga, Simon (1997) *Financial Modeling*. MIT Press, Cambridge, MA.

Berndt, Ernst. R. (1991) *The Practice of Econometrics: Classic and Contemporary*. Addison-Wesley Publishing Co., Reading, MA.

Bernstein, Peter L. (1992) *Capital Ideas: The Improbable Origins of Modern Wall Street*. The Free Press, New York, NY.

Bernstein, Peter L. (1996) *Against the Gods: The Remarkable Story of Risk*. John Wiley&Sons, New York, NY.

Black, Fisher (1972) "Capital Market Equilibrium with Restricted Borrowing." *Journal of Business,* 45, 444-454.

Campbell, John Y. and Luis Viceira (2000) *Strategic Asset Allocation: Portfolio Choice for Long Term Investors*. Book manuscript, Harvard University, November (published by Oxford University Press, 2002).

Chamberlain, Gary (1983) "Funds, Factors, and Diversification in Arbitrage Pricing Models." *Econometrica*, 50, 1305-1324.

Chamberlain, Gary and Michael Rothschild (1983) "Arbitrage, Factor Structure and Mean-Variance Analysis on Large Asset Markets." *Econometrica*, 50, 1281-1304.

Cochrane, John H. (2001) *Asset Pricing*. Princeton University Press, Princeton , NJ.

Dixit, Avinash K. (1990) *Optimization in Economics*, Second Edition. Oxford University Press, New York, NY.

Fama, Eugene (1976) *Foundations of Finance*. Basic Books, New York, NY.

Huang, Chi-Fu and Robert Litzenberger (1987) *Foundations of Financial Economics*. North-Holland, Amsterdam.

Ingersoll, Jonathan (1987) *Theory of Financial Decision-Making*. Rowman&Littlefield, New York, NY.

Jagannathan, Ravi and Ellen McGrattan (1995) "The CAPM Debate." Federal Reserve Bank of Minneapolis *Quarterly Review*, 19, 2-17.

Jensen, Michael (1969) "Risk, the Pricing of Capital Assets and the Evaluation of Investment Portfolios." *Journal of Business*, 42, 167-247.

Lintner, John (1965) "The Valuation of Risky Assets and the Selection of Risky Investments in Stock Portfolios and Capital Budgets." *Review of Economic and Statistics,*47, 13-37.

Luenberger, David (1998) *Investment Science.* Oxford University Press, Oxford, UK.

Markowitz, Henry (1952) "Portfolio Selection." *Journal of Finance*, 7, 77-91.

Markowitz, Henry (1987a) *Portfolio Selection.* John Wiley&Sons, New York, NY.

Markowitz, Henry (1987b) *Mean-Variance Analysis in Portfolio Choice and Capital Markets.* Basil Blackwell, New York, NY.

Merton, Robert (1972) "An Analytical Derivation of the Efficient Portfolio Frontier." *Journal of Financial and Quantitative Analysis,* 7, 1851-1872.

Mossin, Jan (1966) "Equilibrium in a Capital Asset Market." *Econometrica,* 34, 768-783.

Myerson, Roger B. (2002) *Probability and Decision Analysis.* Book manuscript, Department of Economics, University of Chicago, July.

Ross, Stephen (1978) "Mutual Fund Separation in Financial Theory: The Separation Distributions." *Journal of Economic Theory*, 17, 254-286.

Sharpe, William F. (1964) "Capital Asset Prices: A Theory of Market Equilibrium under Conditions of Risk." *Journal of Finance*, 19, 425-442.

Sharpe, William F. (1966) "Mutual Fund Performance." *Journal of Business,*39, 119-138.

Sharpe, William F. (1967) "Portfolio Analysis." *Journal of Financial and Quantitative Analysis,* 2, 76-84.

Tobin, James (1978) "Liquidity Preference as Behavior Towards Risk." *Review of Economic Studies,* 26, 65-86.

3. Expected Utility Approach to Financial Decision-Making

3.1 Introduction

Preferences over consumption bundles are essentially binary relations. For example, consider a world without uncertainty with a set of consumption bundles denoted $\boldsymbol{X}$. Let two consumption bundles be denoted x and y, with $\{x,y\}\in \boldsymbol{X}$, and examine the function $u{:}\boldsymbol{X}\rightarrow \mathbb{R}$, which reads the domain of u is $\boldsymbol{X}$ contained in $\mathbb{R}$, or $\boldsymbol{X}\subset\mathbb{R}$. Then, the ordering

$$x \text{ is preferred to } y \quad \text{if and only if} \quad u(x) > u(y)$$

follows intuitively. The ordering means that consumption bundle x is preferred to consumption bundle y as long as the value of the function u evaluated at x is strictly greater than the value of the function u evaluated at y.

In what follows, our goal is to search for an ordering like the one above, but in the case where the consumption bundles x and y represent risky consumption prospects. What is the effect of risk on the function u? To answer this question, we consider the widely used case where uncertainty is objectively assessed, i.e. exogenously given. In addition, we discuss individual behavior towards risk and a portfolio choice problem that highlight some of the relationships between behavior towards risk and the demand for risky assets. In this context, we examine the quadratic utility index case and provide a simple numerical example. Finally, we apply those concepts to the case of diversification with independent returns.

An appendix to this chapter provides a discussion of the special case, where the randomness is restricted to probability distributions of payoffs in the class of Normal distributions and utility is exponential, and thus characterized by the mean and variance only. In addition, we examine an alternative to the traditional Von-Neumann-Morgenstern expected utility framework, involving non-additive probabilities.

3.2 The Von-Neumann-Morgenstern (VNM) Framework: Probability Distributions over Outcomes

First, we discuss utility in the case where uncertainty is exogenous, basically risk as opposed to uncertainty in the Knightian sense. The VNM expected utility framework is specialized for the case where the uncertain consumption prospect is an outcome, and hence there is a probability distribution over possible outcomes.

We let $\boldsymbol{X}$ be a set of random "prizes" or "outcomes," with $x \in \boldsymbol{X}$ representing a possible "prize" or "outcome," i.e. a random variable. For example, x is the bundle $x=(x_1=10 \text{ cans of soda})$ for $x \in \boldsymbol{X}$ where x is a one element vector in this case.

Next, we let $\boldsymbol{P}$ be a set of probability distributions on the vector of prizes x, with $p \in \boldsymbol{P}$ representing a "lottery" or a "gamble" or a "probability distribution" over the vector of prizes $x \in \boldsymbol{X}$. The lottery (or gamble or probability distribution) space satisfies the following two properties:
(i) There are a finite number of possible outcomes which is a subset of $\boldsymbol{X}$, called the support of $p \in \boldsymbol{P}$, denoted by $\boldsymbol{X}^P$;
(ii) For each $x \in \boldsymbol{X}^P$, there is a number $p(x)>0$ with $\Sigma_{x \in XP}\, p(x) = 1$.
Thus, $\boldsymbol{P}$ is a set of (simple) probability distributions on $x \in \boldsymbol{X}$ that satisfies (i) and (ii).

For example, let $\boldsymbol{X} \to \mathbb{R}^2_+$, or the domain of $\boldsymbol{X} \subset \mathbb{R}^2_+$ is the positive orthant. Consider the bundles $x=(x_1=\text{cans of soda},\ x_2=\text{bottles of water})$. The probability distribution can be defined accordingly as:
• possible outcomes, or the support of $p \in \boldsymbol{P}$, are $\{(x_1=10 \text{ cans of soda},\ x_2=2 \text{ bottles of water}),\ (x_1=4 \text{ cans of soda},\ x_2=4 \text{ bottles of water})\}$;
• the probability distributions are $p(x_1=10,x_2=2)=1/3$, $p(x_1=4,x_2=4)=2/3$, or the bundle $(10,2)$ with probability $1/3$ and bundle $(4,4)$ with probability $2/3$.

For general binary relations under risk, it is common to assume that three axioms are satisfied: the transitivity axiom, the substitution (or independence) axiom, and the continuity (or Archimedean) axiom; see Notes on the Literature. Given the axioms, we can represent a preference relation of the VNM type with an ordering over the "lottery" or "gamble" or "probability distribution." This gives the following so-called expected utility theorem:

Expected Utility Theorem: There is a utility function $u{:}\boldsymbol{X}\rightarrow\mathbb{R}$, such that for two probability distributions, p and q, satisfying (i) and (ii) and the transitivity, substitution (or independence), and the continuity (or Archimedean) axioms, we have

$$p \text{ is preferred to } q \quad \text{if and only if } \Sigma_{x\in X^P}\, p(x)\, u(x) > \Sigma_{x\in X^Q}\, q(x)\, u(x) \quad (3.1)$$

i.e. a preference ordering over probability distributions on the vector of prizes x.

The ordering in (3.1) is based on an expected utility representation of the bundles, or the value of the probability distribution measured by the expected utility level it provides. Moreover, the probabilities over the prizes are given exogenously. An ordering representation of this type is called a VNM expected utility representation. Intuitively, the bundle that has the highest probability, or likelihood, is preferred to the bundle with the lowest probability, translating into more expected utility is preferred to less expected utility.

We are able to further specialize the VNM expected utility representation by examining two properties of the utility function: (i) non-satiation; and (ii) risk aversion.

(i) First, consider the issue of non-satiation. For a lottery (gamble or probability distribution) p, over prizes x,

$$E[x] = \Sigma_{x\in X^P}\, p(x)\, x \quad (3.2)$$

denotes its expected value. Next, let $\boldsymbol{X}$, the set of "prizes" or "outcomes," be some dollar value ($) or wealth, so that $\boldsymbol{X}$ can be thought as the real line or some arbitrary interval of the real line. Then, we are able to define a special "lottery" (gamble or probability distribution) that gives the prize (or outcome) x with probability one, i.e. with certainty. Denote this lottery or ticket δ_x . In this special case, δ_x is a probability distribution with one possible outcome, and using (3.2) we obtain its expected value as

$$E[\delta_x] = \Sigma_{x \in XP}\, p(x)\, x = 1 \times x = x, \tag{3.3}$$

i.e. a realization of prize x. Now, let the function $u{:}\mathbf{X} \rightarrow \mathbb{R}$ be strictly increasing, or $\partial u(x)/\partial x \equiv u'(x) > 0$ all $x \in \mathbf{X}$. Strictly positive marginal utility implies that for a small increase in the prize x, utility increases by u'. Then, for all $\{x,y\} \in \mathbf{X}$ so that $x > y$, the expected utility representation (3.1) implies that $\delta_x > \delta_y$. Intuitively, the expected utility representation (3.1) together with strictly positive marginal utility $u' > 0$, gives non-satiation in preferences, or "more is preferred to less."

(ii) Second, we examine the issue of risk aversion. Consider again two prizes (or outcomes) x and y, $\{x,y\} \in \mathbf{X}$, with respective associated utilities $\{u(x),\ u(y)\}$, and assume the ordering $x > y$, $u(x) > u(y)$. Next, consider some number $\alpha \in [0,1]$ and use expression (3.3) to construct another lottery (gamble or probability distribution), based upon two possible outcomes, as a simple convex (linear) combination of the sure tickets for x and y. This new lottery is a prize given by

$$p = \alpha\, \delta_x + (1 - \alpha)\, \delta_y \tag{3.4}$$

representing a chance α at the sure ticket for x, and a chance $(1-\alpha)$ at the sure y. This is a gamble between the sure tickets for x and y, or a gamble over two possible outcomes. The expected value of the new lottery in (3.4) is, using the linearity of the expectations operator and (3.3), given by

$$E[p] = \alpha\, x + (1 - \alpha)\, y \tag{3.5}$$

where $E[p]$ in (3.5) is just a convex combination of two numbers yielding the prize (or outcome) $E[p]$, for example $E[p] \equiv z \in \mathbf{X}$. Next, we can use (3.2)-(3.3) to consider the sure ticket for the expected value of p, that is the lottery $\delta_{E[p]}$. By (3.3), this is the lottery that yields $E[p]$ with probability one, i.e. for sure, or

$$E[\delta_{E[p]}] = E[p].$$

Therefore, the utility of this lottery is simply

$$u(\delta_{E[p]}) = u(E[p]) = u(\alpha\, x + (1 - \alpha)\, y) \tag{3.6}$$

where the last equality follows from (3.5). In expression (3.6), we have the utility of the expected value of the lottery p, expressed as the utility of a linear combination of each of the possible prizes $\{x,y\}$. The next step is to calculate the expected utility of the lottery p. We use (3.5) to obtain the expected utility of the lottery p, given by (3.4), to obtain

$$E[u(p)] = \alpha\, u(x) + (1 - \alpha)\, u(y) \tag{3.7}$$

a linear combination of the utility of each possible prize $\{x,y\}$. Now, we are ready to compare the utility of the expected value of the lottery p to the expected value of the utility of the same lottery.

If the function u is concave, by the definition of concavity we have that

$$u(\alpha x + (1 - \alpha)\, y) \geq \alpha\, u(x) + (1 - \alpha)\, u(y) \tag{3.8}$$

or the utility of the expected value of the lottery p is no less than the expected value of the utility of the lottery p. Similarly, if the function u is concave, for a random variable p, Jensen's Inequality implies that

$$u(E[p]) \geq E[u(p)] \tag{3.9}$$

which by (3.6) and (3.7) is exactly (3.8). Expressions (3.8)-(3.9) say that, under concavity of the utility function u, the utility of the expected lottery cannot be less that the expected utility of the lottery. In turn, the curvature of the utility function plays a crucial role in determining the value of the lottery for an individual. The final step is to relate this curvature to risk aversion per se.

Consider the risky gamble p in (3.4), i.e. the lottery itself; and the ticket for the prize $E[p]$ for sure, $\delta_{E[p]}$, i.e. the individual's willingness to pay for the lottery. Entertain the possibility that

$$\delta_{E[p]} < p,$$

e.g. the gamble described in (3.4) is strictly greater than obtaining the expected value of that gamble for sure, $\delta_{E[p]}$. This inequality would contradict (3.9) or (3.8) by comparing the utility of $\delta_{E[p]}$ with the expected utility of p by (3.6) and (3.7). Hence, under concavity of the function u, we have that

$$\delta_{E[p]} \geq p \tag{3.10}$$

i.e. the sure ticket for the expected value of the gamble is (weakly) preferred to the gamble per se. We are ready to understand risk aversion in the context of the function u.

Risk Aversion: If expression (3.10) is satisfied for all lotteries (or gambles or probability distributions) on $\boldsymbol{X}$, that is $\delta_{E[p]} \succeq p$ for all $p \in \boldsymbol{P}$, the individual endowed with the utility function u is said to be risk averse. Equivalently, the individual is said to be risk averse if and only if her endowed utility function is concave, consistently with (3.10).

Intuitively, equation (3.10) shows that a risk averse individual is one that, when faced with a choice between a ticket for the expected value of the gamble, that is $E[p]$ for sure, and the gamble per se, p, she will prefer the expected value of the gamble, e.g. the individual prefers the expected value for sure than the risky prospect. In our heuristic presentation above, we considered a gamble with two possible outcomes, however this can be easily generalized for more outcomes, i.e. for any support of the probability distribution. Figure 3.1 illustrates the idea of risk aversion. Note that, in the figure, there is a prize (or outcome), denoted x^*, which, by the intermediate value theorem, satisfies the conditions

$$u(x^*) = \alpha\, u(x) + (1 - \alpha)\, u(y) = E[u(p)] \quad \text{and} \quad x^* \neq E[p]. \qquad (3.11)$$

Thus, by the expected utility representation (3.1) and by (3.10), we have that

$$\delta_{x^*} \sim p$$

where "$\sim$" indicates indifference. Expression (3.11) indicates that the prize x^* makes the individual indifferent between the sure ticket for x^* and the gamble per se. Logically, the ticket δ_{x^*} is called the Certainty Equivalent of the lottery p. This is the prize that would make the individual indifferent between receiving it for sure or gambling. This is analogous to the certainty equivalent of a risky asset studied in Chapter 2.

In the case of risk aversion, we have that (see Figure 3.1)

$$x^* < E[p] \qquad (3.12)$$

in words, the prize that gives the certainty equivalent of the lottery p is strictly less than the expected value of the lottery itself. But, the prize that gives the certainty equivalent, x^*, can be appropriately interpreted as the individual's willingness to pay for the lottery. In turn, risk aversion implies that the individual's willingness to pay to enter a gamble is strictly less that the expected value of the lottery itself.

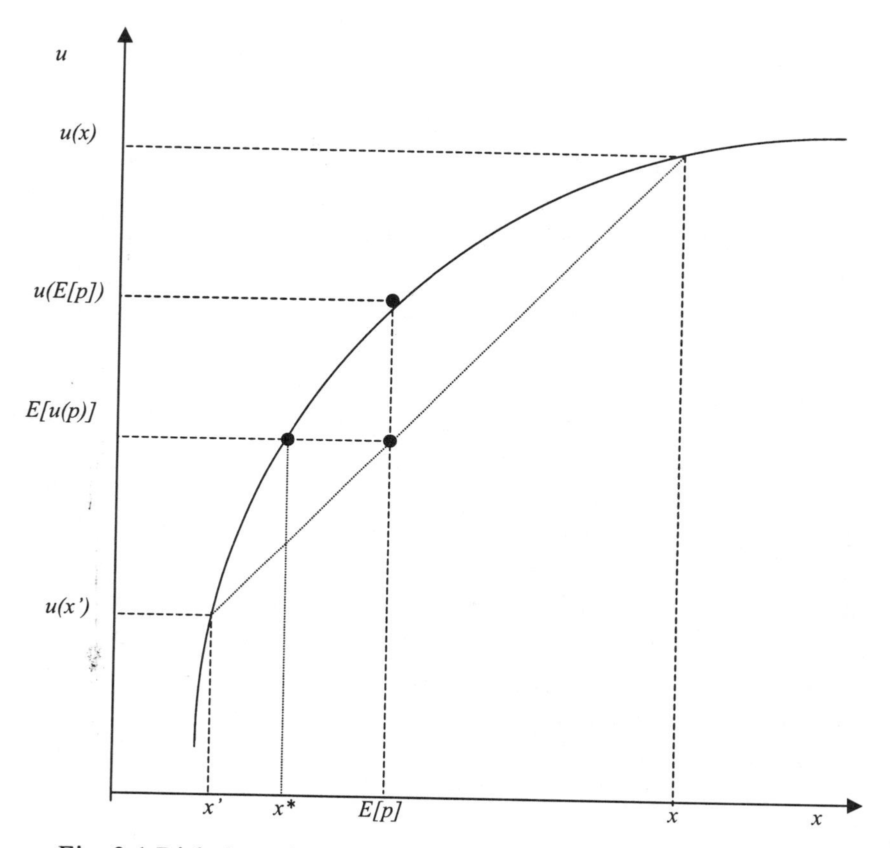

Fig. 3.1 Risk Aversion and the Curvature of the Utility Function

Furthermore, we may define the difference between the expected value of the lottery and the willingness to pay for the lottery, $E[p] - x^*$, as the risk premium:

$$risk\ premium \equiv E[p] - x^*; \qquad (3.13)$$

and for a risk averse individual, $risk\ premium \equiv E[p] - x^* > 0$.

In summary: we note that the second derivative of the utility function, u'', refers to the change in the marginal utility, u', given a small change in the prize; and thus

(i) If the individual is endowed with a utility index u which is <u>strictly</u> concave, $u''<0$, then $u(E[p])>E[u(p)]$ by (3.6) and (3.7), $E[p]>x^*$ by

(3.11), and the *risk premium* $\equiv E[p] - x^{*}>0$ by (3.13). In this case, the individual is said to be risk averse because she prefers the expected value of the risky gamble instead of the gamble per se;
(ii) If the individual is endowed with a utility index u which is linear, $u''=0$, then $u(E[p])=E[u(p)]$ by (3.6) and (3.7), $E[p]=x^{*}$ by (3.11), and *risk premium* $\equiv E[p] - x^{*}=0$ by (3.13); in this case the individual is said to be risk neutral;
(iii) If the individual is endowed with a utility index u which is strictly convex, $u''>0$, then $u(E[p])<E[u(p)]$ by (3.6) and (3.7), $E[p]<x^{*}$ by (3.11), and *risk premium* $\equiv E[p] - x^{*}<0$ by (3.13); in this case the individual is said to be risk loving because exhibits a willingness to accept a risky gamble over its expected value.

We have shown that when the utility index u is twice continuously differentiable, strictly increasing and strictly concave, then $u'>0$ (the change in utility given a small change in the prize) indicates non-satiation, and $u''< 0$ indicates individual aversion towards risk, where $u''= d^2u(x)/dx^2$ (the change in marginal utility given a small change in the prize). We can use the information in these two differentials to 'measure' risk aversion, in particular using the degree of concavity, or curvature of the utility function.

3.3 Measurement of Risk Aversion

We first consider the so-called Arrow-Pratt constant absolute risk aversion measure, defined as

$$CARA(x) \equiv - u''(x) / u'(x) \qquad (3.14a)$$

where *CARA* stands for constant absolute risk aversion. According to this definition, as the absolute value of the second differential of the function u increases locally, or $|u''|$ increases; it gives more curvature to the utility function and the individual becomes more risk averse. It is an absolute measure because it compares the absolute gains and losses of a random prize x. To see this, consider an absolute (small) change in the prize as $x'=x+\epsilon$, for some (small) real number ϵ. Then, compute

$$\partial\, u(x+\epsilon)/\partial\epsilon = [\partial\, u(x+\epsilon)/\partial(x+\epsilon)][\partial(x+\epsilon)/\partial\epsilon] = u'(x+\epsilon)$$

$$\partial^2 u(x+\epsilon)/\partial\epsilon^2 = u''(x+\epsilon)$$

which when substituted into (3.14a), appropriately evaluated at $\epsilon = 0$, yields the well defined measure of absolute risk aversion in (3.14a).

Figure 3.2 illustrates an increase in absolute risk aversion as measured by expression (3.14a). In the figure, we have that the expected value of the lottery p is given by

$$E[p] = \alpha x + (1 - \alpha)\, x', \quad for\ \alpha \in [0,1]$$

and the utilities of the certainty equivalents are

$$u_1(x_1^*) = \alpha\, u_1(x) + (1 - \alpha)\, u_1(x') = E[u_1(p)], \quad for\ \alpha \in [0,1]$$

$$u_2(x_2^*) = \alpha\, u_2(x) + (1 - \alpha)\, u_2(x') = E[u_2(p)], \quad for\ \alpha \in [0,1]$$

where we used expression (3.10).

Hence, the risk premia for the utility index u_1 and for the utility index u_2 obey

$$E[p] - x_1^* < E[p] - x_2^*$$

where x^* is defined in (3.10). It follows that an individual with utility u_2 demands a larger risk premium than u_1, i.e. the individual's willingness to pay for the lottery is smaller. In turn, the utility index u_2 shows more (absolute) risk aversion than the index u_1. Alternatively,

$$|\, u_2''| > |\, u_1''| \quad for \quad x' \leq x^* \leq E[p].$$

The utility index u_2 has more curvature than u_1, hence exhibiting more (absolute) risk aversion.

It is also of interest to understand how absolute risk aversion varies with the prize x. We compute the derivative

$$d\,CARA(x)/dx \equiv CARA'(x) = \{ - u'''(x)\, u'(x) + [u''(x)]^2 \} / [u'(x)]^2, \quad (3.14b)$$

and see that

$$CARA'(x) = \{ - u'''(x)\, u'(x) + [u''(x)]^2 \} / [u'(x)]^2 \leq 0 \Rightarrow u'''(x) > 0,$$

in words, for absolute risk aversion to be non-increasing in the prize, the third derivative of the utility function must be strictly positive, that is $u''' = d^3 u(x) / dx^3 > 0$ (the change in u'' given a small change in the prize is strictly positive) which implies that the marginal utility, u', is strictly convex.

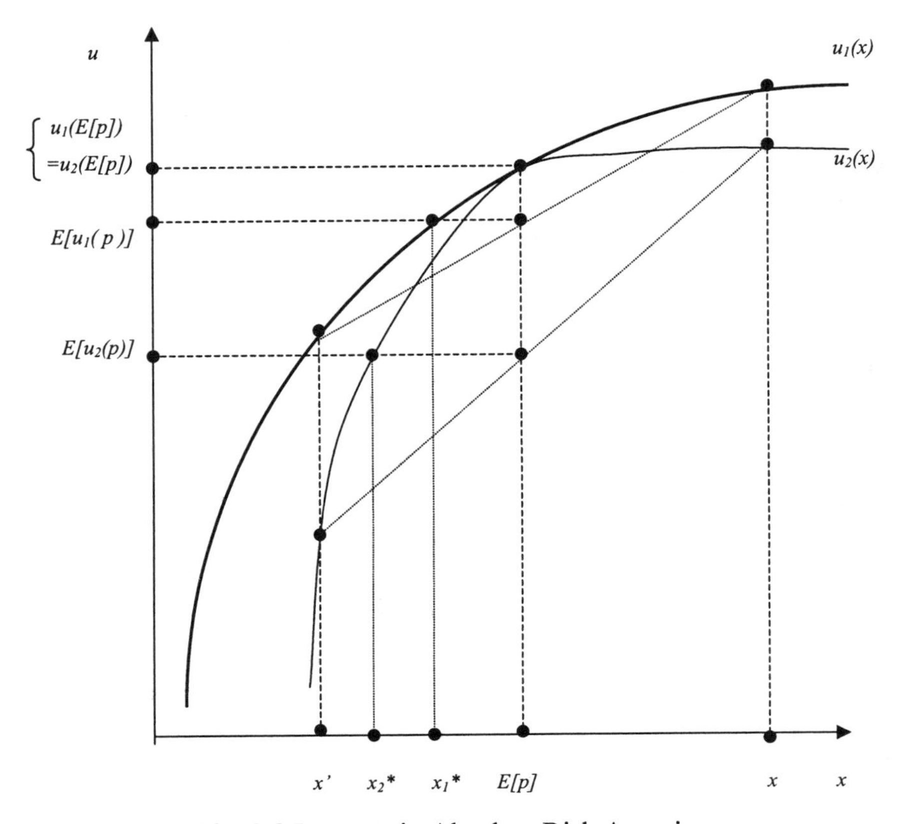

Fig. 3.2 Increase in Absolute Risk Aversion

The implication of the condition on the third derivative is that for a small increase in the prize, the lower the level of absolute risk aversion. Some have related the condition of a strictly positive third derivative of the utility function at a prize x to the concept of (absolute) prudence, analogously to the concept of absolute risk aversion. The analogy follows from the definition of absolute risk aversion in (3.14a) relative to utility level u, and a definition of absolute prudence relative to the marginal utility u' as $\mathcal{P}(x) \equiv -u'''(x)/u''(x)$. The reference to "prudence" relates to the amount of additional saving an individual would be willing to engage when faced with additional random wealth, the so-called precautionary saving motive. In effect, under $\mathcal{P}(x)>0$, the lower (higher) the prize x, the higher (lower) $CARA$ implying positive absolute "prudence." Thus, the condition $u''' > 0$ implies that absolute prudence is strictly positive for a risk averse individual, $u''<0$; see Notes on the Literature.

We can proceed and examine a measure of relative risk aversion along the same lines. Consider the Arrow-Pratt measure of constant relative risk aversion, given by

$$CRRA(x) \equiv - u''(x)\, x \,/\, u'(x) \tag{3.15}$$

where $CRRA$ stands for constant relative risk aversion. As before, according to this definition, as $|\, u''|$ increases locally, it gives more curvature to the utility function and the individual becomes more risk averse. However, in this case, it is a relative measure because it compares the relative gains and losses of a random prize x. To see this, consider a relative (small) change in the prize as $x'=x\epsilon$, for some (small) number τ satisfying, $\epsilon=1+\tau$. Then, compute

$$\partial\, u(x\epsilon)/\partial\epsilon = [\partial\, u(x\epsilon)/\partial(x\epsilon)][\partial(x\epsilon)/\partial\epsilon] = u'(x\epsilon)\, x$$

$$\partial^{2} u(x\epsilon)/\partial\epsilon^{2} = u''(x\epsilon)\, x^{2}$$

which when substituted into (3.14a), appropriately evaluated at $\epsilon = 1$ (or $\tau=0$), yields a measure of relative risk aversion as in (3.15). Expressions (3.14a) and (3.15) imply that

$$CRRA(x) = CARA(x)\, x,$$

and the manner at which relative risk aversion varies with the prize x can be computed similarly to (3.14b).

It is worth mentioning that, in this context, the curvature of the utility function implies the static concept of risk aversion. But, curvature also has an important connection with the concept of intertemporal substitution in a dynamic, deterministic, environment. In effect, risk aversion is inversely related to intertemporal substitution in this framework. This is not a desirable property because risk aversion refers to attitudes towards gambles in a timeless fashion, whereas intertemporal substitution refers to attitudes towards the transfer of consumption bundles across periods in a risk-free fashion. We pursue the consequences and solutions for this problem further in Chapters 7 and 8.

3.4 The Von-Neumann-Morgenstern (VNM) Framework: State Dependent Utility

In the case of probability distributions over possible outcomes discussed is the previous section 3.3, we have not given any explanation for the basic cause of receiving the prize. In other words, we have not explained the causes of the outcomes. In order to introduce a discussion of the basic causes, we need to further develop the mapping between actions and outcomes. As before, let $\boldsymbol{X}$ be a set of random "prizes" or "outcomes," with $x \in \boldsymbol{X}$ representing a possible "prize" or "outcome," i.e. a random variable. Now, introduce a set $\boldsymbol{Z}$ whose elements represent the possible states of nature. We assume that $\boldsymbol{Z}$ is finite. Each $z \in \boldsymbol{Z}$ is a possible state of nature that gives a resolution to uncertainty, where the z states are mutually exclusive and collectively exhaustive. The underlying state of nature is the cause of the prize or outcome. For example, "consumption of sorbet if the temperature, or state of nature, is $\geq 75°$."

Next, using the set of prizes $\boldsymbol{X}$ and the set of states of nature $\boldsymbol{Z}$, we construct a set of pairs $\{x,z\}$ denoted $\boldsymbol{G}$. A lottery (or gamble or probability distribution) $g \in \boldsymbol{G}$ will describe, for each state of nature $z \in \boldsymbol{Z}$, a respective prize $x \in \boldsymbol{X}$ that is awarded when state z occurs. This lottery is denoted $g(z)$ and it is usually called a state contingent claim. The state contingent claim assigns a prize to the lottery holder when the realized state of nature is $z \in \boldsymbol{Z}$.

For example, suppose a game of chess is played between Kasparov (K) and Fisher (F). The possible states of nature are:

$$z_1 = K\ wins,\ \ z_2 = F\ wins,\ \ z_3 = draw$$

where $\{z_1, z_2, z_3\}$ are the possible states of nature for the chess game lottery, and $\{K\ wins,\ F\ wins,\ draw\}$ are the causes of the state. Then, for this example, the g's are given by the mappings

$$g(z_1) = x_1,\ \ prize\ if\ K\ wins,$$

$$g(z_2) = x_2,\ \ prize\ if\ F\ wins,$$

$$g(z_3) = x_3,\ \ prize\ if\ draw$$

the so-called state contingent claims: if state z_1 occurs, may claim prize x_1, etc.

How do we assess the probability of occurrence of a specific state of nature before the resolution of uncertainty? In the VNM tradition, we assume that there is a probability distribution [as in (i)-(ii) of section 3.2], here denoted by the function π, over the set of states of nature $\boldsymbol{Z}$, so that for each state $z \in \boldsymbol{Z}$, there is an objective (exogenous) probability $\pi(z)>0$ that state $z \in \boldsymbol{Z}$ occurs, with $\Sigma_{z \in Z}\ \pi(z)=1$. We are ready to give a VNM representation of the state dependent utility analogously to the expected utility theorem.

<u>Expected State Dependent Utility Theorem</u>: Given a probability distribution π, over states of nature $z \in \boldsymbol{Z}$, a set of prizes $\boldsymbol{X}$, and a set of lotteries $\boldsymbol{G}$, there is a utility function $u: \boldsymbol{X} \times \boldsymbol{Z} \rightarrow \mathbb{R}$, where $\boldsymbol{X} \times \boldsymbol{Z}$ denotes the Cartesian product, or the set of ordered pairs $\{x,z\}\ x \in \boldsymbol{X},\ z \in \boldsymbol{Z}$, such that

g is preferred to g' if and only if

$$\Sigma_{z \in Z}\ \pi(z)\ u(g(z),z) > \Sigma_{z \in Z}\ \pi(z)\ u(g'(z),z) \qquad (3.16)$$

i.e. the lottery g is preferred to the lottery g' as long as the expected utility of the lottery g is greater than the expected utility of the lottery g'. This is sometimes called the extended expected utility approach.

Heuristically, the difference between the expected utility representation (3.1) and the expected state dependent representation (3.16) is the following. In the expected utility theorem, the representation (3.1) gives state independent utility, for which the representation (3.16) implies

g is preferred to g' if and only if $\Sigma_{z \in Z}\ \pi(z)\ u(g(z)) > \Sigma_{z \in Z}\ \pi(z)\ u(g'(z))$

i.e. the function u is the same for every state of nature $z \in \boldsymbol{Z}$. In (3.16), the function u is not the same for every state of nature $z \in \boldsymbol{Z}$, because $u(.,z)$ itself is a function of z, i.e. u changes as z changes. The implications of this difference are important. For example, suppose there are two states of nature, z_1 and z_2, with respective prizes attached, x_1 and x_2, and respective probabilities, $\pi(z_1)$ and $\pi(z_2)$. The state dependent expected utility is, from (3.16), given by

$$\mathbb{U} = \Sigma_{z \in Z}\ \pi(z)\ u(x_z, z) = \pi(z_1)\ u(x_1, z_1) + \pi(z_2)\ u(x_2, z_2).$$

The slope of the indifference curve between the two prizes is given by

$$dx_2/dx_1|_{d\mathbb{U}=0} = -[\pi(z_1)/\pi(z_2)]\ \{u'(x_1, z_1)/u'(x_2, z_2)\} \equiv MRS(z)$$

yielding a marginal rate of substitution across prizes, which is state dependent, *MRS(z)*. For the case of the state independent expected utility given in (3.1), we have expected utility as

$$U^* = \Sigma_{z \in Z}\, \pi(z)\, u(x_z) = \pi(z_1)\, u(x_1) + \pi(z_2)\, u(x_2)$$

and the slope of the indifference curve between the two prizes is given by

$$dx_2/dx_1\,|_{dU^*=0} = -[\pi(z_1)/\pi(z_2)]\,\{u'(x_1)/u'(x_2)\} \equiv MRS$$

yielding a marginal rate of substitution across prizes, which is state independent, *MRS*.

Figure 3.3 further illustrates the issue for a risk averse individual, where a concave utility function yields marginal rate of substitution that is convex towards the origin. The line $x_1 = x_2$ is called the certainty line because it gives the same prize in every state of nature. In the state dependent case, the marginal rate of substitution evaluated along two different points in the certainty line may be different because the utilities may be different, depending upon the different states of nature. Thus, in going from x^* to $2x^*$,

$$-[\pi(z_1)/\pi(z_2)]\,\{u'(x^*, z_1)/u'(x^*, z_2)\} \neq$$

$$-[\pi(z_1)/\pi(z_2)]\,\{u'(2x^*, z_1)/u'(2x^*, z_2)\}.$$

Alternatively, in the state independent case

$$-[\pi(z_1)/\pi(z_2)]\,\{u'(x^*)/u'(x^*)\} =$$

$$-[\pi(z_1)/\pi(z_2)]\,\{u'(2x^*)/u'(2x^*)\} = -[\pi(z_1)/\pi(z_2)]$$

i.e. the marginal rate of substitution is constant along the certainty line. The reader should be aware that this property is analogous to the property of homotheticity (ordinal criteria) in the case of certainty and choice over two goods.

Finally, we note that there is an alternative approach of subjective probabilities, the Savage approach, which very roughly speaking replaces all objective probabilities that we considered with their subjective counterparts. We do not discuss this case here and the interested reader is referred to the Notes on the Literature at the end of this chapter.

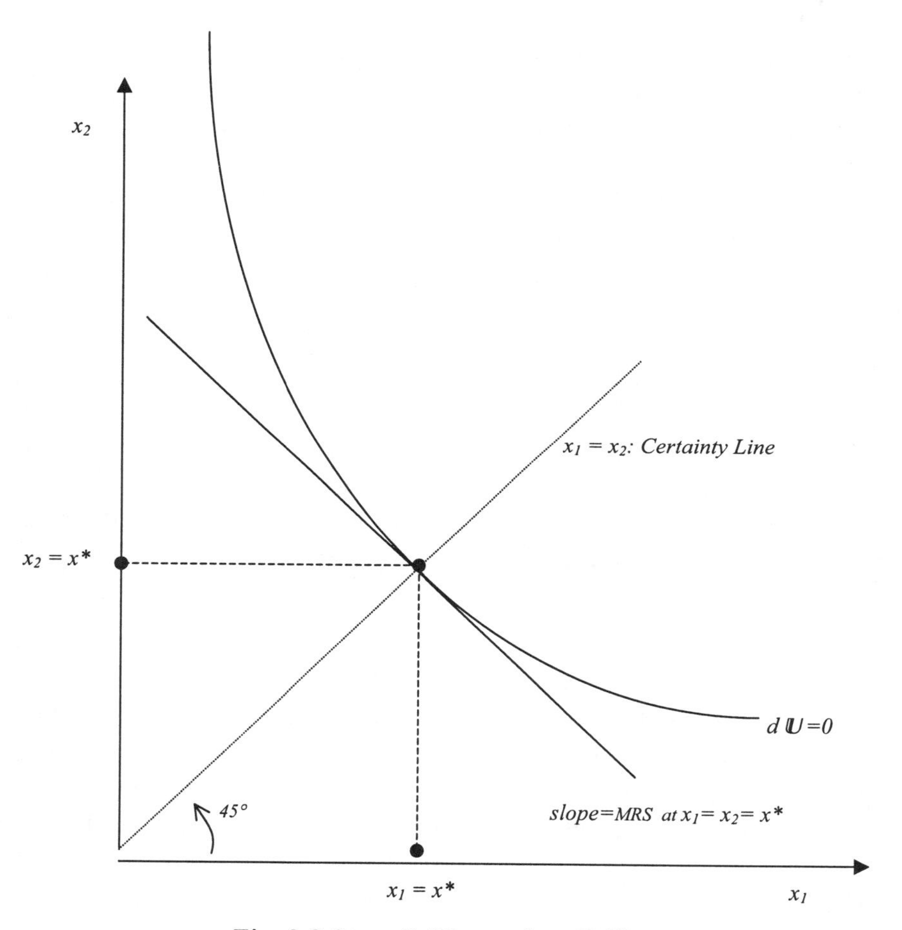

Fig. 3.3 State (In)Dependent Utility

3.5 Portfolio Choice and Comparative Statics with VNM State Independent Expected Utility

We proceed with an analysis of a simple portfolio choice problem using the utility framework presented above.

Suppose there are two assets available for an individual to choose to invest a given amount of wealth. One asset is risky in the following sense. There is a finite state space, denoted $\mathbf{Z}$, so that the gross return on the risky asset is $R(z)$ when state $z \in \mathbf{Z}$ occurs. The finite state space is supported be the probability distribution $p(z)>0$, with $\Sigma_{z \in Z}\, p(z)=1$. The other asset is risk-free, with gross return R_f.

Let the non-stochastic value of initial wealth be denoted W_o. The portfolio choice problem faced by an individual can be written as

$$\underset{\{\alpha\}}{Max}\ \Sigma_{z \in Z}\, p(z)\, u(R_f\, [W_o - \alpha] + R(z)\, \alpha) \qquad (3.17)$$

where α denotes the value of the risky asset in the portfolio and W_o-α denotes the value of the risk-free asset in the portfolio. In this case, we examine the value of investment in each asset and relate the implied demand for each asset to the measure of absolute risk aversion. Alternatively, we could examine the share of each asset in total wealth and relate it to relative risk aversion as well; see e.g. the Appendix to this chapter and Notes on the Literature. The final random wealth of an initial investment is thus

$$W(z) \equiv R_f\, [W_o - \alpha] + R(z)\, \alpha > 0$$

with short selling allowed as long as the inequality is satisfied; in other words final random wealth must be strictly positive and no bankruptcy is allowed. The utility function u is strictly increasing and twice continuously differentiable, with further properties discussed below. According to problem (3.17), an individual maximizes the expected utility of the final random wealth by choice of an investment strategy, α. The necessary first order condition for an interior equilibrium is given by

$$\Sigma_{z \in Z}\, p(z)\ u'(R_f\, [W_o - \alpha] + R(z)\, \alpha)\, [\, R(z) - R_f\,] = 0 \qquad (3.18a)$$

whose solution yields a demand function for the risky asset denoted

$$\alpha^* = \alpha^*(R_f, W_o, p(z)). \qquad (3.18b)$$

The (sufficient) second order condition,

$$\Sigma_{z \in Z}\, p(z)\ u''(R_f\, [\, W_o - \alpha] + R(z)\, \alpha)\, [\, R(z) - R_f\,]^2 < 0 \qquad (3.18c)$$

is guaranteed to hold for a strictly concave utility function, $u''<0$, which implies that the individual is risk averse, e.g. section 3.2. Expression (3.18a) simply equates the marginal benefit and marginal cost of the additional dollar invested in the risky asset. Suppose the individual decides to invest one more dollar in the risky asset. The marginal gain is measured by the marginal utility of the additional dollar, $u'(W(z))$, times the (gross) risky return on the dollar, $R(z)$. However, there is a forgone revenue from the risk-free asset given by the marginal utility, $u'(W(z))$, times the (gross) risk-free return on the dollar, R_f. The efficient allocation is where the expected marginal benefit is equated to the expected marginal cost, or

$$E[u'(W(z))\,R(z)] = E[u'(W(z))R_f]$$

which is exactly expression (3.18a).

It is of interest to examine the comparative statics of the demand function in (3.18b) with respect to changes in the initial level of wealth, W_o. Differentiating the first order condition, (3.18a), with respect to W_o yields the expression

$$\partial\alpha^*/\partial W_o = -\Sigma_{z\in Z}\, p(z)\ u''(R_f\,[W_o-\alpha] + R(z)\,\alpha)\,[R(z) - R_f]\,R_f / \Sigma_{z\in Z}\, p(z)\ u''(R_f\,[W_o-\alpha] + R(z)\,\alpha)\,[R(z) - R_f]^2 \qquad (3.19)$$

which cannot be easily signed because, in the numerator, $R(z) - R_f \gtreqless 0$, depending upon the realization of the state $z\in\mathbf{Z}$. The expression in the denominator, $\Sigma_{z\in Z}\, p(z)\ u''(R_f[W_o-\alpha]+R(z)\alpha)[R(z)-R_f]^2 < 0$, is strictly negative for risk averse individuals, or when $u''<0$, so that the sufficient second order condition is satisfied and an optimum exists.

However, we can relate the numerator to the coefficient of absolute risk aversion in the following way. Recall expression (3.14a), and rewrite it as

$$u'' = -CARA \times u'.$$

Then, we can substitute into the numerator of (3.19) to obtain

$$\Sigma_{z\in Z}\, p(z)\ u'(R_f\,[W_o-\alpha] + R(z)\,\alpha)\,[R(z) - R_f]\,R_f\ CARA(R_f\,[W_o-\alpha] + R(z)\,\alpha) \qquad (3.20a)$$

where the functions are appropriately evaluated at the maximum, α^*. Equation (3.20a) may be suitably compared to the expression

$$\Sigma_{z\in Z}\, p(z)\ u'(R_f\,[W_o-\alpha] + R(z)\,\alpha)\,[R(z) - R_f]\ R_f\ CARA(R_f\,W_o) = 0 \qquad (3.20b)$$

where the equality to zero follows from the first order condition (3.18a), and the fact that the term

$$R_f \, CARA(R_f \, W_o)$$

can be factored out of the expectation because it is not a function of the state z. Based upon a comparison between expression (3.20a)-(3.20b) and absolute risk aversion, we obtain the following characterization of the effect of initial wealth upon asset allocation:

(i) For all $z \in \mathbf{Z}$ so that $R(z) - R_f > 0$, it implies that $\{R_f [W_o - \alpha] + R(z)\alpha\} > \{R_f W_o\}$; if the coefficient of absolute risk aversion is strictly decreasing in wealth, or $CARA' < 0$, then

$$CARA(R_f \, [W_o - \alpha] + R(z)\, \alpha) < CARA(R_f \, W_o),$$

and it implies that expression (3.20b) = $0 >$ (3.20a), or (3.20a) < 0, and

$$\partial \alpha^* / \partial W_o > 0,$$

that is the allocation of the risky asset increases as initial wealth increases;

(ii) For all $z \in \mathbf{Z}$ so that $R(z) - R_f < 0$, it implies that $\{R_f [W_o - \alpha] + R(z)\alpha\} < \{R_f W_o\}$; if the coefficient of absolute risk aversion is strictly decreasing in wealth, or $CARA' < 0$, then

$$CARA(R_f \, [W_o - \alpha] + R(z)\, \alpha) > CARA(R_f \, W_o),$$

but in this case, $R(z) - R_f < 0$, implying that expression (3.20b) = $0 >$ (3.20a), or (3.20a)<0, or

$$\partial \alpha^* / \partial W_o > 0,$$

and again the allocation of the risky asset increases as initial wealth increases.

Ultimately, the key is the dependence of the coefficient of absolute risk aversion on wealth. It is easy to show, by the same reasoning, that if the coefficient of absolute risk aversion is strictly increasing in wealth, or $CARA' > 0$, then

$$\partial \alpha^* / \partial W_o < 0 \quad \textit{for all } z \in \mathbf{Z}.$$

Of course, if the coefficient of absolute risk aversion is independent of wealth, or $CARA' = 0$, then

$$\partial \alpha^* / \partial W_o = 0 \quad \textit{for all } z \in \mathbf{Z}.$$

Table 3.1 gives a summary of this characterization. In particular, the table shows that as initial wealth increases, the demand for the risky asset may increase, in which case the risky asset is a normal good; or decrease in which case the risky asset is an inferior good. The latter case is not very desirable from an economic perspective since it implies that, at higher (lower) levels of initial wealth, the demand for the risky asset decreases (increases). Common wisdom would require the contrary though. In the next section, we discuss this problem with a class of utility functions (quadratic utility) that yields, among other things, this undesirable property.

Table 3.1
Static Demand for Risky Asset and Dependence of Absolute Risk Aversion on Wealth

CARA'	*<0* *RISK AVERSION DECREASING IN WEALTH*	*>0* *RISK AVERSION INCREASING IN WEALTH*	*=0* *RISK AVERSION NOT A FUNCTION OF WEALTH*
$\partial\alpha^*/\partial W_o$	*>0* *Risky Asset is a Normal Good*	*<0* *Risky Asset is an Inferior Good*	*=0* *Independent of Wealth*

Finally, we recall expression (3.14b). Note that the condition on the third derivative, $u'''>0$ is necessary for $CARA'<0$, but not sufficient. A necessary and sufficient condition for $CARA'<0$ is that

$$CARA' = \{-u'''u' + [u'']^2\}/[u']^2 < 0 \Rightarrow u''' > (u'')^2/u' > 0$$

$$\Rightarrow \mathcal{P} = -u'''/u'' > -u''/u' = CARA > 0,$$

in words, the level of absolute prudence, $\mathcal{P}$, must be strictly greater than the level of absolute risk aversion, *CARA*, for the risky asset to be a normal good. Mathematically, it says that the degree of convexity of the marginal utility, u', must be strictly greater than the degree of concavity of the utility function u.

3.6 The Quadratic Utility Function

Consider a utility function of the following form

$$u(W(z)) = W(z)(1 + \varphi W(z)), \quad \varphi \in \mathbb{R}, \, z \in \mathbf{Z} \qquad (3.21)$$

where $W(z)$ denotes the value of state contingent final random wealth, and there is a finite state space, denoted $\mathbf{Z}$. The value of final random wealth is $W(z)$ when state $z \in \mathbf{Z}$ occurs. As before, the finite state space denotes the support of the simple probability distribution $p(z)>0$, with $\Sigma_{z \in Z}\, p(z)=1$. In this special case, the state independent VNM expected utility index is given by

$$E[u(W(z))] = \Sigma_{z \in Z}\, p(z)\, W(z)\, [1 + \varphi W(z)]$$

$$= E[W(z)] + \varphi E[W(z)^2] \qquad (3.22)$$

and recalling that $E[W(z)^2] = var(W(z)) + E[W(z)]^2$, we have that

$$E[u(W(z))] = E[W(z)] + \varphi \{ (E[W(z)])^2 + var(W(z)) \}. \quad (3.23)$$

Expression (3.23) shows that, for a quadratic utility index (3.21), the expected utility only depends upon the first two moments of the distribution of the random wealth, its mean and variance. This makes this class of utility functions appealing for mean-variance analysis, e.g. Chapter 2. The parameter φ denotes the weight of the second moment (risk) on total expected utility.

For this utility index, we compute the first and second differentials as

$$u'(W(z)) = 1 + 2\varphi W(z),$$

$$u''(W(z)) = 2\varphi, \text{ constant.}$$

First, for the utility index to be consistent with risk aversion, we need $u''(W(z))<0$ all $z \in \mathbf{Z}$ which implies the restriction

$$\varphi < 0.$$

This restriction implies that in expected utility (3.22), the individual attaches a negative weight to the second moment (risk) in total expected utility. Thus, the individual likes mean, but dislikes deviation from the mean (risk) as in the mean-variance framework, e.g. Chapter 2.

But, with $\varphi < 0$, non-satiation requires that $u'(W(z))>0$ all $z \in \mathbf{Z}$ which implies that

$$W(z) > -1/2\varphi, \quad \text{all } z \in \mathbf{Z}$$

i.e. we must truncate the probability distribution of the states of nature so that a realization

$$W(z) < -1/2\varphi, \quad all\ z \in \boldsymbol{Z}$$

occurs with probability zero (recall that $\varphi < 0$).

Secondly, the coefficient of absolute risk aversion for this utility index is, by (3.14a),

$$CARA = -2\varphi / [1 + 2\varphi W(z)]$$

so that absolute risk aversion is decreasing in the parameter $\varphi < 0$ given the level of wealth, i.e. the lower φ the more disutility the individual receives from the second moment (risk) and the higher the coefficient of absolute risk aversion. Also,

$$CARA' = 4\varphi^2 / [1 + 2\varphi W(z)]^2 > 0$$

which, using Table 3.1, indicates that $\partial \alpha^* / \partial W_o < 0$, i.e. the risky asset is an inferior good with quadratic utility index (3.21). From expression (3.14b), for the quadratic utility function absolute prudence is null, $\mathcal{P}=0$, and of course the condition that $\mathcal{P} > CARA$ cannot be satisfied.

A simple numerical example using the quadratic utility index is as follows. Suppose there are two possible states of nature, $\{z_1, z_2\}$ with well behaved probability function $p(z)$. Then, expected utility is, using (3.21),

$$E[u(W(z))] = p(z_1)\, W(z_1)\, [1 + \varphi W(z_1)] + p(z_2)\, W(z_2)\, [1 + \varphi W(z_2)].$$

Using the first order condition (3.18a), the solution for the demand for the risky asset α^* is determined by the condition

$$p(z_1)\, \{[1 + 2\varphi (R_f\, W_o + [R(z_1) - R_f]\alpha^*)]\, [R(z_1) - R_f]\} +$$

$$p(z_2)\, \{[1 + 2\varphi (R_f\, W_o + [R(z_2) - R_f]\alpha^*)]\, [R(z_2) - R_f]\} = 0.$$

Letting $p(z_1)=0.25$, $p(z_2)=0.75$, $R(z_1)=1.5$, $R(z_2)=1.05$, $R_f=1.1$, $\varphi=-0.35$, and $W_o=1$; we find that $\alpha^*=1/2$, so that one-half of the wealth is invested in the risky asset and the other half in the risk-free asset (recall that $W_o=1$). In this case, the absolute risk aversion evaluated at $W_o=1$ is $CARA(W_o)=2.3$. Increasing initial wealth to $W_o=1.25$ tilts the allocation to the risk-free asset, $\alpha^*=0.09$ and $W_o-\alpha^*=1.16$ since the risky asset is an inferior good. Notice also that decreasing the parameter of the disutility of risk φ to -0.40, $\varphi = -0.40$, and holding wealth constant, increases absolute risk aversion to $CARA(W(o))=4.0$, and the equilibrium allocation tilts to the risk-free asset so that $\alpha^*=0.22$, and the wealth invested in the risky asset decreases as expected.

Although a quadratic utility index gives an appropriate framework for mean-variance analysis, its drawbacks in terms of truncating the probability distribution of the states, and making the risky asset an inferior good, makes it less appealing for portfolio choice theory.

3.7 Diversification, Risk Aversion and Non-Systematic Risk

In Chapter 2, we discussed in some detail the possibility of full diversification of non-systematic or idiosyncratic risk. The material presented in this chapter provides a good framework to further discuss this issue.

First, consider the diversification problem from an expected utility point of view. Suppose an individual endowed with a well behaved utility function u, faces a fair gamble x with zero expected value $E[x]=0$ and $var(x)=\sigma_x^2$. The individual's willingness to take the fair gamble may be gauged by examining the following relationship

$$u(W_o - P_x(W_o)) \sim E[u(W_o + x)] \qquad (3.24)$$

where W_o is the individual's given wealth. It is simple to understand the term $P_x(W_o)$ in this case. The right-hand side of (3.24) is the expected utility obtained when the individual plays the fair gamble. The left-hand side is the non-stochastic utility of initial wealth minus a payment schedule, $P_x(W_o)$. Thus, given the indifference between the two sides, the payment schedule $P_x(W_o)$ is the individual's willingness to pay to avoid the risk of playing the gamble; in other words $P_x(W_o)$ is the risk premium, e.g. section 3.2.

We proceed by taking a Taylor's series approximation of both sides of the identity (3.24) evaluated at W_o. The approximation is taken up to the first order on the left-hand side because it is a non-stochastic stream, and to the second order on the right-hand side because of the stochastic component, x. The resulting expression becomes

$$u(W_o) - P_x(W_o)\, u'(W_o) \approx E[u(W_o) + x\, u'(W_o) + 1/2\, x^2 u''(W_o)] \qquad (3.25a)$$

$$u(W_o) - P_x(W_o)\, u'(W_o) \approx u(W_o) + 1/2\, \sigma_x^2 u''(W_o) \qquad (3.25b)$$

where (3.25b) is obtained by applying the expectation operator on the right-hand side of (3.25a) and noting that $E[x]=0$. We can use equation (3.14a), to express the willingness to pay as a function of absolute risk aversion, or

$$P_x(W_o) \approx 1/2\, \sigma_x^2\, CARA(W_o). \qquad (3.26)$$

Expression (3.26) implies that the variance of the gamble σ_x^2, and absolute risk aversion $CARA(W_o)$, are directly proportional to the risk premium. The higher (lower) the variance of the gamble σ_x^2, and the higher (lower) absolute risk aversion $CARA(W_o)$, the more (less) the individual is willing to pay to forgo the gamble, i.e. the higher (lower) the risk premium.

Now, we can reinterpret (3.24)-(3.26) in the following manner. Suppose the given initial wealth W_o is invested in a risky project x. Then, the risk premium of the project is given by (3.26). However, suppose alternatively that the individual is able to choose a portfolio of n statistically independent risky projects, with an allocation of equal shares of W_o to each project, i.e. W_o/n in each project. In this case, for statistically independent fair gambles, the risk premium is gauged by examining the analogous to (3.24), given by

$$u(W_o - P_{nx}(W_o)) \sim E[u(W_o + (x/n))]. \qquad (3.27)$$

Using the same methodology as in (3.25), the total risk premium of investing in the n projects is given by

$$n\, P_{nx}(W_o) \approx [1/2\, \sigma_x^2\, CARA(W_o)]\, /\, n \qquad (3.28)$$

which is exactly $1/n$ of the risk premium when only one risky project is engaged as in (3.26). Therefore, under uncorrelated risks and risk aversion we note the same principle seen in Chapter 2, that non-systematic or idiosyncratic risk can be fully diversified away.

The same point can be made using the state independent utility framework or state contingent claims. Suppose one asset pays off Y in the "good" state (g) and zero in the "bad" state (b), with probability function $p(g)=p(b)=1/2$. Investing all resources in this asset yields an expected payoff of $Y/2$, that is: $1/2\times zero + 1/2\times Y$; with variance $(Y/2)^2$,

i.e. $1/2\times (zero\text{-}Y/2)^2 + 1/2\times (Y\text{-}Y/2)^2$. Adding a statistically independent asset to this portfolio, with $1/2$ of the wealth invested in each asset, would imply state contingent claims:
(i) Y if both states are "good" with probability $1/4$;
(ii) $Y/2$ if one state is "good" and the other is "bad" with probability $1/4$;
(iii) $Y/2$ if one state is "bad" and the other is "good" with probability $1/4$; and
(iv) *Zero* if both states are "bad."

The expected return of this portfolio is again $Y/2$, but its variance is $(Y/2)^2/2$, i.e. $1/4\times (zero\text{-}Y/2)^2 + 1/2\times (Y/2\text{-}Y/2)^2 + 1/4\times (Y\text{-}Y/2)^2$. Thus, if n statistically independent assets are added to the portfolio, the variance of the portfolio decreases by $1/n$, exactly as in expression (3.28).

3.8 **Summary**

We have examined how economic theorists model utility derived from final random wealth. From the point of view of portfolio choice, for the mean-variance approach to be well defined, the expected utility of an individual must be expressed as a function of the mean and variance (standard deviation) of random wealth. The portfolio choice problem is one of maximizing expected utility subject to appropriate budget constraints, but the expected utility must ultimately be a function of the first two moments of the distribution of the final random wealth. The important advantage of a well defined utility framework is that risk aversion can be gauged by parameters associated with the individuals utility function.

There are three basic ways to reconcile the assumptions and axioms of expected utility theory of the VNM type with a representation in mean-variance space: (i) restrict the utility index to be quadratic as in section 3.6; (ii) restrict the decision-making to probability distribution of payoffs in the class of Normal distributions, characterized by the mean and variance only, with exponential utility index; (iii) restrict the decision-making to probability distribution of payoffs of the Lognormal distribution class with power utility index.

As we noted, case (i) has two disadvantages [distribution of outcomes is truncated to guarantee $u'>0$ and risky asset is an inferior good]. In the Appendix to this chapter, we examine case (ii) in some detail. Case (iii) is not pursued here (see Notes on the Literature), but the power utility function will be used extensively in the next chapters. In addition, the Appendix provides an analysis of an alternative to the Von-Neumann-Morgenstern expected utility framework.

Problems

1. Let $\boldsymbol{X}$ be a set of random "money prizes" along the real line, with $x \in \boldsymbol{X}$ denoting the "prize". Define a "lottery" (say a gamble or probability distribution) that gives prize x with probability one as δ_x (this is certainty, a "probability distribution" with one possible outcome).

i. Given the Von-Neumann-Morgenstern Expected Utility Representation, for the utility function $U{:}\boldsymbol{X} \rightarrow \boldsymbol{R}$, show that for all $x,y \in \boldsymbol{X}$ such that $x>y$, then $\delta_x > \delta_y$ if and only if the function U is strictly increasing. What does this result mean?

ii. For another lottery p, let $E[p]$ be its expected value. Show that

$$\delta_{E[p]} \geq p$$

if and only if U is concave. What does this result mean?

2. "The way absolute risk aversion covaries with random wealth matters for asset allocation." Comment.

3. Is the quadratic utility function appropriate for expected utility maximization in the mean-variance framework? Why or why not?

4. Explain what an arbitrage opportunity is in the context of this chapter.

5. Explain in your own words the concepts of (i) non-satiation; (ii) risk aversion.

6. Explain how you obtain the solution(s) (3.18) to problem (3.17).

8. Explain the Arrow-Pratt measures of absolute and relative risk aversion.

9. What is a state contingent claim?

10. Consider the portfolio choice problem: $Max\ E[\ u\ (R_f\ \{W(o) - \alpha\} + R(z)\ \alpha)\ ]$ by choice of α.

i. Explain this problem.
ii. Solve for the demand for the risky asset.
iii. Let utility for the final random wealth be quadratic: $u(x) = x(1+\varphi x)$. What is the effect of a change in initial wealth on the demand for the risky asset in this case? Explain.

Notes on the Literature

The material in sections 3.2-3.4 follows closely the presentation of Kreps (1990), Chapter 3. The foundations of choice under uncertainty are in Von Neumann and Morgenstern (1944). The book by Pratt et al (1995) presents a complete and brilliant general introduction to statistical decision theory. A popular writing about the 18th century historical aspects of choice under risk is found in Bernstein (1996), Chapter 6.

Several graduate and a few undergraduate textbooks in microeconomic analysis present useful discussions. For example, Henderson and Quandt (1971), Varian (1992) and more recently Kreps (1990) and Mas-Collel, Whinston and Green (1995) are good graduate sources; Pindyck and Rubinfeld (1995) is a good undergraduate source; and Dixit (1990) is an intermediate source. Gollier (2001) is an excellent book on the general subject of risk analysis, and Gollier et al (1997) is a good example of risk tolerance applied to investment. The concept of prudence and its relation to precautionary saving is due to Kimball (1990), and surveyed in Gollier (2001), Chapter 16. Bultel (1999) provides comparative statics results of changes in risk aversion in choice under uncertainty.

A general review of the literature on preferences and risk is available in Machina (1987); a more advanced but lively discussion is found in Kreps (1988); and Epstein (1992) provides a useful review of risk in decision-making analysis with particular attention to the issue of risk aversion versus intertemporal substitution mentioned in section 3.3 [Campbell and Viceira (2000) present several models of dynamic financial economics where the distinction is made explicitly; see Chapters 7 and 8 below].

A concise presentation of the axioms of expected utility theory is in Eichberger and Harper (1995), and Gollier (2001). Debreu (1959) is the classic reference on state contingent claims.

Measures of risk aversion are due to Arrow (1965) and Pratt (1964).

The subjective approach is presented in Savage (1954) and Kreps (1988, 1990).

A clear and precise treatment of risk aversion, the portfolio choice problem, and the quadratic utility case are given in Huang and Litzenberger (1988), Chapter 1, including the case of constant relative risk aversion utility.

References

Arrow, Kenneth J. (1965) *Aspects of the Theory of Risk Bearing*. Yrjo Jahnssonin Saatio, Helsinki.

Bernstein, Peter L. (1996) *Against the Gods: The Remarkable Story of Risk*. John Wiley&Sons, New York, NY.

Bultel, Dirk (1999) "Increases in Risk Aversion." *International Economic Review*, 40, 63-67.

Campbell, John Y. and Luis Viceira (2000) *Strategic Asset Allocation: Portfolio Choice for Long Term Investors*. Book manuscript, Harvard University, November. (Published by Oxford University Press, 2002).

Debreu, Gerard (1959) *Theory of Value*. Cowles Foundation, Yale University, New Haven, CT.

Dixit, Avinash K. (1990) *Optimization in Economics*, Second Edition. Oxford University Press, New York, NY.

Eichberger, Jurgen and Ian R. Harper (1995) *Financial Economics*.Oxford University Press, Oxford, UK.

Epstein, Larry (1992) "Behavior Under Risk: Recent Developments in Theory and Applications." In Laffont, Jean J., Ed., *Advances in Economic Theory: Sixth World Congress*, Vol. II, 1-63. Cambridge University Press, Cambridge, UK.

Gollier, Christian (2001) *The Economics of Risk and Time*. The MIT Press, Cambridge, MA.

Gollier, Christian, John Lindsey, and Richard Zeckhauser (1997) "Investment Flexibility and the Acceptance of Risk." *Journal of Economic Theory*, 76, 219-241.

Henderson, James M. and Richard E. Quandt (1971) *Microeconomic Theory*. McGraw Hill Book Co., New York, NY.

Huang, Chi Fu and Robert Litzenberger (1988) *Foundations of Financial Economics*. North Holland, Amsterdam.

Kimball, Miles (1990) "Precautionary Saving in the Small and in the Large." *Econometrica*, 58, 53-74.

Kreps, David M. (1990) *A Course in Microeconomic Theory*. Princeton University Press, Princeton, NJ.

Kreps, David M. (1988) *Notes on the Theory of Choice*. Westview Press, Boulder, CO.

Machina, Mark (1987) "Choice under Uncertainty: Problems Solved and Unsolved." *Journal of Economic Perspectives*, 1, 121-154.

Mas-Collel, Andreu, Michael Whinston, and Jerry Green (1995) *Microeconomic Theory.* Oxford University Press, Oxford, UK.

Pindyck, Robert S. and Daniel L. Rubinfeld (1995) *Microeconomics*. Prentice Hall, New York, NY.

Pratt, John W. (1964) "Risk Aversion in the Small and in the Large." *Econometrica*, 32, 122-136.

Pratt, John W., Howard Raiffa and Robert Schlaifer (1995) *Introduction to Statistical Decision Theory.* The MIT Press, Cambridge, MA.

Savage, Leonard J. (1954) *The Foundations of Statistics.* John Wiley&Sons, New York, NY.

Varian, Hal (1992) *Microeconomic Analysis.* Norton, New York, NY.

Von Neumann, John and Oskar Morgenstern (1944) *Theory of Games and Economic Behavior.* Princeton University Press, Princeton, NJ (updated editions are available).

Appendix to Chapter 3

A3.1 Introduction to Mean-Variance Analysis with Expected Utility based on Normal Distribution of Payoffs

Let us restrict the decision-making of an individual by imposing the class of Normal distributions for the final random payoffs with exponential utility index. In this special case, the payoffs are completely characterized by the first two moments of the distribution: the mean and variance.

Recall that in the normal distribution class, a continuous random variable x is said to follow a normal (stable) distribution with mean μ and finite variance σ^2, with notation $x \sim N(\mu, \sigma^2)$, if it has a probability density function given by the formula

$$f(x) = (2\pi \sigma^2)^{-1/2} exp(-(x - \mu)^2 / 2 \sigma^2) \qquad (A3.1)$$

where $\pi = 3.141592...$ is a physical constant and *exp* denotes the exponential operator, e.g. Chapter 1.

A3.2 The Payoffs

We denote the final random payoff or prize to some initial investment by W and let there be a continuum of states of nature, say $W \in \mathbf{R}^1$. Think of the random variable W as 'state-contingent' wealth. For example, wealth is going to fall in the interval $(W+\epsilon, W-\epsilon)$ for a small real number $\epsilon > 0$ with some positive probability. Also, let there be a well defined state independent utility function $u(W)$, strictly increasing and concave.

We assume that final random wealth is normally distributed, or $W \sim N(\mu_W, \sigma_W^2)$. Hence, the Von-Neumann-Morgenstern(VNM) expected utility is given by

$$E[u(W)] = \int_{-\infty}^{\infty} u(W) f(W) dW \qquad (A3.2)$$

where $f(W)$ is the probability density function of W, defined in (A3.1). This distribution allows expected utility to be unbounded below and above (unbound above is plausible, but below is not as plausible!). Substituting (A3.1), the expression for expected utility is

$$E[u(W)] = \int_{-\infty}^{\infty} u(W)\,(2\pi\sigma_W^2)^{-1/2} \exp(-(W-\mu_W)^2/2\sigma_W^2)\, dW. \qquad (A3.3)$$

A3.3 The Specific Functional Form for the Utility Function

Consider the exponential function

$$u(W) = -\exp(-\lambda W), \quad \lambda > 0 \qquad (A3.4)$$

depicted in Figure A3.1. The final random wealth W is allowed to be unboundedly positive and negative [it is fine that utility $u(W)$ is negative because only comparisons matter]. For this function, we have first, second and third derivatives

$$u'(W) = \lambda \exp(-\lambda W) > 0 \qquad (A3.5a)$$

$$u''(W) = -\lambda^2 \exp(-\lambda W) < 0 \qquad (A3.5b)$$

$$u'''(W) = \lambda^3 \exp(-\lambda W) > 0 \qquad (A3.5c)$$

i.e. strictly increasing, strictly concave and the marginal utility is strictly convex. It implies a constant coefficient of absolute risk aversion, denoted $CARA(W)$, given by the parameter $\lambda > 0$, or

$$CARA(W) = -u''(W)/u'(W) = \lambda \qquad (A3.6)$$

which is independent of wealth, or $CARA'(W)=0$. The absolute prudence for this function is

$$\mathcal{P}(W) = -u'''(W)/u''(W) = \lambda \qquad (A3.7)$$

as well, and $\mathcal{P} = CARA$ for this function. Table 3.1 shows that, in this case, we expect the dollar value of the risky asset in the portfolio to be independent of initial wealth, i.e. all change in initial wealth is absorbed by borrowing or lending in the risk-free asset.

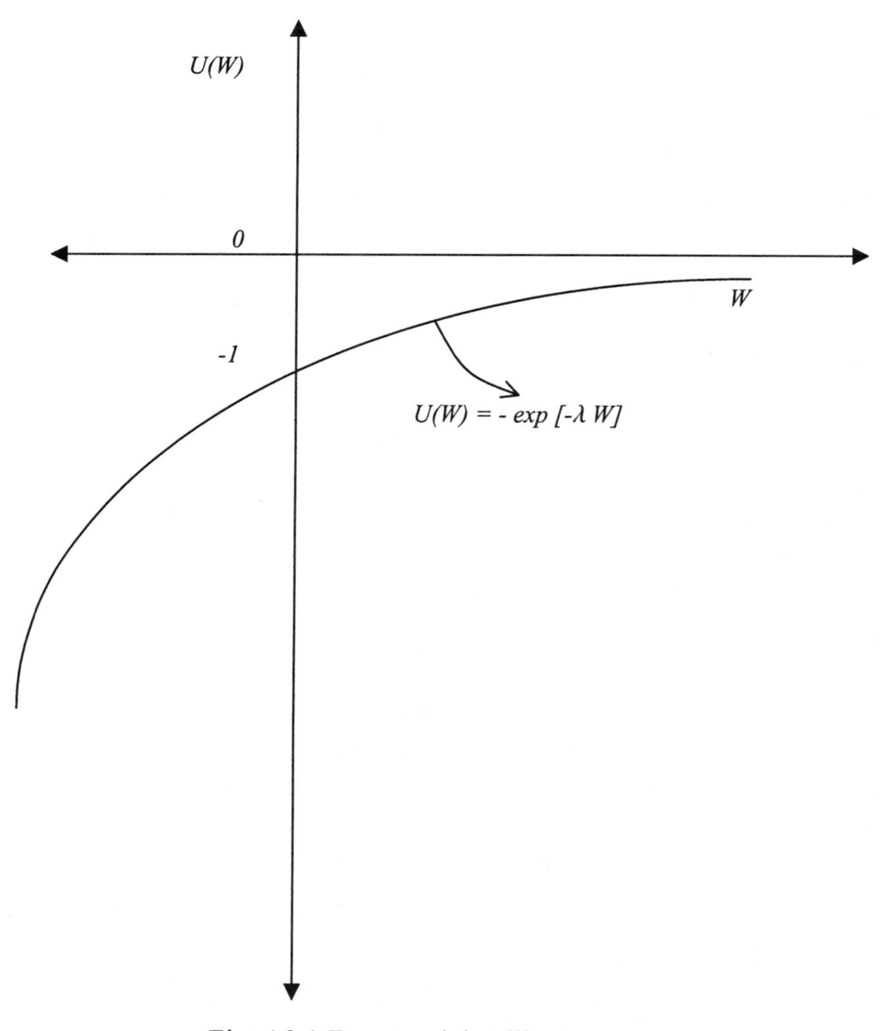

Fig. A3.1 Exponential Utility Function

We can compute the VNM expected utility for the exponential utility index described in (A3.4). Substituting (A3.4) into (A3.3), we obtain the expression

$$E[u(W)]=\int_{-\infty}^{\infty}-exp(-\lambda\ W)(2\pi\sigma_W^2)^{-1/2}exp(-(W-\mu_W)^2/2\sigma_W^2)\ dW. \quad (A3.8)$$

Then, collecting terms in the *exp* operator, we obtain

$$E[u(W)]=\int_{-\infty}^{\infty} - exp(-\{(\lambda\ W)+[(W-\ \mu_W)^2 / 2\ \sigma_W^2]\})\ (2\ \pi\ \sigma_W^2)^{-1/2} dW. \tag{A3.9}$$

Momentarily, concentrate on the term in keys inside the *exp* operator:

$$\{(\lambda\ W) + [(W-\ \mu_W)^2 / 2\ \sigma_W^2]\};$$

expand this term to obtain

$$\{[(\lambda\ W\, 2\ \sigma_W^2) + W^2 - 2\, W\ \mu_W + \mu_W^2] / 2\ \sigma_W^2\};$$

now add and subtract the quantity $(2\ \mu_W\ \lambda\ \sigma_W^2\ - \lambda^2\ \sigma_W^4)$ from the numerator to obtain

$$\{([W - (\mu_W -\ \lambda\ \sigma_W^2)]^2 / 2\ \sigma_W^2) + \lambda\, [\mu_W - (1/2)\, \lambda\, \sigma_W^2]\}$$

and substitute the last expression back into (A3.9) to obtain

$$E[u(W)] = \int_{-\infty}^{\infty} - exp(-\{\, ([W - (\mu_W -\ \lambda\ \sigma_W^2)]^2 / 2\ \sigma_W^2\,) + \lambda\, [\mu_W - (1/2)\, \lambda\, \sigma_W^2]\,\}\,)\ (2\ \pi\ \sigma_W^2)^{-1/2}\ dW. \tag{A3.10}$$

The reason for these steps is that term $-exp(-\lambda\ \{\mu_W - (1/2)\ \lambda\ \sigma_W^2\})$ is not a function of W, thus can be pulled out of the integral, yielding

$$E[u(W)] = -exp(-\lambda\, [\mu_W - (1/2)\ \ \lambda\ \ \sigma_W^2])\ \times$$

$$\int_{-\infty}^{\infty} exp(-\{\, ([W - (\mu_W -\ \lambda\ \sigma_W^2)]^2 / 2\sigma_W^2\,)\,\})\ (2\pi\sigma_W^2)^{-1/2} dW, \tag{A3.11}$$

and we can consider the integral in this last expression as

$$\int_{-\infty}^{\infty}\ exp(-\{\, ([W - (\mu_W -\ \lambda\ \sigma_W^2)]^2 / 2\ \sigma_W^2\,)\,\}\,)\ (2\ \pi\ \sigma_W^2)^{-1/2}\ dW.$$

This is simply the total area under a normal distribution with mean $(\mu_W - \lambda\ \ \sigma_W^2)$ and variance σ_W^2, thus (since area under the curve denotes the probability) this integral adds up to one, or

$$\int_{-\infty}^{\infty} exp(-\{\, ([W - (\mu_W -\ \lambda\ \sigma_W^2)]^2 / 2\ \sigma_W^2\,)\,\}\,)\ (2\ \pi\ \sigma_W^2)^{-1/2}\ dW = 1.$$

Hence, the VNM expected utility for the exponential utility function is given by

$$E[u(W)] =\ - exp\ (-\lambda\, [\mu_W - (1/2)\ \ \lambda\ \ \sigma_W^2]\,) \tag{A3.12}$$

appropriately only a function of μ_W and σ_W^2, given the parameter for absolute risk aversion, $\lambda > 0$. In addition, expected utility varies with the mean and standard deviation according to:

$$\partial\ E[u(W)]/\partial\ \mu_W = \lambda \exp(-\lambda [\mu_W - (1/2)\ \lambda\ \sigma_W^2]) > 0 \quad (A3.13)$$

$$\partial\ E[u(W)]/\partial\ \sigma_W = -\lambda^2 \exp(-\lambda [\mu_W - (1/2)\ \lambda\ \sigma_W^2]) < 0 \quad (A3.14)$$

or expected utility is strictly increasing in μ_W ("more mean is preferred to less") and strictly decreasing in σ_W ("risk aversion as disutility from σ_W").

The expected utility (A3.12) gives an indifference set in $\{\mu_W, \sigma_W\}$ space as:

$$dE[u(W)] = 0 = \lambda \exp[-\lambda [\mu_W - (1/2)\ \lambda\ \sigma_W^2]]\ d\mu_W -$$

$$\lambda^2 \exp[-\lambda [\mu_W - (1/2)\ \lambda\ \sigma_W^2]]\ d\sigma_W \quad (A3.15)$$

with

$$d\mu_W/d\sigma_W|_{dE[u(W)]=0} = \lambda\ \sigma_W > 0 \quad (A3.16)$$

$$d^2\mu_W/d\sigma_W^2|_{dE[u(W)]=0} = \lambda > 0. \quad (A3.17)$$

Figure A3.2 illustrates the indifference map. The indifference curves are strictly increasing and convex. Curves to the northwest indicate higher levels of expected utility. The marginal rate of substitution between mean and standard deviation (the slope) is increasing in both λ and σ_W and equals zero for either $\lambda=0$ or $\sigma_W=0$. In the special case of $\lambda=0$, the individual is risk neutral [see equation (A3.7)].

To sum: in the case of a continuum of random payoffs which are normally distributed, or when $W \sim N(\mu_W, \sigma_W^2)$, the constant absolute risk aversion state independent utility index $u(W) = -exp(-\lambda W)$ yields VNM expected utility which depends only on the two moments of the distribution of payoffs; it is increasing in the mean and decreasing in the standard deviation, hence suitable for the mean-variance approach.

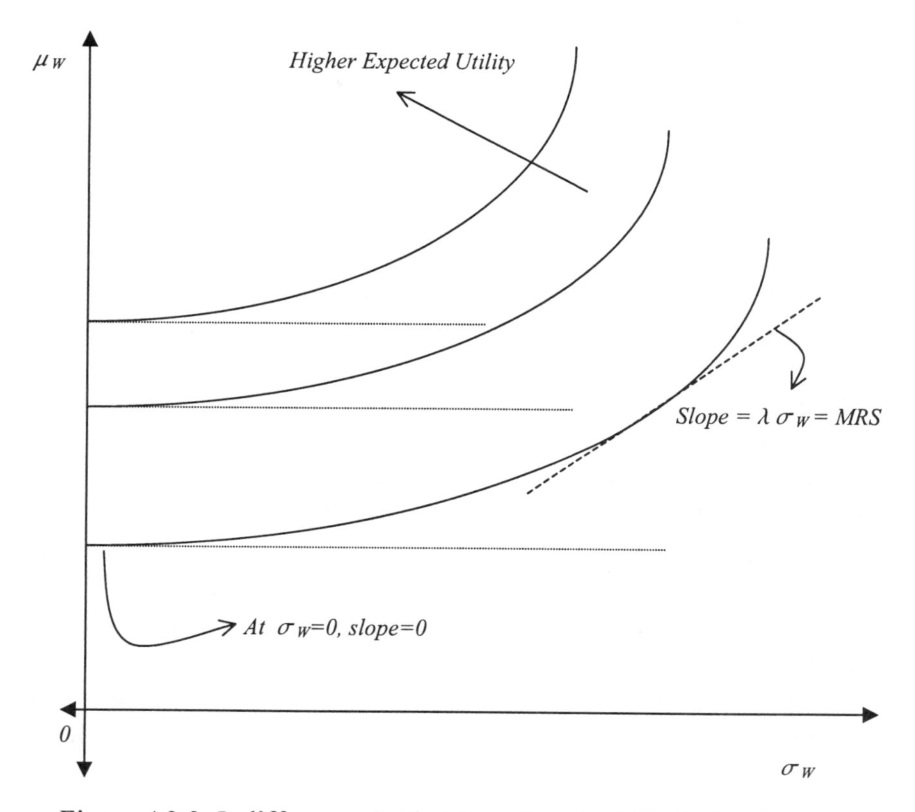

Figure A3.2: Indifference Set in Mean-Standard Deviation Space

A3.4 ***The Individual Budget Constraint***

What are the market opportunities available for this risk averse individual? The answer can be found by noting that by the Two-Fund theorem, e.g. Chapter 2, one can choose portfolios along the capital market line. At the initial instant, the individual has a deterministic given amount of wealth $W_0>0$ to allocate between a portfolio of risk-free assets and a portfolio of risky assets available in the market.

Let the risk-free rate of return be deterministic (zero variance) and denoted by $r_f>0$. Choosing the risk-free portfolio gives a sure total (gross) return of $(1+r_f)W_0$.

Let the random (risky) rate of return on the market portfolio be normally distributed, denoted by r, with first and second moments given by

$$E[r] = r_M, \quad Var(r) = \sigma_M^2 \tag{A3.18}$$

or $r \sim N(r_M, \sigma_M^2)$. These moments are taken as given by the individual. This is the usual atomistic assumption of perfect competition: the individual is always small relative to the market, thus cannot influence market returns. Choosing the risky portfolio gives a mean total return of $(1+r_M)W_O$ with variance $\sigma_M^2 W_O^2$.

Denote by $\alpha \in (0,1)$, the share of initial wealth invested in the risk-free portfolio, in turn $(1-\alpha) \in (0,1)$ is the share invested in the risky portfolio, and no short selling is allowed. You should note in this case that, unlike the problem of sections 3.5-3.6, we are defining α as the share of initial wealth invested in the risk-free asset. Hence, given the utility index (A3.5)-(A3.6a,b)-(A3.7a,b), a change in initial wealth does not change the dollar value of the risky asset in the portfolio, but it does change the share of the total initial wealth invested in the risky asset.

By the Two-Fund theorem, the convex combinations characterize the capital market line. Along this line the final random payoffs or wealth of the individual is given by

$$W = \alpha\,[(1+r_f)W_O] + (1-\alpha)[(1+r)W_O] \tag{A3.19}$$

a simple weighted average of the risk-free and risky returns.

This random payoff has mean

$$E[W] = \mu_W = \alpha\,[(1+r_f)W_O] + (1-\alpha)[(1+r_M)W_O]. \tag{A3.20}$$

and variance and standard deviation

$$E[\{W - E(W)\}^2] = \sigma_W^2 = (1-\alpha)^2\, \sigma_M^2\, W_O^2 \tag{A3.21}$$

$$\sigma_W = (1-\alpha)\, \sigma_M\, W_O \tag{A3.22}$$

respectively. It is important to understand that by varying α, one can obtain combinations of $\{\mu_W, \sigma_W\}$ that represent portfolios between risk-free and risky assets (or portfolios) such that all three variables α, μ_W, and σ_W are simultaneously varying. Hence, the capital market line in the mean-standard deviation space, $\{\mu_W, \sigma_W\}$, is obtained by eliminating α

between (A3.20) and (A3.22). It will be useful to solve first expression (A3.20) for α as a function of μ_W, obtaining

$$\alpha = [(1+r_M)/(r_M - r_f)] - [\mu_W \ / \ (r_M - r_f)W_O], \qquad (A3.23)$$

and then solve expression (A3.22) for α as a function of σ_W, obtaining

$$\alpha = 1 - (\sigma_W \ / \ \sigma_M W_O). \qquad (A3.24)$$

The last expression (A3.24) implies a restriction on the standard deviation, σ_W of the form

$$\sigma_W \in \ (0, \ \sigma_M W_O),$$

and (A3.23) implies a restriction on μ_W of the form

$$(1+r_f) \ W_O \ \le \ \mu_W \ \le (1+r_M) \ W_O$$

both assumed to be satisfied to guarantee that $\alpha \in (0,1)$. Of course you could substitute (A3.24) into (A3.20) directly, but we'll consider expressions (A3.23)-(A3.24) to illustrate some comparative statics issues below. Equating expressions (A3.23)-(A3.24), one obtains the capital market line as a linear relationship in $\{\mu_W, \sigma_W\}$ given by

$$\mu_W = [(1+r_f)W_O] + [(r_M - r_f) / \ \sigma_M] \ \sigma_W. \qquad (A3.25)$$

As long as the risky asset is a favorable bet, or $r_M - r_f > 0$ (consistent with risk aversion, or the price of risk is strictly positive), the slope of the capital market line is positive. Again, each point along this line represents a combination of $\{\mu_W, \sigma_W\}$ with an associated share of risk-free portfolio α. Figure A3.3 illustrates the capital market line. If $\alpha=1$, the portfolio consists of risk-free assets only with sure mean return

$$\mu_W = (1+r_f)W_O$$

and zero standard deviation(variance), denoted point F in Figure A3.3. As α decreases monotonically towards zero, μ_W and σ_W increase monotonically towards $(1+r_M) \ W_O$ and $\sigma_M \ W_O$ respectively. The alternative portfolios along the line include a mix of risk-free and the risky (market) portfolio where the individual trades some additional risk for the higher expected return at the constant positive price of risk, that is where $(r_M - r_f) / \sigma_M > 0$. At $\alpha=0$, the portfolio consists of risky assets only with mean return $\mu_W = (1+r_M)W_O$ and standard deviation $\sigma_W = \sigma_M \ W_O$, denoted point M in Figure A3.3.

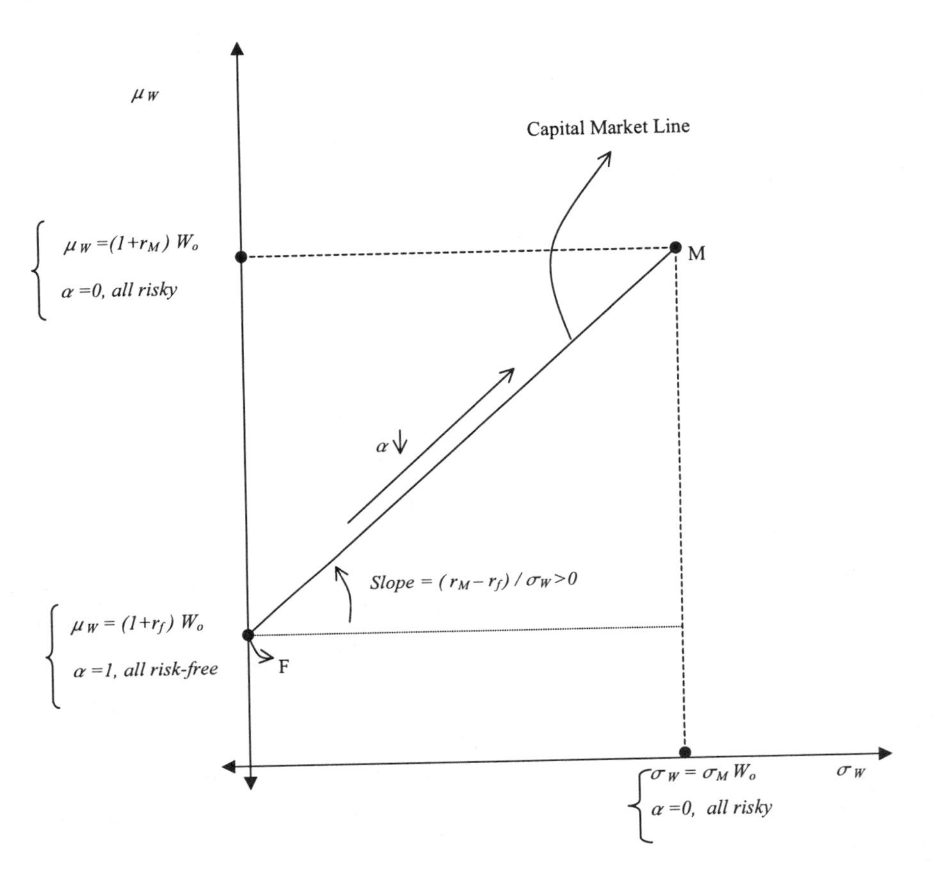

Figure A3.3: The Capital Market Line

A3.5 ***The Equilibrium Allocation***

An individual seeks to maximize expected utility (A3.12) subject to the available market opportunities given by (A3.25) by choice of $\{\mu_W, \sigma_W\}$ and thus implicitly α. We write down this problem as

$$\underset{\{\mu_W, \sigma_W\}}{Max} \quad E[u(W)] = -\exp(-\lambda\{\mu_W - (1/2)\ \lambda\ \sigma_W^{\ 2}\}) \qquad (A3.26)$$

$$subject\ to \qquad \mu_W = [(1+r_f)W_o] + [(r_M - r_f)/\sigma_M]\ \sigma_W$$

given $\{r_M, \sigma_M, r_f, W_O, \lambda\}$. Once a solution for this problem is found, it may be suitably substituted into (A3.23) or (A3.24) to yield a solution for the portfolio share α.

First, we are going to solve this problem by the Lagrangean method, and a direct solution is presented later. Denote the VNM expected utility by

$$V(\mu_W, \sigma_W^2) = E[u(W)] = -\exp(-\lambda\{\mu_W - (1/2)\ \lambda\ \sigma_W^2\})$$

with respective first and second partial derivatives and cross-partials given by

$$V_1 = \exp(-\lambda\{\mu_W - (1/2)\ \lambda\ \sigma_W^2\})\,\lambda > 0$$

$$V_2 = -\exp(-\ \lambda\{\mu_W - (1/2)\ \lambda\ \sigma_W^2\})\,\lambda^2\ \sigma_W < 0$$

$$V_{11} = \exp(-\lambda\{\mu_W - (1/2)\ \lambda\ \sigma_W^2\})\,(-\lambda^2) < 0$$

$$V_{22} = -\exp(-\lambda\{\mu_W - (1/2)\ \lambda\ \sigma_W^2\})\,\lambda^2(\lambda^2\ \sigma_W^2 + 1) < 0$$

$$V_{12} = V_{21} = \exp(-\lambda\{\mu_W - (1/2)\ \lambda\ \sigma_W^2\})\,(-\lambda^3\sigma_W^2) > 0$$

where mean and variance are Edgeworth complements in utility, $V_{12}=V_{21}>0$. Under this representation for expected utility, we write down the Lagrangean function as

$$\mathcal{L} = V(\mu_W, \sigma_W^2) + \gamma[\{[(1+r_f)W_O] + [(r_M - r_f)/\ \sigma_M]\,\sigma_W\} - \mu_W] \quad (A3.27)$$

where $\gamma \geq 0$ is a non-negative Lagrange multiplier associated with the capital market line constraint. The first order necessary conditions for an interior equilibrium are given by

$$\partial\mathcal{L}/\partial\mu_W = V_1 - \gamma = 0 \quad (A3.28)$$

$$\partial\mathcal{L}/\partial\sigma_W = V_2 + \gamma[(r_M - r_f)/\sigma_M] = 0 \quad (A3.29)$$

$$\partial\mathcal{L}/\partial\gamma = \{[(1+r_f)W_O] + [(r_M - r_f)/\sigma_M]\,\sigma_W\} - \mu_W = 0. \quad (A3.30)$$

These give solutions for $\{\mu_W^*, \sigma_W^*, \gamma^*\}$ and implicitly α^*, given the parameters $\{r_M, \sigma_M, r_f, W_O, \lambda\}$. The sufficient second order condition for a constrained maximum requires that the determinant of the bordered Hessian of order two be strictly positive ($|\boldsymbol{H}_2|>0$). This condition is satisfied given that the constraint is linear and the objective function gives indifference curves, which are strictly increasing and convex.

Figure A3.4 illustrates the equilibrium portfolio choice yielding asset allocation at point E. The equilibrium illustrates a Mutual Fund theorem: if $V(\mu_W, \sigma_W^2)$ is the expected utility index, all individuals hold the <u>same</u> ratio of risky to risk-free assets; see Notes on the Literature.

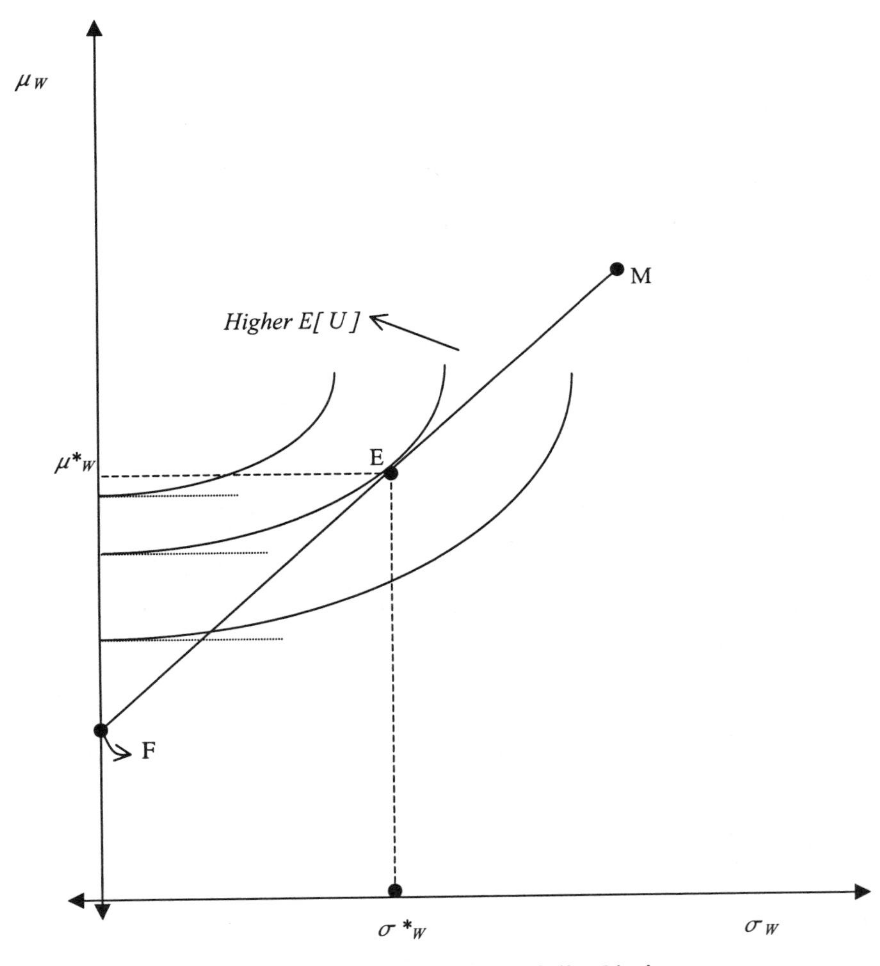

Fig. A3.4 Equilibrium Portfolio Choice

A3.6 **Comparative Statics**

(i) An increase in the expected market rate of return: $\partial\ r_M>0$

The differential of the system (A3.28)-(A3.30) with respect to changes in the expected market rate of return is given by

$$\begin{vmatrix} V_{11} & V_{12} & -1 \\ V_{21} & V_{22} & [(r_M-r_f)/\sigma_M] \\ -1 & [(r_M-r_f)/\sigma_M] & 0 \end{vmatrix} \begin{vmatrix} \partial\mu_W^*/\partial\ r_M \\ \partial\sigma_W^*/\partial\ r_M \\ \partial\gamma^*/\partial\ r_M \end{vmatrix} = \begin{vmatrix} 0 \\ -\gamma/\sigma_M \\ -\sigma_W/\sigma_M \end{vmatrix} \quad (A3.31)$$

The determinant of this system is positive by the second order condition and the comparative statics for $\partial\mu_W^*/\partial r_M$ and $\partial\sigma_W^*/\partial r_M$ are given by

$$\partial\mu_W^*/\partial r_M = \{[(r_M-r_f)/\sigma^2_M][\gamma - V_{12}\sigma^2_W] - [V_{22}\sigma_W/\sigma^2_M]\} / |\mathbf{H}_2| \quad (A3.32)$$

$$\partial\sigma_W^*/\partial r_M = \{[(r_M-r_f)/\sigma_M](V_{11}\sigma_W)+(V_{12}\sigma_W/\sigma_M)+(\gamma/\sigma_M)\}/|\mathbf{H}_2|. \quad (A3.33)$$

where both are ambiguous in sign. Why is this the case? The answer is substitution and wealth effects going in opposite directions.

However, it is plausible to think in the following way. For small values of $(r_M - r_f)$ (the slope of the capital market line is small and flat), an increase in the expected market rate of return, $\partial r_M >0$, induces the individual to take more risk at a higher expected return, $\partial\mu_W^*/\partial r_M>0$, and $\partial\sigma_W^*/\partial r_M>0$, implying that $\partial\alpha^*/\partial r_M<0$ from (A3.23) or (A3.24) above. In this case, the higher the mean return, the lower the share of the risk-free asset in the portfolio. Also note from (A3.32)-(A3.33), that small values of $(r_M - r_f)$ imply that the terms multiplied by $(r_M - r_f)$ become small indicating the likelihood of positive signs for both of the partials, $\partial\sigma_W^*/\partial r_M >0$ and $\partial\mu_W^*/\partial r_M >0$. This also implies that $\partial\alpha^*/\partial r_M<0$ from expressions (A3.23) or (A3.24).

The intuition is that from a relatively flat capital market line with a small price of risk (close to the zero standard deviation indeed), an increase in expected market rate of return induces the individual to take more risk, i.e. the substitution effect outweighs the income or wealth effect, and the share of the safe asset in the portfolio falls. This situation is depicted in Figure A3.5. From the initial equilibrium at point E, to point A describes the pure substitution effect (a pure substitution effect always trades more variance for more mean since the marginal rate of substitution is positive). From points A to E', denotes the pure income or wealth effect which is offsetting. In this case, the final equilibrium is at point E' where the substitution effect outweighs the income effect.

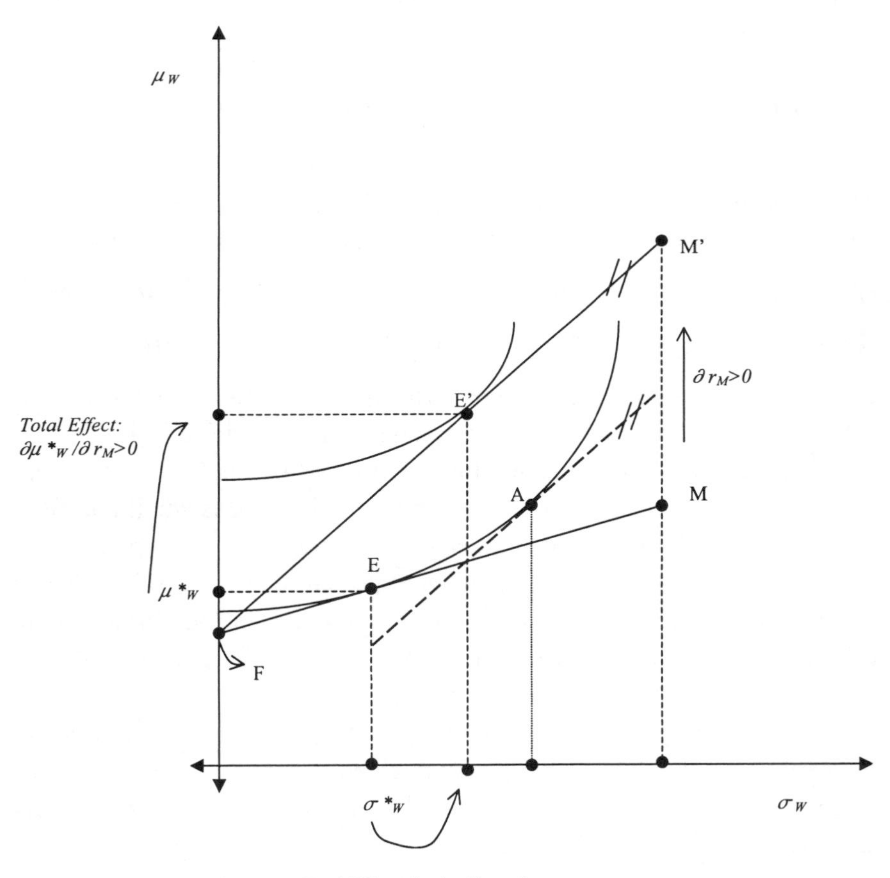

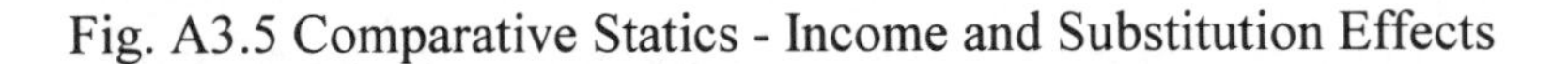

Fig. A3.5 Comparative Statics - Income and Substitution Effects

However, for higher levels of $(r_M - r_f)$ this need not be the case. From a steep or more vertical capital market line far from zero standard deviation where the price of risk is already large, a further increase in mean return can plausibly induce the individual to take less risk, i.e. the wealth effect outweighs the substitution effect, and the individual moves away from the risky portfolio into the risk-free asset.

(ii) An increase in initial wealth: $\partial W_o > 0$

The differential of the system (A3.28)-(A3.30), with respect to changes in initial wealth is given by

$$\begin{vmatrix} V_{11} & V_{12} & -1 \\ V_{21} & V_{22} & [(r_M - r_f)/\sigma_M] \\ -1 & [(r_M - r_f)/\sigma_M] & 0 \end{vmatrix} \begin{vmatrix} \partial\mu_W^*/\partial W_o \\ \partial\sigma_W^*/\partial W_o \\ \partial\gamma^*/\partial W_o \end{vmatrix} = \begin{vmatrix} 0 \\ 0 \\ -(1+r_f) \end{vmatrix} \qquad (A3.34)$$

The comparative statics for $\partial\mu_W^*/\partial W_o$ and $\partial\sigma_W^*/\partial W_o$ are given by

$$\partial\mu_W^*/\partial W_o = -(1+r_f)\{V_{12}[(r_M - r_f)/\sigma^2_M] + V_{22}\} / |\mathbf{H}_2| \qquad (A3.35)$$

$$\partial\sigma_W^*/\partial W_o = \{[(1+r_f)V_{12}] + [(r_M - r_f)/\sigma_M]V_{11}\} / |\mathbf{H}_2| \qquad (A3.36)$$

which are both ambiguous in sign again. Other comparative statics results can be computed in the same fashion.

A3.7 **Closed Form Solutions**

Problem (A3.26) does have a closed form solution. One way to obtain the closed form solution is to substitute the constraint directly into the objective function, and solve the unconstrained problem:

$$\underset{\{\sigma_W\}}{Max} \quad E[u(W)] = -\exp(-\lambda[(1+r_f)W_o + [(r_M - r_f)/\sigma_M] - (1/2)\ \lambda\ \sigma_W^2]). \qquad (A3.37)$$

The first order necessary condition for an interior equilibrium is given by

$$\partial E[u(W)]/\partial\sigma_W = -\exp(.)\{-\ \lambda[(r_M - r_f)/\sigma_M] + \lambda^2\ \sigma_W]\} = 0, \qquad (A3.38)$$

with the second order sufficient condition for a maximum satisfied by

$$\partial^2 E[u(W)]/\partial \sigma_W^2 =$$

$$-exp(.) \{(-\lambda [(r_M - r_f)/\sigma_M] + \lambda^2 \sigma_W]) ^2 + \lambda^2\} < 0. \qquad (A3.39)$$

The optimal choice of σ_W, denoted σ_W*, is the closed form solution from (A3.38) given by

$$\sigma_W* = [(r_M - r_f) / \lambda \sigma_M] > 0. \qquad (A3.40)$$

This is just the standard deviation that equates the marginal rate of substitution to the slope of the capital market line. But note, the formula for the optimal amount of risk in (A3.40) is not an explicit function of wealth. But, this is only one side of the story, recall (A3.36). Along the capital market line, σ_W is a function of wealth through the effect of wealth on α and the total effect can be ambiguous as in (A3.35)-(A3.36) above. This can also be seen by inspecting (A3.23)-(A3.24): for a given σ_W, (A3.24) indicates that an increase in wealth increases α which is consistent with (A3.23) for a given μ_W, however, as α changes μ_W and σ_W also change, being all three simultaneously determined.

The closed form optimal expected return on the portfolio is found by substituting σ_W* into (A3.25) obtaining the optimal choice of μ_W denoted μ_W* given by

$$\mu_W* = (1+r_f)W_o + [(r_M - r_f)^2 / \lambda \sigma_M^2]. \qquad (A3.41)$$

These solutions are the same as the ones obtained in the system (A3.28)-(A3.30) above, say substituting (A3.28) into (A3.29) and solving the resulting equations for $\{\mu_W*, \sigma_W*\}$ gives exactly the same solutions. (You could also solve directly by equating the marginal rate of substitution to the slope of the capital market line, right!).

Thus, the closed form optimal choice for the share of the risk-free asset in the portfolio, denoted $\alpha*$, can be easily obtained by substituting (A3.41) or (A3.40) into (A3.23) or (A3.24) respectively yielding

$$\alpha* = 1 - [(r_M - r_f) / \lambda \sigma_M^2 W_o] \qquad (A3.42)$$

which satisfies $\alpha \in (0,1)$ given $\sigma_W \in (0, \sigma_M W_o)$. In turn, the share of the risky portfolio is given by

$$1 - \alpha^* = [(r_M - r_f) / \lambda \sigma_M^2 W_o]. \qquad (A3.43)$$

The closed form solutions conceal some important wealth and substitution effects though, the reason being why we solved the general problem above. The reader should be aware that the Envelope theorem is an approximation, although it works well in most cases! Finally, note that the dollar value of the initial wealth invested in the risky asset, or

$$(1 - \alpha^*) W_O = [(r_M - r_f) / \lambda \ \sigma_M^2]$$

is indeed independent of the initial level of wealth, since the exponential utility function implies $CARA'(W)=0$. Similarly, the dollar value of the initial wealth invested in the risk-free asset is

$$\alpha^* W_O = W_O - [(r_M - r_f) / \lambda \ \sigma_M^2],$$

thus directly related to initial wealth, i.e. any change in initial wealth is absorbed by borrowing and lending in the risk-free asset.

A3.8 ***Summary I***

The mean-variance problem of portfolio choice between a risk-free portfolio and a risky market portfolio can be embedded in an equilibrium VNM expected utility maximization framework when the payoffs are normally distributed. For the exponential utility index with constant absolute risk aversion, the VNM expected utility is a function of mean and variance only. Comparative statics in given parameters may lead to offsetting income and substitution effects.

The advantages of the assumption of normally distributed payoffs are two fold: (i) Normal random variables benefit from the Central Limit Theorem; (ii) A linear combination of normal random variables is normally distributed as well and this is attractive from the point of view of portfolio (various assets) choice.

However, one problem is that, somewhat unrealistically, wealth is assumed to be unbounded above and below.

A3.9 ***Introduction to Non-Additive Probabilities***

In the remainder of this appendix, we examine cases where utility deviates from the traditional Von-Neumann-Morgenstern expected utility framework studied in Chapter 3. We focus on a discussion of non-additive probabilities and its application to behavior towards risk and expected utility. Then, we present economic examples drawn from risky investment and economic growth. In addition, we consider uncertainty in Knight's sense.

Knightian uncertainty refers to the market participant's absence of knowledge of the probability distribution of the random variables of interest. Under non-additive probabilities, we can relate Knightian uncertainty to a subjective concept of uncertainty aversion. In the presence of Knightian uncertainty, market participants may apply a Bayesian approach attaching a prior probability distribution to the unknown distribution. We show that under uncertainty aversion the economic costs of Knightian uncertainty are of the first order, but under risk aversion the economic costs are of the second order.

We start with an introduction to the concept of non-additive probabilities and the related problem of uncertainty aversion. This framework relates to a subjective assessment of probability as opposed to the objective framework of the traditional VNM type. Let $\boldsymbol{Z}$ be a set of states of nature, and let $\boldsymbol{X}$ be the corresponding set of events. A function $P\colon \boldsymbol{X} \rightarrow [0,1]$ is called a <u>non-additive probability</u>, if it satisfies the following:

(i) The probability of the empty set is zero and the probability of the state space is one,

$$P(\varnothing) = 0, \qquad P(\boldsymbol{Z}) = 1;$$

and (ii) Monotonicity,

$$B \subset A \Rightarrow P(A) \geq P(B) \qquad all\ \{A,B\} \in \boldsymbol{X}.$$

Then, non-additive probability reveals uncertainty if for all events $\{A,B\} \in \boldsymbol{X}$,

$$P(A \cup B) \geq P(A) + P(B) - P(A \cap B)$$

with equality for the additive case. Non-additive probability implies that the probability of the union can be greater that the probability of each event separately minus the intersection (see e.g. section 1.3), providing some additional subjective probability content not captured by the additive case. We can parameterize this uncertainty content by a well defined function c as

$$c(P,A) = 1 - P(A) - P(A^c), \qquad (A3.44a)$$

where A^c is the set of elements not in A. The function c denotes a measure of uncertainty aversion of a probability distribution P at event $A \in \boldsymbol{X}$, i.e. it is the additional subjective probability content not captured by the additive case.

The monotonicity property and the function c show that it is possible for $P(A)+P(A^c)<1$. This specific difference is a measure of uncertainty that the agent attaches to a possible event $A \in \boldsymbol{X}$. The pure additive case gives

$$P(A) + P(A^c) = 1 \quad and \quad c(P,A) \to 0,$$

thus, $c(P,A)$ measures the deviation from the traditional additive case.

Intuitively, non-additive probability allows the term $c(P,A)$ to measure a subjective faith in the measures of likelihood of $P(A)$ and $P(A^c)$. For example, we can think of some subjective assessment in a contractual relationship where the $c(P,A)$ function captures the extent of lack of trust of one of the parties.

We can calculate expected utility for a non-additive probability function as well. Let u be a real valued function $u\colon \boldsymbol{X} \to \boldsymbol{R}$, and let P denote the non-additive probability function discussed above. Then, expected utility $E[u]$ is defined as

$$E[u] \equiv \int_{-\infty}^{0} [P\,(u \geq x) - 1]\, dx + \int_{0}^{\infty} P(u \geq x)\, dx, \qquad (A3.44b)$$

where the term $\int_0^\infty P(u \geq x)dx$ denotes the expectation of the non-negative domain of utility, and $\int_\infty^0 [P\ (u \geq x) - 1]dx$ denotes the expectation over the negative domain. In the additive case, we have that $E[u] = \int_{-\infty}^{\infty} P(u \geq x)\ dx$. You should not be concerned that non-additive probabilities do not add up to one because they are not objective probabilities, but probability assessments based upon subjectivity.

Let's consider a simple example for an individual interested in a risky investment. Suppose an entrepreneur or investor is risk neutral. Then, the reward function is linear, or $u = R$, where $R \geq 0$ stands for the total reward or income on investment. Non-additive (subjective) probabilities are then as follows. Suppose there are two possible bounds for the set of events: $\{L, H\}$ and associated incomes or rewards $\{R_L, R_H\}$, for $R_L < R_H$. An event R falls in the interval $R_L<R<R_H$ symmetrically, with probability $p>0$. Hence, the probability function for non-negative possible outcome is given by

$$P(R \geq R_o) = \begin{cases} 1 \text{ for } R_L > R_o; \\ p \text{ for } R_H > R_o > R_L; \\ 0 \text{ for } R_o > R_H. \end{cases}$$

The non-additive probability function $P(R \geq R_o)$ simply states that the probability of a reward being greater than a given R_o is one, when R_o is strictly grater that the lower bound $R_L > R_o$; it is equal to $p>0$ when a given R_o is between the lower and higher bounds (symmetric); and zero otherwise. Then, from the formula for expected utility in (A3.44b), using the non-negative domain of R, we can compute $E[R]$ as

$$E[R] = \int_0^\infty P(R \geq R_o)\, dR = \int_0^{R_L} 1\, dR + \int_{RL}^{RH} P(R \geq R_o)\, dR + \int_{RH}^\infty 0\, dR$$
$$= R\,\big|_o^{R_L} + p\,R\,\big|_{RL}^{RH}$$
$$= R_L + p\,(R_H - R_L)$$
$$= R_L\,(1-p) + p\,R_H.$$

Notice that, for this probability function, the probability of a reward between R_L and R_H is symmetrically equal to p, so that the measure of uncertainty aversion in this case is

$$c(P,A) = 1 - P(A) - P(A^c) = 1 - 2p.$$

This implies that

$$p = (1 - c)/2, \quad and \quad (1 - p) = (1 + c)/2,$$

which can be appropriately substituted into $E[R]$ to yield

$$E[R] = \int_0^\infty P(R \geq R_o)\, dR = c\,R_L + [\,(1-c)\,(R_L + R_H)/2].$$

The formula for expected utility is the linear sum of two terms. The first term, cR_L, may be described as a "worst case" scenario, R_L, times the extent of the individual uncertainty aversion to this event, c. The second term, $(1- c)(R_L+R_H)/2$ is just the expected reward if the distribution of returns is uniform, $(R_L+R_H)/2$, times $(1- c)$, which is the simple sum of the probabilities of each event, e.g. see the function c above. In other words, the expected utility of the project is a weighted average of the worst case scenario and the "risk neutral" scenario where the distribution of the returns is assumed uniformly distributed. The weight is the parameter c measuring the extent of uncertainty aversion: when $c \rightarrow 0$, or $p \rightarrow 1/2$ the agent is "risk neutral;" the larger c, or the lower the probability p, the more uncertainty aversion the individual attaches to event R_L, the "worst case" scenario.

A3.10 ***The Cost of Knightian Uncertainty with Uncertainty Aversion***

In this section, we present a more detailed economic application of the non-additive probability framework of section A3.9, focusing on the presence of Knightian uncertainty, i.e. unknown probability distribution. In this application, we compute the output costs of Knightian uncertainty. We consider a firm with a technology for the production of final goods that takes the following specific form:

$$Z(L, x_n) = L^{1-\alpha} \Sigma_{n=1}^{N} x_n^{\alpha}, \quad \alpha \in (0,1) \tag{A3.45}$$

where x_n is an intermediate capital good input indexed by $n=1,2...N$, the homogeneous domestic labor input is denoted L, and α is the share of capital in total production.

We assume, without loss of generality, that the intermediate capital good input is produced outside the firm, say in a foreign nation under free trade. Hence, the value of final goods produced (GDP) in this economy is the sum of domestic labor income only, since capital inputs are purchased from abroad.

Progress, in this economy, relates directly to the number of intermediate capital good inputs N. The nth intermediate capital good has respective price P_n. The marginal cost of each capital good is denoted w, assumed to be identical across all capital goods. Adding capital good n to the firm's production process, requires a random sunk cost specific to that capital good, where the sunk cost is taken to be a linear function, $\theta \times n$, for θ a random variable to be defined below.

The timing of events and activities is as follows. At the initial instant, before uncertainty is realized, the firm is operating at full capacity and receives an investment. At the end of the period, uncertainty is revealed and trade and production take place.

The firm is assumed to operate in a perfectly competitive environment. The production function (A3.45) yields a marginal physical product of intermediate capital good input n given by

$$\partial Z / \partial x_n = L^{1-\alpha} \alpha \, x_n^{\alpha-1}, \qquad n=1,2,\ldots,N$$

which can be equated to the respective factor cost P_n, yielding a demand function for the nth capital good as

$$x_n^d = (\alpha / P_n)^{1/(1-\alpha)} L, \qquad n=1,2,\ldots,N. \qquad (A3.46a)$$

The producers of intermediate capital goods outside the firm face a demand function, given in expression (A3.46), with price elasticity $1/(1-\alpha)$. The market for production of intermediate capital goods is (assumed) not competitive, so that a producer charges a mark–up above the fixed marginal cost w, given by

$$P_n = w / \alpha, \qquad all\ n \qquad (A3.46b)$$

for the price of the intermediate capital good input.

Then, adding capital good n to the productive capacity of the competitive firm at the initial instant, leads to random discounted profits (revenue minus cost) given by

$$\pi_n(\theta) = [\,(P_n - w)\, x_n \;/\; (1 + r)\,] - \theta n$$

$$= [(w^{-\alpha'} k L) / (1 + r)] - \theta n, \qquad n=1,2,\ldots,N \qquad (A3.47)$$

for $r>0$ the risk-free market interest rate taken as given, $\alpha' \equiv \alpha/(1-\alpha)$ the price elasticity of demand, and constant $k \equiv [(1-\alpha)/\alpha](\alpha)^{2/(1-\alpha)}$, where we used the demand function (A3.46a), and the mark-up pricing rule (A3.46b), to obtain expression (A3.47). The profit is just the net revenue generated by the additional input, $(P_n - w)\, x_n$, discounted to the initial instant by the given risk-free interest factor $(1+r)$, minus the initial random sunk cost θn.

We assume that the random variable associated with the sunk cost function, θn, takes the specific form:

θ can take two possible values:

Low: $\theta = 1-\delta$

High: $\theta = 1+\delta$

for a non-negative parameter $\delta \geq 0$, where the probability distribution is assumed to be unknown to the firm, i.e. Knightian uncertainty.

The deterministic, no uncertainty, case is when $\delta=0$, and $\theta=1$ takes one known possible value. In general, for $\delta \geq 0$, the precise probability distribution of occurrence of each state is unknown. A foreign individual producer of intermediate capital goods (or shareholder) investing in the firm, faces uncertainty in the Knightian sense. However, in facing this type of uncertainty, the investor can construct possible scenarios based upon the framework of non-additive probabilities studied in section A3.9, relating to uncertainty aversion. Suppose a potential investor considers two statistics:

(i) a "worst case" scenario where profit is denoted $\underline{\pi}$;

(ii) a "middle case" scenario where profit is determined by expected profit with a prior that the expected sunk cost is constant, we denote this by $E[\pi]$, where the uncertainty in this case is ignored, i.e. $\delta=0$ and $\theta=1$.

Therefore, the investor's decision rule is to maximize an average of the two statistics:

$$U = c\ \underline{\pi} + (1 - c)\ E[\pi], \quad c \in [0,1] \tag{A3.48}$$

where c is a parameter reflecting the degree of uncertainty aversion discussed in section A3.9. In this example, we can associate c to the polar cases:

• $c \rightarrow 0 \Rightarrow$ Risk neutral case, the investor in this case ignores the uncertainty;

• $c \rightarrow 1 \Rightarrow$ More ambiguity of information, the investor has excessive degree of uncertainty aversion.

Suppose a foreign individual producing intermediate capital good n considers an investment in the domestic firm. We let profit, in absence of new investment, be some constant π_o. Then, investment is undertaken if it increases utility U in expression (A3.48) by considering the alternative scenarios (i)-(ii). In the "worst case" scenario (i), $\theta=1+\delta$, the high value of the sunk cost (thus lower profit), and the total profit is

$$\underline{\pi} = \pi_o + [(w^{-\alpha'} k\, L)/(1+r)] - (1+\delta)\, n$$

that is, the constant profit π_o plus the discounted profit of the new investment assuming the sunk cost is $(1+\delta) \times n$, i.e. the worst possible outcome. In the "middle case" scenario (ii), $\theta = 1$ $(\delta = 0)$, uncertainty is ignored and expected profit is

$$E[\pi] = \pi_o + [(w^{-\alpha'} k L)/(1+r)] - n;$$

the constant profit π_o plus the discounted profit of the new investment assuming the sunk cost is n, i.e. the neutral case.

Hence, the decision whether to invest requires that the average in expression (A3.48) be strictly greater than the profit without investment, π_o, or

$$c\,\{ \pi_o + [(w^{-\alpha'} k L)/(1+r)] - (1+\delta)\, n \} + (1-c)\, \{\pi_o + [(w^{-\alpha'} k L)/(1+r)] - n\} > c\, \pi_o + (1-c)\; \pi_o = \pi_o. \qquad (A3.49)$$

We can appropriately define an adjusted rate of return $\underline{r}$, as

$$\underline{r} \equiv (1+ r)\,(1+ c\delta) - 1 \cong r + c\,\delta,$$

where the term $c\delta$ denotes an uncertainty premium over the risk–free market rate of return. The rate $\underline{r}$ can be considered the uncertainty-adjusted rate of return. Then, the condition for an investment to take place in expression (A3.49) simplifies to

$$w^{-\alpha'} k L / (1+\underline{r}) > n, \qquad (A3.50a)$$

which states that the net revenue generated by the additional input, discounted by the uncertainty premium adjusted rate of return, must be greater than the riskless sunk cost, n $(\theta = 1)$. Examining expression (A3.47), we see that uncertainty enters through the uncertainty premium adjusted rate of return $\underline{r}$, providing a "certainty equivalent" profit function for the investor's decision. Another way to see the analogy is to consider condition (A3.49) for the risk neutral investor, or $c \rightarrow 0$, where the expression simplifies to

$$w^{-\alpha'} k L / (1+r) > n \qquad (A3.50b)$$

and the individual completely ignores the uncertainty when making a decision whether to invest, i.e. $\theta = 1$ $(\delta = 0)$ and use the risk-free return. Expressions (A3.50a,b) are similar except for the uncertainty-adjusted rate of return.

We are ready to compute the costs of uncertainty aversion for this economy. Assuming that all firms share the same uncertainty aversion index c, then the equilibrium number of intermediate capital goods in the economy with Knightian uncertainty is obtained by solving (A3.50a) with equality yielding:

$$\underline{N} = (w^{-\alpha'} k\, L) / (1+\underline{r}), \qquad (A3.51a)$$

and similarly, without uncertainty, solving (A3.50b) with equality gives

$$N = (w^{-\alpha'} k\, L) / (1+r). \qquad (A3.51b)$$

The output of final goods produced (GDP) in the economy with Knightian uncertainty is the sum of labor income, which by $x_n{}^d$ and P_n in (A3.46a,b) and symmetry yields

$$\underline{y} = (1-\alpha)\, \underline{N}\, L^{1-\alpha} (\alpha^2 / w)^{\alpha'\alpha} \qquad (A3.52a)$$

and similarly, without uncertainty, GDP is:

$$y = (1-\alpha)\, N\, L^{1-\alpha} (\alpha^2 / w)^{\alpha'\alpha} \qquad (A3.52b)$$

Therefore Knightian uncertainty reduces GDP by:

$$(\underline{y}/y) - 1 = (\underline{N}/N) - 1 \cong -c\,\delta \qquad (A3.53)$$

i.e. it reduces the number of the new activities (technological progress) in the economy at the rate $c\delta$. The reduction in GDP is measured by the hurdle rate $c\delta$, which is the product of the ambiguity in information or uncertainty aversion parameter, c, and the range of possible outcomes, δ. If there is no uncertainty aversion, $c=0$, or individuals choose to ignore uncertainty, $\delta = 0$, the effect is negligible. Otherwise, the effect is linear (first order) in range of possible outcomes δ, times the uncertainty aversion c.

A3.11 **Risk Averse Bayesian Behavior**

We consider now an extension relating to Bayesian prior under risk aversion (not risk neutrality as in the last section A3.10), and compare with the results of Knightian uncertainty and uncertainty aversion (non-additive probability) of the last section A3.10. Not surprisingly, we show that risk aversion leads to economic costs of the second order, e.g. section 3.7. Consider a case where the N (foreign) firms are Bayesian in the sense of having uniform priors in the presence of symmetric uncertainty and are endowed with strictly increasing and strictly concave utility, i.e. risk averse.

In this case, total value for firms is given by

$$V = u(I_o) + [u(I_1)]/(1+r) \tag{A3.54}$$

for a well defined function u, assumed to be strictly increasing and strictly concave, $u' > 0$, $u'' < 0$; and $r > 0$ is the risk-free rate. The pair $\{I_o, I_1\}$ denotes the total initial and end-of-period income of the firm respectively. Initially, before uncertainty is realized, the firm's budget constraint is given by

$$I_o = R_o + B - \theta n \tag{A3.55a}$$

where B is the firm initial borrowing in a competitive capital market at the risk-free rate r, and R_0 is the firm's income, before borrowing, if the firm refrains from investing. The end-of-period income, after uncertainty is realized, is given by

$$I_1 = R_1 - B(1+r) + w^{-\alpha'} k L. \tag{A3.55b}$$

where R_1 is the end-of-period firm's income and $w^{-\alpha'} k L = (P_n - w) x_n$ is revenue.

As before, randomness takes the form: θ can take two possible values *Low:* $\theta = 1-\delta$; *High:* $\theta = 1+\delta$; for a non-negative parameter $\delta \geq 0$, where the probability distribution is assumed to be unknown to the firm, i.e. Knightian uncertainty. The difference here is that Bayesian firms will consider the two events symmetrically and assign the probability of each to be *1/2*. Recall from sections A3.9-A3.10 that, with non-additive probability, uncertainty aversion is related to the assigned probability of symmetric events by formula (A3.44a). When Bayesian firms have simple priors assigning a symmetric probability *1/2* to the two unexpected events, expected utility becomes

$$V(N, \delta) \equiv (1/2)\{[u(R_o + B - (1+\delta)N)] + u(R_o + B - (1-\delta)N)\} + \{[u(R_1 - B(1+r) + w^{-\alpha'} kL)] / (1+r)\} \tag{A3.56}$$

where the first term, $(1/2)\{[u(R_o+B-(1+\delta)N)] + u(R_o+B-(1-\delta)N)\}$, is the expected utility before uncertainty is realized (assigning a symmetric probability *1/2* to the two unexpected events), and the second term $\{[u(R_1 - B(1+r) + w^{-\alpha'} kL)] / (1+r)\}$ is the present value of future income after uncertainty is realized.

The optimal borrowing problem for the firm involves choosing a level of borrowing, B, to maximize expected utility. In this example, we derive the investment decision by comparing expected utility under optimal borrowing. First, consider the optimal borrowing problem when there is no risky investment. In this special case, the budget constraints (A3.55a,b) become

$$I_o = R_o + B \qquad (A3.57a)$$

$$I_1 = R_1 - B\,(1+r), \qquad (A3.57b)$$

and, the objective function for the optimal borrowing problem in (A3.56) simplifies to

$$V^* = u\,(R_o + B) + \;[\,u(R_1 - \;B\,(1+r\,)\,)\,/\,(1+r)], \qquad (A3.58)$$

which is maximized by choice of B. We let the solution to this optimal borrowing problem without risky investment be denoted

$$\underset{\{B\}}{argmax}\; V^* = V^*_0 \qquad (A3.59)$$

and V^*_0 is the maximum value under optimal borrowing. The decision to make a risky investment under uncertainty involves choosing the number of capital good (input) producing firms, so that expected utility under risky investment and optimal borrowing in expression (A3.56) is no less than the maximum utility without the risky investment, V^*_0 in (A3.59), or

$$V^*_o \leq V^*\,(N,\,\delta\,). \qquad (A3.60)$$

When expression (A3.60) holds with equality, we obtain an equilibrium number of firms under risky investment, $\underline{N}$, that solves $V^*_o = V^*\,(\underline{N},\,\delta\,)$.

Next, suppose the investment is not risky, or $\delta=0$. Then, the decision to invest involves comparing the expected utility in (A3.56) evaluated at $\delta=0$, $V(N,\,0)$, with the optimal borrowing utility in (A3.59), or

$$V^*_o \leq V^*\,(N,\,0), \qquad (A3.61)$$

and at equality, the expression determines the equilibrium number of firms under certainty, N^*.

We gain insight into the costs of uncertainty in this case, by taking a second order Taylor expansion of the function $V^*\,(\,N,\;\delta\,)$ in (A3.60), evaluated at the point where there is no uncertainty, $V^*(N^*,\,0)$, as

$$V^*\,(\,N,\,\delta\,) \cong \;V^*\,(\,N^*,\,0) + \,V^{*\prime}\,(\,N^*,\,0)\,(\,N - N^*) + \tfrac{1}{2}\,V^{*\prime\prime}(\,N - N^*\,)^2 \cong$$

$$V^*\,(\,N^*,\,0) - u'|_O\,(N - N^*\,) + \tfrac{1}{2}\,u''|_O\,[(\,N^* - N\,)^2 + \;N^2\delta^2\,] \qquad (A3.62)$$

where $u'|_O$ is the first derivative of the function u with respect to N evaluated at $I_o = R_o + B - N^*$, i.e. $\delta=0$ (no uncertainty); and $u''|_O$ is the second derivative evaluated likewise. Thus, using the expansion (A3.62), in equilibrium, the number of capital good producing firms under uncertainty, $\underline{N}$, is determined by the expression

$$0 \cong - u'|_O (\underline{N} - N^*) + \tfrac{1}{2} u''|_O [(N^* - \underline{N})^2 + \underline{N}^2 \delta^2].$$

We can further rewrite this formula, recalling from section 3.3 expression (3.14), that the coefficient of absolute risk aversion is given by $CARA(I_o) = - u''|_O / u'|_O$, to obtain

$$0 \cong N^* - \underline{N} - \tfrac{1}{2} CARA(I_o) [(N^* - \underline{N})^2 + \underline{N}^2 \delta^2]. \qquad (A3.63)$$

We evaluate the loss in the number of firms due to uncertainty, at $N^* = \underline{N}$, to obtain:

$$\underline{N}/N^* - 1 \cong - \tfrac{1}{2} CARA(I_o) N^* \delta^2 \qquad (A3.64)$$

i.e. the loss due to uncertainty depends upon absolute risk aversion, but most importantly upon δ^2, thus it is of second order magnitude in the uncertainty parameter δ. This is not surprising because under risk aversion, costs are of second order as shown in section 3.7.

Hence, when we compare the resulting loss under risk aversion in expression (A3.64), with the resulting loss under uncertainty aversion of the Knightian uncertainty case of expression (A3.53), we note the following. In the case of risk aversion the loss or reduction in the number of activities (technological progress) is at the rate of second order in the range of possible outcomes δ, whereas in the case of uncertainty aversion the loss is at the rate of first order in δ.

Therefore, the presence of uncertainty aversion can have potential first order effects, and can be more damaging to the economy, whereas risk aversion has second order effects only.

A3.12 ***Summary II***

In sections A3.9-A3.11, we pursue some issues relating to utility and behavior towards risk in a framework that deviates from the traditional Von-Neumann-Morgenstern framework. This is an active area of research that has blossomed recently. Some have taken this emerging literature towards a fruitful exploration at the edges of the economics and psychology disciplines. The interested reader may examine the further readings in Notes on the Literature.

Notes on the Literature

The Mean-Variance analysis with expected utility based on the Normal distribution of payoffs follows the material in Sargent (1976), Chapter 7.2, based on Tobin's (1958) classic paper. Other important examples are presented in Sargent's chapter, including the state preference model and the Modigliani-Miller theorem. The material is also available in Hirshleifer and Riley (1982), in particular the Mutual Fund theorem in section A3.5 is in Hirshleifer and Riley (1992), page 79.

The material in sections A3.9-A3.12 follows the paper by Aizenman (1997). Non-additive (subjective) probabilities are studied in Schmeidler (1989). A very useful survey of alternative approaches to the traditional Von-Neumann-Morgenstern framework is found in Camerer and Weber (1992) and also Machina (1987).

Knight's (1921) concept of uncertainty relates to the lack of knowledge of the probability distribution of the states of nature, as opposed to risk where the probability distribution is known. Applications of Knightian uncertainty to financial markets is found in Dow and Werlang (1992a,b) and Epstein and Wang (1994); another application in a principal-agent model is in Rigotti (1998a), and in investment behavior in Rigotti (1998b) and Newman (1995). An axiomatic approach to subjective probability is found in Skiadas (1997).

Models with many intermediate inputs are found in Ethier (1982), Romer (1994), and surveyed in Barro and Sala-i-Martin (1995), Chapter 6.

The recent psychology and economics literature is reviewed in Rabin (1998) and an application is found in Rabin (2000). The book Kahneman and Tversky (2000), presents a collection of the most important papers in this area of research. Some like to refer to this area broadly as behavioral economics and finance. Bernstein (1996), Chapter 16, presents a popular discussion of the psychology-economics research with emphasis on financial markets.

References

Aizenman, Joshua (1997) "Investment in New Activities and the Welfare Cost of Uncertainty." *Journal of Development Economics*, 52, 259-277.

Barro, Robert J. and Xavier Sala-i-Martin (1995) *Economic Growth.* McGraw Hill Book Co., New York, NY.

Bernstein, Peter L. (1996) *Against the Gods: The Remarkable Story of Risk.* John Wiley&Sons, New York, NY.

Camerer, Colin and Martin Weber (1992) "Recent Developments in Modeling Preferences: Uncertainty and Ambiguity." *Journal of Risk and Uncertainty,* 5, 325-370.

Dow, James and Sergio R. C. Werlang (1992a) "Excess Volatility of Stock Prices and Knightian Uncertainty." *European Economic Review*, 36, 631-638.

Dow, James and Sergio R. C. Werlang (1992b) "Uncertainty Aversion, Risk Aversion and the Optimal Choice of Portfolio." *Econometrica,* 60, 197-204.

Epstein, Larry G. and Tan Wang (1994) "Intertemporal Asset Pricing under Knightian Uncertainty." *Econometrica*, 62, 283-322.

Ethier, Wilfred (1982) "National and International Returns to Scale in the Modern Theory of International Trade." *American Economic Review,* 72, 389-405.

Hirshleifer, John and John G. Riley (1992) *The Analytics of Uncertainty and Information.* Cambridge Surveys of Economic Literature, Cambridge University Press, Cambridge, UK.

Kahneman, Daniel and Amos Tversky, Eds. (2000) *Choices, Values and Frames.* Cambridge University Press, Cambridge, UK.

Knight, Frank (1921) *Risk, Uncertainty and Profit.* Houghton Mifflin, Boston, MA.

Machina, Mark (1987) "Choice under Uncertainty: Problems Solved and Unsolved." *Journal of Economic Perspectives*, 1, 121-154.

Newman, Andrew (1995) "Risk-Bearing and 'Knightian' Entrepreneurship." Working paper, Department of Economics, Columbia University, October.

Rabin, Mathew (1998) "Psychology and Economics." *Journal of Economic Literature,* 36, 11-46.

Rabin, Mathew (2000) "Inference by Believers in the Law of Small Numbers." Working paper, Department of Economics, University of California, Berkeley, January.

Rigotti, Luca (1998a) "Imprecise Beliefs in a Principal-Agent Model." Working paper, Tilburg University, The Netherlands, October.

Rigotti, Luca (1998b) "Decisive Entrepreneurs and Cautious Investors." Working paper, Tilburg University, The Netherlands, December.

Romer, Paul (1994) "New Goods, Old Theory and the Welfare Costs of Trade Restrictions." *Journal of Development Economics*, 43, 5-38.

Sargent, Thomas J. (1976) *Macroeconomic Theory.* Academic Press San Diego, CA. (A second edition published in 1987 is also available).

Schmeidler, David (1989) "Subjective Probability and Expected Utility without Additivity." *Econometrica*, 57, 571-587.

Skiadas, Costas (1997) "Subjective Probability under Additive Aggregation of Conditional Preferences." *Journal of Economic Theory,* 76, 242-271.

Tobin, James (1958) "Liquidity Preference as Behavior Towards Risk." *Review of Economic Studies*, 25, 65-86.

4. Introduction to Systems of Financial Markets

4.1 Introduction

In this chapter, we present some fundamental analysis of systems of financial markets in a static, short-term context. The focus is on pricing alternative types of securities under uncertainty and, consequently, pricing the states of nature themselves. Some simple portfolio choice problems are examined with derivation of the consumption-based capital asset pricing (CCAPM) formula. The structure of the asset markets is given proper attention with a thorough discussion of complete versus incomplete asset market regimes. An application to the neoclassical theory of the firm is presented last.

4.2 Pricing Securities in a Linear Fashion

Linear pricing of securities implies the absence of arbitrage opportunities, recall Chapter 2. For example, suppose there are two assets with prices p_1 and p_2 . If the sum of the separate prices, p_1+p_2 is greater than the price of the sum of the two securities, say p, it would be profitable to buy the sum of the two securities at p, divide it in two equal parts and sell them separately at a profit. Hence, with economies populated by rational agents, linear pricing implies no-arbitrage.

For example, a well-known result that is a direct implication of linear pricing is the Modigliani-Miller theorem. It says that firms that have identical streams of future net revenues should have the same market value regardless of their equity-debt capital structure, or similarly the market value of a firm should not depend on its equity-debt capital structure. We study the Modigliani-Miller result in the dynamic setting of Chapter 7.

In order to understand more formally the implications of no-arbitrage linear pricing of securities, we first define a typical security.

Definition: An ordinary security is a random prize that gives some payoff in every state of nature.

Let S' be the finite set of states of nature and $s \in S'$ is a possible state, or $s=1,2...S$, for S the finite number of possible states. Let the ordinary security (random prize) be denoted d. Then, d is a list denoted $(d^1, d^2, d^3, d^4, ..., d^S)$ of state contingent payoffs, d^s, denominated in "money" terms, when uncertainty is resolved at the end of the investment period. Note that sometimes the lists appearing in this section may be easily reinterpreted as a vector and we'll leave it clear when this is the case.

Now consider several ordinary securities available in the financial market. Let there be $i=1,2,...n$ ordinary securities in the market place, denoted d_i, each with list of payoffs $(d_i^1, d_i^2, d_i^3, d_i^4, ..., d_i^S)$. Associated with each ordinary security d_i there is an initial unit price denoted P_i. The price reflects the initial ex-ante value (before uncertainty is realized) for the final random payoff. For example, one share (stock) of i=Intel has a current price, P_i, and a random value when uncertainty is resolved at the end of the investment period given by d_i.

In this case, individuals may form portfolios with different quantities of the available securities. Denote by θ_i the quantity of security i, in perfectly divisible units. Then, a portfolio can be defined as a list $(\theta_1, \theta_2, \theta_3, \theta_4, ..., \theta_n)$ invested in each asset i with random payoff at the end of the period, x, given by the linear sum

$$x = \Sigma_{i=1}^{n} \theta_i d_i \qquad (4.1)$$

A negative θ_i denotes a short sale of asset i. The initial price of this portfolio, denoted P, is, by linear pricing (or no arbitrage), given by the linear sum

$$P = \Sigma_{i=1}^{n} \theta_i P_i \qquad (4.2)$$

where again P_i is the initial price of security i.

4.3 **Optimal Portfolio Choice Problems**

Consider forming a portfolio to maximize the expected utility of final random wealth subject to some initial wealth constraint.

Let the individual be risk averse with utility function u strictly increasing, $u' > 0$, and strictly concave, $u'' < 0$. The deterministic initial wealth, in "money" terms is denoted W_o. The $i=1,2,...n$ ordinary securities available in the financial market are all risky for the moment. The final random payoff of a portfolio is given by equation (4.1). Initially, the current price of this portfolio is given by equation (4.2), taken as parametrically given, which is assumed not to exceed initial wealth, or

$$W_o \geq P.$$

The individual will exhaust her initial wealth because preferences are strictly increasing and monotone. The initial budget constraint for a typical individual may be written as

$$W_o = \Sigma_{i=1}{}^n \, \theta_i P_i \geq 0 \qquad (4.3)$$

where some θ_i may be negative. That is, short sales are allowed but one cannot go short on all assets simultaneously. Bankruptcy is not allowed, or the agent is not allowed to end up with $W_o < 0$.

The optimal portfolio choice problem solved by an individual is to maximize expected utility subject to the budget constraints (4.1) and (4.3) by choice of a portfolio of asset holdings $(\theta_1, \theta_2, \theta_3, \theta_4, ..., \theta_n)$. This is written as

$$\underset{\{x, \theta_i\}_{i=1}{}^n}{Max} \; E[u(x)] \qquad (4.4)$$

$$subject\ to \quad x - \Sigma_{i=1}{}^n \, \theta_i d_i = 0$$

$$W_o - \Sigma_{i=1}{}^n \, \theta_i P_i = 0$$

given $\{W_o, P_i\}$ and the probability distribution of random payoffs d_i.

As a first introductory example, suppose there are only two assets. One is risky and the other is risk-free yielding $d_2 = R > 0$, constant with

$P_2=1$, where R is the risk-free gross rate of return. In this case, equation (4.1) becomes

$$x = \theta_1 d_1 + \theta_2 R$$

and the budget constraint (4.3) becomes

$$W_o = \theta_1 P_1 + \theta_2.$$

The solution to problem (4.4) is obtained straightforward, by substituting the constraints into the objective function, obtaining

$$\underset{\{\theta_1\}}{Max} \; E[u(\theta_1 d_1 + \{W_o - \theta_1 P_1\} R)]$$

with first order necessary condition

$$E[u'(\theta_1 d_1 + \{W_o - \theta_1 P_1\} R) \{(d_1 / R) - P_1\}] = 0$$

and the strict concavity of the utility function and linearity of the constraint satisfy the sufficient second order condition. The solution involves equating the marginal benefit of the additional unit invested in the risky asset, given by $u' d_1$, to the marginal cost (equal to the forgone risk-free gain), given by $u' P_1 R$. The result is that the average return of the risky asset, the payoff discounted by the risk-free return, minus the initial price of the risky asset all adjusted by the marginal utility of final random wealth, should be zero.

The expression above can solve for the demand for the risky asset, θ_1^*, as a function of the probability distribution of the final payoff, initial wealth and the risk-free return, i.e. $\theta_1^*=\theta_1^*(W_o, P_1, R,$ *probability distribution of* $d_1)$. Then, using the budget constraint, the demand for the risk-free asset is $\theta_2^* = W_o - \theta_1^*$.

Alternatively, given the quantity θ_1, we may express the initial price of the risky asset as a function of the probability distribution of the final payoff, the marginal utilities of final random wealth and the risk-free return, or

$$P_1 = E[u'(\theta_1 d_1 + \{W_o - \theta_1 P_1\} R) (d_1 /R)] / E[u'(\theta_1 d_1 + \{W_o - \theta_1 P_1\} R)] .$$

Hence, the initial price of the risky asset is the expected discounted final payoff, d_1 /R, adjusted by the marginal utility of final random wealth, i.e. an asset pricing formula. If there is no risk, or $d_1=d$ constant, then the pricing formula gives $P_1 = d/R$, the initial price is the discounted value of the final payoff, where the discount factor is the risk-free return.

A pricing formula of this type is consistent with the Efficient Market hypothesis: prices should reflect the present discounted value of expected payoffs (a more detailed discussion of the Efficient Market hypothesis is in Chapter 7). In fact, the results obtained in this introductory example provide the seeds of a more general asset pricing formula consistent with the so-called consumption-based capital asset pricing model (CCAPM) studied in the dynamic context of Chapters 7 and 8.

In order to understand the consumption-based capital asset pricing formula, we continue with a solution for the general case, (4.4), when all assets are risky. Substituting the first constraint into the objective function, we form the Lagrangean function for this problem as

$$\mathcal{L} = E[u(\Sigma_{i=1}^{n} \theta_i d_i)] + \lambda [W_o - \Sigma_{i=1}^{n} \theta_i P_i] \qquad (4.5)$$

where $\lambda \geq 0$ is the Lagrange multiplier associated with the initial wealth constraint, which gives the marginal utility of initial wealth; recall the Envelope theorem. The first order necessary conditions for an interior equilibrium are given by

$$\partial\mathcal{L}/\partial\theta_i = E[u'(\Sigma_{i=1}^{n} \theta_i d_i)\, d_i] - \lambda P_i = 0, \quad \textit{for each security } i=1,2,...n \qquad (4.6)$$

$$\partial\mathcal{L}/\partial\lambda = W_o - \Sigma_{i=1}^{n} \theta_i P_i = 0. \qquad (4.7)$$

Equations (4.6)-(4.7) are a total of $n+1$ equations in the $n+1$ unknowns $\{\theta_i^*, \lambda^*\}_{i=1}^{n}$ as a function of $\{W_o, P_i\}$ and the probability distribution of random payoffs d_i. Again, the strict concavity of the utility function and linearity of the constraint satisfy the sufficient second order condition. The final optimal payoff is

$$x^* = \Sigma_{i=1}^{n} \theta_i^* d_i$$

and the first order condition (4.6) may be rewritten as

$$P_i = E[u'(x^*)\, d_i] / \lambda, \qquad (4.8)$$

a typical asset pricing formula interpreted as the current price reflecting the expected discounted value of the payoff. Formula (4.8) is basically the consumption-based capital asset pricing formula where the initial price of the asset is the expected value of dividends discounted by the ratio of the marginal utility of final random wealth to the marginal utility of initial wealth, e.g. valued in terms of the marginal utilities of "consumption." As mentioned above, the Envelope theorem yields

$$\partial\mathcal{L}/\partial W_o = \lambda = u'(W_o),$$

the marginal utility of initial wealth. Then, formula (4.8) can be expressed as

$$P_i = E[\, u'(x^*)\, d_i\,] \,/\, u'(W_o). \tag{4.8'}$$

Notice that expression (4.8') can be written as $P_i = E[\,\{u'(x^*)/u'(W_o)\}d_i\,]$, where the term in keys, $\{u'(x^*)/u'(W_o\,)\}$, appropriately represents the marginal rate of substitution between initial and final wealth, i.e. the utility gain of one unit of final random wealth relative to (or priced at) the utility gain of one unit of initial wealth.

The marginal rate of substitution is an important measure for asset pricing. For example, for an asset with "good" payoff, $x^*>W_o$, and for a strictly concave utility function, $u'(x^*)<u'(W_o)$, and the marginal rate of substitution is strictly less than one. For this reason, the marginal rate of substitution is like a discount factor, which given the randomness of x^*, is stochastic, not deterministic. Also of importance is the simple decomposition of the current price in terms of means and covariances. Using the covariance decomposition, $E[XY]=E[X]E[Y]+cov(X,Y)$, expression (4.8') may be rewritten as:

$$P_i = E[u'(x^*)/u'(W_o)]\; E[\, d_i\,] + cov(\,\{u'(x^*)/u'(W_o)\},\, d_i\,), \tag{4.9}$$

which decomposes the current price in terms of means and covariances of marginal rates of substitution and payoffs. We shall return to expression (4.9) shortly.

The importance of expressions (4.8) or (4.8') is that they are able to price any ordinary security with associated random payoff d_i, $i=1,2,\ldots n$. Naturally, a risk-free security in the portfolio can be priced as well. Suppose one of the securities is risk-free: $d_i=R$ for sure and $P_i=1$ as in the introductory example above. The condition (4.6) applies to this security as well yielding

$$E[u'(x^*)]\, R = \lambda \;\Leftrightarrow\; R = 1\,/\,E[u'(x^*)/u'(W_o)] \tag{4.10}$$

where we have used the Envelope theorem again to substitute for the marginal utility of initial wealth.

In general, formula (4.10) expresses the risk-free return of a \$1 investment as the expected value of the ratio of marginal utilities (or marginal rate of substitution) times the risk-free return, or

$$E[u'(x^*)R / u'(W_o)] = 1. \qquad (4.10')$$

Substituting (4.10) into (4.8) gives the price of the risky payoff d_i as

$$P_i = E[u'(x^*)\,(d_i/R)] \,/\, E[u'(x^*)] \qquad (4.11)$$

exactly as in the introductory example. Hence, with a risk-free security in the portfolio, the wealth forgone is just the expected marginal utility of the investment in the risk-free security, or $E[u'(x^*)]R$. We can use the decomposition in expression (4.9), with (4.10), to obtain

$$P_i = E[\,d_i /R] + cov(\,\{u'(x^*)/u'(W_o)\},\, d_i\,),$$

the price is the expected future payoff discounted by the risk-free return adjusted by the covariance between the marginal rate of substitution and the payoff. Consider the following cases:

(i) If there is no uncertainty, $d_i = d$ constant, and

$$cov(\{u'(x^*)/u'(W_o)\},\, d_i\,)=0,$$

and

$$P_i = d_i / R, \qquad (4.11')$$

which is the usual formula of the price reflecting the discounted value of the sure future gain;

(ii) If $cov(\{u'(x^*)/u'(W_o)\},\, d_i\,)=0$, then the asset price is just

$$P_i = E[d_i] / R$$

the discounted value of the expected random payoff.

You should be thinking this: when an asset yields a "good" ("bad") payoff, x^* (or d_i) is high (low), by the strict concavity of the utility function, the marginal utility, $u'(x^*)$ is low (high) and the marginal rate of substitution is low (high). Thus, we can expect that the covariance between the marginal rate of substitution and the payoff is negative,

$$Cov(\{u'(x^*)/u'(W_o)\},\, d_i\,)<0.$$

This means that the price of the risky asset will be lower than the risk-free price, because by risk aversion, there must be a premium for an individual to hold a risky asset, whose payoff covaries negatively with the marginal rate of substitution (we have seen the risk premium in Chapter 3, and we shall study its properties in more detail in Chapter 7).

For another simple illustration, consider the two asset-two state case, let utility be logarithmic, $u(x)=log\ x$, and let the probability distribution of the states be: in state 1, payoff is d_1^1 with probability $\pi>0$, and in state 2, payoff is d_1^2 with probability $1-\pi>0$. The asset pricing formula (4.11) yields

$$([\{\pi d_1^1/[\theta_1 d_1^1+\{W_o-\theta_1 P_1\} R]\}+\{(1-\pi) d_1^2/[\theta_1 d_1^2+\{W_o-\theta_1 P_1\} R]\}]/$$
$$[\{R\pi/[\theta_1 d_1^1+\{W_o-\theta_1 P_1\} R]\}+\{(1-\pi) R/[\theta_1 d_1^2+\{W_o-\theta_1 P_1\} R]\}])$$
$$=P_1$$

which can be easily solved for price given quantity, P_1 as a function of θ_1, or vice-versa. Hence, solving for the price P_1 as a function of quantity $\theta_1=1$, we obtain an asset pricing formula; and solving for the quantity, θ_1, as a function of the price P_1, we obtain the solution for the portfolio choice problem.

4.4 **Pricing the States of Nature**

Recall $s\in S'$ is the finite state space and the state contingent payoff list for ordinary security d_i is

$$(d_i^1, d_i^2, d_i^3, d_i^4,...,d_i^S),$$

with initial price P_i, $i=1,2,...n$. We consider a special security, called a state contingent claim, defined below.

Definition: An elementary ('Arrow') security is a state contingent claim to a payoff of one unit of account (or purchasing power) in a specific state $s\in S'$, and zero in all other states; denoted e^s with payoff list

$$(0, 0, 0,...,d^{S,e}=1,... 0, 0, 0).$$

Whenever such security exists, it has associated with it a strictly positive current price denoted $\psi^s>0$, which is state dependent.

This is a very special security that functions like insurance to the occurrence of one specific state of nature $s\in S'$. For example, a payoff of one unit of commodity or one dollar if an earthquake occurs, and zero otherwise. The associated state dependent prices are also special in the sense that they artificially "price" the states of nature.

The importance of the elementary security in systems of financial markets follows from a series of subtle results. Momentarily, suppose there are available in the market place as many elementary securities as there are states of nature. For each state of nature there is an associated elementary security, and anyone is able to insure against the occurrence of all possible states of nature. Then, for <u>any</u> other security, say an ordinary security d_i, we are able to create an equivalent portfolio of elementary securities that yields exactly the same payoff as the ordinary security, d_i.

More importantly, the elementary securities in the equivalent portfolio have associated state contingent prices, $\psi^s > 0$. Then, by the no-arbitrage theorem, i.e. linear pricing, we are able to price <u>any</u> other security, say any other ordinary security d_i, as a linear combination of the prices of the elementary securities in the equivalent portfolio. We shall see below that this may be possible even when there are fewer elementary securities than states of nature.

At first this sounds strange but not so when you think about it more carefully. In the real world, there are mostly ordinary securities with a given current price; for example the stock of a company on the NYSE and its current price. The result above indicates that the price of this ordinary security can be expressed as a linear combination of the prices of elementary securities, which are the unknown prices of the states of nature. Hence, under some special conditions, we can invert back and extract the prices of the states of nature from the current prices of the ordinary securities. This is important because it means that we can buy and/or sell insurance to all states of nature at those state prices, basically determined by the observed market prices of ordinary securities.

We shall write down the logical steps of this result using the variables defined above.

First, for each ordinary security d, $(d^1, d^2, \ldots, d^S)$, we create an equivalent portfolio of $s=1,2,\ldots S$ elementary securities as contingent claims to d^s that give exactly the same payoff, that is

$$d = \Sigma_{s\in S'} d^s e^s. \quad (4.12)$$

This formula indicates that the security with payoff d, $(d^1, d^2, ..., d^S)$, can be perfectly substituted by a combination of state contingent claims which pay exactly $1\times d^s$ in state s, and zero in all other states, that is a contingent claim to the payoff d^s. The list of payoffs for the right hand side of (4.12), given by $\Sigma_{s\in S'} d^s e^s$ is just

$$d^1 [e^1, (1,...,0,...,0)] + d^2 [e^2, (0, 1,...,0,...,0)] +...+ d^S [e^S, (0,...,1)] \equiv$$

$$[e^1, (1\times d^1,...,0,...,0)] + [e^2, (0, 1\times d^2,...,0,...,0)] +...+ [e^S, (0,...,0,...,1\times d^S)].$$

Hence, holding either d or the sum $\Sigma_{s\in S'} d^s e^s$ gives exactly the same random payoff, and the portfolio of elementary securities as contingent claims to d^s is equivalent.

Second, by the no-arbitrage theorem on prices, or linear pricing, we price the ordinary security d as a linear combination of the prices of the elementary securities or contingent claims in the equivalent portfolio, $\Sigma_{s\in S'} d^s e^s$. Thus, the price of the payoff d, $(d^1, d^2, ..., d^S)$, is the sum of the price of each contingent claim in the equivalent portfolio, $d^s[e^s, (0,...,1,...,0)]$. But note that the cost of obtaining the contingent claim $d^s[e^s, (0,...,1,...,0)]$ is the price of the elementary security $[e^s, (0,...,1,...,0)]$; that is the state price ψ^s, times the payoff in state s, d^s. Therefore, the initial price or initial cost of buying the equivalent portfolio $\Sigma_{s\in S'} \psi^s d^s$ is

$$P = \Sigma_{s\in S'} d^s \psi^s \quad (4.13)$$

for $\psi^s > 0$. Equation (4.13) is equivalent to a dot or inner product, say $P = d\cdot\psi$ for d and ψ listings as the appropriate row vectors. As before, we obtain (4.13) because ψ^s is the price of one unit at e^s, and $\psi^s d^s$ is the price of 1 times d^s units at e^s.

Therefore, the price of d is equal to the cost of constructing an equivalent portfolio with the associated state contingent claims to d^s. Formula (4.13) is crucial for the understanding of market regimes because, given the initial price P and the probability distribution of the state dependent payoff d^s, one can extract the state prices ψ^s and then "price" all states of nature.

Here is the beauty of this framework. Even if the elementary security does not exist in some states of nature, one can construct portfolios with existing ordinary securities that exactly replicate the elementary security in those states.

For a simple example, consider two states of nature, $s=1,2;$ and two ordinary securities with payoffs: $d_1^1=1$ and $d_1^2=1$, the risk-free security; and security 2 yielding $d_2^1=1$ and $d_2^2=-1$, a risky gamble. Then, we can construct portfolios with quantities of security 1 and 2 as follows. First, the investor goes long $1/2$ units on each security, or $\theta* = (1/2,1/2)$. In another portfolio, the investor goes long $1/2$ units of security 1 and short (borrow) $1/2$ of security 2, or $\theta** = (1/2,-1/2)$. Computing the payoffs of each portfolio, after the state of nature is realized, we obtain:
(i) Portfolio $\theta* =(1/2,1/2)$: in state $s=1$, $(1/2)d_1^1+(1/2)d_1^2=d^1=1$; in state $s=2$, $(1/2)d_1^2+(1/2)d_2^2=d^2=0$; thus yielding the elementary security for state 1, $e^1=(1,0)$;
(ii) Portfolio $\theta** =(1/2,-1/2)$: in state $s=1$, $(1/2)d_1^1-(1/2)d_1^2=d^1=0$; in state $s=2$, $(1/2)d_1^2-(1/2)d_2^2 =d^2=1$; thus yielding the elementary security for state 2, $e^2=(0,1)$.

We conclude from this example that, with two ordinary securities and two states of nature, we can construct portfolios that replicate elementary securities, insure against both states of nature, and price both states of nature with linear pricing as well.

We'll present now, with another example, the exact method and conditions needed to find the elementary securities for the states as a linear combination of ordinary securities. Suppose there are two states, $s=1,2;$ and two ordinary securities, $i=1,2$ with state contingent payoffs given by d_i^s. Security 1 yields $d_1^1=A$ in state of nature 1 and $d_1^2=B$ in state of nature 2, and security 2 yields $d_2^1=C$ and $d_2^2=H$ respectively.

Take the following steps:
(i) Set up a matrix of payoffs with dimension $n \times S$, say $i=1,...n$ rows and $s=1,...S$ columns; in this case $i=1,2$ and $s=1,2$, and call it D, where each row represents the payoffs of security i across different states of nature s, or

$$D = \begin{vmatrix} d_1^1 & d_1^2 \\ d_2^1 & d_2^2 \end{vmatrix} = \begin{vmatrix} A & B \\ C & H \end{vmatrix};$$

(ii) Then, invert the matrix D to obtain, in our 2×2 case,

$$D^{-1} = (1/|D|) \begin{vmatrix} d_2^2 & -d_1^2 \\ -d_2^1 & d_1^1 \end{vmatrix}$$

or

$$D^{-1} = \begin{vmatrix} H/|D| & -B/|D| \\ -C/|D| & A/|D| \end{vmatrix}.$$

The rows of the matrix D^{-1} give the exact quantities of each asset needed to construct the elementary security in the respective state of nature. That is, the first row gives the elementary security for state of nature *1* and the second row gives the elementary security for state of nature *2*. The portfolio of the two assets in the first row of D^{-1}

$$\theta_1 = (H/|D|, -B/|D|)$$

yields a state contingent claim to state *1*:

$$e^{*1} = [H/|D|]\, d_1 + [-B/|D|]\, d_2$$

with payoff in state *1*, recall that $d_1^1=A$ and $d_2^1=C$,

$$(HA-BC)/|D| = 1$$

and payoff in state *2*, recall that $d_1^2=B$ and $d_2^2=H$,

$$(HB-BH)/|D| = 0,$$

therefore the state contingent claim to state *1*, $e^1=(1,0)$.

Similarly, a portfolio of the two assets in the second row of D^{-1}

$$\theta_2 = (-C/|D|, A/|D|)$$

yields a state contingent claim to state *2*, $e^2=(0,1)$; that is

$$e^{*2} = [-C/|D|]\, d_1 + [A/|D|]\, d_2$$

with payoff in state *1*, recall that $d_1^1=A$ and $d_2^1=C$,

$$(-CA+AC)/|D| = 0$$

and payoff in state *2*, recall that $d_1^2=B$ and $d_2^2=H$,

$$(-CB+AH)/|D| = 1,$$

therefore the state contingent claim to state *2*, $e^2=(0,1)$.

Thus, for the example, the equivalent portfolios in (4.12) are

$$d_1 = d_1^1 \{[H/|D|] d_1 + [-B/|D|] d_2\} + d_1^2 \{[-C/|D|] d_1 + [A/|D|] d_2\}$$

$$d_2 = d_2^1 \{[H/|D|] d_1 + [-B/|D|] d_2\} + d_2^2 \{[-C/|D|] d_1 + [A/|D|] d_2\}$$

where the terms in keys are the portfolios that replicate the elementary securities in the respective state.

Returning to the previous simple example, let $A=1$, $B=1$, or the first security is risk-free, and $C=1$, $H=-1$, the second is a risky gamble. The portfolio $\theta_1 = (1/2,\ 1/2)$ yields the contingent claim of one to state *1* and zero in state *2*, $e^1=(1,0)$; and the portfolio $\theta_2 = (1/2,-1/2)$ yields the contingent claim of one to state *2* and zero in state *1*, $e^2=(0,1)$; both of which can be priced as we shall see below. Hence, the equivalent portfolios in (4.12) are

$$d_1 = d_1^1 [(1/2)d_1 + (1/2)d_2] + d_1^2[(1/2)d_1 - (1/2)d_2]$$

$$d_2 = d_2^1 [(1/2)d_1 + (1/2)d_2] + d_2^2[(1/2)d_1 - (1/2)d_2]$$

where the terms in brackets are the portfolios that replicate the elementary securities: $[(1/2)d_1 + (1/2)d_2]$ gives the contingent claim of one to state *1*, $e^1=(1,0)$; $[(1/2)d_1 - (1/2)d_2]$ gives the contingent claim of one to state *2*, $e^2=(0,1)$.

This discussion illustrates a very important property: the matrix of payoffs D is square, $n=S$, and invertible or nonsingular. Hence, we are able to find the combinations of the two ordinary securities that give elementary securities for the two states, i.e. the two states can be fully insured. You probably noticed that there is something important going on here! In fact, the simple example above illustrates crucial issues in systems of financial markets.

Notice from above that the state contingent claims to states *1* and *2* are

$$e^{*1} = [H/|D|] d_1 + [-B/|D|] d_2$$

$$e^{*2} = [-C/|D|]\, d_1 + [A/|D|]\, d_2,$$

or a linear sum of each traded security weighted by real numbers. We can define $a_{11} \equiv H/|D|$; $a_{12} \equiv -B/|D|$; $a_{21} \equiv -C/|D|$; $a_{22} \equiv A/|D|$ and rewrite the state contingent claims as

$$e^{*1} = \Sigma_{i=1}^{2}\, a_{1i}\, d_i$$

$$e^{*2} = \Sigma_{i=1}^{2}\, a_{2i}\, d_i\,.$$

We see that the set of traded securities, d_i, span the set of state contingent claims $e^{*}=(e^{*1},\ e^{*2})$. Basic linear algebra gives us matrix criteria for spanning sets and, in this case, the criteria is exactly that the matrix of payoffs D is square, $n=S$, and invertible or nonsingular. Henceforth, security valuation, state valuation, and spanning are closely related and we move to relate those to the structure of security markets.

4.5 Market Regimes: Complete versus Incomplete

We established that, for each ordinary security $i=1,2,...n$, and states of nature $s=1,2,...S$, there is an equation (4.13) indexed by i, or

$$P_i = \Sigma_{s\in S'}\, d_i^{\,s}\, \psi^s, \qquad \psi^s > 0 \tag{4.14}$$

yielding $i=1,...n$ equations in $s=1,...S$ states of nature. Given the initial prices P_i and the probability distribution of the payoffs d_i^s, equations (4.14) can be inverted to find the state prices $\psi^s>0$. In this sense, equations (4.14) denote a system of $i=1,...n$ equations in $s=1,...S$ state prices, ψ^s to be determined. In fact, those state prices can be determined uniquely if and only if the matrix of payoffs, D, with dimension $n\times S$ is invertible or nonsingular. Expression (4.14) may be compactly written in matrix form as

$$\mathbb{P} = D\,\psi$$

where $\mathbb{P}$ is a $n\times 1$ column matrix (n rows and 1 column) of initial prices, D is a $n\times S$ square matrix (n rows and S columns) of payoffs, and ψ is a $S\times 1$ column matrix (S rows and 1 column) of state prices. Of course, the unique solution for the state prices is

$$\psi = D^{-1} \mathbb{P}$$

where D^{-1} is the inverse of the matrix of payoffs D. Hence, the condition of invertibility (nonsingularity) of the matrix D is natural for the unique determination of state prices. Given linear pricing (no arbitrage), it is no accident that in section 4.4 we obtained the portfolios that replicate the elementary securities from the matrix D^{-1}. The first row of the matrix denoted D^{-1} gives the exact quantities of each asset, $i=1,2,...n$, needed to construct the elementary security for state of nature $s=1$, the second row D^{-1} gives the exact quantities of each asset, $i=1,2,...n$, for the elementary security of state of nature $s=2$, and so on.

Equivalently, one may want to define the state contingent total return of security $i=1,2,...n$ in state $s=1,2,...S$ as

$$R_i^{\,s} = d_i^{\,s} / P_i \tag{4.15}$$

and the system of equations in (4.14) can be rewritten as

$$1 = \Sigma_{s \in S'} R_i^{\,s} \, \psi^s, \qquad \psi^s > 0. \tag{4.16}$$

In this case, state prices are uniquely determined if and only if the matrix of total state contingent returns, say R', with dimension $n \times S$ is invertible or nonsingular. The system (4.16) in matrix form is just

$$1 = R' \psi$$

where 1 is a simple $n \times 1$ column matrix of ones. The equations in (4.16) can be appropriately evaluated for different assets and states of nature providing insight into the relationship between total returns, R_i, state prices, ψ^s, and initial price P_i.

How does it all relate to the market regime? The answer is in the definitions below.

Definition: A system of financial markets is said to be complete if and only if the number of ordinary securities in the market is greater than or equal to the number of states of nature, $n \geq S$, and the matrix of payoffs D (or equivalently of total returns R') has rank equal to S, the number of states of nature. In this case, the state prices $\psi^s > 0$ are uniquely determined.

Intuitively, a system of complete financial markets is consistent with the existence of a market for each and all goods produced and consumed in every possible state of the world, so that all commitments into the spectrum of possible states of nature are unlimited.

Mathematically, the rank of a matrix is the maximum number of linearly independent row (column) vectors. If the rank is equal to the number of states of nature, the payoffs are linearly independent, the matrix of payoffs is nonsingular thus invertible. Therefore, from the initial market prices one can invert the system (4.14) or (4.15) and uniquely determine the state prices $\psi^s>0$.

In effect, the restrictions imposed on the matrix of payoffs D, or equivalently on the matrix of total returns R', in a regime of complete markets are consistent with constructing elementary securities for all states of nature thus providing proper full insurance for all of them. This can be inferred from the method to find the elementary securities for the states of nature in (i)-(ii). The important issue is that there are many available ordinary securities in the market place; not the elementary securities themselves since a linear combination of the ordinary ones can replicate the possible elementary securities.

Equivalently, the criteria for a structure of complete markets can be stated in terms of the spanning characteristic of the set of traded securities. When the set of traded securities, d_i, span the set of state contingent claims e^s *for all* $s \in S'$, that is for the whole state space, then the market structure is complete.

The size of the state space is shown to be of significant importance for the valuation of securities and states. Practically, with complete markets, a unique set of prices can provide valuation to each and all states of nature, and those can be directly inferred from observed prices of ordinary securities in the market place.

When any of the specific conditions provided above fails, we have a different market structure.

Definition: A system of financial markets is said to be incomplete if and only if the number of ordinary securities is less than or equal to the number of states of nature, $n \leq S$, and the matrix of payoffs D, or equivalently of total returns R', has rank *strictly less* than S, the number of states of nature. In this case, the state prices $\psi^s>0$ are not uniquely determined.

Intuitively, a system of incomplete financial markets is so that traded assets for commitments into the spectrum of states of nature are severely limited.

Mathematically, even when $n=S$, if the rank is less than the number of states of nature, the payoffs are linearly dependent, the matrix of payoffs is singular and thus not invertible. In this case, given the market prices, one cannot invert the system (4.14) or (4.15), and one cannot uniquely determine the state prices $\psi^s>0$. Moreover, in a regime of incomplete markets one is unable to construct elementary securities for all states of nature, some states cannot be properly insured, i.e. some "markets" are missing. In terms of the spanning characteristic of the set of traded securities, when the set of traded securities, d_i, does not span the set of state contingent claims e^s *for all* $s\in S'$, the market structure is incomplete.

In the case when there are more securities than states of nature, say $n>S$, portfolios become duplicable and redundant. However, this case is useful for the general analysis of security derivatives in a dynamic context, e.g. Chapter 8.

4.6 Optimal Portfolio Choice and the Price of Elementary Securities (State Prices)

We can use the optimal portfolio choice problem in (4.4)-(4.5) and assign a known probability distribution $\pi^s>0$, with $\Sigma_{s\in S'}\pi^s = 1$, on the payoff d_i. The individual solves a portfolio problem

$$\underset{\{\theta_i\}_{i=1}^n}{Max}\quad \Sigma_{s\in S'}\pi^s\, u(\Sigma_{i=1}^n\,\theta_i\, d_i^{\,s}) \tag{4.17}$$

$$subject\ to\quad W_o - \Sigma_{i=1}^n\,\theta_i\,P_i = 0$$

given $\{W_o,P_i\}$ and the probability distribution of d_i:

$$\pi^s>0,\ with\ \Sigma_{s\in S'}\pi^s = 1.$$

The consumption-based capital asset pricing (CCAPM) formula becomes

$$\Sigma_{s\in S'}\pi^s\, u'(x^{*s})\, d_i^{\,s} = \lambda\, P_i \tag{4.18}$$

for each ordinary security $i=1,...n$ where $\lambda = u'(W_o) \geq 0$ is the marginal utility of initial wealth as before. There is an important result here: for ordinary securities $i=1,...n$, expression (4.18) implies that

$$[\Sigma_{s \in S'} \ \pi^s u'(x*^s) \ d_1^s / P_1] = [\Sigma_{s \in S'} \ \pi^s u'(x*^s) \ d_2^s / P_2] = ... = [\Sigma_{s \in S'} \ \pi^s u'(x*^s) \ d_n^s / P_n] \qquad (4.19)$$

known as the <u>Fundamental Theorem of Risk Bearing for Securities Markets</u>. At the risk bearing optimum, the individual adjusts quantities of each security, or portfolio, up to the point where the initial price of each security becomes proportional to the expected marginal utility derived from the contingent final payoff the security generates. In other words, the expected marginal utility per "money" is equalized across all securities held. This represents trade opportunities across the different securities in order to bear the optimum risk.

In essence, equation (4.19) describes the optimal risk bearing in a market with general ordinary securities. Under a regime of complete markets, we are able to do the same for state contingent claims securities as well. We have seen that, in a regime of complete markets, the state space can be priced uniquely. In this case, there must be an analogous to (4.19) for the state prices ψ^s. A very simple way to understand this analogy is to compare expressions (4.14) and (4.18), for the case of complete markets, obtaining

$$\psi^s = \pi^s u'(x*^s) / \lambda = \pi^s u'(x*^s) / u'(W_0), \quad \textit{for each state } s=1,2,...S. \qquad (4.20)$$

According to this formula the state price is just the marginal utility of the final random wealth in that state, priced at the marginal utility of initial wealth, weighted by the probability of occurrence of that state, i.e. the marginal rate of substitution in state s times the probability of that state. Furthermore, (4.20) implies that for each state of nature $s=1,...S$

$$[\pi^1 u'(x*^1) / \psi^1] = [\pi^2 u'(x*^2) / \psi^2] = ... = [\pi^S u'(x*^S) / \psi^S] \qquad (4.21)$$

known as the <u>Fundamental Theorem of Risk Bearing for State Contingent Claims</u> securities. At the risk bearing optimum, the individual adjusts quantities of each security (portfolio) up to the point where the marginal utility derived from the contingent final payoff priced per "state," adjusted by the probability of occurrence of that state, is equalized across all states of nature. In fact, it represents trade opportunities across different states of nature in order to bear the optimal risk across states.

Another way to derive relationship (4.21), under complete markets, is to allow the individual to trade the uniquely priced states of nature directly by solving the following problem

$$\underset{\{x^s\}_{s\in S}}{Max} \quad \Sigma_{s\in S'} \pi^s u(x^s) \tag{4.22}$$

$$subject\ to \qquad \Sigma_{s\in S'} \psi^s x_o^s - \Sigma_{s\in S'} \psi^s x^s = 0$$

where x_o^s is the given initial endowment (wealth) expressed in terms of state contingent payoff, and the probability distribution $\pi^s > 0$, with $\Sigma_{s\in S'} \pi^s = 1$, of x^s is the same as before, i.e. recall the definition of x in (4.1). More importantly, the choice is over x^s yielding a solution x^{*s} as

$$\psi^s = \pi^s u'(x^{*s}) / \lambda = \pi^s u'(x^{*s}) / u'(W_0), \quad \text{for each state } s=1,2,...S \tag{4.23}$$

where $\lambda \geq 0$ is as before the marginal utility of initial wealth. Formula (4.23) yields the analogous to (4.21), the Fundamental Theorem of Risk Bearing for State Contingent Claims securities. Trading across the states of nature with elementary securities has a clear intuitive interpretation. Suppose that for some state $s \in S'$,

$$\psi^s > \pi^s u'(x^{*s}) / \lambda = \pi^s u'(x^{*s}) / u'(W_0).$$

This means that the price of insurance of state s exceeds the marginal utility of the payoff in that state, adjusted by the probability of its occurrence. Insurance in this state is too expensive. The individual is going to underinsure and bear too much risk in state s. But, since there are complete markets, it must be the case that the individual is overinsured in some other state and there is an opportunity to trade risks across the states of nature and consequently increase expected utility. Hence, it is optimal to explore the trade opportunities offered by the market and spread the risks across states of nature giving the optimal amount of risk bearing.

This is in fact a no-arbitrage argument. Suppose there are two states of nature and examine the equivalence of (4.14) and (4.18) to obtain, for an arbitrary security 1:

$$P_1 = d_1^1 \psi^1 + d_1^2 \psi^2 = [\pi^1 u'(x^{*1}) d_1^1 + \pi^2 u'(x^{*2}) d_1^2]/\lambda = P_1$$

which may be rewritten as

$$d_1^1 \{ [\pi^1 u'(x^{*1})/\lambda] - \psi^1 \} + d_1^2 \{ [\pi^2 u'(x^{*2})/\lambda] - \psi^2 \} = 0.$$

Then, for arbitrary payoffs d_1^1 and d_1^2, the equality can only be satisfied when

$$\psi^s = \pi^s u'(x^{*s}) / \lambda = \pi^s u'(x^{*s}) / u'(W_0) \qquad \textit{for each state } s=1,2$$

otherwise one state is underinsured and the other overinsured. Trade over states of nature allows risk to be optimally spread across states.

It is also important to note that expressions (4.20) or (4.23) relate to the asset pricing formulas (4.8)-(4.8') and (4.10)-(4.10') of the consumption-based capital asset pricing model. For example, using the known probability distribution $\pi^s > 0$, with $\Sigma_{s \in S'} \pi^s = 1$, on the payoff d_i, implies that (4.8') may be written as

$$P_i = [\Sigma_{s \in S'} \pi^s u'(x^*) d_i^s] / u'(W_o), \qquad \textit{for each security } i$$

and using (4.20) or (4.23) yields

$$P_i = \Sigma_{s \in S'} \psi^s d_i^s, \qquad \textit{for each security } i$$

exactly as in (4.13) or (4.14). The initial price can be expressed as the linear sum of all possible payoffs valued by the respective state price, ψ^s, i.e. a market valuation of the security. The consumption-based capital asset pricing (CCAPM) formula is thus generally obtained under an asset market regime of complete markets. For a risk-free asset, $d_i^s = d_i$ all $s \in S'$ and from the last expressions we obtain

$$P_i / d_i = \Sigma_{s \in S'} \psi^s = 1/R$$

where we used expression (4.11') to relate to the risk-free return. Thus, the linear sum of the state prices is just the risk-free discount factor! (You should be able to explain why this is the case, otherwise consult section 4.11 below).

Finally, a common representation of asset pricing is in terms of so-called risk-neutral probabilities defined as

$$\pi^{s*} \equiv [\pi^s u'(x^{*s}) / \lambda] / \{ E[u'(x^{*s})] / \lambda \} = [\pi^s u'(x^{*s}) / \lambda] R = [\pi^s u'(x^{*s}) / u'(W_o)] R$$

where R is the risk-free gross return given, for example, in expression (4.10). The asset pricing formula from (4.8') then becomes

$$P_i = (1/R)\ [\ \Sigma_{s \in S'} \pi^{s*}\ d_i^{\ s}], \qquad \textit{for each security } i$$

reflecting the price as the discounted value of the payoff averaged by the artificial probabilities π^{s*}, just as if all market participants were risk neutral, i.e. discounted by the risk-free return. This is because the initial price, $P_i = (1/R)[\Sigma_{s \in S'}\ \pi^{s*}\ d_i^s]$, would be obtained under linear or risk neutral utility and probability function π^{s*} representing the actual, not artificial, probability function. We can also compare this pricing with the case of certainty, $P = d\ /R$ in (4.11'), and note the similarity. The state prices in this case become, using (4.23)

$$\psi^s = \pi^{s*}\ E\ [\ u'(x^{*\,s})\ /\ \lambda\ ] = \pi^{s*}\ E\ [\ u'(x^{*\,s})\ /\ u'(W_o)\ ], \qquad \textit{for each state } s=1,2,...S.$$

so that the state price is represented by the average marginal rate of substitution, $E[u'(x^{*s})/\ u'(W_o\ )]$, weighted by the risk neutral probability π^{s*}.

Risk-neutral probabilities can be computed either through the portfolio choice problems or directly from the pricing formula in the case of complete market structure examined here.

4.7 Individual Optimal Allocation under Complete Markets Regime: An Example

Consider the following *CRRA* (power) utility function:

$$u(x) \ = \ [x^{(1-\gamma)} - 1]\ /\ (1-\gamma), \qquad \gamma > 0. \tag{4.24}$$

For this function we obtain differentials

$$u'(x) \ = 1\ /\ x^{\gamma} > 0 \tag{4.25}$$

$$u''(x) \ = -\ \gamma\ x^{-(1+\gamma)} < 0 \tag{4.26}$$

i.e. strictly increasing and strictly concave. It implies a constant coefficient of relative risk aversion denoted by *CRRA(x)* given by

$$CRRA(x) \ = -\ u''(x)\ x\ /\ u'(x) = \gamma > 0. \tag{4.27}$$

In the case when $\gamma=1$, applying L'Hopital's rule yields the logarithmic function, $log\ x$, i.e. the logarithmic function has $CRRA=1$. As γ increases, the more curved the utility function becomes indicating higher levels of risk aversion. Figure 4.1 depicts the function for two values of γ. Note that x is allowed to be unboundedly positive without loss of generality. Also, for this utility function, we have

$$u'''(x) = (1+\gamma)\, x^{-2(1+\gamma)} > 0 \Rightarrow \mathcal{P}(x) = (1+\gamma)\, x^{-1} > 0$$

i.e. the third derivative and the measure of prudence are strictly positive, recall section 3.3. The utility index in (4.24) is from the general class of power utility functions.

Of course, risk aversion refers to a static, intratemporal concept, e.g. Chapter 3; but in an intertemporal framework, $1/\gamma$ also determines the elasticity of intertemporal substitution of consumption across periods as we shall see in Chapters 7 and 8.

Solving the optimal portfolio choice problem (4.4)-(4.5) or (4.17) with this utility function yields first order necessary conditions

$$\Sigma_{s\in S'}\, \pi^s\, [d_i^{\,s} / (\Sigma_{i=1}^{n}\, \theta_i\, d_i^{\,s})^{\gamma}] - \lambda P_i = 0 \quad \textit{for each security } i=1,2,...n \qquad (4.28)$$

$$W_o - \Sigma_{i=1}^{n}\, \theta_i P_i = 0. \qquad (4.29)$$

which yield solutions $\{\theta_i^*, \lambda^*\}_{i=1}^{n}$ as a function of $\{W_o, P_i, \pi^s, d_i^{\,s}, \gamma\}$. The optimal portfolio weights, denoted ω_i^*, can be computed as

$$\omega_i^* = \theta_i^*\, P_i / W_o \qquad (4.30)$$

for each security $i=1,...n$, and the state prices can be extracted from expression (4.14).

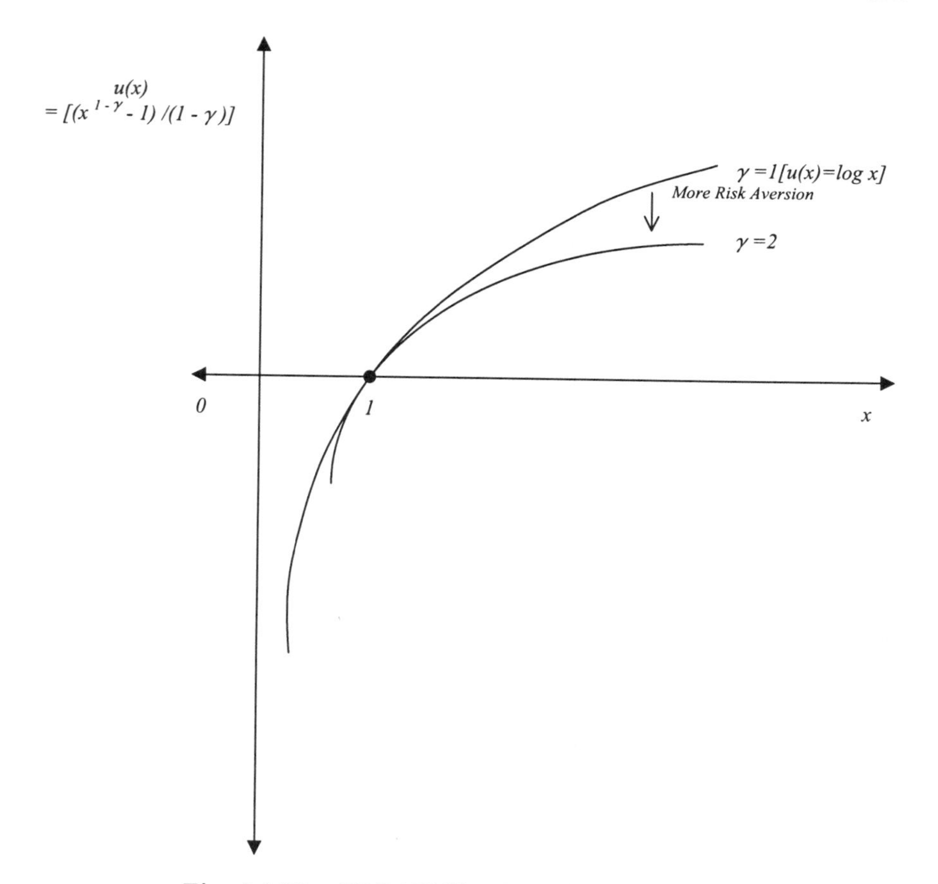

Fig. 4.1 The *CRRA* Utility Function (Isoelastic)

Consider a very simple numerical example. Let there be two states of nature, $s=1,2$ and two ordinary securities with linear independent payoffs, $i=1,2$. Let the base parameter space be

Base Parameter Space:	$\pi^1=1/3,\ \pi^2=2/3,$ $d_1^1=1,\ d_1^2=3,\ d_2^1=2,\ d_2^2=2,$ $P_1=1,\ P_2=1,$ $W_o=20,$ $\gamma=1.$

Thus, security *1* has expected payoff of *2.33* and variance of *0.89*, while security *2* is risk-free with expected payoff of *2*, and the individual has *CRRA=1* (logarithmic utility).

This implies that although security *1* is riskier it provides a favorable bet (a higher expected return). The resulting optimal allocation is obtained by solving the system of equations in (4.28)-(4.29) for θ_1, θ_2 and λ:

$$[\ \pi^1 d_1{}^1/(\ \theta_1 d_1{}^1 +\ \theta_2 d_2{}^1)\,] + [\ \pi^2 d_1{}^2/(\ \theta_1 d_1{}^2 +\ \theta_2 d_2{}^2)\,] - \lambda P_1 = 0,$$

$$[\ \pi^1 d_2{}^1/(\ \theta_1 d_1{}^1 +\ \theta_2 d_2{}^1)\,] + [\ \pi^2 d_2{}^2/(\ \theta_1 d_1{}^2 +\ \theta_2 d_2{}^2)\,] - \lambda P_2 = 0,$$

$$W_o - \theta_1 P_1 - \theta_2 P_2 = 0;$$

then the system of equations (4.30) for ω_1 and ω_2:

$$\omega_1{}^* = \theta_1{}^* P_1 / W_o,$$

$$\omega_2{}^* = \theta_2{}^* P_2 / W_o;$$

and the system of equations (4.14) for ψ^1 and ψ^2:

$$P_1 = d_1{}^1 \psi^1 + d_1{}^2 \psi^2$$

$$P_2 = d_2{}^1 \psi^1 + d_2{}^2 \psi^2;$$

yielding an optimal allocation

$\theta_1{}^* = 13.33$, $\theta_2{}^* = 6.67$, $\lambda^* = 0.05$, $\omega_1{}^* = 0.67$, $\omega_2{}^* = 0.33$, $\psi^1 = \psi^2 = 0.25$.

The mix is unbalanced towards security *1*, which has the higher expected return, but due to risk aversion the share of the risk-free security is nontrivial.

As a first thought experiment, consider from the base parameter space increasing the payoff of security *2* in state *1* to $d_2{}^{1\prime}=2.5$. This makes security *2* risky, the expected returns of the two become the same (*2.33*), but the variance of security *2* (*0.06*) remains lower than the variance of security *1*. The resulting optimal allocation becomes

$d_2{}^{1\prime}=2.5$: $\theta_1{}'^* = 8.89$, $\theta_2{}'^* = 11.11$, $\lambda^* = 0.05$, $\omega_1{}'^* = 0.44$, $\omega_2{}'^* = 0.56$, $\psi^{1\prime} = 0.18$, $\psi^{2\prime} = 0.27$.

The optimal allocation is now tilted towards security *2* that has the same expected payoff with lower variance. Given the *2×2* matrix of payoffs in this example, as the payoff in state *1* increases, the state price falls, whereas the price of state *2* increases. The reason is that in the *2×2* case, state prices are inversely related given a change in payoff. The marginal utility remains the same because neither initial wealth nor the shape of the utility function has changed.

Consider a second thought experiment from the base parameter set, as an increase in the coefficient of relative risk aversion to $\gamma'=2$, say an upward change in the aversion to risk as measured by the curvature of the utility function. This makes the individual more risk averse relative to the same original base parameter space. The resulting optimal allocation becomes

$\gamma'=2$: $\theta_1''^* = 8.04$, $\theta_2''^* = 11.96$, $\lambda^{*''} = 0.0012$, $\omega_1''^* = 0.40$, $\omega_2''^* = 0.60$, $\psi^1 = \psi^2 = 0.25$.

Now, even though security *2* has a lower expected return, the risk aversion is too compelling and there is a substantial shift from the risky to the risk-free security relative to the base optimal allocation. The lesson is that disposition towards risk in the preference set plays an important role in determining the optimal securities and risk allocations. The state prices are unchanged because there were no changes in payoffs for the base parameter space, but the marginal utility decreases because the more curved the utility function (more risk aversion) gives lower marginal utility at initial wealth.

4.8 General Equilibrium under Complete Markets Regime: Full Risk Sharing

Let there be many individuals named $j=1,\ldots J$ with possibly different initial endowments, W_{oj}. We assume a regime of complete markets, full symmetric information across individuals, i.e. π^s, P_i, and d_i^s known to all individuals, competitive markets where individuals take all prices and returns as given, and all individuals have the same preferences. It is well known that under these conditions the general equilibrium is Pareto Optimal, the First Fundamental Theorem of Welfare Economics applies. A planner solving a centralized system can replicate the competitive allocation by solving for the optimal prices and quantities, the Second Fundamental Theorem of Welfare Economics. We shall use this last result to formulate a planning problem as

$$\underset{\{\theta_{ij}\}_{i=1}^{n},_{j=1}^{J}}{Max} \ \Sigma_{j=1}^{J} \eta_j \{ \Sigma_{s\in S'} \pi^s u(\Sigma_{i=1}^{n} \theta_{ij} d_i^{\,s}) \} \qquad (4.31)$$

$$subject\ to \qquad \Sigma_{j=1}^{J} \mu_j [W_{oj} - \Sigma_{i=1}^{n} \theta_{ij} P_i] = 0$$

given $\{W_{oj}, P_i, \pi^s, d_i^{\,s}, \eta_j, \mu_j\}$, where η_j denotes the arbitrary welfare weight attached to each individual, so that $\{\eta_j \geq 0, \Sigma_{j=1}^{J} \eta_j = 1\}$; and μ_j denotes the relative weight attached to each individual in the budget constraint, so that $\{\mu_j \geq 0, \Sigma_{j=1}^{J} \mu_j = 1\}$. The objective function is just the weighted sum of each individual expected utility. The budget constraint is the weighted sum of each individual initial wealth minus the sum of each individual portfolio, which denotes zero excess aggregate initial demand for securities over aggregate initial wealth or the supply. Thus, the budget constraint indicates that individuals are potentially different due to different initial wealth (or endowment), but the wealth is all pooled together in a "world" market to be traded efficiently.

The Lagrangean function for this problem is

$$\mathcal{L} = \Sigma_{j=1}^{J} \eta_j \{ \Sigma_{s\in S'} \pi^s u(\Sigma_{i=1}^{n} \theta_{ij} d_i^{\,s}) \} + \phi\, \Sigma_{j=1}^{J} \mu_j [W_{oj} - \Sigma_{i=1}^{n} \theta_{ij} P_i] \qquad (4.32)$$

where $\phi \geq 0$ is the Lagrange multiplier associated with the aggregate budget constraint. Note that ϕ is the same multiplier for all individuals in the aggregate, given the pooling of resources. Assuming identical welfare and resources weights, or $\eta_j = \mu_j$, *all* $j=1,2,...,J$, the necessary first order conditions for this problem are

$$\partial\mathcal{L}/\partial\theta_{ij} = [\Sigma_{s\in S'} \pi^s u'(\Sigma_{i=1}^{n} \theta_{ij} d_i^{\,s}) \, d_i^{\,s}] - \phi P_i = 0,$$

$$for\ each\ i=1,...n\ and\ each\ j=1,...J \qquad (4.33)$$

$$\partial\mathcal{L}/\partial\phi = \Sigma_{j=1}^{J} \mu_j [W_{oj} - \Sigma_{i=1}^{n} \theta_{ij} P_i] = 0 \qquad (4.34)$$

representing $(n\times J)+1$ equations in the $(n\times J)+1$ unknowns $\{\theta_{ij}, \phi\}_{i=1}^{n},_{j=1}^{J}$.

Formula (4.33) implies that for each asset i, across agents j:

$$[\Sigma_{s\in S'} \pi^s u'(\Sigma_{i=1}{}^n \theta_{i1}\, d_i{}^s) d_i{}^s] = [\Sigma_{s\in S'} \pi^s u'(\Sigma_{i=1}{}^n \theta_{i2}\, d_i{}^s)\, d_i{}^s] = \ldots =$$

$$[\Sigma_{s\in S'} \pi^s u'(\Sigma_{i=1}{}^n \theta_{iJ}\, d_i{}^s) d_i^s]$$

and hence,

$$\theta_{ij} = \theta_i = \Sigma_{j=1}{}^J \theta_{ij} / J \quad \textit{for all } j=1,2,\ldots J. \qquad (4.35)$$

Under complete markets, all individuals have the exact same allocation of wealth regardless differences in their endowments. Individuals are allowed to use capital markets and share risks to the full unlimited extent.

What happens to state prices in this case? From equation (4.14), for each individual $j=1,\ldots J$ and asset $i=1,\ldots n$, we have

$$P_i = \Sigma_{s\in S'} d_i{}^s\, \psi^{sj}$$

so that all individuals face the same state prices, or

$$\psi^{sj} = \psi^s \quad \textit{for all } j=1,\ldots J.$$

The pooling of resources allows individuals to share market risks. For example, consider the numerical example in section 4.7 with two individuals named $j=A,B$. The base parameter space is modified considering that initially individual A is endowed only with security $i=1$, say $W_{oA}=20P_1$, and individual B is endowed only with security $i=2$ (the risk-free one), say $W_{oB}=10P_2$. This is to illustrate an extreme case where one individual is endowed with one security only, the results do not depend on this structure of endowments, but hold as long as the values of the endowments are initially different across individuals. All other parameters remain the same. Solving this problem gives the "world" optimal allocation as

Full Risk Sharing: $\theta_{1A}{}^* = 10,\quad \theta_{1B}{}^* = 10,\quad \theta_{2A}{}^* = 5,\quad \theta_{2B}{}^* = 5$

that yields the same weights for all individuals as in the base parameter space for a single individual: $\omega_1{}^*=0.67;\ \omega_2{}^*=0.33$. Initially individual A has a portfolio that is too risky and is willing to reduce the amount of risk; individual B carries too little risk so is willing to accept some additional risk at favorable odds which is the case in this example. Trade allows both to share the risks by pooling their resources in the complete markets regime, i.e. there is full risk sharing across individuals.

4.9 General Equilibrium under Incomplete Markets Regime

Let there be many individuals named $j=1,...J$ with possibly different initial endowments, W_{oj}. Assume as before full symmetric information across individuals, that is π^s, P_i, and d_i^s known to all individuals. Markets are competitive where individuals take all prices and returns as given, and all individuals have the same preferences. But now assume that the market structure is incomplete, say $n<S$, there are less securities than states of nature. The consequences are awesome.

First, trade across some states of nature is not possible, some markets are "missing", thus some states are uninsurable, the equilibrium is not Pareto Optimal, and the First and Second Fundamental Theorems of Welfare Economics fail.

Second, the state prices become indeterminate, the allocation will depend on prices that can be chosen arbitrarily.

Third, risk sharing opportunities across individuals are limited and not efficient.

For example, suppose there are two states of nature and only one security. Any individual budget constraint gives the feasible allocation:

$$\theta_j^* = W_{oj} / P_i \quad and \quad \omega_j^* = 1,$$

for $j=1,...J$, that is the initial wealth becomes the gamble itself. The set of feasible allocations is constrained to one point, the individual is not able to choose the optimal amount of risk to bear, and it has to bear it all. Hence, the equilibrium is not Pareto Optimal, and the opening of new markets may be Pareto improving; even though this may not always be the case, see Notes on the Literature. Also, it is not possible to price the states of nature uniquely, the current price of the only security is

$$P = d^1 \psi^1 + d^2 \psi^2, \qquad \psi^s > 0,$$

i.e. one equation in two unknowns, which cannot be inverted to obtain state prices uniquely. Moreover, one cannot add the budget constraints. Each individual will hold all the own initial wealth in the security:

$$\theta_A{*} = W_{oA} / P_i, \quad \theta_B{*} = W_{oB} / P_i, \ldots, \quad \theta_J{*} = W_{oJ} / P_i.$$

The are no risk sharing opportunities available. The problem of this example is illustrated in Figure 4.2: the individual cannot trade across states of nature, the market is "missing".

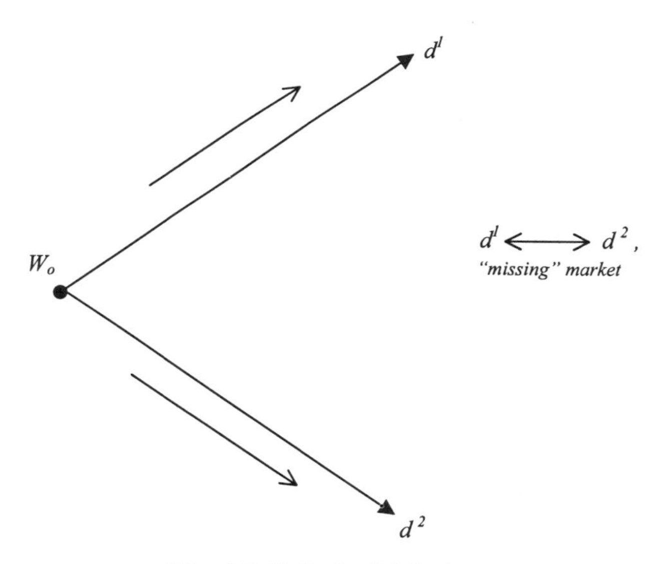

Fig. 4.2 "Missing" Markets

In particular, an incomplete markets regime has important consequences for the pricing of assets themselves. Using the same framework as in the previous section 4.8, under incomplete markets each individual named $j=1,...J$ solves the problem

$$\underset{\{\theta_{ij}\}_{i=1}^{n}}{Max} \ \Sigma_{s \in S'} \pi^s u(\Sigma_{i=1}^{n} \theta_{ij} d_i^{\,s}) \qquad (4.36)$$

$$subject\ to \qquad W_{oj} - \Sigma_{i=1}^{n} \theta_{ij} P^*_i = 0 \qquad each\ j=1,2,...J$$

given $\{W_{oj}, P_i, \pi^s, d_i^s\}$. The Lagrangean function for this problem, for individual j, is

$$\mathcal{L}^* = [\,\Sigma_{s\in S'} \pi^s u(\Sigma_{i=1}^n \theta_{ij} d_i^s)\,] + \phi_j [\,W_{oj} - \Sigma_{i=1}^n \theta_{ij} P^*_i\,], \quad \text{each } j=1,2,...J \tag{4.37}$$

where $\phi_j \geq 0$ is the Lagrange multiplier associated with the individual budget constraint. The necessary first order conditions for this problem are

$$\partial\mathcal{L}^* / \partial\theta_{ij} = [\Sigma_{s\in S'} \pi^s u'(\Sigma_{i=1}^n \theta_{ij} d_i^s) d_i^s] - \phi_j P^*_i = 0, \quad \text{for each } i=1,...n; \text{ each } j=1,2,...J \tag{4.38}$$

$$\partial\mathcal{L}^* / \partial\phi_j = W_{oj} - \Sigma_{i=1}^n \theta_{ij} P^*_i = 0 \qquad \text{each } j=1,2,...J \tag{4.39}$$

representing $n+1$ equations in the $n+1$ unknowns $\{\theta_{ij}, \phi_j\}_{i=1}^n$, each $j=1,2,...J$.

Adding up formula (4.38) across all agents $j=1,2,...J$ and dividing by J yields the price of security i, in the incomplete markets case, as

$$P^*_i = \{(1/J)\, \Sigma_{s\in S'} \pi^s [\Sigma_{j=1}^J u'(\Sigma_{i=1}^n \theta_{ij} d_i^s) d_i^s]\} / \phi, \quad \text{for each } i=1,...n \tag{4.40}$$

where $\phi \equiv \Sigma_{j=1}^J \phi_j / J$, is the marginal utility of initial wealth averaged across J, and thus identical to the marginal utility of initial aggregate wealth in the complete markets problem (4.31). We can compare the price under incomplete markets in (4.40) with the price under complete markets from (4.33) rewritten as

$$P_i = \quad [\Sigma_{s\in S'} \pi^s u'([1/J]\, \Sigma_{i=1}^n \Sigma_{j=1}^J \theta_{ij} d_i^s) d_i^s] / \phi, \quad \text{for each } i=1,...n \tag{4.41}$$

where we used the fact that under complete markets,

$$\theta_{ij} = \theta_i = \Sigma_{j=1}^J \theta_{ij} / J$$

as in (4.35).

When the marginal utility is convex, $u''' \geq 0$, and payoffs are nonnegative, $d_i^s \geq 0$, we can use Jensen's inequality to obtain

$$(1/J)\, \Sigma_{j=1}^J u'(\Sigma_{i=1}^n \theta_{ij} d_i^s) \geq u'([1/J]\, \Sigma_{i=1}^n \Sigma_{j=1}^J \theta_{ij} d_i^s) \quad \text{for all } s\in S' \tag{4.42}$$

implying that, from (4.40)-(4.41),

$$P^*_i \geq P_i, \quad \text{for each } i=1,...n \tag{4.43}$$

where the inequality is strict if we assume that the marginal utility is strictly convex, that is $u'''>0$. Notice that in section 3.3, we relate the third derivative of the utility function to the notion of prudence, and $u'''>0$, implies $\mathcal{P}\equiv -u'''/u''>0$. The result here is that under a regime of incomplete markets, when $u'''>0$, the price of securities is higher relative to the complete markets case and, consequently, expected returns are lower. Incomplete markets are thus a possible cause for lower expected returns in the basic consumption-based capital asset pricing model. This is a useful result for the discussion of excess returns in Chapter 7.

4.10 A Simple Geometrical Illustration of Market Regimes

We give here a simple graphical illustration of complete and incomplete market regimes.

(i) Complete Markets - Suppose there are two securities with two states of nature, possible final random wealth is $x=\{x^1, x^2\}$, and we solve problem (4.22). Figure 4.3 illustrates the equilibrium in $\{x^1, x^2\}$ space. The budget constraint is the negatively sloped linear relationship

$$\Sigma_{s\in S'} \psi^s x_o^s = \psi^1 x^1 + \psi^2 x^2 \qquad (4.44)$$

with slope $-\psi^1/\psi^2<0$. Points along this line are combinations of $\{x^1, x^2\}$ achieved by trading across the states of nature in the market. If short sales restrictions are imposed, the set of feasible trade along this line is restricted close to the corners. The 45° line is the certainty line where $x^1=x^2$, i.e. same total final wealth in all states. Along a given indifference curve, the marginal rate of substitution (MRS) is

$$MRS = dx^2/dx^1|_{E[U]\ const.} = -(\pi^1/\pi^2)\ [u'(x^1)/u'(x^2)] < 0. \qquad (4.45)$$

The equilibrium with complete markets is the solution given in equation (4.23) or simply where the MRS is equal to the slope of the budget constraint

$$\psi^1/\psi^2 = (\pi^1/\pi^2)\ [u'(x^1)/u'(x^2)] \qquad (4.46)$$

a result reminiscent from the Fundamental Theorem of Risk Bearing for State Contingent Claims. Expression (4.46) together with the budget

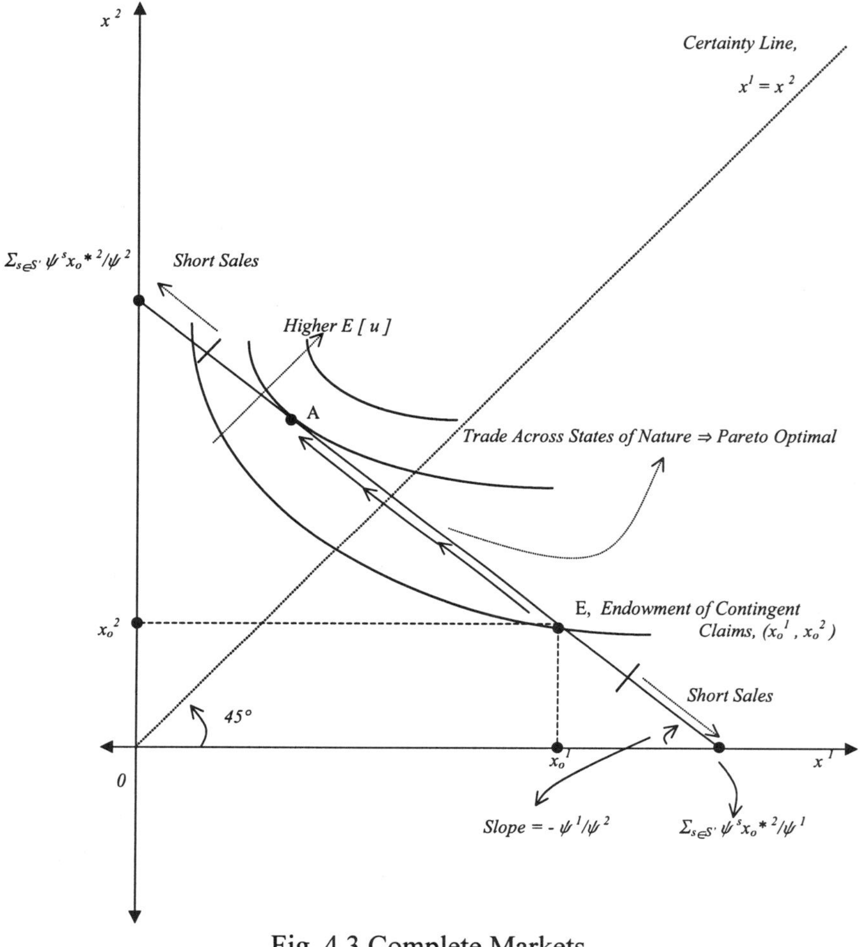

Fig. 4.3 Complete Markets

constraint (4.44), solve for the optimal state contingent wealth $\{x^{*1}, x^{*2}\}$, say point A in the figure. Thus, from the initial endowed state contingent claims in states *1,2*, at point E, trade across states of nature allocates risks efficiently to point A.

(ii) Incomplete Markets - Now suppose there is only one security with two states of nature and possible final random wealth $x=\{x^1, x^2\}$. In Figure 4.4, the initial single security endowed state contingent claims in this case are:

$$x^1 = W_o d^1 / P, \qquad x^2 = W_o d^2 / P$$

at point I. This is the set of feasible allocations, constrained to one point in $\{x^1, x^2\}$ space.

There is no feasible trade across states of nature. The state prices are undetermined because it is not possible to invert from the initial prices to find the state prices uniquely. As a consequence, at point I, any line with indeterminate $-\psi^1/\psi^2$ slopes passes through.

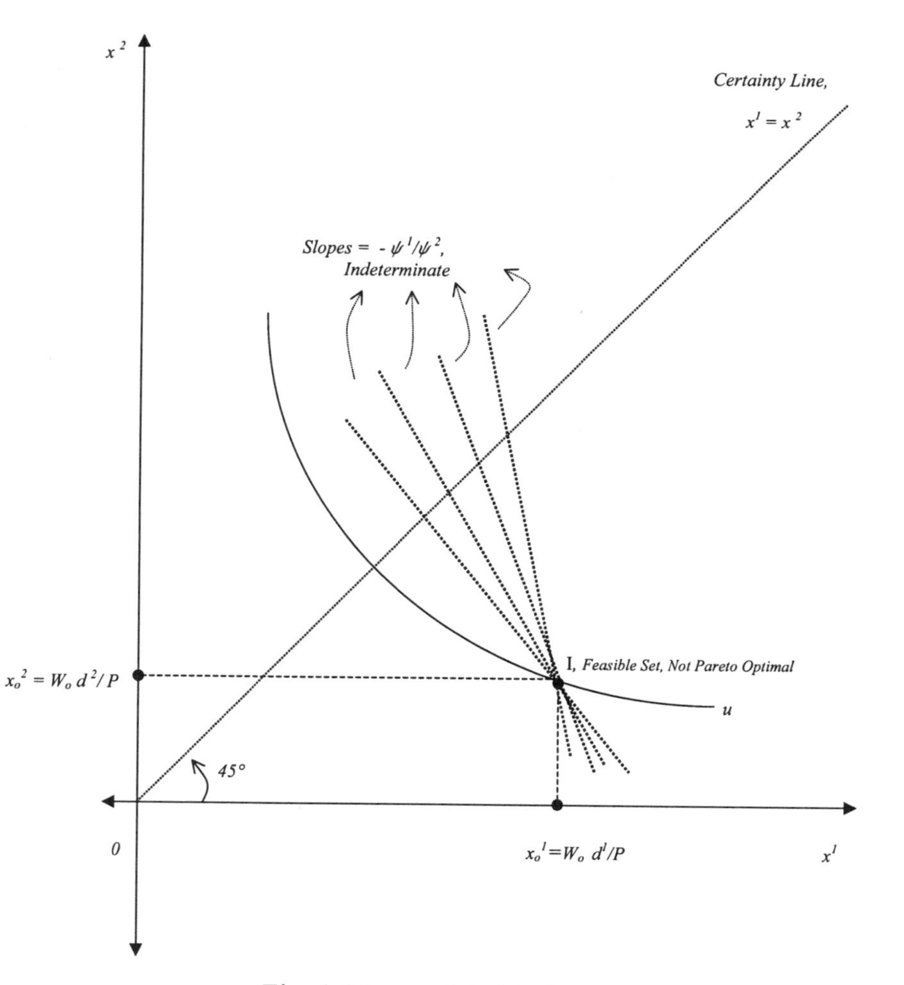

Fig. 4.4 Incomplete Markets

The indifference curve that passes through that point will not in general correspond to the most efficient one. The equilibrium is not Pareto Optimal. If the market across states of nature was not "missing", it would be possible to move along the determined budget constraint, to the optimal risk bearing bliss.

4.11 Application to the Neoclassical Theory of the Firm

We examine the neoclassical theory of the firm focussing on the production decision problem. Financial decisions of the firm are studied in the dynamic setting of Chapter 7.

The neoclassical theory of the firm emphasizes the technology of production. A single product firm, under conditions of certainty, has a typical production function:

$$y = f(x_1, x_2, x_3, \dots x_n) \tag{4.47}$$

where y is the output produced using the inputs $(x_1, x_2, x_3, \dots x_n)$ according to the function f that represents the state of technology. Usually, f is assumed to be nonnegative, strictly increasing and strictly concave in its arguments (diminishing marginal physical products), and constant returns to scale overall. Prices of inputs are taken as given and denoted $(r_1, r_2, r_3, \dots, r_n)$. For simplicity, assume that $n=2$, i.e. there two inputs. The total cost of producing is linear $r_1 x_1 + r_2 x_2$, and given an exogenous level of output, y_o, the firm solves the minimization problem, $Min\ (r_1 x_1 + r_2 x_2)$ by choice of inputs (x_1, x_2) subject to the production technology $y_o = f(x_1, x_2)$. Considering alternative levels of output y, the solutions generate a total cost curve $c(y)$ from which we can deduce an average cost curve, $c(y)/y$ and a marginal cost curve, $dc(y)/dy = c'(y)$.

Under competitive markets, firms maximize profits or income, $py - c(y)$, by choice of output y where p is the market price of output, taken as given. An equilibrium output is then determined by the usual marginal cost pricing formula, $p^* = c'(y^*)$. Under competitive markets and certainty, firms take prices as given and find it in their best interest to do so.

The introduction of risk in the production technology substantially changes the firm's problem. Suppose the single product firm has a stochastic technology

$$y^s = f(x_1, x_2, x_3, \dots x_n; s) \qquad (4.48)$$

for a well defined state space $s \in S'$, where production takes place at the beginning of the period. We let p^s denote the state contingent price of the single good produced when sold at the end of the period, i.e. the spot price. The inputs are contracted at the beginning of the period, before the state is realized, at the sure cost $(r_1, r_2, r_3, \dots r_n)$ for the respective input $(x_1, x_2, x_3, \dots x_n)$.

In each state s, the net (capitalized) income of the firm is given by

$$\Pi^s = p^s y^s - R\, \Sigma_{i=1}^n r_i x_i \qquad (4.49)$$

where $p^s y^s$ is the firm's revenue and $R\, \Sigma_{i=1}^n r_i x_i$ is the total deterministic linear cost of inputs capitalized by the risk-free gross interest rate, for example given in (4.10). Ownership of the firm indicates a claim to its proceeds given by the list of payoffs

$$\Pi = (\Pi^1, \Pi^2, \dots, \Pi^S).$$

Under a regime of complete markets, we can construct an equivalent portfolio of elementary securities that replicates the list as

$$\Pi = \Sigma_{s \in S'} \Pi^s e^s \qquad (4.50)$$

where e^s is the elementary security of state s, analogous to the case examined in (4.12). Under conditions of no-arbitrage, linear pricing implies that the market value (or price) of the firm is given by

$$\Sigma_{s \in S'} \psi^s \Pi^s = \Sigma_{s \in S'} \psi^s [p^s y^s - R\, \Sigma_{i=1}^n r_i x_i] \qquad (4.51)$$

where ψ^s is the price of state s as in (4.23). The market value of the firm in (4.51) represents the full set of market opportunities for the firm valued at the state prices, ψ^s. Under a regime of complete markets, shareholders of the firm agree to choose inputs in order to maximize the firm's market value given in (4.51), subject to the stochastic technology (4.48). The problem is

$$\underset{\{x_i\}_{i=1}^n}{Max} \ \Sigma_{s \in S'} \psi^s [p^s y^s - R\, \Sigma_{i=1}^n r_i x_i] \qquad (4.52)$$

$$subject\ to \quad y^s = f(x_1, x_2, x_3, \dots x_n; s)$$

given $\{r_i, p^s, \psi^s, R$ *and the probability distribution of* $s\}$. The necessary first order conditions for this problem are

$$\Sigma_{s \in S'} \psi^s [p^s f_i(\dots;s) - R\, r_i] = 0, \quad for\ each\ i=1,\dots n \qquad (4.53)$$

where $f_i\ (\dots;s) \equiv \partial f\ (x_i,\dots;s)/\partial\ x_i > 0$ is the marginal physical product of input i. The sufficient second order condition is satisfied by the strict concavity of the technology f. The set of first order conditions can be solved for the input demands as a function of r_i, p^s, ψ^s, R and the probability distribution of s. We can rewrite (4.53) as

$$\Sigma_{s \in S'} \psi^s p^s f_i(\dots;s) - r_i R\, \Sigma_{s \in S'} \psi^s = 0, \quad for\ each\ i=1,\dots n$$

and note that $R\ \Sigma_{s \in S'}\ \psi^s = 1$, because the sum of the state prices over the state space yields the expected marginal rate of substitution, say from (4.23)

$$\Sigma_{s \in S'} \psi^s = \Sigma_{s \in S'} \pi^s u'(x^*)/\lambda = E[u'(x^*)/\lambda]$$

and using (4.10) we obtain $R\ \Sigma_{s \in S'}\ \psi^s = 1$. This is not surprising since $1/R$ is the price of a sure gain of unity in every possible state, thus the linear sum of the prices of all individual states, e.g. section 4.6. The firm's first order condition becomes

$$\Sigma_{s \in S'} \psi^s p^s f_i(\dots;s) - r_i = 0 \quad for\ each\ i=1,\dots n \qquad (4.54)$$

which are the usual production efficiency condition equating marginal products to their respective rewards.

Under a regime of complete markets, state prices and spot output prices, ψ^s and p^s, respectively are identical for all market participants, and state prices are uniquely determined thus providing basis for market valuation of all feasible projects for the firm. In this case, firm shareholders agree unanimously on the market value maximization as an objective for the firm in (4.53) and it is in their best interest to do so.

Under a regime of incomplete markets, the simple and elegant framework of (4.52)-(4.54) is not feasible. As we examined in section 4.9, the state prices are undetermined under incomplete markets rendering the market value of the firm in (4.51) undetermined as well. There will not be unanimity and agreement among shareholders about the valuation of projects for firms. Given the indeterminacy of state prices, shareholders would disagree on how to adjust production because each would differ in their set of risky endowments. The risks to be insured for one individual would differ from risks to be insured for another individual generating disagreement on how to use the production choices available. In particular, under incomplete markets, the state prices ψ^{sj} will be different for each j. This leads to problems of disagreement among shareholders because they may perceive firm valuation differently.

Unless there is a mechanism for firms to have state prices announced to them, there is no general shareholder support for market value maximization. However, in a very special case, there can be agreement among shareholders even if the state prices ψ^{sj} are different for each j, i.e. incomplete markets. Suppose the technology in (4.48) has the special multiplicative form

$$y^s = A(s)\, f(x_1, x_2, x_3, \ldots x_n) \tag{4.55}$$

where $A(s)$ represents the stochastic total factor productivity of the inputs. The property of this technology is that the risk enters multiplicatively in the production process. Under technology (4.55), it is possible that the vector (in the case of many firms) $p^s \times A(s) \times f_i(\ldots)$ is spanned by the set of traded securities in the market place, e.g. sections 4.4-4.5. When this is the case, the product (or inner product in the case of many firms) $\psi^{sj} \times p^s \times A(s) \times f_i(\ldots)$ is identical for all individuals j, regardless that ψ^{sj} depends upon each j. The reason for this result is that under technology (4.55), the production function f changes the scale of the output produced, but not the risk profile. All risk is separate in the term A, and the firm's efficiency condition may be written as

$$\Sigma_{s\in S'} \psi^{sj} p^s A(s) f_i(\ldots) - r_i = 0 \quad \textit{for each } i=1,\ldots n, \textit{ independent of } j. \tag{4.56}$$

In effect, the special technology in (4.55) makes that firms facing incomplete markets do not affect the span of the set of traded securities available in the market place. This allows investors to agree upon value maximization and circumvent the incomplete markets problem. However, the extent to which we can assume that firms do not affect the span of traded securities is limited.

4.12 Summary

Systems of financial markets are crucial for the efficient allocation of risk across individuals and across states of nature. Market regimes can be complete or incomplete depending on the financial structure of the system relative to the possible (finite) states of nature. Complete markets deliver optimal trade opportunities for efficient risk bearing and the equilibrium is Pareto Optimal. Incomplete markets inhibit trade opportunities for efficient risk bearing, the equilibrium is not Pareto Optimal.

Questions and Problems

1. What is an Elementary security?

2. How is it that a portfolio of Elementary securities can be equivalent to an ordinary security?

3. Suppose you have a system with two ordinary securities, d_1 and d_2. Let there be two states of nature, $s=(1,2)$ with payoffs $d_1^1=x$, $d_1^2=y$, $d_2^1=g$, $d_2^2=h$. Can you construct portfolios of the ordinary securities that replicate the Elementary security for each state? How and under what conditions?

4. In a general system of financial markets, explain and characterize an asset structure of Complete Markets.

5. Under an asset structure of complete markets, explain the Fundamental Theorem of Risk Bearing.

6. What are the risk sharing properties of a system of financial markets under an asset structure of complete markets?

7. What are the consequences for a system of financial markets to be characterized by an asset structure of Incomplete Markets?

8. What are the risk sharing properties of a system of financial markets under an asset structure of incomplete markets?

9. "Even if the elementary security does not exist in some states of nature, one may (or may not) be able to construct portfolios with existing ordinary securities that replicate the elementary security in those states of nature." Comment.

10. What is the market valuation of a firm under complete markets for production choices?

Notes on the Literature

The linear pricing representation of systems of financial markets is well explained in Luenberger (1996), Chapter 9. An intermediate and lucid presentation is in LeRoy and Warner (2001), Chapter 2. A more deep mathematical formulation is in Duffie (1996), Chapter 1 and the reader is referred to his Notes, page 17; see also Duffie (1988), Chapter 1. A straightforward discussion at a basic level is in Hirshleifer and Riley (1992), Chapters 2 and 4.

The Modigliani-Miller theorem is in Miller and Modigliani (1958), the efficient markets hypothesis is in Fama (1991), and we discuss both in the dynamic setting of Chapter 7. The theorems of risk bearing and a discussion of asset markets structure can also be found in Hirshleifer and Riley (1992), Chapters 2 and 4.

Cochrane (2001) presents asset pricing in terms of marginal rates of substitution in great detail. The elementary security and basic approach in this chapter are due to Arrow (1965), and the articles and comments in Diamond and Rothschild (1989), Part II are most useful. The risk sharing foundations are found in Arrow (1965) and Debreu (1959), and a good application of the material is in Obstfeld and Rogoff (1995), Chapter 5.

The incomplete markets general equilibrium model is studied in Diamond (1967); and the Pareto improvement question under incomplete markets is considered in Hart (1975). Models of financial innovation under incomplete markets are studied in Allen and Gale (1988), and a brief survey is in Duffie (1992). Asset pricing implications of incomplete markets are surveyed in Duffie (1992) and Scheinkman (1989). The example presented in section 4.9 is from Duffie (1992). An advanced presentation of the theory of incomplete markets is in Geanakopolos (1990), and Magill and Quinzii (1996). Rios-Rull (1994) presents a quantitative assessment of the importance of market completeness.

Bid-Ask spreads and other transactions costs are common characteristics in equilibrium capital markets, a discussion is found in LeRoy and Warner (2001), Chapter 4 and Glosten and Milgrom (1985) present an analysis with information asymmetries.

The theory of the firm under alternative market regimes in section 4.11 is surveyed by Duffie (1992), and also Grinols (1987) who presents a nice application to the theories of international trade. The special case of spanning with the multiplicative technology in (4.55) is due to Diamond (1967).

References

Allen, Franklin and Douglas Gale (1988) "Optimal Security Design." *Review of Financial Studies*, 1, 229-263.

Arrow, Kenneth J. (1965) *Aspects of the Theory of Risk Bearing*. Yrjo Jahnssonin Saatio, Helsinki.

Cochrane, John (2001) *Asset Pricing*. Princeton University Press, Princeton, NJ.

Debreu, Gerard (1959) *Theory of Value*. Cowles Foundation, Yale University, New Haven, CT.

Diamond, Peter (1967) "The Role of the Stock Market in a General Equilibrium Model with Technological Uncertainty." *American Economic Review,* 57, 759-776.

Diamond, Peter and Michael Rothschild (1989) *Uncertainty in Economics: Readings and Exercises,* Second Edition. Academic Press, Boston, MA.

Duffie, Darrell (1988) *Security Markets: Stochastic Models*. Academic Press, Boston, MA.

Duffie, Darrell (1992) "The Nature of Incomplete Security Markets." In Laffont, J. J., Ed., *Advances in Economic Theory: Sixth World Congress*, Vol. II, 263-288. Cambridge University Press, Cambridge, UK.

Duffie, Darrell (1996) *Dynamic Asset Pricing Theory*, Second Edition. Princeton University Press, Princeton, NJ.

Fama, Eugene (1991) "Efficient Capital Markets II." *Journal of Finance*, 46, 1575-1618.

Geanakopolos, John (1990) "An Introduction to General Equilibrium with Incomplete Asset Markets". *Journal of Mathematical Economics*, 19, 1-38.

Glosten, Lawrence R. and Paul Milgrom (1985) "Bid, Ask and Transactions Costs in a Specialist Model with Heterogeneously Informed Agents." *Journal of Financial Economics,* 14, 71-100.

Grinols, Earl L. (1987) *Uncertainty and the Theory of International Trade.* Harwood Academic Publishers, London, UK.

Hart, Oliver (1975) "On the Optimality of Equilibrium when the Market Structure is Incomplete." *Journal of Economic Theory,* 11, 418-443.

Hirshleifer, John and John G. Riley (1992) *The Analytics of Uncertainty and Information.* Cambridge Surveys of Economic Literature, Cambridge University Press, Cambridge, UK.

LeRoy, Stephen and Andrew Warner (2001) *Principles of Financial Economics*. Cambridge University Press, Cambridge, UK.

Luenberger, David (1998) *Investment Science*. Oxford University Press, Oxford, UK.

Magill, Michael and Martine Quinzii (1996) *Theory of Incomplete Markets*. The MIT Press, Cambridge, MA.

Miller, Merton and Franco Modigliani (1958) "The Cost of Capital, Corporation Finance and the Theory of Investment." *American Economic Review*, 48-261-297.

Obstfeld, Maurice and Kenneth Rogoff (1995) *Foundations of International Macroeconomics*. The MIT Press, Cambridge, MA.

Rios-Rull, Jose V. (1994) "On the Quantitative Importance of Market Completeness." *Journal of Monetary Economics*, 34, 463-496.

Scheinkman, Jose A. (1989) "Market Incompleteness and the Equilibrium Valuation of Assets." In, Battacharya, S. and G. Constantinides, Eds. *Frontiers of Financial Research*, Vol. I, 45-51. Rowman and Littlefield, NJ.

5. Contracts, Contract Design, and Static Agency Relationships

5.1 Introduction to Bilateral Relationships and Contracts

In starting this chapter, we give an introduction to alternative theories of the firm. The treatment is from the point of view of the organizational structure and the relationship specific arrangements that may arise within the firm. We briefly discuss agency, transaction costs and property rights within the firm. Then, we discuss asymmetric information in several partial equilibrium problems of bilateral relationships including adverse selection, signaling and moral hazard.

5.2 Theories of the Firm: Agency, Transactions Costs and Property Rights

The neoclassical theory of the firm studied in Chapter 4 has several strengths, such as its focus on the technology of the firm given by

$$y = f(x_1, x_2, x_3, \ldots x_n; s). \tag{5.1}$$

However, it presents several weaknesses: (i) One is that it ignores incentive problems within the firm due to the absence of specific bilateral relationships. In particular, the technology is assumed to operate perfectly and everyone does as they are told; (ii) Another is that the theory ignores internal organization issues such as hierarchies, delegation, monitoring etc; (iii) And, the theory does not attack the problem of the expansion or contraction of the firm. According to the neoclassical theory, the firm has no reason to expand its boundaries.

Thus, it does not explain mergers and acquisitions, additional hiring of skilled versus unskilled workers, etc.

The introduction of agency relationships attacks the problem of incentives within the firm, weakness (i). Suppose we allow one input of the firm, for example x_2, have quality that is endogenous to the firm as opposed to the exogeneity assumption of the neoclassical theory. Imagine that another firm supplies x_2. Let the quality of x_2 be a function of the effort of the manager-owner of the other firm, denoted e, and a random component, ϵ. Hence, the quality of x_2 can be expressed as

$$q_{x2} = q\ (e, \epsilon) \qquad (5.2)$$

for a well-behaved function q, increasing in both arguments.

This simple modification introduces a bilateral relationship between the seller of the input and the original firm, now the buyer of the input. The two parties may write enforceable contracts on the value of q_{x2}. Suppose the buyer's utility from the interaction with the seller is u_B, defined as revenue minus unit cost: $u_B(q_{x2}, p_2^*) = r(q_{x2}) - p_2^*$; where $r(q_{x2})$ is a strictly increasing well behaved revenue function and p_2^* is the unit payment for $x_2=1$. The buyer is assumed to be risk neutral. The seller's utility from the interaction with the buyer is u_S, defined as revenue minus unit cost as well: $u_S\ (p_2^*, e)$; where u_S is a well behaved separable function, strictly increasing in p_2^*, and strictly decreasing in effort e. Hence, the seller is risk averse and dislikes effort.

Solutions for this simple bilateral problem have several important consequences for the theory of the firm. First, optimal or full risk sharing is consistent with the buyer bearing all the risk and the seller bearing no risk, because the former is risk neutral and the latter risk averse. Second, there are key issues of observability and verifiability of the effort by the buyer. If the buyer has full information and can fully observe and verify the effort of the seller, a plausible contract is to offer the following: "Buyer agrees to pay a fixed lump sum amount p_2^* for an arbitrary effort e^*." The fixed payment provides full risk sharing, i.e. the buyer bears all the risk associated with the random component ϵ. If the buyer cannot observe and verify effort, the contract above is infeasible because the seller does not like effort and will choose $e=0$. On the other hand, the buyer can observe the quality of the product it receives, q_{x2}. Another contract involves a variable payment schedule as a function of the quality, $p_2^* = p_2^*(q_{x2})$, i.e. a so-called incentive scheme. In this case, the incentive scheme involves a tradeoff between risk sharing and incentive: let the risk averse seller bear some, but not all of the risk associated with the random component ϵ.

Incentive schemes are powerful mechanisms that fully deal with weakness (i) of the neoclassical theory of the firm. However it has little to say about (ii) and nothing to say about (iii). Towards an attack on (ii) and (iii), consider the transactions cost approach.

The motto of the transactions costs approach is that "writing a good contract is itself costly." In effect, writing about events or describing events is very different than observing events. In the agency theory above, a contract $p_2^* = p_2^*(q_{x2})$ includes all possible states of nature and there is no reason to renegotiate the terms. A contract that includes all possible states of nature is said to be comprehensive or complete. However, in reality it is not simple and easy to write and describe all possible states of nature in a contract. In fact, contracts are written in a simplistic manner with some basic standard features. Contracts can be renegotiated all the time and intent usually becomes a crucial variable in case of enforcement problems. All indicates that, in reality, contracts may not include all the possible states of nature and contingencies. Such limited contracts are called incomplete contracts, and indicate the impossibility to write or describe all possible states of nature precisely in a cost-effective manner. In particular, writing a complete contract becomes prohibitively expensive. Written contracts are thus incomplete due to transaction costs. For example, a contract between the buyer and the seller may not specify what happens in the event there is an unusual catastrophe that destroys the seller's plant, and so on.

The economic implications of incomplete contracts are awesome. First, ambiguity may arise in the contract leading to disputes involving courts, lawyers, etc, thus further increasing costs. Second, it encourages ex-ante relationship specific investments of the sort: "make an investment, then if value is created, continue relationship, if value is not created, split relationship." Third, when unwritten or indescribable events occur, the firm may use the option of expanding its boundaries. For example, if an unusual catastrophe destroys the seller's plant, the buyer may want to expand its own plant to produce the needed input. Thus, transaction costs and the consequent incomplete contracts give a sharp explanation for the expansion of the boundaries of the firm, a clear improvement upon the neoclassical framework.

However, the weakness of the transaction cost theory is that it does not explain what changes after the boundaries expand. In the example above, if the buyer decides to buy the seller due to some unforeseen event, what is it that changes? According to the property rights approach, ownership is a source of power when there are transaction costs and incomplete contracts. When unforeseen events happen, ownership determines control over the usage or management of the emerging asset, thus providing a source of potential gain for its owners.

5.3 **Summary I**

We presented above a heuristic introduction to specific relationships among parties that deviate from the atomistic assumptions found in the traditional neoclassical theory of the firm. In the neoclassical world, market relationships are anonymous. However, bilateral trading relationships, contracts among parties, etc., provide other important dimensions to the problem of allocation of resources. Under bilateral relationships, only small deviations in terms of the information structure can generate differences in outcomes that are worth studying in much more detail. We move towards this direction in the next sections below.

5.4 **Introduction to Adverse Selection**

Consider cases where the Pareto Optimal complete markets allocation fails due to asymmetric (private) information among the parties involved. In the next few sections, the focus is on the case of Adverse Selection: private information takes the form of hidden knowledge of one's own characteristic. The private information is about who the individual is, as opposed to hidden actions influencing outcomes, which would be moral hazard; to be seen later. This private information has an effect on the utilities of all parties involved.

In general, the strategies examined in adverse selection models are summarized by the relationship between two parties, say A and B. For example, party A may be uninformed about some characteristic of party B whereas party B is informed about her own characteristic. The strategic interaction is that the uninformed party A moves first in an attempt to try to induce the informed party B to follow and reveal the private information. The mechanism used in this interaction is a simple contract offer and a response.

In this and the next sections, we focus on asymmetric information problems in a partial equilibrium framework, in the sense that no aggregate resource constraint is imposed. In Chapter 6, we consider asymmetric information in a general equilibrium framework.

5.5 The Principal-Agent Relationship

A more general framework for the study of asymmetric information in the context of the two party interactions mentioned above is called a principal-agent relationship. In this case, the principal, say party A, is uninformed about the characteristic of party B, the agent. The principal moves first by offering a contract to the agent. The agent examines the contract and follows a "take it or leave it" strategy. In this sense, a principal-agent relationship is analogous to a Stackelberg game: the leader proposes a contract, the follower will take it or leave it.

The rationale for relationships of the principal-agent type hinges on the problem of lack of Pareto optimality. In the absence of Pareto optimality, one can obtain a set of constrained Pareto optimal allocations by maximizing the expected utility of one party, holding the expected utility of all other parties constant. Hence, the principal-agent framework presents itself as a simple enough bargaining structure for an introduction to the study of asymmetric information.

5.6 A Simple Example in Insurance Markets

The insurance industry is one of the main areas of the economy where problems of adverse selection (and moral hazard) emerge. Here, we focus on the adverse selection problem. In the context of the principal-agent relationship, let the principal be the insurance company and the agent be the consumer willing to demand insurance. Let there be two possible states of nature: state 1 denotes the occurrence of a loss, say an accident occurs; state 2 denotes no loss. Consider now the agent and the principal separately.

The agent - A typical consumer is assumed to be risk averse with total wealth denoted $W > 0$. The agent is willing to accept (demand) an insurance policy (contract) for a loss of total wealth W. The source of asymmetric information is that there are two types of consumers, and this is private information. Consumer type 1 has a high probability of loss, denoted $\pi^1 > 0$. Consumer type 2 has a low probability of loss, denoted π^2 with ordering

$$0 < \pi^2 < \pi^1.$$

Hence, for consumer type 1 the probability of state 1, the loss, is high (the probability of state 2, no loss, is low) whereas for type 2 the probability of state 1, the loss, is low (the probability of state 2, no loss, is high). Due to risk aversion, both consumers are willing to demand full coverage of their wealth W_i, $i=1,2$ at a fair price. In order to understand the decision-making of the agent, we shall examine the contract offered by the insurance company, the principal.

The Principal - A typical insurance company is risk neutral. In this example, companies will be operating in competitive markets, thus will have zero expected profits. The contract will consist of an infinite supply of insurance sold at a premium rate per unit of insurance. The insurance company cannot observe the type of individual for whom it is selling an insurance policy that is the agent's private information. However, companies can make a general assessment of the proportion of individuals type 1 in the population, denoted by $\theta > 0$. By risk neutrality, the insurance company pools all individuals with an assessment of $\pi^i \in (0,1)$, $i=1,2$, for the specific types in the population, as

$$\theta_p = \theta \pi^1 + (1 - \theta)\, \pi^2 \qquad (5.3)$$

yielding an assessment of the average risk of the population, θ_p, as a linear combination of the specific unobserved types, as assessed by the insurance company measure π^i. In turn, the state-contingent profits ("utility") of the insurance company for full coverage W will be:

$$\text{state 1, loss} \Rightarrow y^1 = - W + \rho W \qquad (5.4)$$

$$\text{state 2, no loss} \Rightarrow y^2 = \rho W \qquad (5.5)$$

where $\rho > 0$ is the premium rate per unit of insurance to be chosen by the principal. In state 1, the profit is just the payoff for the incurred loss $-W$ plus the premium received ρW. In state 2, the profit is just the premium received, ρW.

The insurance company contract offer will be some premium rate, $\rho*$, so that expected profits are zero under competitive markets. Thus, $\rho*$ will solve

$$E[Profits] = \theta_p y^1 + (1 - \theta_p)\, y^2 = 0 \qquad (5.6)$$

given y^1 and y^2 in (5.4)-(5.5) above. The solution is

$$\rho* = \theta_p \qquad (5.7)$$

and the contract offer is a premium rate equal to the average assessed probability of loss in the pool of individuals of different types. The solution in (5.7) is the actuarially fair price for insurance. If the probability of loss is higher (lower), the premium should be more (less) expensive. The contract is indeed designed for the average risk in the population. At this price for insurance of state 1, the company will sell insurance in any amount for any individual willing to pay.

The Equilibrium - The offered contract, $\rho*$ in (5.7), and the assessment in (5.3), imply that the equilibrium is one where

$$\pi^1 > \rho* = \theta_p = \theta \pi^1 + (1 - \theta)\, \pi^2 > \pi^2 \qquad (5.8)$$

This is an important condition. First, examine the inequality on the left-hand side. It says that individuals with a high probability of loss, π^1, face an insurance price, $\rho* < \pi^1$, which is less than the price they are willing to pay based on their private information. For type 1, the price is too low and all of them will buy full insurance. Now, on the right-hand side, individuals with low probability of loss, π^2, face an insurance price, $\rho^* > \pi^2$, which is more than the price they are willing to pay based on their private information. For type 2, the price is too high and they will either buy partial insurance or no insurance at all.

The equilibrium is one where the contract offered by the principal adversely selects those types who represent a high risk in the population. The asymmetric information between principal and agent leads the former to design a contract based on the average risk of the population. Without any other incentive mechanism, the implication is clear: the price of insurance in state 1 is unbalanced relative to the willingness to insure state 1 by both types, there is a violation of the Fundamental Theorem of Risk Bearing (recall Chapter 4) and the equilibrium is not Pareto Optimal.

Under full information, the principal would be fully informed. Then, it could perfectly segment the market by offering different contracts for different types. Type 1 would be offered insurance at a higher price and choose to fully insure at $\rho^{*1} = \pi^1$. Type 2 would be offered insurance at a lower price and choose to fully insure as well at $\rho^{*2} = \pi^2$. This would be first-degree price discrimination and first best for the principal.

The asymmetric information examined above is embedded in a more general problem of contract or mechanism design. It is a common problem in many industries, for example the health insurance industry is a case in point. The principal must design an incentive mechanism that induces individuals to reveal their types truthfully. We consider this optimal contract design in another example below.

5.7 Mechanism Design: A Problem of Price Discrimination

Consider a discrete case problem of first and second-degree price discrimination, so-called vertical differentiation. In this problem, there is a market for some commodity, say sophisticated "cheese" in a locality. The principal-agent relationship is understood as the principal being the producer/seller of the commodity and the agent the consumer willing to buy the commodity.

The Agent - The consumer is assumed to be risk neutral and is willing to buy a fixed quantity, say one unit, of the commodity (for example, a piece of "cheese"). The consumer derives utility from a measure of the quality of the commodity weighted by her own taste for quality, which is her own private information. The consumer also derives disutility from the price it has to pay for the unit of the commodity. The utility function of the risk neutral agent is

$$U(\theta, q, t) = \theta q - t \tag{5.9}$$

where $\theta > 0$ is the individual taste for the "good" quality of the commodity which is the individuals private information and will be allowed to differ across individuals, $q>0$ is the quality of one unit of the commodity consumed, and $t>0$ is the price paid for one unit of the commodity. Hence, θ gives the individual's expertise in recognizing and understanding the quality of the commodity, and the product $\theta \times q$ gives the total enjoyment from consuming one unit of quality q.

This utility function gives a constant marginal rate of substitution between price and quality as

$$MRS = dt/dq|_{U\ const.} = \theta > 0 \tag{5.10}$$

for each given taste $\theta > 0$. The linear indifference set of a consumer, in $\{t,q\}$ space is shown in Figure 5.1 with higher utility denoting lines out to the southeast.

The source of asymmetric information is the individual's expertise in recognizing the quality of the commodity, which is their private information. Let there be two types of individuals (agents) with different levels of expertise (different characteristics) about the quality of the commodity: type 1 is a non-sophisticated type, who does not know much about the quality of the commodity, with taste parameter denoted by θ_1; type 2 is a sophisticated type who knows more about the quality of the commodity, with taste parameter denoted $\theta_2 > \theta_1$. The indifference set (5.9)-(5.10) implies that for all $\theta_2 > \theta_1$, given t or q,

$$\partial\ [U(\theta_2, q, t) - U(\theta_1, q, t)]/\partial\ q > 0. \tag{5.11}$$

Intuitively, as the quality of the commodity increases, the individual with the higher taste derives more utility from quality than the individual with lower taste. In other words, the sophisticated consumer is willing to

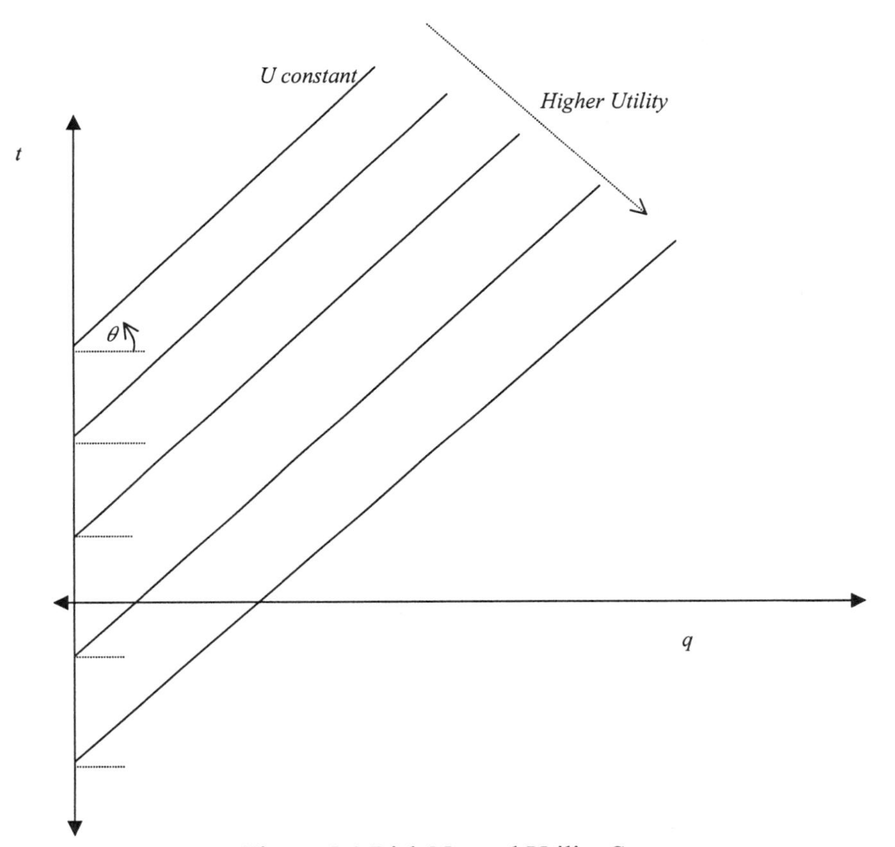

Figure 5.1 Risk Neutral Utility Set

pay more for quality than the non-sophisticated consumer. Figure 5.2 illustrates this property in *{t,q}* space. Notice that this property ultimately allows the principal to segment the market on quality!

The Principal - The producer/seller has a monopoly on the local market for the commodity (this assumption is not crucial, but simplifies the problem) and is also risk neutral. Since the expertise of agents is private information, the principal cannot observe directly the type of

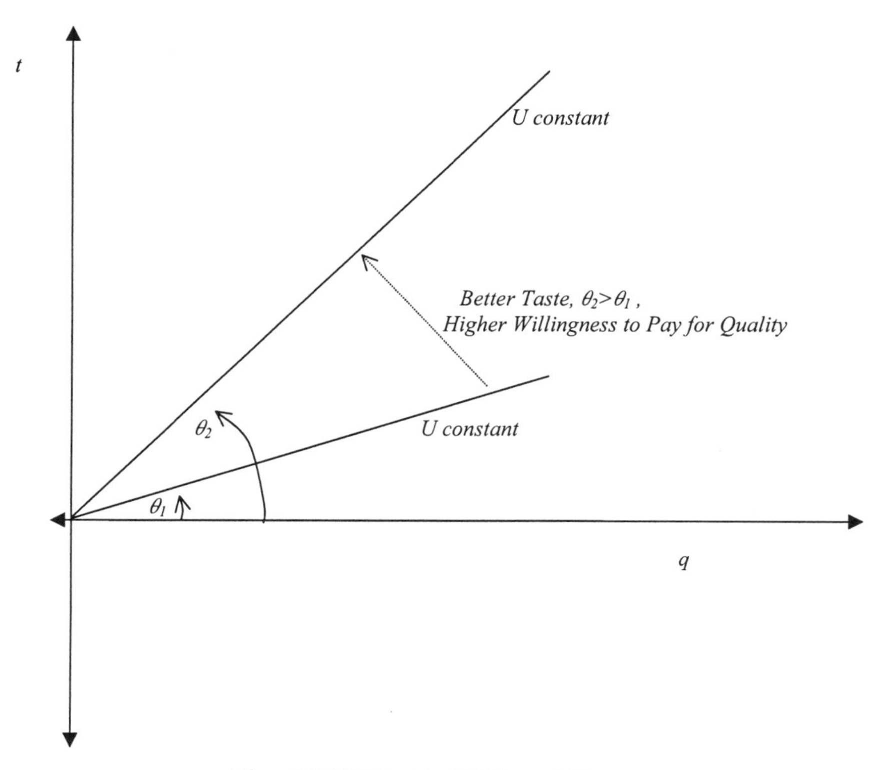

Fig. 5.2 Utility for Different Types

individual willing to buy the commodity. It does have a prior belief or assessment of the proportion of type 1 (the non-sophisticated) in the population, denoted by $\pi > 0$. The production technology consists of the ability to produce one unit of the commodity of quality q at a cost function denoted $c(q)$, which is strictly increasing and convex,

$$c'(q)>0, \qquad c''(q)>0.$$

The profit ("utility") function of the principal is just

$$PROFIT(t,q) = t - c(q) \tag{5.12}$$

and the iso-profit set is given by

$$dt/dq|_{PROFIT\ const.} = c' > 0, \quad d^2t/dq^2|_{PROFIT\ const.} = c'' > 0 \tag{5.13}$$

depicted in Figure 5.3 with curves to the northwest denoting higher profits.

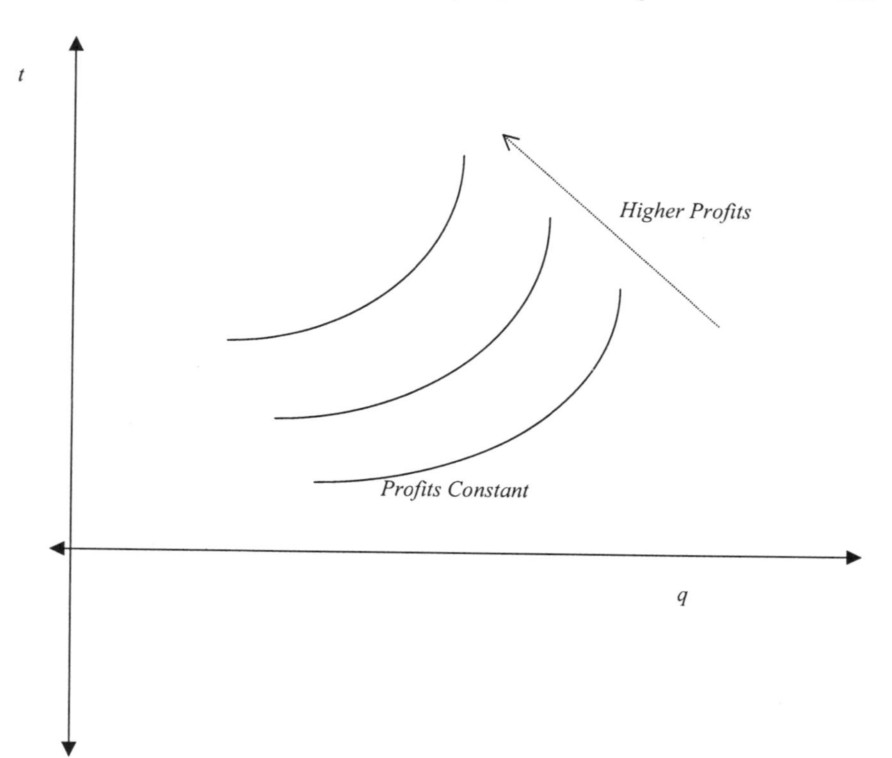

Fig. 5.3 Isoprofit Set

There are several ways to characterize the principal-agent relationship depending on the existence or not of private information and depending on the contract design offered by the principal. In the following, we consider three cases with three alternative contracts: (i) Symmetric (full) information; (ii) Asymmetric information; (iii) Optimal contract design.

(i) Symmetric Information - In this case, the principal is able to perfectly observe types θ_i, $i=1,2$, and there is no private information. The study of this case is useful as a benchmark for the other two cases. It yields first degree price discrimination and is the first best for the principal. The principal moves first in order to design a contract offer to the agent.

The contract is a pair $\{q,t\}$ obtained by maximizing profits subject to a constraint that takes into account the utility of the agent. We write this problem as

$$\underset{\{q_i, t_i\}_{i=1}^{2}}{Max} \quad PROFIT(t_i, q_i) = t_i - c(q_i), \qquad for\ i=1,\ 2 \qquad (5.14)$$

$$subject\ to \qquad U(\theta_i,\ q_i,\ t_i) = \theta_i q_i - t_i \geq 0, \qquad for\ i=1,2$$

given θ_i for $i=1,2$. The constraints

$$U(\theta_i, q_i, t_i) = \theta_i q_i - t_i \geq 0$$

are called the participation or individual rationality constraints. The individual rationality constraint states that the principal gives an incentive to the agent in its contract offer of at least non-negative utility. Of course, in the case of symmetric information, this constraint is binding since the principal has no reason to share any of its own surplus with the agent. A solution for this problem involves writing the Lagrangean function

$$\mathcal{L} = [t_i - c(q_i)] + \lambda\,(\theta_i q_i - t_i) \qquad (5.15)$$

where $\lambda \geq 0$ is the Lagrange multiplier associated with the participation constraint and may be thought of as the marginal benefit of "quality." The first order necessary conditions for an interior equilibrium are given by

$$\partial\mathcal{L}/\partial q_i = -\,c'(q_i) + \lambda\,\theta_i = 0, \quad for\ i=1,2 \qquad (5.16)$$

$$\partial\mathcal{L}/\partial t_i = 1 - \lambda = 0, \quad for\ i=1,2 \qquad (5.17)$$

$$\partial\mathcal{L}/\partial\lambda = \theta_i q_i - t_i = 0, \quad for\ i=1,2. \qquad (5.18)$$

These are five independent equations in the five unknowns $\{q_i^*, t_i^*, \lambda^*\}_{i=1}^{2}$ given $\{\theta_i\}_{i=1}^{2}$. The second order sufficient conditions are satisfied by the convexity of the iso-profit function and the linearity of the constraint. The first best solutions for the principal are given by

$$c'(q_i^*) = \theta_i, for\ i=1,2 \qquad (5.19)$$

$$\lambda^* = 1 \qquad (5.20)$$

$$\theta_i q_i^* = t_i^*, \quad for\ i=1,2. \qquad (5.21)$$

The principal offers a pair of contracts to each type separately as

$$\{q_i^*(\theta_i),\ t_i^*(\theta_i)\},\ for\ i=1,2;\ q_2^*(\theta_2) > q_1^*(\theta_1)\ and\ t_2^*(\theta_2) > t_1^*(\theta_1) \qquad (5.22)$$

where θ_i is fully observed by the principal. The equilibrium is presented in Figure 5.4. The principal, taking advantage of full information, is able to extract the entire surplus from each type by perfectly discriminating the price and quality. For type 1, the non-sophisticated, the principal only offers $\{q_1*(\theta_1), t_1*(\theta_1)\}$, low quality and low price, and the agent accepts it without any other choice or any notice that there are possibly other contracts available, denoted point A in Figure 5.4. For type 2, the sophisticated, the principal only offers $\{q_2*(\theta_2), t_2*(\theta_2)\}$, high quality and high price, and the agent accepts it without any other choice or any notice that there are possibly other contracts available, denoted point B in Figure 5.4.

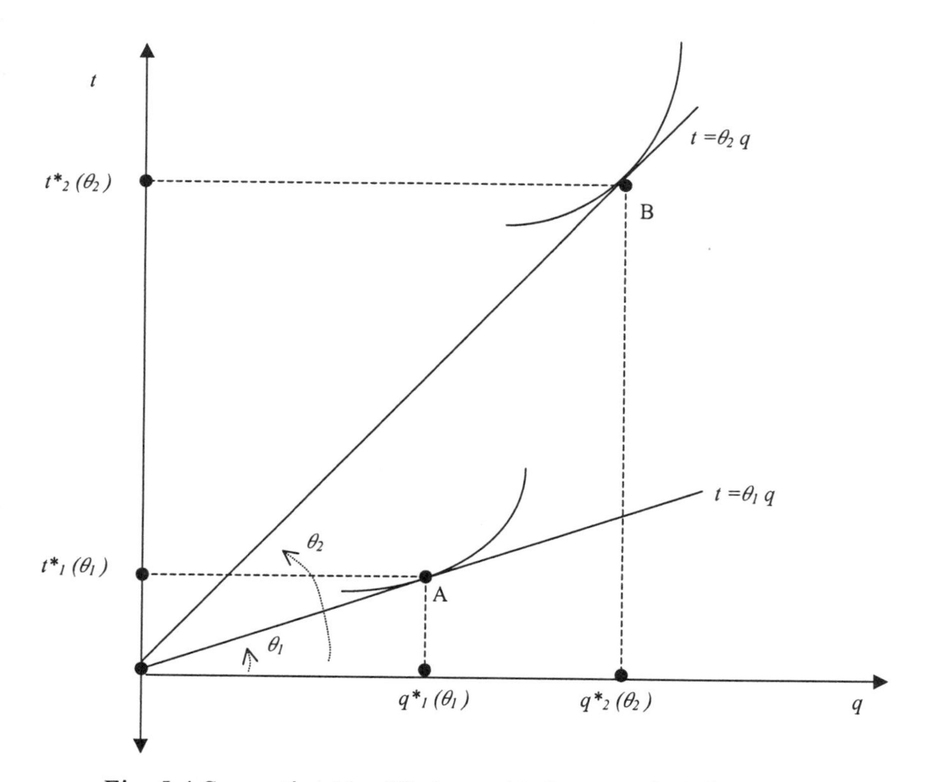

Fig. 5.4 Separating Equilibrium with Symmetric Information

This is first degree price discrimination and is the first best for the principal. Of course, this full discrimination practice is forbidden by rule of law: one cannot offer two consumers different deals without choice!

(ii) Asymmetric Information - In this case, the principal cannot observe types perfectly. It makes an assessment of the proportion of type 1 in the population, denoted $\pi>0$. The θ_i 's for $i=1,2$ are agent's private information. The principal has a menu of contracts to offer, and one possibility would be to offer the first best contract, seen in (i) above, to all regardless of the type, i.e. make both contracts available simultaneously. What would be the equilibrium in the case the principal offers $\{q_i*(\theta_i), t_i*(\theta_i)\}$ from (5.22)?

The answer is obtained by trying out each contract offered for the two types of agents and determine who is going to accept what. First, consider the choice of agent type 1, the non-sophisticated. Trying out the contract designed for type 1 yields utility

$$U(\theta_1, q_1*, t_1*) = \theta_1 q_1* - t_1* = 0 \tag{5.23}$$

which is the principal's first best. Now, we let agent type 1 try out the contract offered to the sophisticated type 2, $\{q_2*(\theta_2), t_2*(\theta_2)\}$, yielding utility

$$U(\theta_1, q_2*, t_2*) = \theta_1 q_2* - t_2* = (\theta_1 - \theta_2) q_2* < 0 \tag{5.24}$$

where the second equality is obtained from the first best solution given by $t_2*=\theta_2 q_2*$ in (5.21), and the inequality comes from the definition of the specific types. Hence, the non-sophisticated agent is better off taking the contract designed for her, and will do so in these circumstances. Second, consider the choice of agent type 2, the sophisticated one. Trying out the contract designed for type 2 yields utility

$$U(\theta_2, q_2*, t_2*) = \theta_2 q_2* - t_2* = 0 \tag{5.25}$$

which is the principal's first best. Now, we let agent type 2 try out the contract offered to the non-sophisticated type 1, $\{q_1*(\theta_1), t_1*(\theta_1)\}$, yielding utility

$$U(\theta_2, q_1*, t_1*) = \theta_2 q_1* - t_1* = (\theta_2 - \theta_1) q_1* > 0 \tag{5.26}$$

where the second equality is obtained from the first best solution given by $t_1*=\theta_1 q_1*$ in (5.21), and the inequality from the definition of the types.

Surprisingly, the sophisticated agent is better off taking the contract designed for the non-sophisticated type 1, and will do so in these circumstances.

The equilibrium with private information and first best contract offered by the principal is one where the sophisticated type misrepresents and pretends to be type 1, a classic adverse selection problem. Both types choose the contract $\{q_1^*, t_1^*\}$ and only low quality-low price commodity is sold. The first best contract adversely selects type 2 as type 1 because of private information. It does not give type 2 any incentive to reveal her true characteristic. By pretending to be type 1, the sophisticated agent is able to extract surplus from the principal in the amount $(\theta_2 - \theta_1)\, q_1^* > 0$ given in (5.26). This situation illustrates the case where uncertainty about the quality of a commodity may impede the functioning of a specific market, in this case it closes down the market for the high quality commodity.

Figure 5.5 illustrates the equilibrium with private information and adverse selection. By moving from the first best point B to A, the sophisticated type 2 pretends to be type 1 and can move to a higher indifference line where B'>B and B'~A, hence A>B (by transitivity), and type 2 extracts the surplus from the principal who moves to a lower iso-profit curve. Agent type 2 takes full advantage of her own private information.

This equilibrium outcome should make you think whether there is a contract that the principal could offer that would give the right incentives for both types to reveal themselves truthfully. The answer is yes, and, in fact, there are many contracts that do exactly that, but only one is optimal.

(iii) The Optimal Contract - The principal has to design a contract that is optimal in the sense of making individuals reveal their types truthfully, the so-called mechanism design. Since the specific types are private information, the principal uses its own assessment of the proportion of type 1 in the population, $\pi > 0$, to compute its own expected profits. At the same time, the principal should give the right incentives for each type

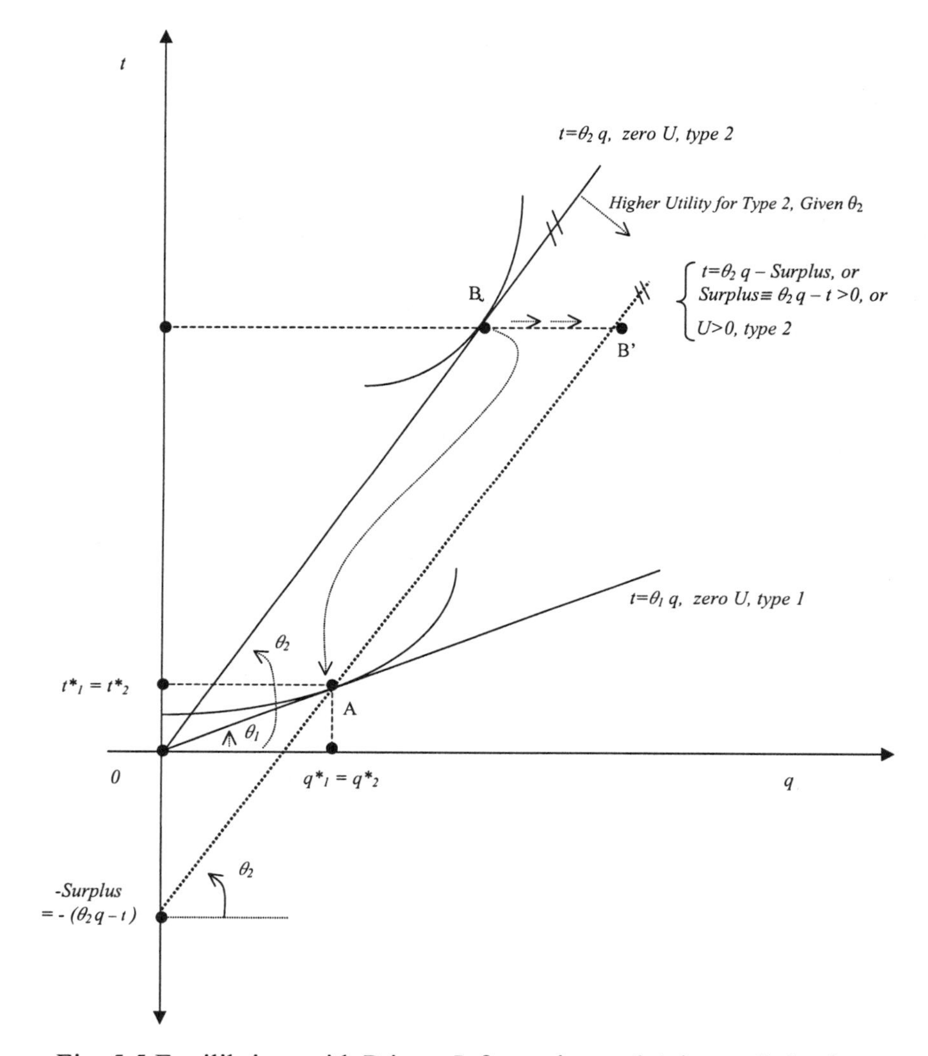

Fig. 5.5 Equilibrium with Private Information and Adverse Selection

to accept the contract that is designed for each. We shall write down the optimal contract problem and then discuss it in detail. The principal will maximize expected profits subject to a set of constraints or

$$\underset{\{q_i,t_i\}_{i=1}^{2}}{Max} \quad \pi[t_1 - c(q_1)] + (1-\pi)[t_2 - c(q_2)] \tag{5.27}$$

$$\text{subject to} \quad U(\theta_i, q_i, t_i) = \theta_i q_i - t_i \geq 0, \quad \text{for } i=1,2$$

$$\theta_1 q_1 - t_1 \geq \theta_1 q_2 - t_2$$

$$\theta_2 q_2 - t_2 \geq \theta_2 q_1 - t_1$$

given $\pi > 0$, and the unobserved θ_i for $i=1,2$. As in problem (5.14), the constraints

$$U(\theta_i, q_i, t_i) = \theta_i q_i - t_i \geq 0$$

are called the participation or individual rationality constraints. In addition, there are the constraints

$$\theta_1 q_1 - t_1 \geq \theta_1 q_2 - t_2 \quad and \quad \theta_2 q_2 - t_2 \geq \theta_2 q_1 - t_1$$

which are called incentive compatibility constraints.

Consider the first incentive compatibility constraint $\theta_1 q_1 - t_1 \geq \theta_1 q_2 - t_2$: it states that the utility of agent type 1 derived from the contract $\{q_1, t_1\}$ designed by the principal for agent type 1, has to be at least equal, if not more than the utility the agent type 1 would obtain if she tried out the contract with offer $\{q_2, t_2\}$ designed by the principal for agent type 2. This gives the right incentive for agent type 1 to take the contract that was designed for her. The other incentive compatibility constraint for agent type 2, given by $\theta_2 q_2 - t_2 \geq \theta_2 q_1 - t_1$, states the same logic. The utility of agent type 2 derived from the contract $\{q_2, t_2\}$, designed by the principal for agent type 2, has to be at least equal, if not more than the utility the agent type 2 would obtain if she tried out the contract $\{q_1, t_1\}$ designed by the principal for agent type 1, giving the right incentive for agent type 2 to take the contract that was designed for her.

This problem has to be attacked using the Kuhn-Tucker theory, e.g. Chapter 1. We form the Lagrangean function

$$\mathcal{L} = \pi[t_1 - c(q_1)] + (1-\pi)[t_2 - c(q_2)] + \gamma_1[\theta_1 q_1 - t_1] + \gamma_2[\theta_2 q_2 - t_2]$$
$$+ \lambda_1[\theta_1 q_1 - t_1 - \theta_1 q_2 + t_2] + \lambda_2[\theta_2 q_2 - t_2 - \theta_2 q_1 - t_1] \tag{5.28}$$

where $\{\gamma_i \geq 0, \lambda_i \geq 0\}_{i=1}^2$ are non-negative Lagrange multipliers attached to the participation and incentive compatibility constraints. The set of first order necessary conditions for $\{q_i, t_i, \gamma_i, \lambda_i\}_{i=1}^2$ is given by

$$\partial\mathcal{L}/\partial q_1 = -\pi c'(q_1) + \gamma_1\theta_1 + \lambda_1\theta_1 - \lambda_2\theta_2 \leq 0, \quad if < 0,\ q_1 = 0 \quad (5.29)$$

$$\partial\mathcal{L}/\partial q_2 = -(1-\pi)\, c'(q_2) + \gamma_2\theta_2 - \lambda_1\theta_1 + \lambda_2\theta_2 \leq 0, \quad if < 0,\ q_2 = 0 \quad (5.30)$$

$$\partial\mathcal{L}/\partial t_1 = \pi - \gamma_1 - \lambda_1 + \lambda_2 \leq 0, \quad if < 0,\ t_1 = 0 \quad (5.31)$$

$$\partial\mathcal{L}/\partial t_2 = (1-\pi) - \gamma_2 + \lambda_1 - \lambda_2 \leq 0, \quad if < 0,\ t_2 = 0 \quad (5.32)$$

$$\partial\mathcal{L}/\partial\gamma_1 = \theta_1 q_1 - t_1 \geq 0, \quad if > 0,\ \gamma_1 = 0 \quad (5.33)$$

$$\partial\mathcal{L}/\partial\gamma_2 = \theta_2 q_2 - t_2 \geq 0, \quad if > 0,\ \gamma_2 = 0 \quad (5.34)$$

$$\partial\mathcal{L}/\partial\lambda_1 = \theta_1 q_1 - t_1 - \theta_1 q_2 + t_2 \geq 0, \quad if > 0,\ \lambda_1 = 0 \quad (5.35)$$

$$\partial\mathcal{L}/\partial\lambda_2 = \theta_2 q_2 - t_2 - \theta_2 q_1 + t_1 \geq 0, \quad if > 0,\ \lambda_2 = 0 \quad (5.36)$$

which give eight equations in the eight unknowns $\{q_i, t_i, \gamma_i, \lambda_i\}_{i=1}^2$ given $\{\theta_i, \pi\}_{i=1}^2$. The second order sufficient conditions are taken care of by the increasing and strictly convex iso-profit curves and the linear constraints.

The conjectured optimal contract offered by the principal that yields the right incentive for the agent to reveal its type truthfully has the following four properties:

(a) $\gamma_1 > 0$ thus $\theta_1 q_1 - t_1 = 0$, the non-sophisticated type 1 obtains zero utility;

(b) $\gamma_2 = 0$ thus $\theta_2 q_2 - t_2 > 0$, the sophisticated type 2 obtains some positive utility, to be determined below;

(c) $\lambda_1 = 0$ thus $\theta_1 q_1 - t_1 - \theta_1 q_2 + t_2 > 0$, the non-sophisticated type 1 obtains the incentive to choose the contract that has been designed for her;

(d) $\lambda_2 > 0$ thus $\theta_2 q_2 - t_2 - \theta_2 q_1 + t_1 = 0$, the sophisticated type 2 is indifferent between the two contracts offered, since already has the incentive to take the contract designed for her from condition (b).

In the case of properties (a)-(d), the first order necessary conditions (5.29)-(5.36) become

$$\partial\mathcal{L}/\partial q_1 = -\pi c'(q_1) + \gamma_1\theta_1 - \lambda_2\theta_2 = 0, \quad (5.37)$$

$$\partial \mathcal{L} / \partial q_2 = - (1 - \pi)\, c'(q_2) + \gamma_2\, \theta_2 = 0, \qquad (5.38)$$

$$\partial \mathcal{L} / \partial t_1 = \pi - \gamma_1 + \lambda_2 = 0, \qquad (5.39)$$

$$\partial \mathcal{L} / \partial t_2 = (1 - \pi) - \lambda_2 = 0, \qquad (5.40)$$

$$\partial \mathcal{L} / \partial \gamma_1 = \theta_1\, q_1 - t_1 = 0, \qquad (5.41)$$

$$\partial \mathcal{L} / \partial \lambda_2 = \theta_2\, q_2 - t_2 - \theta_2\, q_1 + t_1 = 0 \qquad (5.42)$$

which give six equations in the six unknowns $\{q_i, t_i, \gamma_1, \lambda_2\}_{i=1}^{2}$ given $\{\theta_i, \pi\}_{i=1}^{2}$, and the solution is completed by $\gamma_2 = \lambda_1 = 0$. The solution for the conjectured optimal contract (*OC*) in equations (5.37)-(5.42) is recursive.

First, by (5.40)

$$\lambda_2^{OC} = (1 - \pi) \qquad (5.43)$$

where the superscript denotes optimal contract. Next, by (5.39) and (5.43) obtain

$$\gamma_1^{OC} = 1. \qquad (5.44)$$

Then using (5.43)-(5.44), (5.37) becomes

$$- \pi\, c'(q_1) + \theta_1 - (1 - \pi)\, \theta_2 = 0,$$

rearranging and adding and subtracting $(1 - \pi)\, \theta_1$, gives a solution for q_1^{OC} as

$$c'(q_1^{OC}) = \theta_1 - [(1 - \pi)/\pi]\, (\theta_2 - \theta_1) \qquad (5.45)$$

and since the right-hand side is smaller in magnitude than the first best solution θ_1, from (5.19) it implies that

$$q_1^{OC} < q_1^* \qquad (5.46)$$

i.e. the optimum quality offered to the non-sophisticated type 1 is less than the quality offered in the first best contract.

Continuing with the recursive solution, by (5.38), including the solutions already obtained, yields

$$c'(q_2^{OC}) = \theta_2 \;\Rightarrow\; q_2^{OC} = q_2^* \qquad (5.47)$$

and the sophisticated type 2 is offered the quality consistent with the first best principal's solution. Condition (5.41) then yields

$$t_1^{OC} = \theta_1\, q_1^{OC} \qquad (5.48)$$

which by (5.46) implies that

$$t_1^{OC} < t_1^* \tag{5.49}$$

and the price offered for the non-sophisticated type 1 is also lower than the first best. Finally, by (5.42), and the solutions already obtained, we get

$$t_2^{OC} = \theta_2 q_2^{OC} - q_1^{OC} (\theta_2 - \theta_1) < t_2^* = \theta_2 q_2^* \tag{5.50}$$

and the price offered for the sophisticated type 2 is less than the first best price.

To sum, the design of the optimal contract offer is the following: $\{q_1^{OC}, t_1^{OC}\}$, $\{q_2^{OC}, t_2^{OC}\}$; and relative to the first best:

$$q_1^{OC} < q_1^*, \qquad t_1^{OC} < t_1^*;$$

$$q_2^{OC} = q_2^*, \qquad t_2^{OC} < t_2^*.$$

Does this mechanism induce the two types to reveal truthfully? The answer is yes, and hence it is the optimal contract. The strategy is to make the contract to the non-sophisticated type 1, $\{q_1^{OC}, t_1^{OC}\}$, consistent with very low price and quality. This induces the sophisticated type 2 to reveal its type and buy the high price and high quality offer, $\{q_2^{OC}, t_2^{OC}\}$. Type 1 is better off taking the contract designed for her because if she were to take the contract designed for the sophisticated type 2, she would get negative utility as condition (c) shows. Type 2 has no incentive to misrepresent because it is getting a surplus by choosing the contract designed for her.

The equilibrium is illustrated in Figure 5.6 where points A^{OC} and B^{OC} denote the equilibrium of both agents under the optimal contract. The indifference line of type 2 passes through point A^{OC} due to condition (d). The surplus obtained by the sophisticated type 2,

$$q_1^{OC} (\theta_2 - \theta_1) > 0,$$

is some rent derived from its own characteristic, which is private information. The principal is willing to give up a small amount of profits in order to offer the right incentive for type 2 to reveal truthfully.

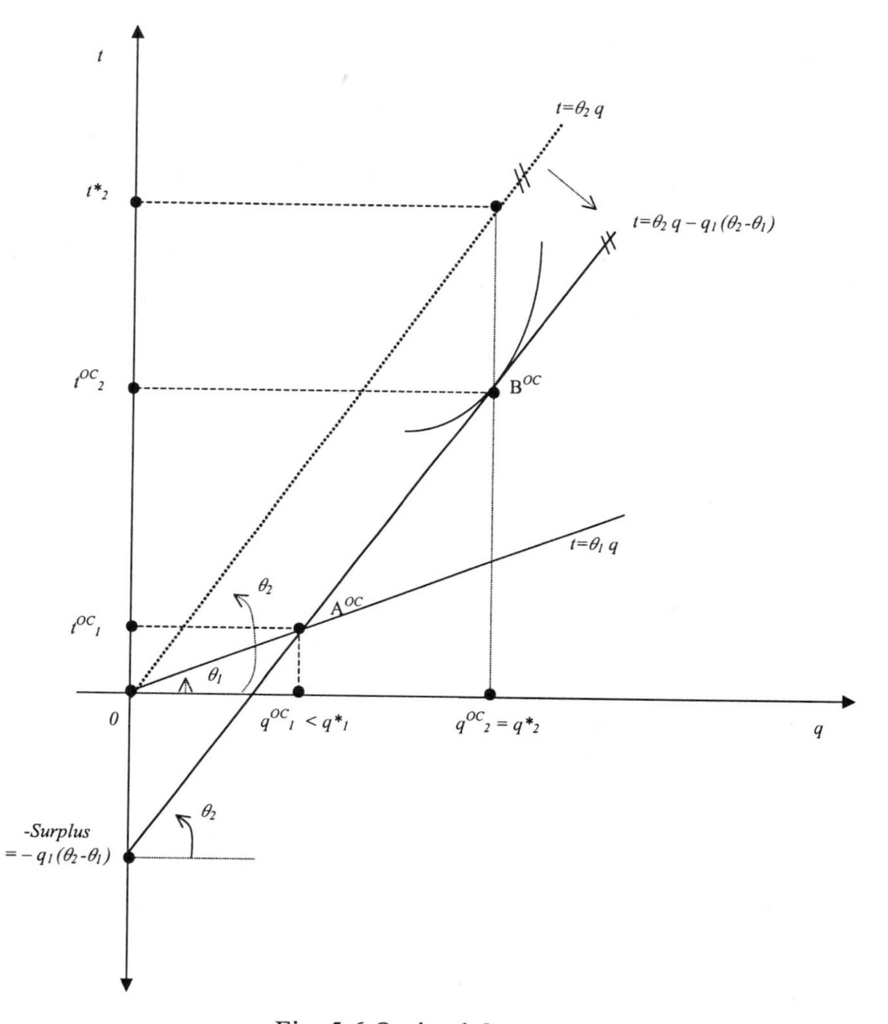

Fig. 5.6 Optimal Contract

Thus, a precise contract design can mitigate the adverse selection problem. The optimal contract induces both agents to reveal their private information truthfully, of course at a price! In effect, the principal can separate both types with simultaneous contract offers opened to all individuals, and for this reason the equilibrium obtained is sometimes called a separating equilibrium.

5.8 Adverse Selection in Credit Markets and the Possibility of Credit Rationing

The analysis in this specific area is not a pure principal-agent relationship where the principal provides incentives for the agent to reveal its type. Here, it describes a potential situation that emerges in credit markets due to the adverse selection induced by private information, and goes beyond in drawing some potential consequences for the behavior of lenders in credit markets.

The Borrowers (or Agents) - There is a continuum of potential investors, say on the real line between zero and one, all assumed to be risk neutral. Each has a prospect of a risky project that requires an initial loan of size $D>0$, taken to be indivisible. There is a finite number of such projects indexed by $i=1,2,...$, but all of them, regardless of their risk prospect, yield an identical expected total return, denoted $R>0$, which includes principal plus interest on the investment.

Therefore, the lender is able to screen projects on their average (expected) return, but not on their specific risk characteristic. This point will be made clear below. Even though all projects yield the same expected total return, they may differ in their risk prospects. We assume there are two states of the world regarding the risk prospects of projects:
(i) the good state (success) yields, as an outcome of the investment, a total return R_i^G, for each project $i=1,2,...$ with probability $p_i>0$;
(ii) the bad state (not success) yields, as an outcome of the investment, a total return R_i^N (assumed to be zero without loss of generality, $R_i^N=0$), for each project $i=1,2,...$ with probability $(1 - p_i)>0$.

Since both have the same expected return, this is given by

$$R = p_i \ R_i^G. \tag{5.51}$$

Each agent knows its own probability of success $p_i>0$, but this is private information because only the agent knows the risk and prospect of its own project.

Agents demand a loan of size $D>0$ in the form of a <u>Standard Debt Contract</u>: "borrow $D>0$ and pay back the state contingent amounts, $(1+r)D>0$ in the good state, or *zero* in the bad state." Here, $r>0$ is the interest rate on the loan charged by the financial institution, the lender or principal in this case. The bad state is just bankruptcy with no collateral required, but this assumption is not crucial to the results obtained.

It is assumed that the investment in the good state is a favorable bet, or

$$R_i^G > (1 + r)\, D > 0. \tag{5.52}$$

Without any collateral, the agent's expected profit ("utility"), P, from project i is given by

$$E[P,\ agent\ i] = p_i\ \ [R_i^G - (1 + r)\, D] = R - p_i(1 + r)D \tag{5.53}$$

where the second equality uses expression (5.51). Hence, the agent is assumed to be risk neutral with linear utility. Since the agent has an available alternative of not entering the standard debt contract and getting zero profits, the participation or individual rationality constraint is given by

$$E[P,\ agent\ i] = R - p_i(1 + r)D \geq 0 \tag{5.54}$$

i.e. the agent will participate as long as she obtains non-negative expected profits.

It is easy to see that the constraint (5.54), when it holds with equality, determines a value for the interest rate on the loan, r, which is critical for the agent's decision to enter the contract or not. The thresholds are as follows: for any interest rate charged by the principal that satisfies

$$r\ \leq (R - p_i) / D\, p_i,$$

the agent is willing to enter the contract since in this case $E[P, agent\ i] \geq 0$; whereas for interest rates

$$r > (R - p_i) / D\, p_i,$$

the agent is not willing to enter the contract since, in this case, the expected value $E[P,\ agent\ i] < 0$. Hence, for each interest rate r, there is an associated proportion of individuals in the population that will accept the standard debt contract and enter the relationship. We denote this proportion as a well defined function of the interest rate r:

$$P(r) > 0.$$

More importantly, this function is strictly decreasing in r,

$$dP(r) / dr < 0.$$

The intuition is simple: the higher the interest rate, the marginal project becomes non-profitable and the fewer individuals are willing to enter the contract.

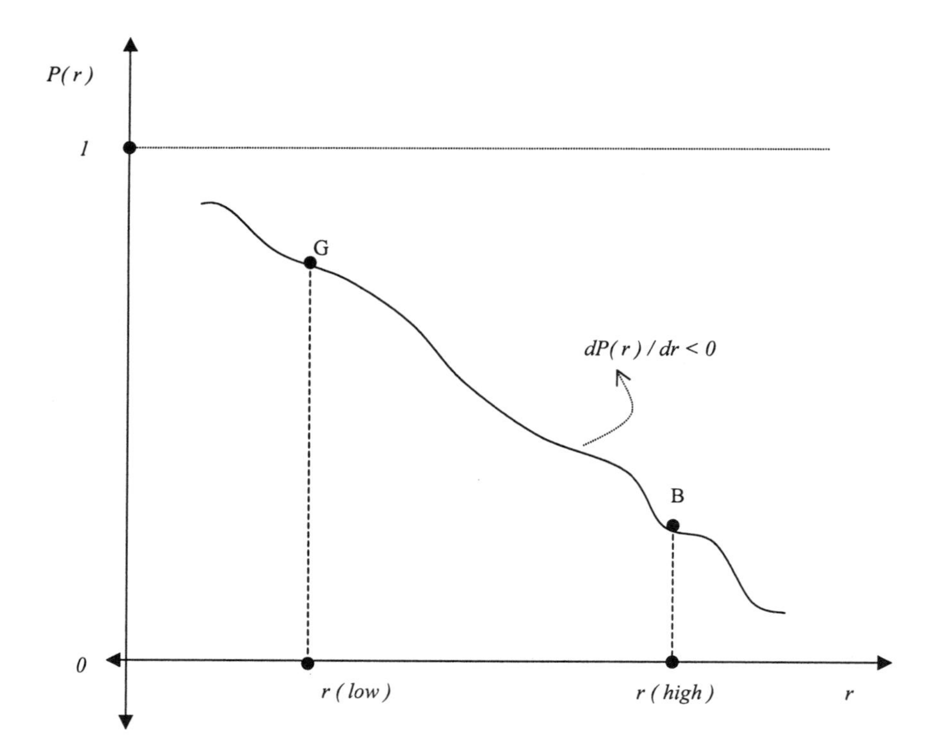

Fig. 5.7 Adverse Selection in Credit Markets: Demand for Investment

This simple "demand" for investment relationship is depicted in Figure 5.7. Points along this downward sloping curve have a very important property. We compute the derivative of the expected profit (5.53) with respect to the probability of success p_i to obtain

$$dE[P,\ agent\ i]/dp_i = -(1 + r)D < 0 \tag{5.55}$$

since $dR/dp_i=0$ because the average total return is taken as given, or only the risk is private information. Thus, the consequence of (5.55) is that, for a given r, as p_i increases and projects become less risky, the expected profit falls and individuals in the class of high p_i's (the low risk one's) will be willing to pay less for the loan. In turn, for a given r, as p_i decreases and projects become more risky, the expected profit increases and individuals in the class of low p_i's (the high risk one's) will be willing to pay more for the loan. This is the adverse selection of types that emerges from the focus on the average returns.

In Figure 5.7, at point G, the interest rate is low and the participation rate is high. This low interest rate increases the probability of success of the projects and attracts more good types, the less risky ones to the pool of investors. However, at point B, the interest rate is too high and it decreases the probability of success of the projects; the low risk types are driven out of the market, and the high-risk types are the ones that will largely populate the pool of investors. This is the adverse selection problem!

The Lender (or Principal) - The financial institution or bank is risk averse with respect to the interest rate charged, $r>0$. We assume that it does not operate in a competitive market, but it could as well, this is not crucial here. It is willing to offer the loan $D>0$, as a standard debt contract for a chosen interest rate $r>0$. The principal is able to screen projects based on their average returns, $R>0$, but cannot observe the p_i's directly. It does know the probability distribution of the p_i's though. Let the p_i's follow a continuous probability distribution with probability density function, $f(p_i)$, taken as given by the principal. The principal cannot affect this distribution, and it is public information. The principal offers a standard debt contract at an interest rate $r>0$ that maximizes its expected profits subject to its own assessment of the proportion of potential investors that will participate, denoted $P_p(r)$, where the principal recognizes that $dP_p(r)/dr <0$, i.e. it recognizes the adverse selection problem. Given all agents in the continuum from zero to $P_p(r)$ that participate in the contract, the principal's expected profit function is given by

$$E[P, principal] = (1 + r)\, D \int_0^{P_p(r)} p_i f(p_i)\, dp_i \qquad (5.56)$$

which denotes the total revenue received from those who participate and are expected to succeed and pay back the loan.

The key to understand how adverse selection leads to the possibility of credit rationing is to note that the expected profit function in (5.56) is strictly concave in the interest rate r for some relevant region in the domain of the function, i.e. the principal is risk averse.

Thus, a maximum for this function will be an interest rate $r*$ that satisfies the first order necessary condition

$$dE[P, principal]/dr =$$

$$D \int_0^{P_p(r)} p_i f(p_i)\, dp_i + (1 + r)\, D\, [dP_p(r)/dr]\, P_p(r)\, f(P_p(r)) = 0 \qquad (5.57)$$

where the derivative

$$d[\int_0^{P_p(r)} p_i f(p_i)\, dp_i]/dr = [dP_p(r)/dr]\, P_p(r)\, f(P_p(r))$$

is obtained by applying Liebniz's rule, e.g. Chapter 1. The sufficient second order condition is satisfied by the strict concavity of the expected profit function in the relevant domain of the interest rate. Expression (5.57) gives the intuition for the strict concavity of the expected profit function together with the adverse selection cost. The first term on the right-hand side,

$$D \int_0^{P_p(r)} p_i f(p_i)\, dp_i,$$

represents the positive marginal revenue obtained by the principal if the interest rate increases by one hundred basis points. The second term is the adverse selection cost:

$$(1 + r)\, D\, [dP_p(r)/dr]\, P_p(r)\, f(P_p(r))$$

is negative because $dP_p(r)/dr < 0$, and it represents the marginal cost of increasing the interest rate in terms of the lower number of participants in a mix of (consequently) more risky types, thus making the pool more risky to the lender. In effect, more risky types will be willing to pay the higher interest rate driving out the less risky types, analogous to "Gresham's Law" in monetary theory ("bad money drives out good money!"). An increase in the interest rate has these two opposing effects: the gain in revenue, but the loss of the good risks that are driven out by the bad risks due to adverse selection. The strict local concavity of the expected profit function comes from these opposing effects: at a low interest rate, the marginal revenue outweighs the marginal cost and the expected profit is increasing in r; at the maximum marginal revenue equals marginal cost; and at high levels of the interest rate the marginal cost exceeds the marginal revenue decreasing expected profits. Figure 5.8 depicts the "supply" of credit by the principal under these conditions.

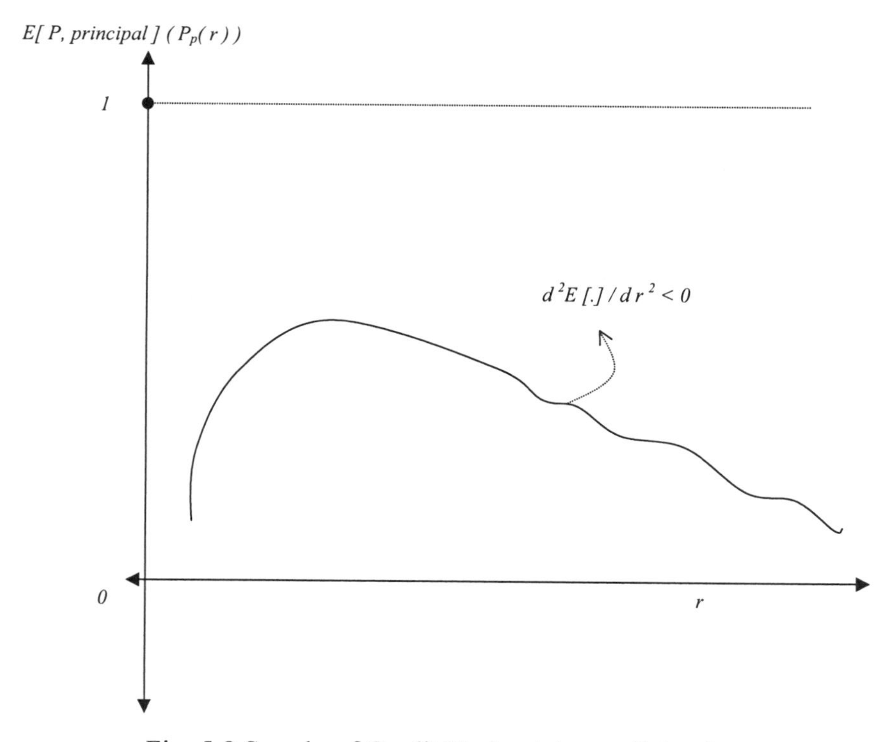

Fig. 5.8 Supply of Credit Under Adverse Selection

The possibility of credit rationing is depicted in Figure 5.9a,b where the "demand" and "supply" for credit are depicted. Figure 5.9a shows that it is possible, given some probability density function for the distribution of the types, $f(p_i)$, that the maximum expected profit for the principal is at point A, to the left of the equilibrium of "demand" and "supply" at point B. In this case, the principal offers the contract at $r*$ that solves (5.57) to all interested in making the investment up to $P_p(r*)$. However, at $r*$ the participation rate of agents, from (5.54), is at point A' and the distance A-A' is an excess demand for loans. Hence, the principal rations credit up to $P_p(r*)$ without any discrimination of types, i.e. one individual may be funded and another individual with the same project and type may not be funded. Figure 5.9b shows this result in the continuum of agents.

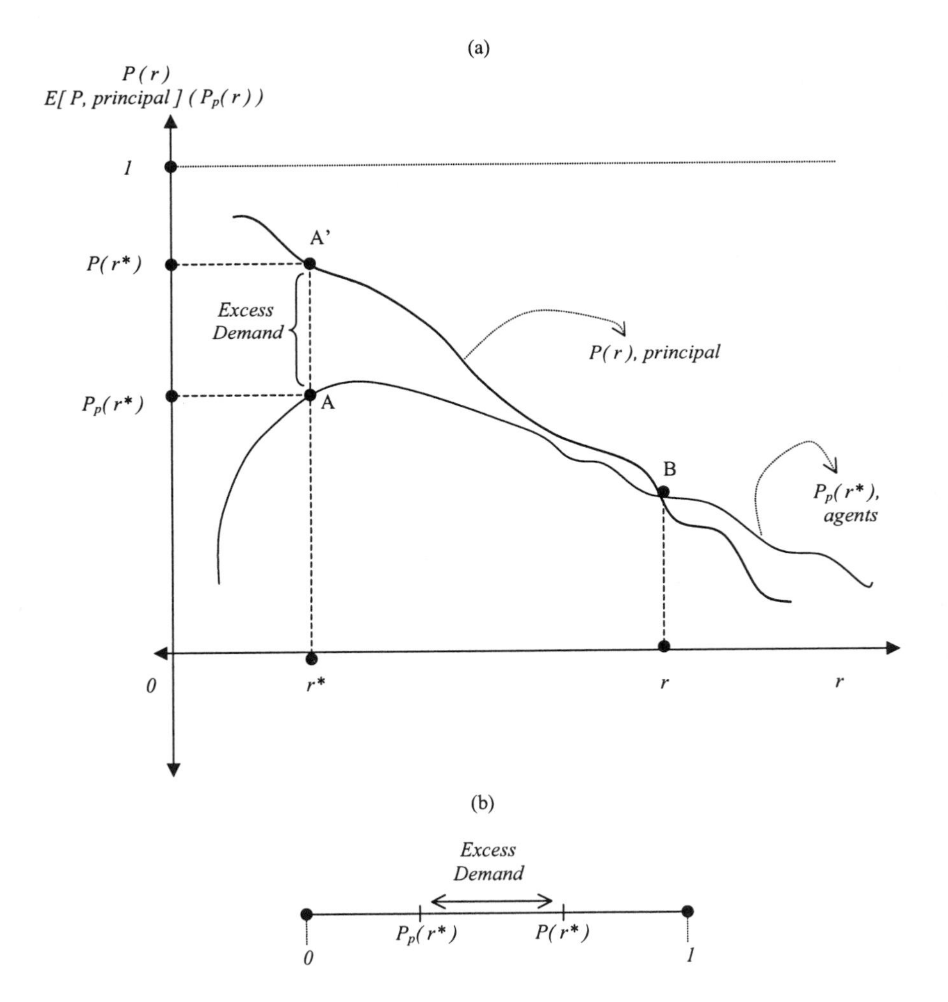

Fig. 5.9 Credit Rationing under Adverse Selection

You may be asking yourself: why wouldn't the principal increase interest rates and offer loans to all potential investors? That's simple. By increasing the interest rate, the bad risks will drive out the good ones in detriment to the principal's expected profits. The principal will be better off rationing at a lower interest rate and maintaining the good risks in the pool.

5.9 Signaling

We examined adverse selection models above where party *A* is uninformed about some characteristic of party *B,* whereas party *B* is informed about her own characteristic. We assumed that the uninformed party *A* moves first in an attempt to try to induce the informed party *B* to follow and reveal the private information truthfully.

Alternatively, the signaling equilibrium involves the informed party *B* moving first to send a signal to the uninformed party *A* revealing some of the private information. Then, the uninformed party *A* uses the signal to respond. The signaling problem may be embedded in the adverse selection problem. The classic signaling problem with adverse selection is the used car market, known as the "lemons" problem.

The lesson is that uncertainty about the quality of a commodity may impede the functioning of a specific market exactly as we've seen in the price discrimination problem of section 5.7.

The "lemons" problem goes as follows. Suppose there are two types of used cars in the market place:

(i) Good cars ("peaches");

(ii) Bad cars ("lemons").

There is a risk neutral seller with a finite supply of cars who is better informed about the quality of the used car for sale, but this may be private information of the seller. There are many (an infinite number of) potential risk neutral buyers who may not be able to observe the quality of the used car for sure, but do have an assessment of the proportion of good cars in the market, denoted $q>0$. Both buyers and sellers value the cars according to the following:

(i) A good car ("peach") is worth $\$g$ to the seller and $\$G$ to the buyer with $G > g$, so that G-g is the seller's profit on the good car;
(ii) A bad car ("lemon") is worth $\$l$ to the seller and $\$L$ to the buyer with $L > l$, and L-l is the seller's profit on the bad car.

Given (i)-(ii) and the infinite number of potential buyers, both cars should be traded and, if there is no private information, good cars would sell for G and bad cars would sell for L, the simple market equilibrium.

Another possibility would be if both parties were uninformed, then the equilibrium price of any car would be the average value

$$q\,G + (1 - q)\,L$$

given risk neutrality and infinite number of potential buyers.

But, the interesting case is the asymmetric information case. Suppose the seller knows the quality of the used car for sale and this is private information. The buyer only assesses the proportion of good cars in the market $q>0$, but cannot perfectly distinguish a good car (peach) from a bad car (lemon).

What is the equilibrium market price, that is $p>0$, in this case? First, the informed seller will only offer good cars if the market price is $p > g$, otherwise she would make a loss in selling good cars. Second, uniformed buyers, with assessment of proportion of good cars $q>0$, will infer that if the price of used cars is too low, say $p < g$, then all cars in the market are lemons, i.e. why would one be selling such a good car for such low price? In this case, buyers would trade only if the price is $p \leq L$, where L is the value the agent attaches to the bad cars. Thus, there are three possible "equilibria" all illustrated in Figure 5.10:
(a) If the market price is

$$L < p < g,$$

no cars will be sold and the market is blocked, private information impedes the market to function since the buyer would only buy if $p \leq L$, i.e. the price is too low for "peaches" and too high for "lemons;"
(b) If the market price is

$$p \leq L,$$

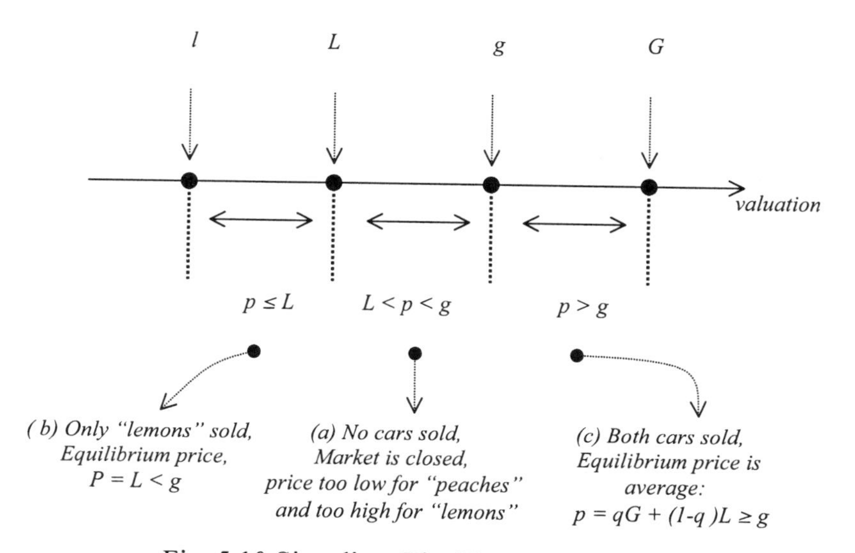

Fig. 5.10 Signaling: The "Lemons" Problem

then the seller only sells bad used cars (lemons), no good cars will be sold and the market price equilibrium will be

$$p = L < g,$$

i.e. the adverse selection case;
(c) If the market price is

$$p > g,$$

then the price is high and both cars are put up for sale; the market price (by risk neutrality) will be

$$p = q\,G + (1 - q)\,L \geq g$$

and both cars will be sold as if both the principal and the agent were uninformed, i.e. no revelation mechanism at work.

The disfunctioning of the market comes from the inability of the seller to signal the right quality of the used car for sale to the buyer. One possible contract offer designed to mitigate the problem would be for the seller to signal the quality of the used car by offering a warrantee, but signals and messages of this sort come at a cost. The optimal contract must balance these benefits and costs as we examined in section 5.7.

5.10 **Summary II**

In sections 5.4-5.9, we show how asymmetric information regarding an individual characteristic can obstruct the regular functioning of markets. Mitigating the adverse selection problem involves mechanism design through simple principal-agent relationships. We examine below the alternative case where asymmetric information is about an individual's action, the so-called moral hazard problem.

5.11 **Introduction to Moral Hazard**

The moral hazard problem deals with information about "what the agent does; the decision the agent takes." It may fit into the principal-agent framework as well but with some qualifications. The agent takes an "action" (an activity for example) that may be private information due to the lack of perfect monitoring. This action may have an effect on the agent's own utility, and on the utility of the principal who may not be able to observe perfectly the action of the agent.

The action of the agent leads to an "outcome" that can be fully observed by both the principal and the agent. The key to the solution for the moral hazard problem is the link between the action and the outcome. The principal can only observe an imperfect and noisy signal of the agent's action by inference based upon the outcome. The nature of the contract offered by the principal will be a transfer. The transfer could be positive, say a premium, or negative (a penalty), but that will depend on the observed outcome, and eventually on the link between action and outcome.

In the adverse selection model seen before, the principal offers a contract giving an incentive to the agent to reveal her private information. In that case, the source of asymmetric information is at the time of the contract offer; for example at the time the trade of a commodity takes place. In the moral hazard case, the incentive comes from the principal making the transfer depend specifically on the observed outcome. In this sense, the moral hazard problem brings about a tension or tradeoff between risk sharing and incentives to be fully elaborated below. Moreover, the action will occur after the contract has been offered so that the principal will have an interest to guarantee, with the right incentives, that the agent takes the preferred action. Figure 5.11 sums up some of the issues discussed above.

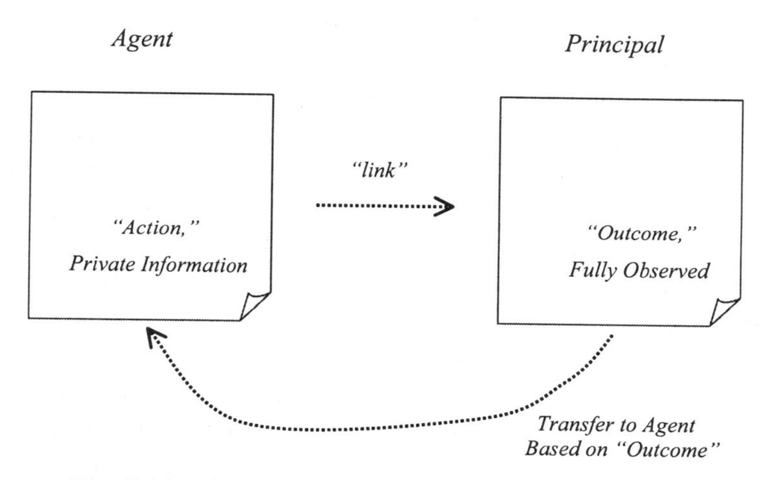

Fig. 5.11 Principal-Agent Flows under Moral Hazard

One of the main qualifications of the moral hazard problem, in the context of the principal-agent relationship, is the connection to game theory. The principal offers a contract to the agent who will take it or leave it. If the agent decides to take the contract, then it may choose its own action under private information. There is simultaneity though. When the principal offers the contract, it takes the unobserved action of the agent as given. When the agent accepts the contract, then it takes the contract as given and chooses its action optimally. The optimal action chosen is assumed to be consistent with the incentive structure provided in the principal's contract. Thus the equilibrium concept is one of Subgame Perfect Nash Equilibrium: the agent's action is the one that is agreeable with the incentive scheme provided in the principal's contract. Subgame perfect Nash equilibrium guarantees that the action and decision of the agent are both consistent with the incentive structure.

The issues to be examined below are as follows. First, we let the space of actions and outcomes be finite. In this set up, we start with the case where the principal and the agent are both risk neutral. The moral hazard due to private information leads the principal to design a contract that induces the agent to internalize the externality of her private information. This is done by forcing all the risk of the outcome to be borne by the agent.

Next, we examine the case where the agent is risk averse and the principal is risk neutral with a finite number of actions. In this case, there will be partial risk sharing in the design of the optimal contract and the tension between risk sharing and incentives due to moral hazard becomes crystal clear. The optimal contract will balance the need for efficient risk sharing with the incentive to mitigate moral hazard. We examine, in some detail, the link or technology between action and outcome, and also consider a case with an infinite number of possible outcomes on a continuum.

Then, in the context where the agent is risk averse and the principal is risk neutral, we change the space of actions by considering the case of an infinite number of possible actions on a continuum.

This is an appropriate framework for a discussion of the First Order Approach: the agent's incentive compatibility constraint will be substituted for the first order necessary condition of a choice of action that maximizes the agent's expected utility, taking all other relevant variables as given. First, we examine this case in the context of a simple insurance market example with finite number outcomes. Second, we generalize the example of section 5.13 (with a finite number of actions) to infinite number of actions and outcomes.

5.12 Finite Number of Actions and Outcomes, Principal and Agent both Risk Neutral

Consider the simplest discrete case of two actions and two possible outcomes. The most common example of problems of this type is in situations of effort in a supervisor-labor context. For example, in a sports team context the principal is the owner of the team and the agent is the manager of the team; in an attorney-client context the principal is the client and the attorney is the agent; in a company with publicly held stock the principal is the shareholder and the agent is the CEO, and the list of examples goes on.

Here we assume both principal and agent are risk neutral. The principal is assumed to hire the agent for one specific task at some reward denoted

$$w > 0.$$

The two possible outcomes of this specific task are assumed to yield value to the principal denoted:

$$V > 0 \;\; if\, success;$$

$$V = 0 \;\; if\, failure.$$

The key element of private information is that the probability of success of the outcome depends on the agent's action. In this case, the specific action is the agent's level of effort. Consider first the agent's problem.

The Agent - Assumed to be risk neutral with action (effort) being her private information. The agent may accept the contract offered by the principal to perform a specific task, at a possibly state contingent reward $w>0$. The next best alternative, in case of refusal, is an outside offer denoted,

$$\underline{\omega} > 0.$$

We express the agent's action, in this case the work effort assumed equal to the cost of effort, as

$$a \;\geq 0.$$

The two possible actions are:

(i) $a > 0$ indicates high effort with an implied probability of success in the outcome given by $p > 0$;

(ii) $a = 0$ indicates low effort with an implied probability of success in the outcome given by $q > 0$, where $0 < q < p$.

Hence, the probability of success, given high effort, is larger than the probability of success given low effort, i.e. the key link between action and outcome.

The agent's utility consistent with risk neutrality is just linear, increasing in the reward and decreasing in the effort:

$$Agent's\ utility = w - a.$$

The Principal - Also assumed to be risk neutral, only observes the outcome of the agent's action, not the effort itself. In other words, there is no perfect monitoring technology available to the principal. In effect, the principal can only design a reward scheme based on the outcome observed, and a simple state contingent scheme is the following:

offer the contract $\{w^1, w^2\}$ so that:

(a) $w^1 > 0$ if the observed outcome is success ($V > 0$);

(b) $0 < w^2 < w^1$ if the observed outcome is failure ($V=0$).

The unobserved action of the agent is taken as given by the principal under the assumption that the agent acts rationally, and chooses the course of action consistent with the incentive provided by the principal in the reward scheme. This is in accord with the subgame perfect equilibrium concept discussed above.

How does the principal design the optimal contract? The answer is that the principal has to take into account the agent's participation and incentive compatibility constraints. In taking these constraints into account, the principal can give incentives for the agent to choose high effort. Alternatively, the principal could offer a different scheme, and let the agent shirk and choose low effort.

First, we examine the case where the principal gives incentives for the agent to choose high effort. The agent's participation constraint, for the choice of high effort is given by

$$[p\,w^1 + (1 - p)\,w^2] - a \geq \underline{\omega} \tag{5.58}$$

where the left-hand side is the agent's expected reward out of high effort (the term in brackets) minus the cost of high effort, yielding the agent's expected utility for high effort with contract offer $\{w^1, w^2\}$, which should be no less than the next best alternative given by $\underline{\omega}$. The agent's incentive compatibility constraint for a choice of high effort is given by

$$[p\,w^1 + (1 - p)\,w^2] - a \geq q\,w^1 + (1 - q)\,w^2 \tag{5.59}$$

which says that, given the contract offer $\{w^1, w^2\}$, the agent's expected utility out of high effort should be no less than the agent's expected utility out of low effort ($a=0$).

The principal's expected utility (U), under the induced action consistent with the participation and incentive compatibility constraints above being satisfied, is given by

$$E[U, principal] = p\,V - [\,p\,w^1 + (1 - p)\,w^2] \qquad (5.60)$$

i.e. the expected revenue from success under high effort, $p \times V$, minus the expected cost of contract offer, $\{w^1, w^2\}$, under success $[\,p\,w^1 + (1 - p)\,w^2]$.

The solution for the optimal contract, denoted $\{w^{1*}, w^{2*}\}$ is obtained by the maximization of the principal's expected utility in (5.60) subject to the participation and incentive compatibility constraints, (5.58)-(5.59), taking $\{p, V, q, a, \underline{\omega}\}$ as given.

The simplest solution is one where the participation and incentive compatibility constraints, (5.58)-(5.59), hold with equality. In this case, the agent is indifferent between the outside offer and the low effort action, and an additional assumption must be made that the agent breaks the tie in favor of high effort. It turns out that this is in the best interest of the agent as well. In this case, the optimal solution $\{w^{1*}, w^{2*}\}$ that solves the two constraints with equality is

$$w^{1*} = \underline{\omega} + a\,[(1 - q)/(p - q)] \qquad (5.61)$$

$$w^{2*} = \underline{\omega} - a\,[q/(p - q)]. \qquad (5.62)$$

The optimal contract $\{w^{1*}, w^{2*}\}$ consists of a premium

$$a\,[(1 - q)/(p - q)]$$

over the base $\underline{\omega}$ if the outcome is success, and a penalty

$$a\,[q/(p - q)]$$

subtracted from the base in case of failure. This optimal scheme implies that the agent's reward depends fully upon the observed outcome. Consequently, the principal is able to fully transfer all the risk to the agent. In effect, the scheme makes the agent internalize all the gains from high effort, in other words the agent internalizes the externality from its own hidden action which is private information. It thus induces the agent to choose the high effort action.

The contract offer depends fundamentally on the probabilities p and q, or the link between action and outcome. In particular, given a level of effort, an increase in the probability of success p (under high effort) unambiguously decreases the premium, $a\,[(1 - q)/(p - q)]$, in case of an observed success. An increase in the probability of success $q < p$ (under low effort) unambiguously increases the penalty, $a[q/(p - q)]$, in case of an observed failure. This optimal contract may be substituted into the principal's expected utility (5.60) yielding

$$E[U, principal]^* = p\,V - (\underline{\omega} + a) \qquad (5.63)$$

where the expected cost from success is the base $\underline{\omega}$ plus the cost of effort a. The optimal contract mitigates the moral hazard problem because it provides an incentive for the agent to pursue high effort, by giving the agent a premium over the base when $a>0$.

Now, we examine the case where the principal is not interested in designing the optimal contract to mitigate the moral hazard. Suppose the principal is willing to offer full insurance to the agent and bears all the risk of the agent's action. The logical contract offer would be a fixed amount in all states,

$$\{w^1, w^2\} = \{\underline{\omega}, \underline{\omega}\}.$$

Under this fixed amount, the agent is allowed to shirk as she finds fit. The principal's expected profit from this contract offer would be, using (5.60),

$$E[U, principal]^{\underline{\omega}} = q\,V - \underline{\omega} \qquad (5.64)$$

the expected revenue from success with low effort minus the fixed cost.

The difference between (5.63) and (5.64) is

$$E[U, principal]^* - E[U, principal]^{\underline{\omega}} = V\,(p - q) - a. \qquad (5.65)$$

We note that, for a given $\{p,q,a\}$, the larger the success yield V the more likely the principal is willing to offer the optimal contract $\{w^{1*},w^{2*}\}$ to induce agent to choose the high effort action, i.e. expression (5.65) is more likely positive.

There are some basic problems with the framework presented above though. One is that under the optimal contract, and in case of failure, the penalty may be too harsh, say $w^{2*}<0$ in expression (5.62), and such transfer from the agent to the principal would be difficult to enforce. In general, the full enforcement of contractual relationships is an important problem. Another restrictive assumption is the one-task performance. We can easily find multi-task performances that would give more complicated payment schemes. Another restriction worth noting is the absence of monitoring, but this is a less important restriction since a monitor could always collude with the agent and misrepresent.

Finally, an important assumption of the framework presented is risk neutrality. If the agent were to be risk averse instead of risk neutral, she would not be willing to bear all the risk as the optimal contract dictates, some risk sharing must occur. We move to this case next.

5.13 Variations with Finite Number of Actions, Principal is Risk Neutral, Agent is Risk Averse

Consider again the simplest case of two actions and a finite number of possible outcomes (discrete case). The situation is of effort in a supervisor-labor context for one specific task. The agent is risk averse and can choose two possible actions, taken to be its own private information, denoted a_i, $i=1,2$, with

$$a_1 > 0 \text{ denoting high effort;}$$

$$0 < a_2 < a_1 \text{ denoting low effort.}$$

An action yields one outcome from a finite set of outcomes or states of nature. The finitely many possible outcomes, fully observed by all in the economy, are denoted x^j, $j=1,2,...,m$. The possible outcomes are assumed to be equal to the monetary payoff to the principal for the completion of the task, with an ordering given by

$$0 < x^m < ... < x^3 < x^2 < x^1$$

so that x^1 yields the highest payoff and x^m yields the lowest payoff, all strictly positive. The relationship between action and outcome, in this case, is more involved referring to an information structure with noise. The key link or technology between action and outcome is the following:

$$p[x^j, a_i] > 0, \quad i=1,2, j=1,2,...,m,$$

denotes the probability of observing a possible outcome x^j, $j=1,2,...m$, when the agent chooses action a_i, $i=1,2$.

These probabilities will be further qualified below. Momentarily, we think intuitively that they represent the likelihood that when the principal observes an outcome, x^j, the action chosen by the agent was a_i. That is, for each action we have a probability distribution of the outcomes.

The contract offered by the principal may depend on the outcome x^j, $j=1,2,...,m$. Hence, if the principal observes an arbitrary outcome x^j, it pays the agent a state contingent reward w^j, and the principal's utility will be, in this case,

$$\text{Principal's utility} = (x^j - w^j), \quad j=1,2,...,m,$$

since the principal is assumed to be risk neutral, i.e. linear reward.

The agent is assumed to be risk averse towards the reward prospect. Let the reward be given by $w^j, j=1,2,...,m$, and effort a_i, $i=1,2$, so that the agent's utility, assumed to be separable in reward and effort, is given by

$$\text{Agent's utility} = [U(w^j) - a_i], \quad j=1,2,...,m,\ i=1,2$$

where U is strictly increasing and strictly concave, $U' > 0$, $U'' < 0$. Next, we examine the decision-making of the agent and the principal.

The Agent - Assumed to be risk averse. The agent's action, or effort is private information, and she may accept the contract from the principal to perform the specific task at the state contingent reward $w^j, j=1,2,...,m$. The next best alternative, in case of refusal of the principal's contract offer, is an outside offer that gives some exogenous utility denoted, $\underline{U}>0$. Given the wage prospect $w^j, j=1,2,...,m$, the agent can choose a level of effort a_i, $i=1,2$, which yields expected utility

$$E[U, agent] = \{\Sigma_{j=1}^{m} p[x^j, a_i] U(w^j)\} - a_i, \quad \text{for } i=1,2 \qquad (5.66)$$

where $a_i>0$, $i=1,2$ may be suitably interpreted as the cost of the agent's effort in terms of utility.

The Principal - Assumed to be risk neutral. The principal chooses to offer a contract based on the observed outcome, taking the action of the agent as given. The right and preferred incentives for the agent to perform high effort, $a_1>a_2$, must take into account the agent's participation and incentive compatibility constraints. Thus, for action a_1

to be preferred to action a_2, the agent's participation constraint is given by

$$\{\Sigma_{j=1}^{m} p[x^j, a_1] U(w^j)\} - a_1 \geq \underline{U} \qquad (5.67)$$

i.e. the expected utility derived from level of effort a_1, must be no less than the utility from the next best alternative, $\underline{U}$. The incentive compatibility constraint for action a_1 to be preferred to action a_2 is given by

$$\{\Sigma_{j=1}^{m} p[x^j, a_1] U(w^j)\} - a_1 \geq \{\Sigma_{j=1}^{m} p[x^j, a_2] U(w^j)\} - a_2 \qquad (5.68)$$

denoting that the expected utility from high effort, a_1, must be no less than the expected utility from low effort, a_2.

The principal's expected utility, with induced high effort a_1, is given by

$$E[U, principal] = \Sigma_{j=1}^{m} p[x^j, a_1] (x^j - w^j) \qquad (5.69)$$

which denotes the principal's expected gain when the agent chooses high effort a_1. The principal offers an optimal contract, $\{w^j\}, j=1,2,...,m$, so that it solves the problem

$$\max_{\{w^j\}_{j=1}^{m}} E[U, principal] = \Sigma_{j=1}^{m} p[x^j, a_1] (x^j - w^j) \qquad (5.70)$$

$$\text{subject to} \qquad \{\Sigma_{j=1}^{m} p[x^j, a_1] U(w^j)\} - a_1 \geq \underline{U}$$

$$\{\Sigma_{j=1}^{m} p[x^j, a_1] U(w^j)\} - a_1 \geq \{\Sigma_{j=1}^{m} p[x^j, a_2] U(w^j)\} - a_2,$$

taking as given $\{p[x^j, a_i], x^j, a_i, \underline{U}\}, i=1,2, j=1,2,...,m$, where the two constraints are the participation and incentive compatibility constraints from (5.67)-(5.68).

This problem may be solved using Khun-Tucker's theory. We form the Lagrangean function

$$\mathcal{L} = \Sigma_{j=1}^{m} p[x^j, a_1] (x^j - w^j) + \mu \{[\Sigma_{j=1}^{m} p[x^j, a_1] U(w^j)] - a_1 - \underline{U}\}$$

$$+ \lambda \{[\Sigma_{j=1}^{m} p[x^j, a_1] U(w^j)] - a_1 - [\Sigma_{j=1}^{m} p[x^j, a_2] U(w^j)] + a_2\} \qquad (5.71)$$

where $\{\mu \geq 0, \lambda \geq 0\}$ are non-negative Lagrange multipliers attached to the participation and incentive compatibility constraints respectively. The set of first order necessary conditions for $\{w^j, \mu, \lambda\}_{j=1}^{m}$ is given by

$$\partial \underline{\mathcal{L}}/\partial w^j = -p[x^j, a_1] + \lambda p[x^j, a_1]\, U'(w^j) - \lambda p[x^j, a_2]\, U'(w^j) + \mu p[x^j, a_1]\, U'(w^j) \leq 0, \quad if < 0,\ w^j = 0, \quad for\ j=1,2,...,m \quad (5.72)$$

$$\partial \underline{\mathcal{L}}/\partial \mu = \{\Sigma_{j=1}^{m} p[x^j, a_1]\, U(w^j)\} - a_1 - \underline{U} \geq 0, \quad if > 0,\ \mu = 0 \ (5.73)$$

$$\partial \underline{\mathcal{L}}/\partial \lambda = \{\Sigma_{j=1}^{m} p[x^j, a_1]\, U(w^j)\} - a_1 - \{\Sigma_{j=1}^{m} p[x^j, a_2]\, U(w^j)\} + a_2 \geq 0, \quad if > 0,\ \lambda = 0 \quad (5.74)$$

which yields $m+2$ equations in the $m+2$ unknowns $\{w^j, \mu, \lambda\}_{j=1}^{m}$ given $\{p[x^j, a_i], x^j, a_i, \underline{U}\}$, $i=1,2$, $j=1,2,...,m$. The second order sufficient conditions are satisfied by the linearity of the objective function with the strictly concave constraints.

We can now consider two possible contract offers, from the solutions (5.72)-(5.74). The first yields efficient risk sharing and is the first best, in the context of a risk neutral principal and a risk averse agent, but it does not mitigate the moral hazard problem. The second best involves less than efficient risk sharing but mitigates the moral hazard problem. The two possible offers illustrate the tension between risk sharing and incentives under moral hazard.

(i) Efficient Risk Sharing (First Best) - This contract involves setting

$$\mu > 0, \quad \lambda = 0,$$

thus the participation constraint (5.73) is binding, and the incentive compatibility constraint (5.74) is not binding, thus making high effort preferred to low effort on the incentive compatibility constraint. Substitution of these conditions into the first order necessary condition (5.72) yields

$$[1/U'(w^1)] = [1/U'(w^2)] = ... = [1/U'(w^m)] = \mu > 0, \ constant \quad (5.75)$$

and the contract offer is given by

$$w^1 = w^2 = ... = w^m = w^{constant}. \quad (5.76)$$

Since the agent is risk averse and the principal is risk neutral, the efficient risk sharing is to transfer all the risk to the principal, i.e. the principal provides full insurance to the agent. This is achieved by offering a constant reward independent of the outcome. But, this contract does not give any incentive for the agent to choose high effort due to the private information and consequent moral hazard; as we have seen in section 5.15 as well. This is the conflict between risk sharing and incentives. If there were no private information, contract (5.76) would be the optimal risk sharing scheme, i.e. the principal provides full insurance to the agent. This is because the principal is risk neutral, hence should be indifferent to bearing risk. But, with private information, this optimal risk sharing scheme may work in detriment to the principal because it lacks the right incentive for the agent to choose the high effort action.

(ii) Partial Risk Sharing (Second Best) - This contract involves setting

$$\mu > 0, \quad \lambda > 0,$$

thus the participation constraint and the incentive compatibility constraint are both binding. Substitution of these conditions into the first order necessary condition (5.72) above yields

$$[1/U'(w^j)] = \mu + \lambda\,[1 - (p[x^j,\, a_2\,] \,/\, p[x^j,\, a_1\,])] \quad \text{for } j=1,2,...,m \qquad (5.77)$$

and the contract offer is $\{w^{j*}\}, j=1,2,...,m$, that satisfy condition (5.77).

The general form of the optimal contract $\{w^{j*}\}$ from (5.77) is some base reward, $\mu>0$, plus a transfer $\lambda[1-(p[x^j,\, a_2\,]/p[x^j,\, a_1\,])]$. It is clear that the optimal contract depends crucially on the key link or technology between action and outcome, through the information content of the probabilities ratios $p[x^j,\, a_2\,]/p[x^j,\, a_1\,]$. The ratio refers to the probability of observing a possible outcome, x^j, when the agent chose low effort action a_2, $p[x^j, a_2\,]$; divided by the probability of observing a possible outcome, x^j, when the agent chose high effort action a_1 , $p[x^j,\, a_1\,]$. Suppose we observe an arbitrary outcome, x^j. Then, if the ratio is greater than one, $p[x^j,a_2\,]/p[x^j,a_1\,]>1$, it implies that there is a higher probability that the agent took the low effort action a_2. Expression (5.77) implies that the reward is the base $\mu>0$, minus a penalty

$$\lambda\,[1 - (p[x^j,\, a_2\,]/p[x^j,\, a_1\,])] < 0 \quad \Leftrightarrow \quad p[x^j,\, a_2\,] \,/\, p[x^j,\, a_1\,] > 1.$$

Alternatively, if the ratio is less than one, $p[x^j,\, a_2\,]/p[x^j,\, a_1\,]>1$, it implies that there is a higher probability that the agent took the high effort action a_1 and expression (5.77) implies that the reward is the base $\mu>0$, plus a premium

$$\lambda\,[1 - (p[x^j,\, a_2\,]/p[x^j,\, a_1\,])] > 0 \quad \Leftrightarrow \quad p[x^j,\, a_2\,] \,/\, p[x^j,\, a_1\,] < 1.$$

Thus, the partial risk sharing contract delivers a consistent incentive scheme for the agent to choose the high effort action, mitigating moral hazard.

We proceed with a more thorough discussion of the probability ratios. For example, what happens when the probability ratios equal to one? There are several related theoretical, intuitive and practical explanations for these ratios and we present a few of them.

First, we may plausibly draw an analogy between the incentive problem of the principal and classical statistical inference. This requires caution because there is no statistical inference going on here, the principal has the knowledge that the offer gives the right incentive for the agent to choose high effort. In classical statistical analysis, the probability ratios, $p[x^j, a_2]/p[x^j, a_1]$, are suitably called Likelihood Ratios which, in our context, reflect a comparison between the probability distribution of possible outcomes associated with alternative actions chosen by the agent. One would be willing to statistically infer the unknown action, say a_i, $i=1,2$, by observing one of the outcomes (x^j, $j=1,2,...,m$). In classical statistical inference, one would estimate the parameter "a_i" from a data sample "x^j" by computing the maximum likelihood estimator of "a_i." The maximum likelihood estimate (MLE), or "a_i;" thus maximizes the probability $p[x^j, a_i]$ of the action a_i given the outcome x^j. Under maximum likelihood estimation, we expect that

$$p[x^j, a_k] / p[x^j, a_i] \leq 1, \quad \text{for all } j=1,2,...,m; \ i, k=1,2, \ k \neq i. \qquad (5.78)$$

When the outcome x^j is likely to have come from action a_i, for example, $i=1$, high effort action, we expect x^j to be large and close to x^1. Alternatively, when $i=2$, the low effort action is chosen, we expect x^j to be small and close to x^m. Condition (5.78) is familiar from Likelihood Ratio tests in statistics. The denominator denotes the probability under the MLE, "a_i," so that it yields the largest unconstrained probability, conditional on the observed sample. The numerator denotes the probability under maximum likelihood estimation constrained by the action a_k, $k \neq i$, conditional on the observed sample. Since the denominator is the unconstrained global maximum, relationship (5.78) is appropriate. Therefore, from (5.77) and (5.78), an observation of x^j large, close to x^1, yields a reward $\{w^{j*}\}$ that is greater than in the case of an observation of x^j that is small and close to x^m.

Another related interpretation is that the likelihood ratio reflects a signal that the observed sample was picked from one distribution relative to another. In this case, the likelihood ratio

$$p[x^j, a_k] / p[x^j, a_i], \quad \text{for all } j=1,2,...,m; \; i,k=1,2, \; k \neq i \qquad (5.79)$$

denotes how strongly the observed sample comes from the distribution associated with action a_k, i.e. $p[x^j, a_k]$; relative to the distribution associated with action a_i, $p[x^j, a_i]$.

Then, for example, for $i=1$ and $k=2$, $p[x^j, a_2]/p[x^j, a_1]$ denotes how strongly the observed sample comes from the distribution associated with low effort a_2, i.e. $p[x^j, a_2]$; relative to the distribution associated with high effort a_1, $p[x^j, a_1]$.

Bayes' theorem is useful in this context. We denote by $p(a_1)$ the unconditional prior probability of high effort, and hence $1-p(a_1)$ is the unconditional prior probability of low effort. The link or technology between action and outcome is given by $p[x^j, a_i]$, $i=1,2; j=1,2,...,m$, the conditional probability of x^j given chosen action a_i. Let $p[a_1, x^j]$, $j=1,2,...,m$ be the posterior or updated probability of high effort a_1 given that the observed outcome is x^j. Bayes' theorem, e.g. Chapter 1, yields

$$p[a_1, x^j] = p[x^j, a_1]\, p(a_1) / \{p[x^j, a_1] p(a_1) + p[x^j, a_2][1-p(a_1)]\} \qquad (5.80a)$$

which implies that

$$p[x^j, a_2] / p[x^j, a_1] = p(a_1)\{1-p[a_1, x^j]\} / \{p[a_1, x^j][1-p(a_1)]\}. \qquad (5.80b)$$

Expressions (5.80a,b) have the following interpretation. When the observation of an outcome, x^j, provides basis for a revision of the prior belief about the high effort action a_1 upward; or in mathematical terms when

$$p[a_1, x^j] > p(a_1)$$

then

$$p[x^j, a_2] / p[x^j, a_1] < 1$$

and the signal received is that the distribution over outcomes is the high effort one. In this case, the agent receives a premium in the state contingent wage (5.77), over and above the base reward $\mu > 0$.

Alternatively, when the observation of an outcome, x^j, provides basis for a revision of prior beliefs about the high effort action a_1 downwards, or when

$$p[a_1, x^j] < p(a_1)$$

then

$$p[x^j, a_2] / p[x^j, a_1] > 1,$$

and the signal received is that the distribution is the low effort one. In this case, the agent is penalized in the state contingent reward (5.77). Of course, when

$$p[a_1, x^j] = p(a_1)$$

there is no useful information in the sample, and the principal will not have statistical basis to deviate from the full insurance solution.

A third related interpretation is the following. Intuitively, because the principal is giving the right incentive for the agent to choose high effort, we expect the likelihood ratios to be small. But, the signal includes noise, hence it is like a "noise to signal ratio." For example, when

$$p[x^j, a_2] / p[x^j, a_1] < 1,$$

it implies a higher probability of high effort relative to low effort, i.e. in the sample x^j, the "noise to signal ratio" is low for the principal. In this case, the principal will provide a premium above the base $\mu > 0$. This is because the principal is confident that the action chosen was high effort and there is substantive statistical evidence that supports deviation from the full insurance solution. However, there can be mistakes sometimes, i.e. there is "noise." For some other sample x^j, the "noise to signal ratio" is high for the principal, and the likelihood ratio is high. In this alternative case, the principal is not confident that the action chosen was high effort and believes not to have induced much effort on the agent. There is no substantive statistical evidence that supports deviation from the full insurance solution, i.e. $1 - (p[x^j, a_2]/p[x^j, a_1])$ is low.

One important issue remains though. None of the related interpretations offered above guarantees that the likelihood ratios vary monotonically with respect to changes in the outcomes. The question is: how exactly does the likelihood ratio $p[x^j, a_2]/p[x^j, a_1]$ vary with respect to changes in x^j? A condition that guarantees that the likelihood ratio is monotonically decreasing in the outcomes is given by

$$p[x^h, a_2]/p[x^h, a_1] > p[x^j, a_2]/p[x^j, a_1] \quad \text{for all } j<h;\ j, h=1,2,...,m \qquad (5.81)$$

known as the <u>Monotone Likelihood Ratio Condition</u>. This condition states roughly that when we observe an outcome of higher payoff, or a higher value of x^j, it indicates that it is more plausible that the agent took a high effort, a_1, action. Assuming that the monotone likelihood ratio condition (5.81) holds, the second best optimal contract $\{w^{j*}\}, j=1,2,...,m$ can be easily computed. From expression (5.77), one obtains, by risk aversion of the agent,

$$dw^j / d(p[x^j, a_2]/p[x^j, a_1]) = \lambda (U')^2 /(U'') < 0 \quad \text{for } j=1,2,...,m, \qquad (5.82)$$

which states that the reward schedule varies inversely with the likelihood ratio. An increase in the likelihood ratio indicates more "noise" and is associated with a decrease in the reward offered by the principal.

In this case, the optimal contract has the following guidelines:

(i) If observed sample is high payoff:

$$[1/U'(w^{j*})] = \mu + \lambda [1 - (p[x^j, a_2]/p[x^j, a_1])]$$

yields a large w^{j*}, because $p[x^j, a_2]/p[x^j, a_1]$ is low, or close to zero; the principal observes an outcome with low "noise to signal ratio" and thus gives a higher reward to the agent;

(ii) If observed sample is low payoff:

$$[1/U'(w^{j*})] = \mu + \lambda [1 - (p[x^j, a_2]/p[x^j, a_1])]$$

yields a small w^{j*}, because $p[x^j, a_2]/p[x^j, a_1]$ is high; the principal observes an outcome that has a high "noise to signal ratio" and thus must penalize the agent.

The optimal contract is of the general form:

(i) If observed sample is high payoff: w^{j*};

(ii) If observed sample is low payoff: $w^{h*} < w^{j*}$; *for* $j<h;\ j, h=1,2,...,m.$

The second best solution involves some partial risk sharing giving the right incentive for the agent to choose high effort so that the principal can mitigate the moral hazard problem. In other words, the loss in efficiency due to partial risk sharing is balanced by the right incentive for the agent to choose high effort thus avoiding the moral hazard problem.

A very simple case that provides further insight is when there are only two possible outcomes, $m=2$, $j=1,2$, for example $j=1$ denotes success and $j=2$ denotes failure, with an ordering $x^1 > x^2$. As in section 5.12, in this case, the solution to (5.70) is obtained directly by solving the individual rationality and incentive compatibility constraints, both with equality, which are appropriately linear in utility $U(w^j)$. The solutions are given by

$$U(w^{1*}) = \underline{U} + \{(1-p[x^1, a_2])/(p[x^1, a_1]-p[x^1, a_2])\}a_1 - \{(1-p[x^1, a_1])/(p[x^1, a_1]-p[x^1, a_2])\}a_2,$$

$$U(w^{2*}) = \underline{U} - \{p[x^1, a_2]/(p[x^1, a_1]-p[x^1, a_2])\}a_1 + \{p[x^1, a_1](1-p[x^1, a_1])/(p[x^1, a_1]-p[x^1, a_2])\}a_2,$$

and the Lagrange multipliers are given by

$$\lambda^* = \{p[x^1, a_1](1-p[x^1, a_1])/(p[x^1, a_1]-p[x^1, a_2])\} \times \{[U'(w^{2*})-U'(w^{1*})]/U'(w^{1*})\, U'(w^{2*})\}$$

$$\mu^* = \{U'(w^{1*})(1-p[x^1, a_1]) + p[x^1, a_1]\, U'(w^{2*})\}/\, U'(w^{1*})\, U'(w^{2*}).$$

It is easy to see that: (i) $\lambda^*>0$, for $w^{1*} > w^{2*}$ and $p[x^1,a_1] > p[x^1,a_2]$, giving the partial risk sharing second best contract; and (ii) $\lambda^*=0$ for $w^{1*} = w^{2*}$, giving the full risk sharing contract; (iii) $\mu^*>0$ everywhere. In this special case, (5.78) suffices for an ordering of the probability ratios since events are mutually exclusive and collectively exhaustive, i.e. the monotone likelihood ratio condition is not needed.

For a numerical example, a utility function that presents risk aversion for the agent is $U(w)=2w^{1/2}$, with $U'=w^{-1/2}>0$ and $U''= -(1/2)w^{-3/2}<0$. In this case, the optimal contract will be of the form

$$w^{j*} = \{\mu + \lambda\,[1 - (p[x^j, a_2]/p[x^j, a_1])]\}^2 \quad for\ j=1,2.$$

Assuming the probability distributions are given by

$$p[x^1, a_1] = 3/5, \quad p[x^1, a_2] = 1/5;$$

and that the exogenous outside utility and effort are

$$\underline{U} = 1;\ a_1 = 1;\ a_2 = 0;$$

direct computation yields rewards and prices

$$w^{1*} = 4; \quad w^{2*} = 1/16; \quad \lambda^* = 21/20; \quad \mu^* = 19/20.$$

The high payoff state provides a large reward relative to the low payoff state in order to give the right incentive for the agent to choose high effort.

In this case, the first best solution would be

$$w'^{1} = w'^{2} = \lambda^{*2} = (21/20)^2 = 1.1025,$$

denoting a fixed reward schedule where there is no incentive for the agent to choose high effort.

The welfare of the agent, $U(w)=2w^{1/2}$, can be directly computed as

$$U(w^{1*}) = 4 > U(w'^{1}) = U(w'^{2}) = 21/10 > U(w^{2*}) = 1/2,$$

so that it is in the best interest of the agent to choose high effort when the contract is second best. Profits for the risk neutral principal are computed as

$$x^1 - 4; \quad x^2 - 1/16; \quad x^i - 1.1025, i=1,2; \quad x^1 > x^2$$

so that the second best solution involves giving up some profit in return for high effort. In this example, the premium for high effort is large and the penalty for low effort is substantive due partly to the agent's risk aversion.

In general, for partial risk sharing problems of this sort, schemes based on the observed outcome, x^j, involve a linear sharing rule of the form:

$$w^{1*} = BASE(\underline{U}) + SHARE(p[x^1, a_2]/p[x^1, a_1])\,(x^1 - x^2)$$

$$w^{2*} = BASE(\underline{U})$$

where the $SHARE(p[x^1, a_2]/p[x^1, a_1]) \in (0,1]$ is some fraction of the observed gain (x^1-x^2) so that the principal engages in some profit sharing to provide the right incentive for the agent to perform, i.e. mitigate moral hazard.

Finally, a variation on the other extreme is when there are an infinite number of possible outcomes (states of nature) on a continuum. To accommodate for the continuum of states of nature, we let the bounded set of possible outcomes in the continuum be denoted by

$$\boldsymbol{X} = \{x \geq 0: x \in [x^1, x^m], x^1 > x^m\}.$$

As before, there are two possible actions, $a_1 > a_2$ for high and low effort respectively. The conditional probability density functions of the outcomes, assumed to be public information, are denoted by

$$f_i(x), \quad i=1,2; \quad x \in \mathbf{X}$$

for high ($i=1$) and low ($i=2$) effort respectively. Then, the principal's problem, (5.70) can be recast as

$$\underset{\{w(x)\}}{Max} \quad E[U, principal] = \int_{x \in \mathbf{X}} f_1(x)\,[x - w(x)]\,dx \qquad (5.70')$$

$$subject\ to \quad \{ \int_{x \in \mathbf{X}} f_1(x)\, U(w(x))\, dx\} - a_1 \geq \underline{U}$$

$$\{ \int_{x \in \mathbf{X}} f_1(x)\, U(w(x))\, dx\} - a_1 \geq \{ \int_{x \in \mathbf{X}} f_2(x)\, U(w(x))\, dx\} - a_2,$$

taking as given $\{f_i(x), x, a_i, \underline{U}\}$, $i=1,2, x \in \mathbf{X}$, where the two constraints are the participation and incentive compatibility constraints and $w(x)$ is the state contingent wage schedule. The second best solution, obtained exactly as above with $\{\mu > 0, \lambda > 0\}$ the Lagrange multipliers attached to the participation and incentive compatibility constraints respectively, yields

$$[1/U'(w(x))] = \mu + \lambda\,[1 - (f_2(x)\,/\,f_1(x)\,)\,], \quad x \in \mathbf{X} \qquad (5.77')$$

where the likelihood ratio is interpreted as above: if

$$f_2(x)\,/\,f_1(x) < 1, \quad x \in \mathbf{X}$$

it implies that the sample speaks strongly for the high effort action; if

$$f_2(x)\,/\,f_1(x) > 1, \quad x \in \mathbf{X}$$

it implies that the sample speaks strongly for the low effort action; and if

$$f_2(x)\,/\,f_1(x) = 1, \quad x \in \mathbf{X}$$

it does not provide useful information. The monotone likelihood ratio condition guarantees that

$$f_2(x)\,/\,f_1(x) \text{ is decreasing in } x, \quad x \in \mathbf{X}.$$

Thus, the optimal contract is as above: a sample with observation of large (high payoff) outcomes yields a premium over the base; and a sample with observation of low (low payoff) outcomes yields a penalty.

5.14 **Infinite Number of Actions and First Order Approach, Principal is Risk Neutral, Agent is Risk Averse**

We now consider cases with an infinite number of possible actions in a continuum. This is more interesting because it impacts substantively on the agent's incentive compatibility constraints. First, we examine an insurance problem and then we examine a supervisor-labor problem.

(i) Here, we examine the simple case of a finite number of states of nature or outcomes, with two possible outcomes; in a situation of variable, and continuum of effort. The problem is an insurer-insuree relationship in the context of insurance markets. The agent is risk averse and the principal is risk neutral so that the optimal contract involves some partial risk sharing and incentives. The principal is the insurance company and the agent is willing to buy (demand) insurance. In this case, the infinite number of possible actions by the agent requires a solution based upon the first order approach. We shall examine it in full detail below.

The Agent - Assumed to be risk averse, the agent demands insurance of her own wealth, denoted $W > 0$. There are two possible states of the world:

$$\textit{loss of wealth, } W = 0;$$

$$\textit{no loss of wealth, } W > 0.$$

Let $s \in [0,1]$ be defined as the share of wealth, W, to be insured at a premium $\rho > 0$ per unit of wealth W. The premium, ρ, is parametrically given to the agent. The agent may take an action, $a > 0$ in a continuum bounded by the interval $a \in [\underline{a}, a^h]$, for $a^h > \underline{a}$, where $\underline{a}$ indicates the lowest possible effort and a^h the highest possible effort, which are all agent's private information, i.e. the bounded set of possible actions is

$$\boldsymbol{A} = \{a\text{: } a \in [\underline{a}, a^h],\ a^h > \underline{a}\}.$$

The main difference relating to the model of insurance seen before under adverse selection, section 5.6, is that, given the private information of the agent's action (effort or self-protection), the probability of a loss is assumed to be a function of the level of the agent's effort. A simple function that captures this property is given by

$$\pi = \pi(a) > 0, \quad \pi' < 0, \quad \pi'' < 0, \qquad a \in A \qquad (5.83)$$

where π denotes the probability of loss of wealth by the agent. This function is assumed strictly decreasing and strictly convex implying that more effort yields a lower probability of loss. An example of such function is

$$\pi(a) = b - a^2$$

for some real number b, so that $a^2 < b < 1 + a^2$ for a well defined probability and $\pi' = -2a < 0$, $\pi'' = -2 < 0$.

The agent's utility function is, as before,

$$Agent's\ utility = U - a,$$

where the function U is strictly increasing and strictly concave due to risk aversion, $U' > 0$, $U'' < 0$. The agent's expected utility is then given by

$$E[U, agent] = \pi(a)\, U(s\, W - \rho s\, W) + [1 - \pi(a)]\, U(W - \rho s\, W) - a \quad (5.84)$$

where the term $U(sW - \rho sW)$ is the net utility obtained in case of loss, and the term $U(W - \rho sW)$ is the net utility obtained in case of no loss. The expected utility in (5.84) is just the expected gain out of the utility of the wealth minus the cost of effort.

The agent's participation constraint is now familiar. The expected utility should be no less than some given bound consistent with no purchase of insurance, for example $\underline{U} > 0$, or

$$E[U, agent] = \pi(a)\, U(s\, W - \rho s\, W) + [1 - \pi(a)]\, U(W - \rho s\, W) - a \geq \underline{U}. \qquad (5.85)$$

However, in this case, the incentive compatibility constraint is more complicated due to the infinite number of possible actions that the agent may take. This is because it would require comparing the agent's expected utility to an infinite number of alternative actions in the interval $[\underline{a}, a^h]$, an impossible task.

One clever way out of this analytical difficulty is to substitute the usual incentive compatibility constraint for another constraint that is somewhat equivalent. We let the agent maximize its own expected utility by choice of action $a \in A$ taking $\{W, s, \rho\}$ as given, or

$$\underset{\{a \in A\}}{Max}\ E[U, agent] = \pi(a)\ U(s\,W - \rho s\,W) + [1 - \pi(a)]\ U(W - \rho s\,W) - a \qquad (5.86)$$

yielding a first order necessary condition

$$1/\pi'(a) = U(s\,W - \rho s\,W) - U(W - \rho s\,W) < 0 \qquad (5.87)$$

with second order sufficient condition satisfied by the convexity of the π function.

The first order condition (5.87) is the constraint to be substituted for the incentive compatibility constraint, a procedure known as the First Order Approach to the principal-agent problem. The first order approach involves substituting the optimality condition of the agent's action choice as the incentive compatibility constraint in the optimal contract problem. The first order condition implies that, given ρ,

$$as\ s \rightarrow 1,\ \ 1/\pi' \ \rightarrow 0,\ \ \pi' \ \rightarrow -\infty\ ,\ \ and\ \ a\ \rightarrow \underline{a},$$

i.e. the action converges to the agent's "minimum" effort or minimum self-protection as the share of wealth to be insured approaches unity, $s \rightarrow 1$. Hence, if the principal offers full coverage, $s=1$, it gives no incentive for the agent to choose a substantive level of effort or care for the insured wealth. Since the agent is risk averse and the principal is risk neutral, full coverage is the efficient (first best) risk sharing contract, i.e. it transfers all risk to the principal. However, to give some incentive for the agent to choose a substantive level of effort, i.e. higher than the minimum level of effort; the principal must offer partial coverage, or $s \in (0,1)$, ultimately making s a function of the level of effort a. Here again we observe the usual tradeoff between risk sharing and incentives under moral hazard.

The Principal - Assumed risk neutral operates in a competitive market so that the zero expected profit condition

$$E[P, principal] = \rho\, s\, W - \pi(a)\, s\, W = 0 \qquad (5.88)$$

yields a competitive pricing formula

$$\rho^* = \pi(a) \qquad (5.89)$$

given the agent's action, $a \in A$. Expression (5.89) yields the actuarially fair price of insurance as in (5.7). In this case, the higher (lower) the effort, the lower (higher) the probability of loss and the lower (higher) the insurance premium.

Substituting (5.87) and (5.89) into (5.85), yields an optimal supply of coverage schedule, as a function $s_*(a)$ which solves

$$U[W - \pi(a)\, s_*(a)\, W] \geq \underline{U} + a - [\pi(a)/\pi'(a)] > 0. \qquad (5.90)$$

This formula yields a function $s_*(a)$ when insurance is available at the competitive price $\rho_*=\pi(a)$, given the optimal action that satisfies the first order necessary condition (5.87), and the participation constraint (5.85). We may easily express the optimal supply schedule as a function of the continuous effort, a, by computing

$$ds_*/da \leq -\{[\pi\pi''/(\pi')^2]+U'(W - \pi s_*W)\, \pi' s_*W\}/U'(W - \pi s_*W)\, \pi W > 0. \qquad (5.91)$$

Hence, for the (participation) constraint (5.90) holding with equality, the optimal supply schedule is increasing in a, or $s_*'(a)>0$. Simply, as the level of the agent's effort a increases, the principal is willing to supply more insurance per unit of wealth, and s_* increases. At the same time, as the level of the agent's effort a increases, the competitive market price ρ_* decreases, but the expected profit remains zero. The intuitive reason for the principal to offer more insurance is that even though the expected (ex-ante) profit remains zero, the total ex-post revenue out of a higher effort state, $s_* W$, increases and ex-post realized profits may increase.

The optimal schedule in (5.90)-(5.91) involves partial risk sharing giving the incentive for the agent to choose a level of effort that is above the "minimum" level, $\underline{a}$, thus mitigating the moral hazard problem.

For example, one way to implement this mechanism in the insurance industry is for the principal to fix the level of coverage at some $s_*(.)=s_{**}$, possibly equal to one, and offer a deductible or co-payment scheme so that:

(a) For a large deductible taken by the agent, the premium rate is low, and the implied accepted level of effort or self-protection is high;

(b) For a small deductible taken by the agent, the premium rate is high, and the implied accepted level of effort or self-protection is low.

(ii) Second, we examine the situation of effort in a supervisor-labor context for one specific task as in section 5.13, but here we use the first order approach. The agent is risk averse and can choose actions smoothly (an infinite number of possible levels of effort, with private information), with actions denoted $a > 0$ in a continuum bounded by the interval $a \in [\underline{a}, a^h]$, for $a^h > \underline{a}$, where $\underline{a}$ indicates the lowest possible effort and a^h the highest possible effort, i.e. the bounded set of possible actions is

$$\boldsymbol{A} = \{a: a \in [\underline{a}, a^h],\ a^h > \underline{a}\}.$$

There are an infinite number of possible outcomes or states of nature on a continuum. The bounded set of possible outcomes in the continuum is denoted by

$$\boldsymbol{X} = \{x \geq 0: x \in [x^l, x^m],\ x^l > x^m\}.$$

The probability density function of the outcomes conditional on action a, assumed to be public information, is denoted by

$$f(x,a), \quad x \in \boldsymbol{X},\ a \in \boldsymbol{A}.$$

Then, the agent's problem is to maximize expected utility [recall expression (5.66)] by choice of a given level of effort, or

$$\underset{\{a \in \boldsymbol{A}\}}{Max}\ E[U, agent] = \{\int_{x \in \boldsymbol{X}} f(x,a)\, U(w(x))\, dx\} - a, \qquad x \in \boldsymbol{X}. \tag{5.92}$$

where $w(x)$ is the state contingent wage schedule as in section 5.13. This yields a first order necessary condition,

$$\int_{x \in \boldsymbol{X}} f_a(x,a)\, U(w(x))\, dx = 1, \quad x \in \boldsymbol{X}, \quad a \in \boldsymbol{A}. \tag{5.93}$$

where $f_a(x,a) = \partial f(x,a)/\partial a$, $a \in \boldsymbol{A}$. The second order sufficient condition is satisfied by the assumed concavity of the probability density function with respect to a in the domain $a \in \boldsymbol{A}$. Equation (5.93) is the appropriate incentive compatibility constraint to be used in the principal's contract design.

The principal's problem is to give the incentive for the agent to choose the induced action. Therefore, the principal chooses $\{w(x),a\}$, $x \in \boldsymbol{X}$, $a \in \boldsymbol{A}$, so that:

$$\underset{\{w(x),a\}}{Max}\ E[U, principal] = \int_{x \in \boldsymbol{X}} f(x,a)\, [x - w(x)]\, dx \tag{5.94}$$

$$\text{subject to} \quad \{\int_{x \in \boldsymbol{X}} f(x,a)\, U(w(x))\, dx\} - a \geq \underline{U},$$

$$\{\int_{x \in \boldsymbol{X}} f_a(x,a)\, U(w(x))\, dx\} - 1 = 0,$$

taking as given $\{f(x,a), x, \underline{U}\}$, $x \in \boldsymbol{X}$, $a \in \boldsymbol{A}$, where the two constraints are the participation and incentive compatibility constraints. The solution, with $\{\mu > 0, \lambda > 0\}$ denoting the Lagrange multipliers attached to the participation and incentive compatibility constraints respectively, yields first order necessary conditions

$$[1/U'(w(x))] = \mu + \lambda\,[f_a(x,a) / f(x,a)], \quad x \in \boldsymbol{X}, \quad a \in \boldsymbol{A} \qquad (5.95)$$

$$\int_{x \in \boldsymbol{X}} f_a(x,a)\,[x - w(x)]\,dx + \lambda\,\{\int_{x \in \boldsymbol{X}} f_{aa}(x,a)\,U(w(x))\,dx\} = 0,$$
$$x \in \boldsymbol{X}, \quad a \in \boldsymbol{A} \qquad (5.96)$$

where $f_{aa}(x,a) = \partial^2 f(x,a)/\partial a^2$, $a \in \boldsymbol{A}$. Expressions (5.95)-(5.96) together with the participation and incentive compatibility constraints solve for $\{w^*(x), a^*, \mu^*, \lambda^*\}$, given $\{f(x,a), x, \underline{U}\}$, $x \in \boldsymbol{X}$, $a \in \boldsymbol{A}$. The second order sufficient conditions are appropriately satisfied given the concavity of the probability density function with respect to a in the domain $a \in \boldsymbol{A}$ and the strict concavity of U. In this case, notice that the likelihood ratio is

$$f_a(x,a) / f(x,a), \quad x \in \boldsymbol{X}, a \in \boldsymbol{A}$$

which appropriately denotes the partial derivative of the logarithm of the likelihood function, $log\, f$, with respect to a:

$$\partial\,[\log f(x,a)]/\partial\, a = f_a(x,a)/f(x,a), \; x \in \boldsymbol{X}, a \in \boldsymbol{A}.$$

Under the maximum likelihood interpretation, the closer $f_a(x,a)$ / $f(x,a)$ to one, or

$$f_a(x,a) / f(x,a) \to 1 > 0,$$

it indicates that $f(x,a)$ is barely sensitive to a, so that the sample has information that the agent took an action consistent with the principal's incentive, i.e. high effort. On the other hand, as $f_a(x,a)/f(x,a)$ deviates below one, or for example

$$f_a(x,a) / f(x,a) < 0,$$

it indicates that $f(x,a)$ is more sensitive to a, so that the sample has information that the agent did not take an action consistent with the principal's incentive, i.e. the agent chose low effort. When

$$f_a(x,a) = 0,$$

there is no useful information in the sample.

The monotone likelihood ratio condition in this case is

$$f_a(x,a) \,/\, f(x,a) \text{ is increasing in } x, \quad x \in \mathbf{X},\ a \in \mathbf{A}.$$

Since, from expression (5.95), $w(x)$ is increasing in $f_a(x,a)/f(x,a)$, i.e. U is strictly concave; we conclude that the wage schedule is increasing in x. Hence, for $f_a(x,a)/f(x,a)>0$, the offer is the base plus a premium with the implied low "noise to signal ratio" as in section 5.13; and for $f_a(x,a)/f(x,a)<0$, the offer is the base minus the penalty, with the implied high "noise to signal ratio."

Reviewing the example in section 5.13, with utility function $U(w)=2w^{1/2}$, the optimal contract will be of the form

$$w^*(x) = \{\mu^* + \lambda^* \,[\, f_a(x,a^*) \,/\, f(x,a^*) \,]\}^2, \quad x \in \mathbf{X},\ a \in \mathbf{A}.$$

Assuming an exponential probability distribution with mean $a \in \mathbf{A}$, the probability density function is given by the formula

$$f(x,a) = (1/a)\ exp[-(x/a)], \quad x \in \mathbf{X}$$

where *exp* is the exponential operator. By direct computation, we obtain a wage schedule as

$$w^*(x) = \{\mu^* + \lambda^* \,[\, (x - a^*) \,/\, a^{*\,2} \,]\}^{\,2}$$

where $\lambda^*>0$ is obtained by solving (5.96), and $\mu^*>0$ and a^*, by solving the participation and incentive compatibility constraints. The wage schedule is strictly increasing and strictly convex in x as depicted in Figure 5.12. At $x = a$, the average of the distribution of outcomes, we obtain the first best solution,

$$w'^{**}(x) = \mu'^{**\,2}.$$

As before, if $x>a$, or $f_a(x,a)/f(x,a)>0$, the offer is the base plus a premium with the implied low "noise to signal ratio;" whereas if $x<a$, or $f_a(x,a)/f(x,a)<0$, the offer is the base minus the penalty, with the implied high "noise to signal ratio."

5.15 **Summary III**

In sections 5.11-5.14, we examined the moral hazard problem emerging from asymmetric information. In this case, we may find a contract that mitigates moral hazard, and the tradeoff between efficient risk sharing and incentives emerges naturally. We examine below the effects of moral hazard on asset returns.

5.16 Asset Returns and Moral Hazard

The final part of this chapter presents and discusses the effects of moral hazard on asset returns in the context of the consumption-based capital asset pricing model (CCAPM) with particular attention to issues regarding idiosyncratic versus systematic risk, e.g. Chapter 2, and Zero-level pricing.

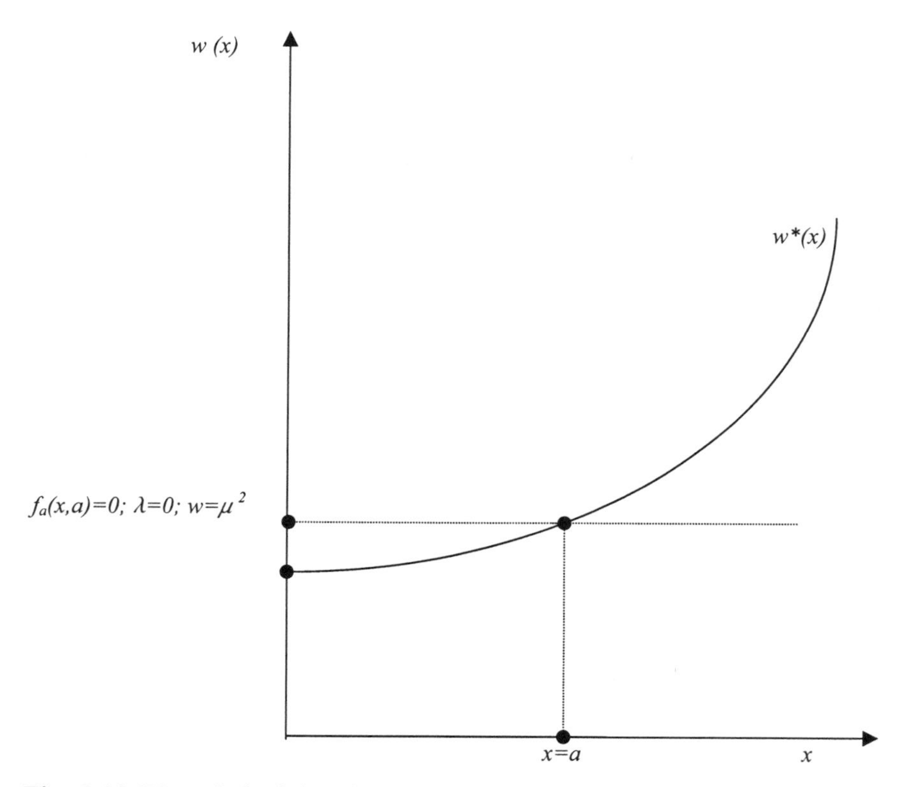

Fig. 5.12 Wage Schedule with Infinite Number of Actions and Outcomes

5.17 **Basic Model**

Here, we consider economies populated by a large number of individuals with each individual denoted by the name i. Each individual named i may have private information about their work effort in the production of a single good. In particular, there may be an infinite number of actions/effort taken by the individual so that the action/effort is unobserved by the general public. Individuals are assumed to be risk averse.

The basic model pays particular attention to the distinction between idiosyncratic and systematic risk as discussed in Chapter 2. Systematic or aggregate risk is denoted by the aggregate state variable, θ, which affects the distribution of the agent's output, conditional upon the unobserved work effort. Individuals can observe the aggregate state, θ, before their choice of work effort, thus making work effort contingent upon the aggregate state.

The idiosyncratic or non-systematic risk is introduced as follows. Each individual named i chooses one state contingent action/effort, denoted $a^i(\theta) \geq 0$, in order to operate a production technology that yields individual stochastic output denoted y^i. The individual output has two possible outcomes,

$$y^i_\gamma, \text{ for } \gamma = g \text{ ("good"), } b \text{ ("bad").}$$

The probability distribution of the individual output is binomial, and assumed to depend upon the state contingent work effort as:

$$y^i_g \text{ with probability } p(a(\theta)),$$

$$y^i_b \text{ with probability } 1 - p(a(\theta));$$

$$y^i_g > y^i_b \text{ all } i$$

where $p(a(\theta)) \geq 0$ is a strictly increasing and strictly concave function of effort, $a(\theta)$, contingent on the aggregate state, or $p' > 0$, $p'' < 0$. Intuitively, the larger (lower) the work effort for a given aggregate state, the larger (lower) the probability of higher output produced by the individual. Also, given the aggregate state, the individual output in the "good" state, y^i_g, is uniformly higher than in the "bad" state, y^i_b.

Aggregate or systematic risk is assumed to follow a binomial distribution as well. The average or aggregate per individual output is denoted Y_θ, for $\theta = G,B$. The probability distribution of average output is given by

$$Y_G \text{ with probability } \pi,$$

$$Y_B \text{ with probability } 1-\pi;$$

$$Y_G > Y_B$$

where $\pi \geq 0$ is a simple probability measure.

Therefore average (aggregate per individual) and individual outputs are related by the expressions

$$Y_G = p(a(G))\, y^i_g + [1-p(a(G))]\, y^i_b \qquad (5.97a)$$

$$Y_B = p(a(B))\, y^i_g + [1-p(a(B))]\, y^i_b. \qquad (5.97b)$$

Expressions (5.97a,b) simply state the average output as a linear combination of individual outputs. In particular, (5.97a,b) show that average output is an increasing function of the effort supplied by the individual. We can easily compute the mean and variance of aggregate output in this economy as

$$E[Y] = \pi Y_G + (1-\pi)\, Y_B$$

$$Var(Y) = \pi\,(1-\pi)\,(Y_G - Y_B)^2$$

where we used the law of large numbers to express the dependence on the individual type $\gamma = g,b$ only through Y_θ, given in (5.97a,b).

An individual derives utility from consumption of the single good and disutility from effort according to an additively separable function

$$U = u(c^i_\gamma(\theta)) - v(a(\theta)) \qquad (5.98)$$

where $c^i_\gamma(\theta)$ is the individual state contingent consumption, u strictly increasing and strictly concave, $u' > 0$, $u'' < 0$, and v is a (dis)utility function of effort, assumed strictly increasing and convex, $v' > 0$, $v'' \geq 0$. Thus, an individual is assumed to be risk averse regarding the consumption prospect and the disutility of labor. The function v could be linear without much loss of generality, indicating the special case where the individual is risk neutral regarding the disutility of labor. The expected utility, over individual type γ and aggregate state θ, of a consumer i is given by

$$E[U^i] = \pi\{p(a(G))\, u(c^i_g(G)) + [1-p(a(G))]\, u(c^i_b(G)) - v(a(G))\} + (1-\pi)\{p(a(B))\, u(c^i_g(B)) + [1-p(a(B))]\, u(c^i_b(B)) - v(a(B))\}. \quad (5.99)$$

The basic setup is simple enough to accommodate a linear sharing rule. Because there are two aggregate and two idiosyncratic states (2×2 case), a 'principal' can basically offer a linear sharing rule contingent upon the final observable outcomes y^i_γ and Y_θ, see e.g. sections 5.13-5.14. In particular, we can postulate a linear consumption rule for an individual i, of the form

$$c^i_\gamma(\theta) = \alpha_\theta\, y^i_\gamma + (1-\alpha_\theta)\, Y_\theta, \quad \text{each } \theta = G,B;\ \gamma = g,b \qquad (5.100)$$

where α_θ is a choice variable in the 'principal's' problem of maximizing the expected utility of the agent subject to the appropriate incentive compatibility and participation constraints (more on this below!).

However, the level of effort of the individual is assumed to be continuous. We thus apply the First Order Approach discussed in section 5.14 to obtain the appropriate incentive compatibility constraint.

The appropriate incentive compatibility constraint is the one that maximizes the expected utility, over individual type γ, by choice of the level of effort $a^i(\theta)$ of an individual, subject to consumption following the linear sharing rule given in (5.99), all conditional on the aggregate state θ, and taking the principal's share parameter α_θ as given. The problem may be written as

$$\underset{\{a^i(\theta)\geq 0\}}{Max}\ p(a^i(\theta))\, u(c^i_g(\theta)) + [1-p(a^i(\theta))]\, u(c^i_b(\theta)) - v(a^i(\theta)) \quad (5.101)$$

subject to (5.100),

with $\{\alpha_\theta, Y_\theta\}$ taken as given by the individual agent. Substituting the constraint into the objective function and computing the necessary first order conditions of the resulting unconstrained problem yields the expression

$$p'(a^i(\theta))[u(\alpha_\theta\, y^i_g + (1-\alpha_\theta)\, Y_\theta) - u(\alpha_\theta\, y^i_b + (1-\alpha_\theta)\, Y_\theta)] = v'(a^i(\theta)), \quad \text{each } \theta \qquad (5.102)$$

with the sufficient second order condition for the unconstrained maximum appropriately satisfied by the strict concavity of the p function.

The solution in (5.102) is familiar from asymmetric information problems where the tradeoff between incentives and risk sharing arises (sections 5.13-5.14). The first best would be to provide full insurance to the risk averse individual over the idiosyncratic (non-systematic) risk, i.e. efficient or full risk sharing. Full insurance for the idiosyncratic risk in this context implies that $\alpha_\theta = 0$ for all θ. This is because with $\alpha_\theta = 0$ in (5.100), consumption is just the average across individuals, $c^i_\gamma(\theta) = Y_\theta$, for each $\theta = G,B$ and all $\gamma = g,b$. This implies, from expression (5.102) that

$$v'(a^i(\theta))/p'(a^i(\theta)) \to 0, \text{ all } \theta \iff v' \to 0 \text{ or } p' \to \infty. \qquad (5.103)$$

Hence, expression (5.103) shows that under the first best or full insurance for the idiosyncratic risk, the effort is minimum, $a^i \to 0$, since the agent has no incentive to provide effort, i.e. the classic moral hazard problem. The solution is for the 'principal' to provide partial insurance to the individual idiosyncratic risk. In the second best with partial insurance, the agent has an incentive to provide effort above the minimum level, and we observe the usual tradeoff between incentive and risk sharing.

The second best solution that mitigates moral hazard is for the 'principal' to choose $\alpha_\theta > 0$ inducing the agent to provide an arbitrary level of effort $a^i > 0$. It suffices, for our purposes here, to note that the first order conditions (5.102) allow the 'principal' to choose a linear share parameter $\alpha_\theta > 0$, contingent on the aggregate state θ, that induces the desired and arbitrary level of effort $a^i > 0$, given that Y_θ is appropriately substituted from expressions (5.97a,b). For an arbitrary level of effort $a^{i*} > 0$, $\alpha_\theta^* > 0$ must then solve

$$p'(a^{i*}(\theta))\,[\,u(\alpha_\theta^*\, y^i_g + (1-\alpha_\theta^*)\{p(a^{i*}(\theta))\, y^i_g + [1-p(a^{i*}(\theta))]\, y^i_b\}) -$$
$$u(\alpha_\theta^*\, y^i_b + (1-\alpha_\theta^*)\{p(a^{i*}(\theta))\, y^i_g + [1-p(a^{i*}(\theta))]\, y^i_b\})\,] = v'(a^{i*}(\theta)),$$
$$\text{each } \theta.$$

Such a solution mitigates the moral hazard problem, because the principal provides a share $\alpha_\theta > 0$, which induces an arbitrary state contingent effort $a^{i*}(\theta) > 0$, above the minimum level. In effect, $\alpha_\theta \geq 0$, measures the extent of partial risk sharing between the principal and the agent. If $\alpha_\theta = 0$, from the linear sharing rule (5.100), the agent only faces aggregate risk and the principal bears all the idiosyncratic risk. When $\alpha_\theta > 0$, the agent bears some of the idiosyncratic risk thus having an incentive to provide effort above the minimum level.

5.18 **Equilibrium Asset Returns**

Given the α_θ^* chosen to induce the desired level of effort by the agent, we can now describe optimal state contingent consumption in terms of a portfolio of assets. We let an individual hold a portfolio with three assets.

First, we have a private asset subject to the own individual risk, then we have two other macro assets. The characteristics of the assets are as follows: (i) the private risky asset yields the return y^i_γ representing the own individual risk; (ii) one aggregate risky asset yields the average per individual output, Y_θ , representing aggregate macro risk; and (iii) the other macro asset is a risk-free asset yielding the risk-free total return R_f. Therefore, the state contingent individual consumption is replicated by holding a portfolio with the three assets expressed as

$$c^i_\gamma(\theta) = \alpha_\theta \ y^i_\gamma + (1-\alpha_\theta)\,[\,s\,Y_\theta + (1-s)\,R_f]\ ,\ each\ \theta{=}G,B;\ \gamma=g,b \qquad (5.104)$$

where $s \in [0,1]$ is the share of the risky asset in the portfolio of macro assets. Indeed, $\alpha_\theta \geq 0$, measures the extent of partial risk sharing between the principal and the agent.

The important question here is the extent to which the introduction of moral hazard affects asset returns. Towards an answer, first consider the risk-free return, R_f. We obtain the endogenous value for R_f using the Zero-Level pricing method: maximize expected utility of an individual by choice of the share of macro assets, s, taking the risk-free return, R_f as given (we apply this method in Chapter 8 as well). Then, we set the share of the risk-free macro asset on the portfolio to zero, i.e. $s=1$, and find the appropriate R_f that makes the risk-free asset in zero net supply, i.e. the value of the asset is the appropriate shadow value that makes the agent indifferent between owning it or not. The specific problem for an individual i may be written as

$$Max_{\{s\}}\ \pi\ \{ p(a(G))\ u\ (c^i_g(G)) + [1-p(a(G))]\ u\ (c^i_b(G)) - v(a(G)) \} +$$
$$(1-\pi)\ \{ p(a(B))\ u\ (c^i_g(B)) + [1-p(a(B))]\ u\ (c^i_b(B)) - v(a(B)) \}$$
$$\text{subject to (5.104)}, \qquad (5.105)$$

taking $\{\alpha_\theta, a(\theta), R_f\}$ as given. As before, the solution is easily obtained by substituting the constraint into the objective obtaining necessary first order condition as

$$\pi\ (1-\alpha_G)(Y_G - R_f)\ \{ p(a(G))\ u'\ (c^i_g(G)) + [1-p(a(G))]\ u'\ (c^i_b(G))\} +$$
$$(1-\pi)\ (1-\alpha_B)(Y_B - R_f)\{ p(a(B))\ u'\ (c^i_g(B)) + [1-p(a(B))]\ u'\ (c^i_b(B)) \}$$
$$= 0 \qquad (5.106)$$

where $c^i_\gamma(\theta)$ is given in (5.104), and the sufficient second order condition is satisfied by the strict concavity of the function u. Applying the zero-level pricing method involves solving the first order condition (5.106) for R_f, and evaluating at $s=1$, yielding

$$R_f^* = E[\ u'\ (c^i_\gamma(\theta))\ (1-\alpha_\theta)\ Y_\theta] \,/\, E[u'\ (c^i_\gamma(\theta))\ (1-\alpha_\theta)] \qquad (5.107)$$

where the expectation is over γ and θ. The economy with idiosyncratic risk shows a risk-free return that depends upon the marginal utility of individual consumption and the sharing parameter, besides the usual macro factors, see e.g. section 4.3.

Next, we compute the expected total return of the average (aggregate per individual) output as

$$E\,[Y] = \pi\, Y_G + (1-\pi)\ Y_B \qquad (5.108a)$$

with variance

$$Var\ (Y) = \pi\,(1-\pi)\ (Y_G - Y_B)^2 \qquad (5.108b)$$

where by the law of large numbers the expected return depends on γ only through Y_θ as in (5.97).

We are ready to compute the risk premium in the economy with moral hazard: $E[Y]-R_f^*$. Using expressions (5.107), (5.108a), and the usual covariance decomposition formula, we obtain

$$E[Y] - R_f^* = -\ cov\ (\ u'\ (c^i_\gamma(\theta))\ (1-\alpha_\theta),\ Y_\theta) \,/\, E\,[u'\ (c^i_\gamma(\theta))\ (1-\alpha_\theta)] \qquad (5.109)$$

This is a familiar formula for the risk premium except for the presence of the term relating to the principal's share parameter $(1-\alpha_\theta)$. When there is no private information, optimal full insurance (*FI*) yields $\alpha_\theta=0$, and evaluating at $s=1$ or $c^i_\gamma(\theta)=Y_\theta$, the risk premium takes the familiar form (as seen in Chapter 4)

$$E[Y] - R_f^* |_{FI} = - cov\,(u'(Y_\theta), Y_\theta) / E\,[u'(Y_\theta)]. \qquad (5.110)$$

In practice, the covariance of the marginal utility with the risky asset is negative since in the good state consumption is high and the marginal utility is low, i.e. the risk premium is positive. Thus, the higher the covariance (the less negative), the lower the risk premium.

Next, we examine the effect of moral hazard on the risk premium. We understand, say from section 5.17, that $\alpha_\theta>0$ is necessary to induce effort. But, at the same time, $\alpha_\theta>0$ affects the asset returns as seen in (5.109). With moral hazard, there is partial insurance for the idiosyncratic risk and agents bear some of their own idiosyncratic risk in consumption. Intuitively, it is analogous to the case when individuals develop a need for saving, in other words when a 'precautionary' saving motive emerges (see section 3.3 on the precautionary savings motive based on the positive third derivative of the utility function, $u'''>0$). In particular, this implies that, in addition to aggregate risk, individuals must take into account idiosyncratic risk thus making the marginal utility more variable relative to the case of full insurance.

For example, when there is full insurance to the idiosyncratic risk, or $\alpha_\theta=0$, the risk premium is given in expression (5.110), or equal to the term $-cov(u'(Y_\theta),Y_\theta)/E[u'(Y_\theta)]$ (as in the full information case) because individual consumption $c^i_\gamma(\theta)=Y_\theta$ for all γ, and all variation in the marginal utility comes from the aggregate risk. When there is partial insurance to the idiosyncratic risk ($\alpha_\theta>0$), there is the additional variation of the individual idiosyncratic risk in consumption, because $c^i_\gamma(\theta)\neq Y_\theta$, thus increasing the variation of the marginal utility. In this case, the risk premium is given in (5.109), and we have that

$$- cov\,(u'(c^i_\gamma(\theta))\,(1-\alpha_\theta), Y_\theta) / E\,[u'(c^i_\gamma(\theta))\,(1-\alpha_\theta)] >$$
$$- cov(u'(Y_\theta),Y_\theta)/E[u'(Y_\theta)] \qquad (5.111)$$

exactly because of the additional variation of the marginal utility in the second best.

Hence, the main result here is that the risk premium, in the presence of optimal contracts that mitigate moral hazard, is potentially larger relative to the case of full information. Given that one of the puzzles of general equilibrium asset pricing models is the small magnitude of the risk premium, the presence of moral hazard seems to work in the right direction in solving the puzzle. We shall examine the issue of risk premiums and excess returns more closely for the dynamic case in Chapter 7.

5.19 **Summary IV**

Sections 5.16-5.18 show that asymmetric information can have important effects on asset valuation and returns. In particular, we have shown that mitigating moral hazard can increase the risk premium in the CCAPM.

Problems and Questions

1. Explain exactly how the traditional neoclassical theory of the firm can be changed to explain incentives within the firm.

2. Explain exactly how the traditional neoclassical theory of the firm can be changed to explain the private ownership and boundaries of firms.

3. Explain the concept of a Constrained Pareto Optimal Equilibrium?

4. What is the difference between Adverse Selection and Moral Hazard?

5. Derive the equilibrium price of insurance when the agent is risk averse and the principal is risk neutral under adverse selection. Does your answer relate to the Fundamental Theorem of Risk Bearing? How?

6. What is the Signaling problem in reference to the Adverse Selection case examined in section 5.12? Explain.

7. Derive an equilibrium incentive scheme in a simple insurance problem under Moral Hazard.

8. Derive the effect of Moral Hazard on Asset Returns? Explain.

Notes on the Literature

A recent book by Laffont and Martimort (2002) is a brilliant and up-to-date presentation of models of adverse selection, signaling and moral hazard of sections 5.4-5.15. The reader is encouraged to consult their book for a more broad presentation and set of bibliographic references.

The material in sections 5.1-5.3 is based on the monograph by Hart (1995) Chapters 1, 2, and 3. The reader is encouraged to consult his book and the references therein. A recent survey of incentives in the theory of the firm, including career concerns is in Prendergast (1999). The problem of incomplete contracts is covered in a Special Issue (1999) of the *Review of Economic Studies*, Salanie (1997), Chapter 7, and more recently in Battigalli and Maggi (2002). Valuable surveys on financial contracting are found in Harris and Raviv (1992), and Hart (2001).

Sections 5.4-5.6 on adverse selection, principal-agent relationships, and insurance markets examples, can be found in Laffont and Martimort (2002), Chapter 3; Salanie (1997), Chapter 2; Laffont (1980, 1989); Eichberger and Harper (1995), Chapter 6; and Macho-Stadler and Perez-Castrillo (1997). Dixit (1999) presents an important extension with disclosure clauses; and Prendergast (2000) provides important empirical evidence on the tradeoff between risk and incentives under agency.

Section 5.7 on price discrimination is based on a classic paper by Mussa and Rosen (1978), and is also found in Salanie (1997), Chapter 2; Macho-Stadler and Perez-Castrillo (1997), Chapter 4.

Section 5.8 on credit markets is based on the classic paper by Stiglitz and Weiss (1981) and may be found in Blanchard and Fischer (1989), Chapter 9; De Meza and Webb (1987); Freixas and Rochet (1998), Chapter 5 present a synthesis.

Section 5.9 on signaling is based on the classic paper by Akerlof (1970) and is found in Salanie (1997), Chapter 4; Macho-Stadler and Perez-Castrillo (1997), Chapter 5. A classic paper on signaling is Spence (1973) which is an application to labor markets. Riley (2001) is an excellent survey of the literature on signaling and asymmetric information. Jones and Manuelli (2000) present comparative statics results of increases in risk under signaling.

Sections 5.11-5.15 on moral hazard is based on the classic paper by Holmstrom (1979) and is found in Laffont and Martimort (2002), Chapter 4; Salanie (1997), Chapter 5; Hart and Holmstrom (1987) is a very complete and useful survey; and Macho-Stadler and Perez-Castrillo (1997), Chapter 3. The concept of subgame perfect Nash equilibrium is presented in Friedman (1986). Likelihood ratio tests in statistics are found in Godfrey (1988). The monotone likelihood ratio condition is presented in Hart and Holmstron (1987) and Laffont and Martimort (2002), Chapter 4.

Rogerson (1985) and Jewitt (1988) are classic references on the first-order approach of section 5.14; a useful survey is found in Laffont and Martimort (2002), Chapter 5; and a recent paper by Araujo and Moreira (2001) show a method to solve for the moral hazard problem with a continuum of actions giving a general characterization of the first order approach. Laffont and Martimort (2002), Chapter 7, present a survey of mixed models of moral hazard and adverse selection. Extensions to a dynamic setting are provided in Chapter 7 of this book, in Laffont and Tirole (1987), Hart and Tirole (1988), and a survey is in Laffont and Martimort (2002), Chapter 8.

Sections 5.17-5.19 are based on Kahn (1990), and the Zero-level pricing method can be found in Chapter 8 of this book and Luenberger (1998), Chapter 16.

References

Akerlof, George (1970) "The Market for Lemons: Quality Uncertainty and the Market Mechanism." *Quarterly Journal of Economics,* 89, 488-500.

Araujo, Aloisio and Humberto Moreira (2001) "A General Lagrangian Approach for Non-Concave Moral Hazard Problems." *Journal of Mathematical Economics*, 35, 17-39.

Battigalli, Pierpaolo and Giovanni Maggi (2002) "Rigidity, Discretion and the Costs of Writing Contracts." *American Economic Review*, 92, 798-817.

Blanchard, Olivier and Stanley Fischer (1989) *Lectures on Macroeconomics.* The MIT Press, Cambridge, MA.

De Meza, David and David C. Webb (1987) "Too Much Investment: A Problem of Asymmetric Information." *Quarterly Journal of Economics,* 102, 281-292.

Dixit, Avinash (1999) "Adverse Selection and Insurance with 'Uberrima Fides'." Working paper, Department of Economics, Princeton University, October.

Eichberger, Jurgen and Ian R. Harper (1995) *Financial Economics.* Oxford University Press, Oxford, UK.

Freixas, Xavier and Jean-Charles Rochet (1998) *Microeconomics of Banking.* The MIT Press, Cambridge, MA.

Friedman, James (1986) *Game Theory with Applications to Economics.* Oxford University Press. Oxford, UK.

Godfrey, Leslie G. (1988) *Misspecification Tests in Econometrics: The Lagrange Multiplier Principle and Other Approaches.* Econometric Society Monographs by Cambridge University Press, Cambridge, UK.

Harris, Milton and Arthur Raviv (1992) "Financial Contracting Theory." In Laffont, Jean J., Ed., *Advances in Economic Theory: Sixth World Congress, Vol. II.* Cambridge University Press, Cambridge, UK.

Hart, Oliver (2001) "Financial Contracting." NBER Working Paper No. 8285, May. Forthcoming *Journal of Economic Literature.*

Hart, Oliver (1995) *Firms, Contracts, and Financial Structure.* Claredon Press, Oxford, UK

Hart, Oliver and Jean Tirole (1988) "Contract Renegotiation and Coasian Dynamics." *Review of Economic Studies,* 60, 509-540.

Holmstrom, Bengt (1979) "Moral Hazard and Observability." *Bell Journal of Economics*, 10, 74-91.

Holmstrom, Bengt and Oliver Hart (1987) "The Theory of Contracts." In Bewley, Truman, Editor: *Advances in Economic Theory: The Fifth World Congress*. Cambridge University Press, Cambridge, UK.

Jewitt, Ian (1988) "Justifying the First-Order Approach to Principal-Agent Problems." *Econometrica*, 56, 1177-1190.

Jones, Larry and Rodolfo Manuelli (2000) "Increases in Risk and the Probability of Trade." Working paper, Department of Economics, University of Minnesota and University of Wisconsin-Madison, March.

Kahn, James A. (1990) "Moral Hazard, Imperfect Risk Sharing, and the Behavior of Asset Returns." *Journal of Monetary Economics*, 26, 27-44.

Laffont, Jean J. (1980) *Essays in the Economics of Uncertainty*. Harvard University Press, Cambridge, MA.

Laffont, Jean J. (1989) *The Economics of Uncertainty and Information*. The MIT Press, Cambridge, MA.

Laffont, Jean J. and Jean Tirole (1987) "Comparative Statics of the Optimal Dynamic Incentive Contract." *European Economic Review*, 31, 901-926.

Laffont, Jean. J. and David Martimort (2002) *The Theory of Incentives - The Principal-Agent Model*. Princeton University Press, Princeton, NJ.

Luenberger, David (1998) *Investment Science*. Oxford University Press, Oxford, UK.

Macho-Stadler, Ines and J. David Perez-Castrillo (1997) *An Introduction to the Economics of Information: Incentives and Contracts*. Oxford University Press, Oxford, UK.

Mussa, Michael and Sherwin Rosen (1978) "Monopoly and Product Quality." *Journal of Economic Theory*, 18, 301-317.

Prendergast, Canice (1999) "The Provision of Incentives in Firms." *Journal of Economic Literature*, 37, 7-63.

Prendergast, Canice (2000) "The Tenuous Tradeoff Between Risk and Incentive." NBER Working Papers Series No. 7815, July.

Riley, John (2001) "Silver Signals: Twenty-Five Years of Screening and Signaling." *Journal of Economic Literature*, 39, 432-478.

Rogerson, William (1985) "The First-Order Approach to Principal-Agent Problems." *Econometrica*, 53, 1357-1368.

Salanie, Bernard (1997) *The Economics of Contracts: A Primer*. The MIT Press, Cambridge, MA.

Special Issue: Contracts (1999) *Review of Economic Studies*, 66, 1-199.

Spence, Michael (1973) "Job Market Signaling." *Quarterly Journal of Economics*, 87, 355-379.

Stiglitz, Joseph and Andrew Weiss (1981) "Credit Rationing in Markets with Imperfect Information." *American Economic Review*, 71, 393-410.

6. Non-convexities and Lotteries in General Equilibrium

6.1 Introduction

A simple general equilibrium aggregative model is the so-called "Robinson Crusoe" framework. In this setting, an individual produces, given the available technology, and consumes, by a choice of consumption and labor-leisure that maximizes (possibly expected) utility, subject to the technology and level of capital stock and resources constraint. This simple model can be refined in several directions, one of them being the decentralization of consumption and production decisions by identical individuals and firms.

In this chapter, first we use the decentralized version of the model to study a more subtle issue that turns out to be of great importance in our study of asymmetric information under general equilibrium. In the classical general equilibrium framework, the optimal allocation is based on a feasible set that is convex. The question we pursue is the extent to which this is an important assumption. Alternatively, are non-convexities important? The answer is yes. We present an example in the context of a decentralized competitive economy where the individual's labor supply is indivisible. Then, we extend the partial equilibrium framework of Chapter 5, to a classical general equilibrium framework under asymmetric information. Lotteries and the aggregate resources constraint play a key role in determining the properties and feasibility of equilibrium. Edward C. Prescott and Robert Townsend have developed the basic framework of analysis in this chapter. In this part, we pay particular attention to the fundamental theorems of welfare economics in environments with asymmetric information.

We recall that the First Fundamental Theorem of Welfare Economics states that competitive markets lead to Pareto Optimal allocations, and the Second Fundamental Theorem states that a Pareto Optimal allocation can be decentralized by appropriate competitive prices. Finally, we present an example of the macroeconomic and asset allocation effects of alternative unemployment insurance regimes under asymmetric information, in the context of interactions between countries or regions.

6.2 A Static Decentralized Competitive Framework

The specific setting here is one of full information based on the contribution of Richard Rogerson. There is a continuum of identical individuals whose names are over the unit interval $[0,1]$, or name denoted $g \in [0,1]$. All activities take place in a single period. Each agent is endowed with one unit of time and one unit of capital. Time is assumed to be indivisible, this being the source of non-convexity. Hence, either the individual supplies the entire time for labor (and has no leisure) or the individual does not supply any labor at all (entire time for leisure). The labor supply indivisibility is an exogenous constraint faced by agents in the economy. We assume that the individual stock of capital is fixed at k_o, and that labor, or n for the fraction of the unit of time supplied as labor, is the only variable input in the production process, i.e. $n \in \{0,1\}$. The aggregate technology is denoted $f(K,N)$ which is assumed to be constant returns to scale (strictly increasing and strictly concave in each of its arguments), and K and N are the aggregate capital and hours worked in the economy.

Individuals are endowed with identical utility index, assumed to be separable in consumption and labor, taking the form (familiar from Chapter 5)

$$\textit{Individual's utility} = u(c) - v(n) \qquad (6.1)$$

where $c \geq 0$ is individual consumption of the single good produced, and $n \in \{0,1\}$ is the indivisible labor supply. The utility function u is assumed to be twice continuously differentiable, strictly increasing and concave. The v function is so that the only values that matter are

$$v(0)=0, \quad v(1)=m>0 \tag{6.2}$$

for a given constant real number $m>0$. When $n=0$, the individual is not supplying any labor and the disutility of labor is zero. When $n=1$, the individual is supplying all labor, the disutility is a constant denoted m. If labor supply, n, were perfectly divisible, or $n \in [0,1]$, then the v function would be smooth, and the individual would be able to choose a divisible amount of labor supply in the open interval $[0,1]$.

The underlying consumption set for the representative agent in this setting is given by

$$\boldsymbol{X} = \{(c,n,k_o) \in \mathbb{R}^3 : c \geq 0,\ n \in \{0,1\},\ k_o = 1\} \tag{6.3}$$

which is non-convex because $n \in \{0,1\}$.

6.3 The Competitive Equilibrium

We let w denote the market wage rate and r denote the market rental rate. The competitive equilibrium for each individual $g \in [0,1]$ is:

(i) $\{c(g), n(g), k_o, K, N\}$ is a feasible allocation, i.e. $\{c(g), n(g), k_o\} \in \boldsymbol{X}$, and $K,N \geq 0$;

(ii) Given w and r, $\{c(g), n(g)\}$ is a solution for the household's problem

$$\underset{\{c,n\}}{Max}\ [u(c) - v(n)] \tag{6.4}$$

$$\begin{aligned} subject\ to \quad & c \leq n\,w + r\,k_o \\ & c \geq 0 \\ & n \in \{0,1\} \\ & k_o = 1, \end{aligned}$$

and $\{N,K\}$ is a solution to the firm's problem

$$\underset{\{N,K\}}{Max}\ [f(K,N) - w\,N - r\,K] \tag{6.5}$$

$$subject\ to \quad K,N \geq 0;$$

(iii) The resources constraints are satisfied, or supply equals demand in the goods, labor, and capital market

$$f(K,N) = \int_0^1 c(g)\, dg \tag{6.6}$$

$$N = \int_0^1 n(g)\, dg \tag{6.7}$$

$$K = \int_0^1 k_o\, dg = 1. \tag{6.8}$$

There are two important problems with the allocation of the competitive equilibrium in (i)-(iii). First, identical individuals may receive different allocation in equilibrium. In particular, the consumption of an individual supplying labor, $n=1$, is

$$c^{n=1} = w + r \tag{6.9}$$

whereas the consumption of those not supplying labor, $n=0$, is

$$c^{n=0} = r. \tag{6.10}$$

We can characterize the equilibrium by defining the proportion of individuals in the total population that supply labor as $x \geq 0$, so that for those who supply labor

$$[c^{n=1}(g),\ n(g)=1,\ k_o=1] \quad for \quad g \in [0, x\,], \tag{6.11}$$

and for those who do not supply labor

$$[c^{n=0}(g),\ n(g)=0,\ k_o=1] \quad for \quad g \in [\,x\,, 1]. \tag{6.12}$$

However, if the choice set $\boldsymbol{X}$ were convex, all individuals would receive the same allocation, and $x=1$, for some $n \in [0\,, 1]$.

The second problem is more subtle. Suppose one is able to divide the consumption bundle evenly across all individuals in the following manner: give an average consumption

$$c^{\,average} = \phi\, c^{\,n=1}(g) + (1 - \phi)\, c^{\,n=0}(g) \tag{6.13}$$

to each individual, and let each individual hold a lottery ticket with (fair) probability $\phi \geq 0$ to be employed, i.e. probability ϕ that $n=1$. This allocation would be strictly preferred to the original one because it gives a higher expected utility (we show this formally in the next section). However, this alternative allocation does not belong to the feasible set $\boldsymbol{X}$ in (6.3) and is not attainable in the economy with non-convexity.

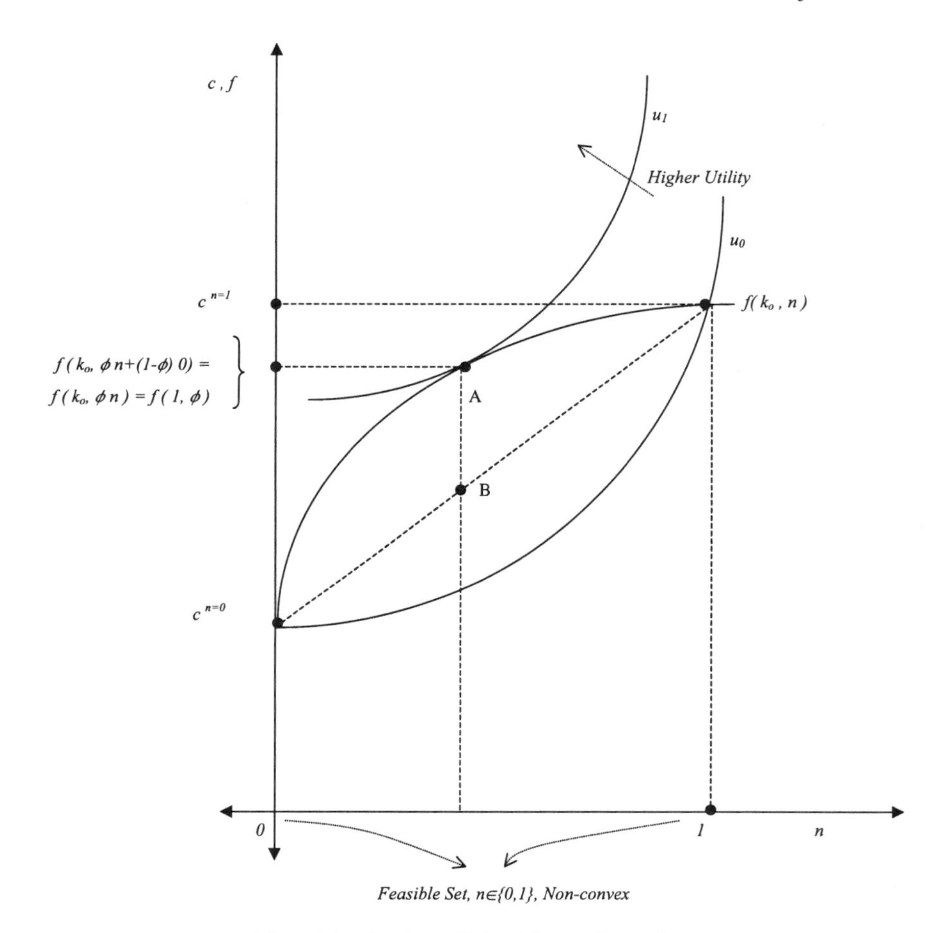

Fig. 6.1 The Feasible Allocations Set

The two problems are illustrated in Figure 6.1. The consumption set is non-convex, and the different allocations are $c^{n=0}$ and $c^{n=1}$. Providing average consumption of (6.13), will take us to the higher utility level at point B. In effect, for strictly concave utility in consumption, any randomized bundle provides higher expected utility. But, in our original setting with indivisible labor, point B is outside the feasible consumption set. The bliss point with divisible labor is point A, and this could be obtained if we were able to introduce trade in particular labor supply lotteries. The main issue is that if trade in lotteries is introduced, individuals will be better off.

6.4 Trade in Lotteries

In order to introduce trade in lotteries, we make changes on the consumption set and preferences. We seek to define an allocation set for all individuals, supplying labor or not, that is convex. First, let the set of allocations of individuals supplying labor be

$$\boldsymbol{X}_1 = \{(c,n,k_o) \in \boldsymbol{X} : n = 1\}. \qquad (6.14a)$$

Next, let the set of allocations of individuals not supplying labor be

$$\boldsymbol{X}_2 = \{(c,n,k_o) \in \boldsymbol{X} : n = 0\}. \qquad (6.14b)$$

Now, define the allocation set

$$\underline{\boldsymbol{X}} = \boldsymbol{X}_1 \times \boldsymbol{X}_2 \times [0,1] \qquad (6.15)$$

as the Cartesian product of $\boldsymbol{X}_1$ and $\boldsymbol{X}_2$ and $[0,1]$, where the latter is the probability measure that $\boldsymbol{X}_1$ is realized. This set is convex. Denoting the probability measure as $\phi \in [0,1]$, a number indicating the probability of $\boldsymbol{X}_1$, an element of the set $\underline{\boldsymbol{X}}$ is the 3-tuple $\{(c^{n=1}, 1, k_o), (c^{n=0}, 0, k_o), \phi\}$.

The Von-Neumann-Morgenstern (VNM) expected utility of individuals holding the lottery is then

$$\textit{Individual's expected utility} = \phi\,[\,u(c^{n=1}) - m] + (1-\phi)\,u(c^{n=0}). \qquad (6.16)$$

In this economy with trade in lotteries, ex-ante (before uncertainty is realized) all individuals are identical and hold a lottery ticket ϕ as the equal opportunity of employment. A competitive equilibrium for the economy with trade in lottery is a list $\{c^{n=1}(g), c^{n=0}(g), \phi(g), k_o, K, N, w, r\}$ where for each individual $g \in [0,1]$:

(i) $\{c^{n=1}(g), c^{n=0}(g), \phi(g), k_o\}$ is a feasible allocation, i.e. $\{c^{n=1}(g), c^{n=0}(g), \phi(g), k_o\} \in \underline{\boldsymbol{X}}$, and $K, N \geq 0$;

(ii) Given w and r, $\{c^{n=1}(g), c^{n=0}(g), \phi(g)\}$ is a solution for the household's problem

$$\max_{\{c^{n=1}, c^{n=0}, \phi\}} \{\phi[u(c^{n=1}) - m] + (1-\phi)\, u(c^{n=0})\} \tag{6.17}$$

$$\text{subject to} \quad \phi c^{n=1} + (1-\phi)c^{n=0} \leq \phi w + r k_o$$

$$c^{n=1}, c^{n=0} \geq 0$$

$$\phi \in [0,1]$$

$$k_o = 1,$$

i.e. ϕ is a choice variable that allows the individual to maximize the probability of employment, and $\{N,K\}$ is a solution to the firm's problem

$$\max_{\{N,K\}} [f(K,N) - wN - rK] \tag{6.18}$$

$$\text{subject to} \quad K,N \geq 0;$$

(iii) The resources constraints are satisfied, or supply equals demand in the goods, labor, and capital market

$$f(K,N) = \int_0^1 \{\phi(g)\, c^{n=1}(g) + [1-\phi(g)]\, c^{n=0}(g)\}\, dg \tag{6.19}$$

$$N = \int_0^1 \phi(g)\, dg \tag{6.20}$$

$$K = \int_0^1 k_o\, dg = 1. \tag{6.21}$$

For this competitive equilibrium, identical individuals receive identical consumption bundles. To see this, note that the first order necessary conditions for an interior equilibrium for problem (6.17) for $\{c^{n=1}, c^{n=0}\}$ are respectively

$$\phi\, u'(c^{n=1}) - \phi\theta = 0 \tag{6.22}$$

$$(1-\phi)\, u'(c^{n=0}) - (1-\phi)\,\theta = 0 \tag{6.23}$$

where $\theta \geq 0$ is the Lagrange multiplier on the budget constraint to problem (6.17). Hence, as long as $\phi \in (0,1)$, then consumption is identical, or $c^{n=1} = c^{n=0}$ by expressions (6.22) and (6.23).

Also, introducing trade in the lottery makes agents better off. To see this, consider an allocation in the indivisible labor economy (6.4)-(6.8), $\{c(g), n(g), k_o, K, N, w, r\}$, and an allocation in the indivisible labor economy with lottery (6.17)-(6.21), $\{c^{n=1}(g), c^{n=0}(g), \phi(g), k_o, K, N,$

$w, r\}$. Welfare, for an individual g, in the indivisible labor economy (without lottery) is

$$\{u[c(g)] - n(g)\, m\},$$

and welfare in the indivisible labor economy with lottery is

$$\{\phi(g)\,[u(c^{n=1}(g)) - m] + (1 - \phi(g))\, u(c^{n=0}(g))\}.$$

Now, define an allocation for the indivisible labor economy with lottery as

$$c^{n=1}(g) = c^{n=0}(g) = c^* = \int_0^1 c(g)\, dg$$

$$\phi(g) = \phi^* = \int_0^1 n(g)\, dg$$

where $\{c(g), n(g)\} \in \boldsymbol{X}$ are allocations in the indivisible labor economy (without lottery), (6.4)-(6.8). The defined allocation gives an average consumption-labor supply bundle to all individuals. This allocation is feasible and aggregate welfare can be compared using Jensen's inequality as

$$\begin{aligned}
\int_0^1 \{u(c(g)) - n(g)\, m\}\, dg &\le \\
&\int_0^1 \{\phi(g)[u(c^{n=1}(g)) - m] + (1-\phi(g))\, u(c^{n=0}(g))\}\, dg \\
&= \int_0^1 \{\phi^* [u(c^*) - m] + (1-\phi^*)\, u(c^*)\}\, dg \\
&= \int_0^1 \{u(c^*) - \phi^* m\}\, dg \\
&= u(c^*) - m\,\phi^* \\
&= u\left(\int_0^1 c(g)\, dg\right) - m \int_0^1 n(g)\, dg
\end{aligned}$$

where the inequality is strict as long as $n(g)$ is not constant for all g. Hence, aggregate utility in the economy with trade in lottery,

$$\int_0^1 \{\phi(g)[u(c^{n=1}(g)) - m] + (1-\phi(g))\, u(c^{n=0}(g))\}\, dg,$$

is at least as large as the utility in the economy without lottery,

$$\int_0^1 \{u(c(g)) - n(g)\, m\}\, dg$$

(you may also see Figure 6.1 and relabel the horizontal axis for ϕ).

The allocation in the indivisible labor economy with trade in lottery in (i)-(iii) above is in fact identical to the allocation of a convex economy. The introduction of the lottery in the labor supply does indeed convexify the consumption set and virtually transforms the economy with a non-convexity into a conventional competitive economy as if the non-convexity did not exist. In particular, this economy satisfies the First and Second Fundamental Theorems of Welfare Economics, that is the competitive equilibrium is Pareto Optimal and a social planning problem can replicate the competitive equilibrium optimal allocation. In addition, in economies with trade in lotteries, the hypothetical representative agent has preferences that are linear in the labor supply n, because the extent of the labor supply is measured by the lottery ϕ that enters linearly in the expected utility characterization.

6.5 Implications for the Elasticity of Labor Supply

The individual problem examined above can be compactly written for the cases of divisible labor, $n\in[0,1]$, and trade in lottery, $\phi\in[0,1]$, as

$$\underset{\{c,\, I\}}{Max}\ [\, u(c) - v(I)], \quad for\ I=n\ or\ \phi \tag{6.24}$$

$$\begin{aligned} subject\ to \quad & c \leq f(I), \quad for\ I=n\ or\ \phi \\ & c \geq 0 \\ & n\in[0,1],\ \phi\in[0,1], \\ & k_o = 1, \end{aligned}$$

where $I=\{n,\phi\}$ is an indicator. The Lagrangean function for this problem is

$$\mathcal{L} = [\, u(c) - v(I)] + \theta\, [f(I) - c] \tag{6.25}$$

where $\theta\geq 0$ is the Lagrange multiplier attached to the resources constraint.

The first order necessary conditions for an interior equilibrium (sufficient second order conditions are satisfied by concavity of utility and strict convexity of the production set) give an equilibrium labor supply condition as

$$- v'(I) + u'(f(I))f'(I) = 0, \quad for\ I=\{n,\ \phi\}. \tag{6.26}$$

Let utility take the form $u(c)=c^{\alpha}$, for $\alpha\in(0,1)$, and the disutility of labor:
(i) For the divisible labor economy, $I=n$, $v(n)=m\, n^{\beta}$, $\beta\geq 1$, $f(n)=nw+r$;

(ii) For the indivisible labor economy with lottery, $I=\phi$, $v(\phi)=m\ \phi$, $f(\phi)=\phi\, w+r$.

We can substitute the appropriate differentials in (6.26), and compute the elasticity of labor supply for the two economies as

$$1 + (dn/dw)\ (w/n)|_{convex} = f(n)/[f(n)(\beta - 1)+(1-\alpha)nw] \qquad (6.27)$$

$$1 + (d\phi/dw)\ (w/\phi)|_{lottery} = f(\phi)/(1-\alpha)\phi\, w \qquad (6.28)$$

and for $\beta >1$, the elasticity of labor supply in the indivisible labor economy with lottery is larger than the elasticity in the convex (no indivisibility) economy. Of course, if $\beta=1$, both economies have linear preferences in n and the elasticities are identical. However, notice that even when $\beta >1$, the associated economy with trade in lotteries gives expected utility linear in the lottery, e.g. section 6.4.

The discrepancy in the elasticity of labor supply is an important characteristic of the economy with trade in lotteries. It has important implications for quantitative issues regarding aggregate fluctuations driven by stochastic technology disturbances, e.g. the stochastic Ramsey model presented in section 1.8. In these quantitative models, stochastic technology disturbances propagate in the economy through vibrations in the labor supply and the magnitude of the elasticity of labor supply becomes a critical parameter, see Notes on the Literature.

6.6 Summary I

In sections 6.1-6.5, we examine the importance of non-convexities for general equilibrium allocations under full information and an indivisibility constraint. In the next sections, we extend this framework to include asymmetric information.

6.7 General Equilibrium Approach to Asymmetric Information

The problems of asymmetric information that we examined in Chapter 5 are cast in a partial equilibrium structure. Here, we examine some of those same problems in a classical general equilibrium structure as proposed by Edward C. Prescott and Robert Townsend. The difficulty is that incentive constraints under asymmetric information introduce non-convexities in the consumption possibilities set. We use lotteries to overcome the non-convexities, as in sections 6.1-6.6. In essence, lotteries make the environments with asymmetric information satisfy certain convexity assumptions necessary for classical general equilibrium analysis, analogously to the introduction of lotteries in the case of an indivisibility. In particular here, we are interested in the First and Second Fundamental Theorems of Welfare Economics applied to economies with asymmetric information.

After presenting a general framework for general equilibrium analysis with asymmetric information, we study an insurance (adverse selection) example with private information where the First Fundamental Theorem is satisfied, but the Second Fundamental Theorem fails. Then we present an application for a consumption and labor supply problem with private information and adverse selection where both the First and Second Fundamental Theorems are satisfied.

6.8 Basic Structure, Pareto Optimality and Decentralized Competitive Equilibrium

The basic structure of the general equilibrium framework with private information is as follows. Individuals in the economy are indexed by type i, where i is finite, $i=1,2,\ldots,I$ all elements of the set $\boldsymbol{I}$. There is a continuum of each individual type i in the economy. Each type i is the agent's own private information and not public information. The fraction of agents of type i in the total I is denoted by λ_i and this is usually public information. The commodity space is assumed to be of finite dimension and linear, denoted $\boldsymbol{L}$ with respective n elements. The consumption possibility set is denoted $\boldsymbol{X} \subset \boldsymbol{L}$, closed and possibly convex for each agent of type i.

The utility function of each individual type i is defined as $u_i : \boldsymbol{X} \rightarrow \mathbb{R}$, concave and possibly linear. The endowment is assumed the same for all agents, independent of i, denoted $\omega \in \boldsymbol{L}$. Let $x_i \in \boldsymbol{L}$ be an allocation of the consumption bundle x to an agent of type i.

The characterization of the aggregate resources constraints in this framework is crucial. In general, there can be several aggregate resources constraints. For example, expressions (6.6)-(6.7)-(6.8) constitutes a set of aggregate resources constraints. We index the several constraints by $k=1,\ldots,K$ [(6.6)-(6.7)-(6.8) indicate $K=3$ in sections 6.2-6.5], and express each constraint weighted by a real valued linear function on the commodity space $\boldsymbol{L}$, denoted $r_{i\,k}$ for each individual $i=1,2,\ldots I$. Hence, the aggregate resources constraints can be expressed as

$$\Sigma_{i=1}^{\;\;I} \lambda_i\, r_{i\,k} \,(x_i - \omega) \leq 0, \quad k=1,\ldots,K \qquad (6.29)$$

where the weights r_{ik} represent a particular characterization of the production possibilities set of the economy.

In this setting, an *I-tuple* $x=(\,x_i\,)$, $x \in \boldsymbol{L}$, is a feasible and implementable allocation if:

(i) The consumption vector belongs to the consumption possibility set, or

$$x_i \in \boldsymbol{X}, \quad i=1,\ldots,I; \qquad (6.30)$$

(ii) The following inequalities hold

$$u_i\,(x_i) \geq u_i\,(x_j), \quad i=1,\ldots,I;\, j=1,\ldots,I;\; i \neq j \qquad (6.31)$$

or each individual type i weakly prefers x_i to all other $x_{j's}$, which implies that it is not in the best interest of any individual to claim to be of another type when faced with choices of different consumption bundles (notice that these are the incentive compatibility constraints that are of vital importance for economies with private information); and

(iii) The resources constraints (6.29) are satisfied.

Because the commodity space $\boldsymbol{L}$ is of finite dimension, for a linear function we may use the dot product as, for example,

$$r_{ik} \cdot x_i = \Sigma_{\ell \in L} \; r_{ik\ell} \; x_{i\ell}$$

where ℓ indexes the components of r and x in the commodity space.

The introduction of private information and incentive compatibility constraints of the form (6.31) present a potential problem for classical general equilibrium analysis. To illustrate the problem, let there be, without loss of generality, two individual types indexed by $i=1,2$. The incentive compatibility constraints in (6.31) are then

$$u_1 (x_1) \geq u_1 (x_2) \tag{6.32a}$$

$$u_2 (x_2) \geq u_2 (x_1). \tag{6.32b}$$

The issue is that whenever u_i is strictly concave, the above incentive compatibility constraints do not define a consumption possibility set that is convex. This is a problem because classical general equilibrium analysis requires consumption and production sets to be convex. In the case of (6.32a)-(6.32b), for different types, for example different degrees of risk aversion for each type, the set of allocations that satisfy the first constraint does not coincide with the set of allocations that satisfy the second constraint. Hence, a linear convex combination of two allocations that satisfy the constraints separately will violate one of the constraints. Incentive compatibility constraints in the context of private information introduce a problematic non-convexity in the general equilibrium framework. Figure 6.2 illustrates the problem of the non-convexity.

We have seen in the previous sections, in the context of a labor supply indivisibility that the introduction of trade in lotteries, in effect, convexifies the consumption possibility set. The same is true in the context of private information. For example, let the vector ϕ_i, $i=1,2,$ be a random assignment (lottery) to each agent of type i, where $\phi_i(x)$ is the probability of obtaining a consumption allocation x for type i. The random allocation can be achieved if

$$\Sigma_{x \in X} \; u_1 (x) \; \phi_1 (x) \;\; \geq \Sigma_{x \in X} \; u_1 (x) \; \phi_2 (x) \tag{6.32c}$$

$$\Sigma_{x \in X} \; u_2 (x) \; \phi_2 (x) \;\; \geq \Sigma_{x \in X} \; u_2 (x) \; \phi_1 (x) \tag{6.32d}$$

which are the randomized analogous to (6.32a)-(6.32b). Constraint (6.32c) indicates that agent of type *1* weakly prefers the lottery ϕ_1 to ϕ_2 because it gives greater than or equal expected utility. Constraint (6.32d) indicates that agent of type *2* weakly prefers the lottery ϕ_2 over ϕ_1

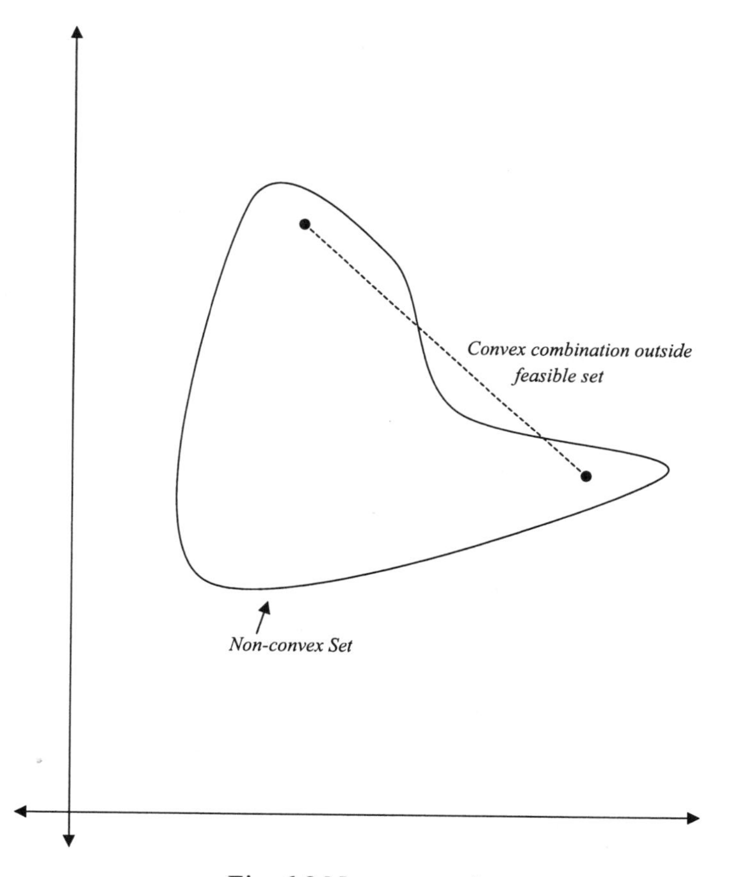

Fig. 6.2 Non-convexity

because it gives greater than or equal expected utility, e.g. recall the expected utility representation of section 3.2. More importantly, these constraints are linear in the lotteries ϕ_i and thus define a set over consumption lotteries that is convex and suitable for classical general equilibrium analysis.

In effect, lotteries will play a crucial role in assuring Pareto Optimality in economies with private information, exactly as in the case of indivisibilities examined in the previous sections.

(i) Pareto Optimal Allocations

We derive Pareto Optimal allocations for the general structure with private information presented above, with appropriate convex consumption possibility sets. This is obtained by maximizing a weighed average of the utilities of the different individual types. In this context, let the set of possible welfare weights be defined as

$$\boldsymbol{\Gamma} = \{\gamma \in \mathbb{R}^{I} : \gamma_i \geq 0, \ \Sigma_{i \in I} \gamma_i = 1\}. \tag{6.33}$$

Next, for welfare weights $\gamma \in \boldsymbol{\Gamma}$, let $\chi(\gamma)$ denote the set of allocations that solve the compactly written program

$$\underset{\{x=(x_i)\}}{Max} \ \Sigma_{i \in I} \gamma_i \ (u_i \cdot x_i) \tag{6.34}$$

$$subject\ to \quad x_i \in \boldsymbol{X}, \quad i=1,\ldots,I$$

$$u_i \cdot x_i \geq u_i \cdot x_j, \quad i=1,\ldots,I; j=1,\ldots,I; \ i \neq j$$

$$\Sigma_{i=1}^{I} \lambda_i r_{ik} (x_i - \omega) \leq 0, \quad k=1,\ldots,K$$

where $u_i \cdot x_i = \Sigma_{\ell \in L} \ u_{i\ell} \ x_{i\ell}$ is the dot product. There is a solution for this program that is Pareto Optimal, i.e. there is an $x \in \chi(\gamma)$ that is Pareto Optimal (see Notes on the Literature). The program in (6.34) is a general framework with incentive compatibility constraints of the form (6.32c,d), in dot product: $u_i \cdot x_i \geq u_i \cdot x_j$. The x's can be appropriately interpreted as the probability of obtaining the consumption bundle x.

Thus, under the regular conditions of convexity of consumption possibility sets, the First Fundamental Theorem of Welfare Economics is satisfied in economies with private information. A natural question is whether or not these Pareto Optimal allocations can be decentralized by a competitive price system, or whether the Second Fundamental Theorem of Welfare Economics holds in economies with private information. The answer is that, under certain specific conditions on the production possibility set, the Pareto Optimal allocation can be decentralized by an appropriate price system.

(ii) Competitive Equilibrium

The decentralized Competitive Equilibrium is as follows (notice that the economy studied in sections 6.2-6.5 is a decentralized competitive equilibrium). Let the consumption and production decisions be decentralized with a production-intermediation set denoted by $\boldsymbol{Y}$, with finite elements $y \in \boldsymbol{Y}$; and a price vector belonging to the commodity space, denoted by $p \in \boldsymbol{L}$. As we shall see below, the form of the production set $\boldsymbol{Y}$ holds the key to one being able to decentralize the Pareto Optimal allocations.

First, we define a competitive equilibrium as an *(I+1)-tuple* $((x^*_i), y^*) \in \boldsymbol{L}$, so that

(i) Utility maximization by individuals holds, or given prices, each x^*_i maximizes $u_i(x)$ over the feasible set $\{x \in \boldsymbol{X}: p^* \cdot (x - \omega) \leq 0\}$;

(ii) Net revenue maximization by firms holds, or given prices, y^* maximizes $p^* \cdot y^*$ over the production set $\boldsymbol{Y}$;

(iii) The resources constraint,

$$\Sigma_{i=1}^{I} \lambda_i (x_i - \omega) = y, \qquad y \in \boldsymbol{Y} \tag{6.35}$$

is satisfied, where we recall that λ_i is the fraction of agents of type *i,* in the total population *I*.

The reason the form of the production set is of crucial importance for the validity of the Second Fundamental Theorem of Welfare Economics is that, in centralized and decentralized equilibria, the resources constraint takes two forms that must be consistent with each other, or (6.29)

$$\Sigma_{i=1}^{I} \lambda_i r_{ik} (x_i - \omega) \leq 0, \quad k=1,\ldots,K$$

and (6.35)

$$\Sigma_{i=1}^{I} \lambda_i (x_i - \omega) = y, \quad y \in \boldsymbol{Y}$$

respectively. This requires that the weights r_{ik} characterize a production possibility set invariant to the specific individual type *i*, i.e. $r_{ik} = r_k$ not a function of the individual type *i*. If not, the production possibilities set is not sufficiently smooth to guarantee the separation between consumption and production possibilities, and price decentralization fails.

Hence, if the production possibility set is of the form

$$\boldsymbol{Y} = \{y \in \boldsymbol{L} : \Sigma_{\ell \in \boldsymbol{L}} r_{k\ell}\, y_\ell = (r_k \cdot y) \leq 0,\ k=1,\ldots,K\}, \tag{6.36}$$

the two resources constraints are consistent with each other and the consumption and production sides separate appropriately. The Pareto Optimal allocation is appropriately decentralized by a price system giving the market competitive equilibrium.

However, if the production possibility set is of the form

$$\boldsymbol{Y} = \{y \in \boldsymbol{L} : \Sigma_{i \in I} \lambda_i \Sigma_{\ell \in L} r_{ik\ell}\, y_\ell = \Sigma_{i \in I} \lambda_i (r_{ik} \cdot y) \leq 0,\ k=1,\ldots,K\} \tag{6.37}$$

i.e. r_{ik} is a function of the individual type i, then the two resources constraints are inconsistent and the Second Fundamental Theorem of Welfare Economics fails. In this case, the specific type i do affect the resources constraint making the separation between the consumption and production sides infeasible.

We shall examine below two different environments where this difference is highlighted. First, in an insurance problem with adverse selection, we shall see that the production possibility set is of the form of (6.37) and price decentralization fails. In a consumption and labor supply problem with private information and adverse selection, the production possibility set can be shown to be of the form of (6.36) and thus the Pareto Optimal allocation can be appropriately decentralized by a competitive price system.

6.9 **An Insurance Problem with Adverse Selection**

Consider an economy with a continuum of households in the unit interval where households are willing to buy insurance against a potential endowment shock. We cast this version of the adverse selection problem in the general equilibrium framework; a partial equilibrium version is in Chapter 5. Recall that adverse selection problems are associated with who the individual is, i.e. its type in the population.

Here, we assume household of type i receives a random endowment of the form

$$\omega_b \text{ with probability } \pi_i, \quad \omega_g \text{ with probability } 1-\pi_i, \text{ for } i \in \boldsymbol{I} \tag{6.38}$$

where $0 < \omega_b < \omega_g$ is public information. The interpretation is that ω_b represents the bad state of the world, a small endowment or a loss with probability $\pi_i \geq 0$, which is individual's i private information. Similarly, ω_g represents the good state, a large endowment or a gain with probability $1\text{-}\pi_i$. Without loss of generality, we assume that there are two types of households differentiated by risk class:

$$i \in \boldsymbol{I} = \{1,2\} \;\; with \;\; 0 < \pi_1 < \pi_2 \qquad (6.39)$$

i.e. household type $i=2$ has a higher probability of loss than household of type $i=1$. The agent type is own private information. But, even though one cannot distinguish another exactly by its type, the fraction of the various types in the population is public information. In turn, the economy presents idiosyncratic risk, but there is no aggregate risk.

Households engage in consumption, $c \in \mathbb{R}_+$, with utility function $u\colon \mathbb{R}_+ \rightarrow \mathbb{R}$ where u is strictly increasing, strictly concave, and continuously differentiable with $u'(0)=\infty$. Let the collection of consumption allocations form a consumption set denoted $\boldsymbol{C}$, with $c \in \boldsymbol{C}$. The endowment pair belongs to the consumption set, $\{\omega_b, \omega_g\} \in \boldsymbol{C}$. In case of a low state endowment, ω_b, the respective household is assigned the consumption bundle c_b; whereas in the case of a good state endowment, ω_g, the consumer is assigned the consumption bundle c_g.

We have learned in this chapter [see expressions (6.32a)-(6.32b)] that a framework that involves asymmetric information with strictly concave utility function gives a consumption possibility set that is not convex. Thus, we introduce lotteries in the consumption space in order to obtain a consumption possibility set in the space of lotteries that is convex. Because the commodity space is assumed to be of finite dimension and linear (denoted $\boldsymbol{L}$ in the basic structure, section 6.8) with respective n elements, the consumption set $\boldsymbol{C}$ is also of finite dimension with respective n elements. Then, lotteries in the consumption space will be n-dimensional vectors, which specify each bundle in $\boldsymbol{C}$, denoted

$$\phi = \{\phi(c)\}_{c \in \boldsymbol{C}}, \qquad \phi(c) \geq 0, \qquad \Sigma_{c \in \boldsymbol{C}}\, \phi(c) = 1 \qquad (6.40)$$

where ϕ_b is the lottery ticket for the bad state, $\omega_b \rightarrow c_b$; and ϕ_g is the lottery ticket for the good state, $\omega_g \rightarrow c_g$. With this specification, the expected utility for an individual of type i, denoted W_i, is given by

$$W_i = \pi_i \Sigma_{c\in C}\, u(c)\,\phi_b(c) + (1-\pi_i)\Sigma_{c\in C}\, u(c)\,\phi_g(c) \qquad (6.41)$$

where $\Sigma_{c\in C}\, u(c)\phi_b(c)$ is the expected utility in the case of a bad state, and similarly $\Sigma_{c\in C}\, u(c)\phi_g(c)$ for the good state, so that $\pi_i\Sigma_{c\in C}\, u(c)\phi_b(c)$ is the probability weighted expected utility of the bad state, and similarly $(1-\pi_i)\Sigma_{c\in C}\, u(c)\phi_g(c)$ for the good state. Notice that W_i is linear in the lottery and strictly concave in c, so that the consumption possibility set is convex.

Now, we are ready to analyze this problem relative to the basic structure. The commodity space is finite dimensional, with consumption vector elements (x) which are well defined probability distributions. Hence, $\boldsymbol{L}=\mathbb{R}^{2n}$, where the first n components, x_b, are given by

$$x_{bc} = \phi_b(c) \quad for\ c\in \boldsymbol{C}$$

and the second n components, x_g, given by

$$x_{gc} = \phi_g(c) \quad for\ c\in \boldsymbol{C}.$$

Then, the convex consumption possibility set in the space of lotteries is given by

$$\underline{\boldsymbol{X}} = \{x \in \boldsymbol{L} : x_{jc} \geq 0,\ \Sigma_{c\in C}\, x_{jc} = 1 \ \ for\ j=b,g\}. \qquad (6.42)$$

The single ($k=K=1$) resource constraint is given by

$$\Sigma_{i\in I}\lambda_i\{\pi_i\,[(\Sigma_{c\in C}\, x_{ibc}\, c) - \omega_b] + (1-\pi_i)[(\Sigma_{c\in C}\, x_{igc}\, c) - \omega_g]\} \leq 0 \quad (6.43)$$

where the triple indexed bundle x_{ijc}, $i=1,2$, $j=b,g$, indicates the fraction of individual type i in state j consuming bundle $c\in \boldsymbol{C}$.

The constraint (6.43) implies that average consumption cannot exceed average endowment. However, in the resource constraint (6.43), the weights r_{ik} have $2\times n$ components of the respective form $\pi_i\, c$ and $(1-\pi_i)c$, a linear function of the private information of the individual of type i, π_i.

Thus, the resources constraint (6.43) is of the form (6.29) and the Pareto Optimal allocation cannot be decentralized in this application, i.e. the Second Fundamental Theorem fails.

The First Fundamental Theorem holds and we can find the Pareto Optimal allocations for this economy. We denote the average endowment by

$$\underline{\omega} = \Sigma_{i\in I}\, \lambda_i\, \{\pi_i\ \omega_b + (1-\pi_i)\ \omega_g\}. \qquad (6.44)$$

The solution for the set of Pareto Optimal allocations, given weights $\gamma \in \Gamma$ [as in (6.33)], is the solution to the linear program

$$\underset{\{x_{ijc}\}_{i=1,2;\, j=b,g}}{Max} \ \Sigma_{i\in I}\ \gamma_i\ \Sigma_{c\in C}\ u(c)\ \{\pi_i\ x_{ibc} + (1-\pi_i)\ x_{igc}\} \qquad (6.45)$$

subject to

$$\Sigma_{c\in C}\ u(c)\ [\pi_2(x_{1bc} - x_{2bc}) + (1-\pi_2)\ (x_{1gc} - x_{2gc})] \le 0 \qquad (6.46a)$$

$$\Sigma_{c\in C}\ u(c)\ [\pi_1(x_{2bc} - x_{1bc}) + (1-\pi_1)\ (x_{2gc} - x_{1gc})] \le 0 \qquad (6.46b)$$

$$\Sigma_{i\in I}\ \lambda_i\ \{\pi_i\ \Sigma_{c\in C}\ x_{ibc}\ c + (1-\pi_i)\Sigma_{c\in C}\ x_{igc}\ c\} \le \underline{\omega} \qquad (6.46c)$$

$$\Sigma_{c\in C}\ x_{ijc} = 1, \qquad x_{ijc} \ge 0,\ for\ i=1,2;\ j=b,g \qquad (6.46d)$$

where (6.46a)-(6.46b) are the incentive compatibility constraints so that each household type weakly prefers the lottery attached to its own type. For example, by (6.46a) type $i=2$ weakly prefers x_2 to x_1. Expression (6.46c) is the resources constraint as in (6.43), and (6.46d) are the appropriate probability measures. In the expected utility (6.46), the triple indexed x_{ijc}, $i=1,2,\ j=b,g$, indicates the probability of type i consuming the bundle $c\in \boldsymbol{C}$ conditional on state j. The first order necessary conditions are the usual Kuhn-Tucker conditions. Letting $\mu_z \ge 0$ be the Lagrange multiplier attached to the constraints (6.70a,b,c,d) for $z=1,2,3,41b,41g,42b,42g$ respectively, the conditions are: for x_{1bc},

$$\gamma_1 u(c)\ \pi_1 - \mu_1 u(c)\ \pi_2 + \mu_2 u(c)\ \pi_1 - \mu_3 \lambda_1\ c\ \pi_1 + \mu_{41b} \le 0,$$
$$if <, \quad x_{1bc} = 0 \qquad (6.47a)$$

and similarly

$$\gamma_1 u(c)\ (1-\pi_1) - \mu_1 u(c)\ (1-\pi_2) + \mu_2 u(c)\ (1-\pi_1) - \mu_3 \lambda_1\ c\ (1-\pi_1) + \mu_{41g}$$
$$\le 0,\ if <,\ x_{1gc}=0 \qquad (6.47b)$$

$$\gamma_2 u(c)\ \pi_2 + \mu_1 u(c)\ \pi_2 - \mu_2 u(c)\ \pi_1 - \mu_3 \lambda_2\ c\ \pi_2 + \mu_{42b} \le 0,$$
$$if <,\ x_{2bc} = 0 \qquad (6.47c)$$

$$\gamma_2 u(c)\ (1-\pi_2) + \mu_1 u(c)\ (1-\pi_2) - \mu_2 u(c)\ (1-\pi_1) - \mu_3 \lambda_2\ c\ (1-\pi_2) + \mu_{42g}$$
$$\le 0,\ if <,\ x_{2gc} = 0 \qquad (6.47d)$$

$$\Sigma_{c\in C}\ u(c)\ [\pi_2(x_{1bc} - x_{2bc}) + (1-\pi_2)\ (x_{1gc} - x_{2gc})] \le 0,$$
$$if <,\ \mu_1 = 0 \qquad (6.47e)$$

$$\Sigma_{c \in C}\ u(c)\ [\pi_1 (x_{2bc} - x_{1bc}) + (1-\pi_1)(x_{2gc} - x_{1gc})] \leq 0,$$

$$\text{if} <,\ \mu_2=0 \qquad (6.47f)$$

$$\Sigma_{i \in I}\ \lambda_i\ \{\pi_i \Sigma_{c \in C}\ x_{ibc}\ c + (1-\pi_i)\Sigma_{c \in C}\ x_{igc}\ c\} \leq \underline{\omega},$$

$$\text{if} <,\ \mu_3=0 \qquad (6.47g)$$

and (6.70d) which holds with equality, i.e. $\mu_{41b}, \mu_{41g}, \mu_{42b}, \mu_{42g} > 0$.

The set of Pareto Optimal allocations is obtained as follows. In general, when we consider the first order conditions (6.47), note that the weak inequalities in (6.47a)-(6.47d) must hold with equality for some bundle (c), otherwise if, for example, $x_{1bc}(c)=0$ for all $(c) \in \boldsymbol{L}$, then it would violate the probability measure in (6.47d).

Hence, we can find points in the consumption possibility set where the first order conditions hold with equality. For example, rewrite (6.47a) as

$$[\pi_1 (\gamma_1 + \mu_2) - \mu_1 \pi_2]\ u(c) - \mu_3 \lambda_1\ \pi_1 c + \mu_{41b} \leq 0,$$

$$\text{if} < 0,\ x_{1bc}=0 \qquad (6.47a')$$

where the left-hand side is a function of c. Call this function $g(c,.)$ and compute

$$\partial g(c,.)/\partial c = [\pi_1 (\gamma_1 + \mu_2) - \mu_1 \pi_2]\ u'(c) - \mu_3 \lambda_1 \pi_1 \qquad (6.48a)$$

$$\partial g^2(c,.)/\partial c^2 = [\pi_1 (\gamma_1 + \mu_2) - \mu_1 \pi_2]\ u''(c) \qquad (6.48b)$$

and recall that u is strictly increasing and strictly concave. Next, from expressions (6.48a)-(6.48b), if $[\pi_1 (\gamma_1 + \mu_2) - \mu_1 \pi_2] \leq 0$ then $\partial g(c,.)/\partial c < 0$ and $\partial g^2(c,.)/\partial c^2 \geq 0$, the condition (6.47a') is strictly decreasing and convex in c, i.e. attains a maximum at $c=0$ which is ruled out by $u'(0)=\infty$. It remains that if $[\pi_1 (\gamma_1 + \mu_2) - \mu_1 \pi_2] > 0$ then $\partial g^2(c,.)/\partial c^2 < 0$, the condition (6.47a') is strictly concave in c, i.e. attains a unique maximum at a single point $c > 0$. Thus, $x_{1bc} = 1$ for some $c \in \boldsymbol{C}$, given $\gamma \in \boldsymbol{\Gamma}$, and zero otherwise.

Using the same argument, $x_{ijc} = 1$ for some $c \in \boldsymbol{C}$, given $\gamma \in \boldsymbol{\Gamma}$, and zero otherwise for all $i=1,2;\ j=b,g$. Since the probability measures place all their mass on a single point, we can denote it c_{ij}, $i=1,2;\ j=b,g$, where

$$c\, x_{ijc} \Leftrightarrow c_{ij} \quad \text{and} \quad u(c)\, x_{ijc} \Leftrightarrow u(c_{ij}). \qquad (6.49)$$

Then, to find the Pareto Optimal allocations we have to find these points in the consumption set, c_{ij}, $i=1,2; j=b,g$. There are in fact three feasible points that give Pareto Optimal allocations.

The first is the average endowment $\underline{\omega}$, i.e. $c_{ij}=\underline{\omega}$, $i=1,2; j=b,g,$. When all households consume $\underline{\omega}$, with certainty, it is easy to see that the allocation satisfies all the constraints and is a solution to (6.46).

Another Pareto Optimal allocation is more interesting and goes as follows. Let the household of type $i=2$ receive a non-stochastic consumption allocation,

$$c_{2b} = c_{2g} = c_2 < \underline{\omega}. \tag{6.50}$$

In this case, constraint (6.46b) is slack and by condition (6.47f) $\mu_2=0$. Household of type $i=1$ receives a lottery so that $c_{1b} \neq c_{1g}$ and (6.46a) is binding so that by (6.47e), $\mu_1>0$. Also, let the resources constraint bind so that by (6.47g), $\mu_3>0$. Hence, (6.46a) and (6.46c) both hold with equality and using (6.49) yields

$$\pi_2 u(c_{1b}) + (1-\pi_2) u(c_{1g}) = u(c_2) \tag{6.46a'}$$

$$\lambda_1 \{\pi_1 c_{1b} + (1-\pi_1) c_{1g}\} + \lambda_2 c_2 = \underline{\omega}. \tag{6.46c'}$$

These form a pair of equations in (c_{1b}, c_{1g}) as a function of c_2, π_1, π_2, λ_1, λ_2 and $\underline{\omega}$. Figure 6.3 presents the solution. Because, $0 < \omega_b < \omega_g$, the preferred solution is at point A where $c^*_{1b} < c_2 < c^*_{1g}$, where c_2 is in the 45° line between points B and B'. Intuitively, households type $i=1$ must be given a lottery that makes their expected utility exceed the expected utility of receiving the average endowment $\underline{\omega}$, and exceed the expected utility of type $i=2$ implicitly. Thus, given the solution for (c^*_{1b}, c^*_{1g}), the first order necessary conditions (6.47a,b,c,d), and the consistency of the welfare weights in (6.33), using (6.49), yield

$$\gamma_1 u'(c^*_{1b}) - \mu_1 u'(c^*_{1b}) \pi_2 - \mu_3 \lambda_1 \pi_1 = 0 \tag{6.51a}$$

$$\gamma_1 u'(c^*_{1g}) (1-\pi_1) - \mu_1 u'(c^*_{1g}) (1-\pi_2) - \mu_3 \lambda_1 (1-\pi_1) = 0 \tag{6.51b}$$

$$\gamma_2 u'(c_2) + \mu_1 u'(c_2) - \mu_3 \lambda_2 = 0 \tag{6.51c}$$

$$\mu_3 (1 - \lambda_1 - \lambda_2) = 0 \tag{6.51d}$$

$$\gamma_1 + \gamma_2 = 1. \tag{6.51e}$$

These five relationships give four independent equations in $(\mu^*_1, \mu^*_2, \gamma^*_1, \gamma^*_2)$ as a function of (c^*_{1b}, c^*_{1g}) and are implicit functions of c_2, $\pi_1, \pi_2, \lambda_1, \lambda_2$ and $\underline{\omega}$. Since the 4-*tuple* $(\mu^*_1, \mu^*_2, \gamma^*_1, \gamma^*_2)$ must be non-negative, the condition

$$\{(\lambda_2/\lambda_1)(\pi_2 - \pi_1)/[\pi_1(1-\pi_1)]\} \geq$$
$$u'(c_2)[u'(c^*_{1b}) - u'(c^*_{1g})]/[u'(c^*_{1b})\,u'(c^*_{1g})] \qquad (6.52)$$

must be satisfied. This condition is satisfied for $c_{1b} \neq c_{1g}$ and the Pareto Optimal equilibrium exists.

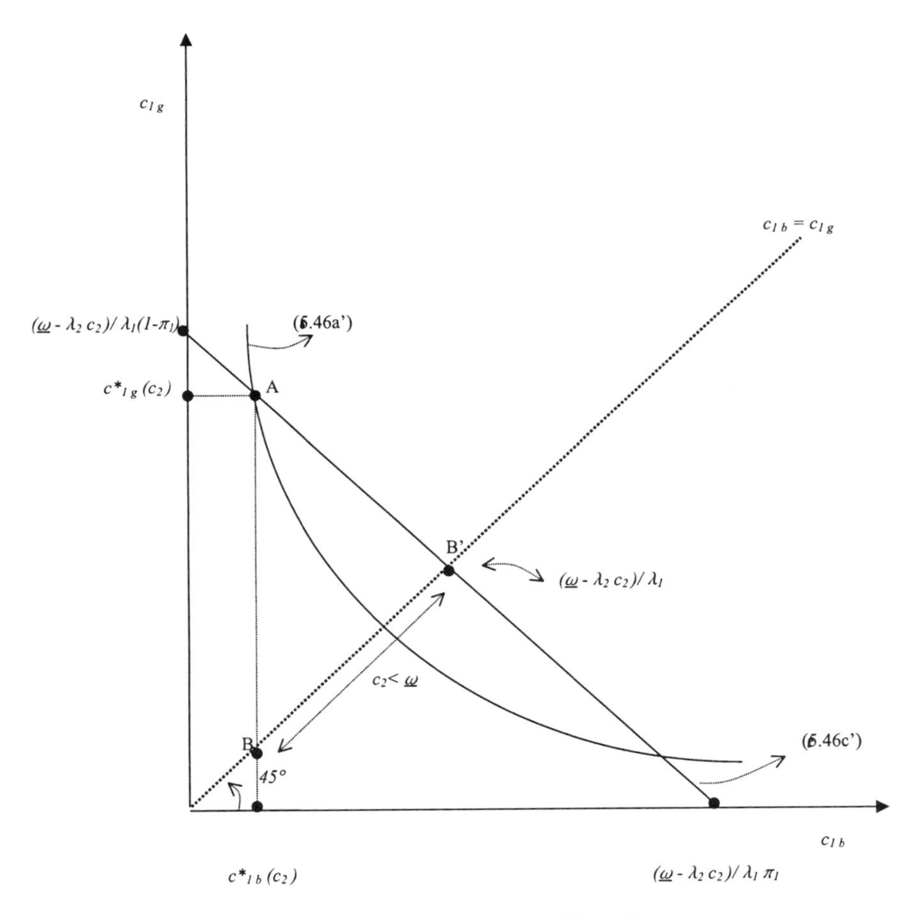

Fig. 6.3 Pareto Optimal Allocation

In this allocation, household of type $i=1$ bears all the uncertainty and is virtually offered a contract with a lottery that gives higher expected utility and higher expected consumption relative to type $i=2$. The introduction of the lottery in consumption allows the two types to be separated and reveal their type truthfully, i.e. a separating equilibrium.

A parameterized example of this Pareto Optimal allocation is worth examining. Let utility be logarithmic, $u(c)=log\ c$, the state dependent endowments be $\omega_b=2$, $\omega_g=6$, the fraction of types be $\lambda_1=\lambda_2=½$ as before, and the probabilities be $\pi_2 = ¾ > \pi_1 = 1/3$. Then, by (6.44), $\underline{\omega}=3.83$. Choose a nonrandom allocation for type 2 as $c_{2b}=c_{2g}=c_2=3 < \underline{\omega} = 3.83$, then (6.46a') and (6.46c') give

$$c^*_{1b}=2.41 < c_2 = 3 < c^*_{2g} = 5.80.$$

Finally, a solution for (6.51a-e) gives

$$\mu^*_1 = 0.24,\ \mu^*_2=0.26,\ \gamma^*_1 = 0.84,\ \gamma^*_2=0.16.$$

The expected utility of type $i=1$ is

$$\pi_1 \log c_{1b} + (1-\pi_1) \log c_{1g} = 1.46$$

which is greater than the expected utility of the nonrandom endowment, $log\ \underline{\omega}=1.34$, which is greater than the expected utility of type $i=1$, $log\ c_2 = 1.10$. Also, the incentive constraint (6.46a) binds so that type $i=2$ is indifferent between the lottery and the nonrandom allocation c_2, whereas the incentive constraint (6.46b) is slack, $(-0.37<0)$, so that type 1 strictly prefers the lottery.

Finally, another Pareto Optimal allocation can be analogously constructed where household of type $i=2$ bears all the uncertainty, $c_{2b} \neq c_{2g}$, and household of type $i=1$ is given a nonrandom bundle $c_{1b}=c_{1g}=c_1 < \underline{\omega}$. Then, in this alternative allocation, household of type 2 bears all the uncertainty, but receives higher expected utility and higher expected consumption relative to type 1.

This adverse selection insurance example yields Pareto Optimal allocations and it satisfies the First Fundamental Theorem of Welfare Economics. However, the Competitive Equilibrium does not obtain in this case, i.e. price decentralization fails. The reason is that the resources constraint is of the general form (6.29) [see (6.42)-(6.43)] and the candidate for production-intermediation possibility set in this economy is

(6.37), which is a function of the individual type i, and thus depends upon individual's private information. In this case, the resources constraint retains the private information differences among types and production and consumption decisions cannot be appropriately separated. In this adverse selection economy, the Second Fundamental Theorem of Welfare Economics fails.

6.10 **Summary II**

In sections 6.7-6.9, we extended problems of asymmetric information studied in Chapter 5 to the classical general equilibrium framework. We examined an insurance problem, and have shown that the First Fundamental Theorem holds, but there are problems with the Second Fundamental theorem. In this case, Pareto Optimal allocations cannot be decentralized by an appropriate price system. The reason is that in this specific environment, the private information enters the resources constraint in a manner that makes the separation of consumption and production possibilities infeasible.

This failure can be shown to exist in a signaling environment as well, see Notes on the Literature. In the next few sections, we present a consumption-labor allocation problem with private information where the First and Second Fundamental Theorems of Welfare Economics hold and ex-ante Pareto Optimal allocations can explain ex-post inefficiencies in the labor market.

6.11 **Unemployment Insurance, Asset Returns and Adverse Selection**

In the last sections of this chapter, we consider a simple general equilibrium model with endogenous labor supply and potential for adverse selection based on private information of individual preferences. We use the model to draw conclusions about asset allocation. In particular, we study the implications of consumption and labor allocations with ex-ante efficiency and possibly ex-post inefficiency for international/inter-regional portfolio diversification. The results are shown to depend crucially on the market regime relative to unemployment insurance. We examine the specific asset allocation that replicates infinitesimally small random deviations from a nonstochastic equilibrium in the presence of private information in a multi-agent general equilibrium model with endogenous labor supply.

In sections 6.2-6.5, we examined an example of applications of lotteries in the context of consumption and labor supply choice when there was indivisibility in the time schedule and full information. Here, we examine a similar framework but assume that the time schedule is smooth. The labor supply is perfectly divisible, but there is private (asymmetric) information in preferences.

In this economy, many important results can be shown. First, the introduction of randomness in the labor supply of certain individuals allows one to obtain ex-ante Pareto Optimal allocations with private information that are consistent with ex-post efficiency in the consumption-labor allocations. Second, and more surprisingly, it can be shown that, under certain conditions on preferences, the ex-ante Pareto Optimal allocations with private information are consistent with ex-post inefficiency in the consumption-labor allocations. This second result is important because it gives a plausible explanation for certain phenomena of otherwise apparent market failure from a well-functioning general equilibrium competitive market system with the amendment of private information. Third, relative to the basic structure and the example of sections 6.7-6.9, the Pareto Optimal allocation can be decentralized by a price system and both the First and Second Fundamental Theorem of Welfare Economics hold for the consumption-labor supply private information economy. The key is that, in this application, the production possibility set is of the general form (6.36). We do not show this last result formally here, but it can be obtained using the methods of sections 6.7-6.8, see Notes on the Literature.

In summary, in a consumption-labor allocation problem with private information, the First and Second Fundamental Theorems of Welfare Economics hold and ex-ante Pareto Optimal allocations can explain ex-post inefficiencies in the labor market. This type of result can also be shown to hold in other private information problems such as the standard moral hazard issue of unobservable actions studied in Chapter 5, e.g. Notes on the Literature.

6.12 **Basic Structure**

The model in this section is a one good-model cast in a static framework with a finite number of units indexed by $j \in \mathfrak{J}$: $j=1,2,\dots,J$. Each unit may be referred as an island, a region or a country inhabited by a large (countably infinite) number of identical individuals. Hence, a priori there may be no differences within units, but potential differences across units. A typical unit has a production technology given by

$$y_j = z_j f(n_j) \tag{6.53}$$

where y_j is the output produced by unit j, n_j are the number of hours spent on the production of the good, f is a strictly increasing and strictly concave function identical for all j ($f'>0$, $f''<0$), and z_j is the productivity level of the technology, i.e. total factor productivity. The differences in productivity across units may be potentially unobservable, however we assume the existence of an organized asset market that reveals the market relevant information of the unit's productivity. In what follows, we assume that $f(n_j)$ takes the specific form

$$f(n_j) = (n_j)^{\alpha} \tag{6.54}$$

for $\alpha \in (0,1)$ with returns to the variable and fixed factors given by

$$w_j = \alpha y_j / n_j \tag{6.55a}$$

$$e_j = (1-\alpha) y_j \tag{6.55b}$$

respectively. Capital markets are perfectly integrated across units, labor is assumed immobile but labor income is assumed to be tradable.

Decisions by representative individuals in each unit are taken according to Figure 6.4. When there is private information in individual preferences, ex-ante equilibrium and ex-post outcomes may differ

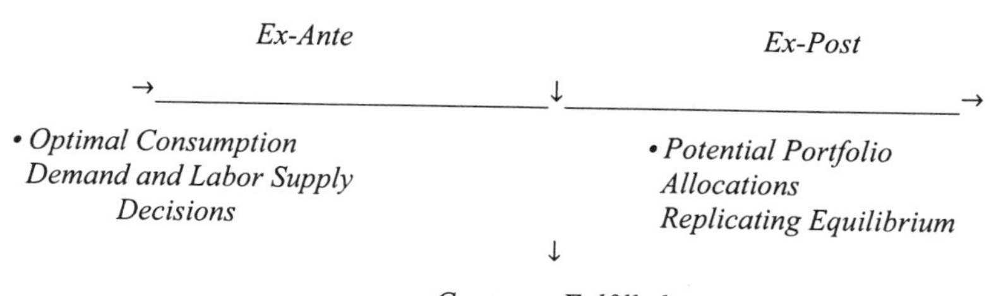

Fig. 6.4 Information and Decision Flows

relative to the point where all contracts are fulfilled, private information is revealed and production technology is realized. All decisions regarding consumption and labor supply are ex-ante, whereas potential portfolio allocations that replicate equilibrium can be taken possibly ex-post. A typical unit utility function is given by

$$w^j(c_j, n_j) \equiv u^j(c_j) + v^j(n_j) \tag{6.56}$$

where c_j is the level of consumption of representative individual in unit j. The function u is assumed to be strictly increasing and strictly concave ($u^j_1>0,\ u^j_{11}<0$), and the function v is assumed to be strictly decreasing and concave ($v^j_1<0,\ v^j_{11} \leq 0$). The main assumption in (6.56) is that consumption and labor are separable in utility. The source of private information in this model will be regarding the parameter $v^j_{11} \leq 0$, i.e. the concavity of the utility function with respect to the labor supply.

The possible randomness in the model will be infinitesimally small. We shall consider infinitesimally small deviations from a nonstochastic equilibrium generated by z_j, i.e. dz_j, from $z_j = 1$. In particular, dz_j is generated randomly and enters the economic equilibrium linearly. This method takes into account up to second order derivatives of the functions above associated with first order moments of the distributions (means), but ignores third order derivatives associated with second order moments (variances), see Notes on the Literature.

In the absence of heterogeneity across units, there is no distinction between ex-ante and ex-post allocations, and $w^j=w$. A Pareto efficient allocation can be obtained by maximizing the social welfare function subject to the resources constraints, or

$$\underset{\{c_j,\, n_j\}}{Max}\ E\left[\Sigma_j\ \omega_j\ w(c_j, n_j)\right] \tag{6.57}$$

$$subject\ to \qquad \Sigma_j \pi_j \left[c_j - z_j\, f(n_j)\right] \leq 0$$

where ω_j are arbitrary welfare weights satisfying $\{\omega_j\!: \omega_j \geq 0, j=1,2,\ldots,J, \Sigma_j\, \omega_j=1\}$, π_j are resource weights to account for possible differences in size of units satisfying

$$\{\pi_j : \pi_j \geq 0, j=1,2,\ldots,J,\ \Sigma_j \pi_j = 1\},$$

and E is the expectation operator. In this framework, a solution to (6.57) yields Pareto Efficiency ex-ante and ex-post, or

$$[-v_1\ (n_j)\,/\,u_1\ (c_j)] = z_j f'(n_j) \qquad all\ j=1,2\ldots J \tag{6.58}$$

i.e. the marginal rate of substitution between consumption and work in utility is equal to the marginal rate of transformation in production. Furthermore, if $\omega_j=\pi_j=1/J$ all j, then $c_j=c$ and $n_j=n$ all j, the perfectly pooled equilibrium.

6.13 Heterogeneity, Efficiency, and Market Completeness

We proceed examining alternative cases relating to the information structure about potential differences in preferences and market regimes.

(i) Full Information with Heterogeneous Types

Consider heterogeneity in preferences across units, but full information, i.e. the heterogeneity is public information. Each unit has only one type and differences across units reflect differences across types. With perfect information, ex-ante and ex-post allocations are identical across units. Again, the Pareto efficient allocation is obtained by maximizing the social welfare function subject to the resources constraints, or

$$\underset{\{c_j,\, n_j\}}{Max}\ E\,[\,\Sigma_j\ \omega_j\ w^j\,(\,c_j\,,\,n_j\,)] \qquad (6.59)$$

$$subject\ to \qquad \Sigma_j \pi_j\,[\,c_j - z_j\ f(\,n_j)\,]\le 0.$$

A solution to (6.59) yields Pareto Efficiency ex-ante and ex-post, or

$$[\,-\,v_1{}^j\,(\,n_j\,)\,/\,u_1{}^j\,(\,c_j\,)] = z_j f'(\,n_j\,) \qquad all\,j=1,2\ldots J \qquad (6.60)$$

i.e. the marginal rate of substitution between consumption and work in utility for unit type j is equal to the marginal rate of transformation in production for that type.

For example, let there be two units or types $J=2$, with $\omega_j=\pi_j$, and preference structure:

$$\bullet\ j=1:\ v_{11}{}^1\,(\,n_1\,) = 0; \qquad (6.61)$$

$$\bullet\ j=2:\ v_{11}{}^2\,(\,n_2\,)<0.$$

Thus, for both types $j=1,2$, preferences are separable in consumption and labor supply, with identical forms in the consumption argument. However, for type $j=1$ it is linear in the labor supply and for type $j=2$ it is strictly concave in labor supply. This difference in preferences implies that for type $j=1$ the elasticity of substitution between consumption and leisure is larger than for type $j=2$. In particular, the linearity in labor supply implies that $j=1$ is 'risk neutral' in labor supply whereas the strict concavity for $j=2$ implies 'risk aversion' in labor supply. Figure 6.5 illustrates this difference: the more risk averse individual, $v_{11}{}^2(n_2)<0$ (the concave curve), will prefer to work more hours at point B than a lottery that would give expected disutility at point A, whereas the risk neutral type, $v_{11}{}^1(n_1)=0$ (straight line), is indifferent between the certain and the gamble both at point A. According to this preference structure, the optimal allocation across units takes the form:

$$c_1 = c_2\ and\ n_1 < n_2\,.$$

Both types consume the same amount, but individuals in unit $j=2$ work more hours since they are risk averse in labor supply whereas individuals in unit $j=1$ are risk neutral and work less. It is optimal for type $j=2$ to work more hours and shift risk to type $j=1$.

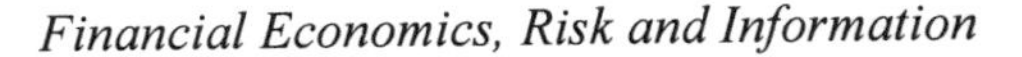

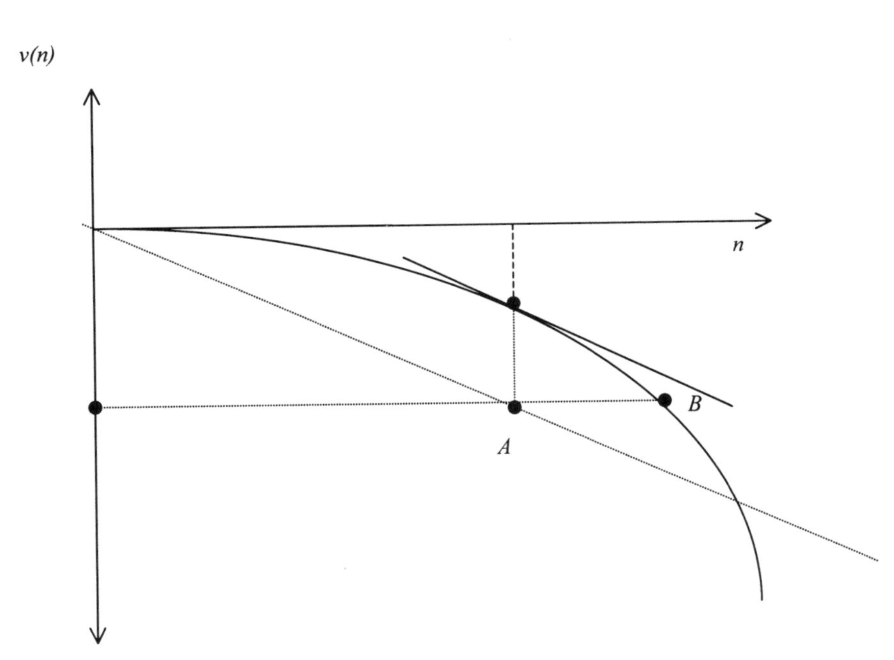

Fig. 6.5 Preferences for Labor

(ii) Private Information in Individual Preferences with Heterogeneous Types and Ex-Post Efficiency: Complete Markets

Now, we introduce private information into the model in (i). In this case, individual differences across units are private information of the specific unit and there may be differences in allocation ex-ante versus ex-post. A feasible and implementable allocation requires incentive compatibility constraints of the form

$$w^j(c_j, n_j) \geq w^j(c_i, n_i) \qquad \text{for all } j, i \in \mathfrak{J},\ i \neq j \qquad (6.62)$$

i.e. individual of unit j when faced with alternative consumption-labor supply bundles will have an incentive to reveal its true type, that is an incentive not to misrepresent its preferences towards labor. It is clear that, with private information, the model in (i) above is such that the incentive compatibility constraints will be violated ex-ante because both types consume the same amount and type $j=2$ works more. Therefore, type $j=2$ will have an incentive to misrepresent as type $j=1$. This is a classic adverse selection problem. Technically, for u^j strictly concave in consumption, the consumption-labor supply possibility set is not convex.

The revelation mechanism used here to avoid the adverse selection problem is to introduce a lottery scheme that convexifies the consumption-labor supply possibilities set as in the basic framework of section 6.8. Denoting the consumption-labor supply bundles $(c,n)\in\mathcal{L}$, where $\mathcal{L}$ is the consumption-labor possibility set, the analogous to the incentive compatibility constraints (6.62) with the introduction of the lottery scheme are

$$\Sigma_{(c,n)\in\mathcal{L}}\ \phi_j(c,n)\,w^j(c,n) \geq \Sigma_{(c,n)\in\mathcal{L}}\ \phi_i(c,n)\,w^j(c,n) \qquad \text{for all } j,i\in\mathfrak{J},\ i\neq j \tag{6.63}$$

where $\phi_j(c,n) \geq 0$, $\Sigma_j\,\phi_j(c,n)=1$ is the lottery for bundle (c,n). Hence, the incentive compatibility constraints in (6.63) are linear in the lottery and yield a convex consumption-labor possibility set. A consequence of introducing the revelation mechanism through the lottery scheme is that ex-ante and ex-post allocations may differ. Ex-ante, all individuals in all units are identical in expectations but ex-post the relevant differences are realized.

For example, consider $J=2$, with $\omega_j=\pi_j$, and preference structure as in (6.61). The revelation mechanism consists of introducing a lottery in the labor supply of individuals of unit $j=1$ to make it unattractive to individuals of unit $j=2$ who are risk averse, while not affecting the decisions of $j=1$ who are risk neutral. Let the lottery for $j=1$ be a contract with the firm with the following terms:

- *with probability* $(1-\phi)$: $n_1 = 0$;
- *with probability* ϕ: $n_1 = \underline{n}>0$;

for $\phi\in(0,1)$, and $\underline{n}>0$ given. The lottery ticket gives every holder full unemployment insurance, thus there are complete markets in unemployment insurance. The effective hours worked will be $\phi\,\underline{n}$ and every individual of unit $j=1$ will receive ex-post a full wage $z_1 f'(\phi\,\underline{n})$ whether working or not.

The expected utility for $j=1$, ex-ante, is $w^1(c_1, \phi \underline{n})$ and $j=1$ maximizes expected utility by choice of probability ϕ. Ex-ante allocations are obtained as solutions to the social problem

$$\underset{\{c_1, c_2, \phi, n_2\}}{Max} \{\omega_1 w^1(c_1, \phi \underline{n}) + \omega_2 w^2(c_2, n_2)\} \tag{6.64}$$

$$\text{subject to} \quad \pi_1[c_1 - z_1 f(\phi \underline{n})] + \pi_2[c_2 - z_2 f(n_2)] \leq 0$$

where in the resources constraint ϕ enters as the proportion of individuals of unit $j=1$ who actually work. Notice that the social planner knows the location of individuals across units, but due to private information must give individuals the right incentive to reveal truthfully.

Hence, the contract is offered to all across units, and gives the right incentive for all to reveal truthfully. In effect, there is no a priori intra-unit differences, and, ex-ante, no inter-unit differences as well. However, ex-post both intra-unit and inter-unit differences will arise. A solution to (6.64) yields Pareto Efficiency ex-ante for all units, or

$$[-v_1{}^1(\phi \underline{n}) / u_1{}^1(c_1)] = z_1 f'(\phi \underline{n}) \tag{6.65a}$$

$$[-v_1{}^2(n_2) / u_1{}^2(c_2)] = z_2 f'(n_2) \tag{6.65b}$$

i.e. ex-ante, the marginal rate of substitution between consumption and work in utility for each unit or type is equal to the marginal rate of transformation in production for that unit or type. Indeed, this ex-ante allocation is identical to the full information allocation with heterogeneity in the model of (i) above when we set

$$n_1 = \phi \underline{n}. \tag{6.65c}$$

Hence, ex-ante Pareto efficiency holds and the lottery makes everyone better off in expectations.

The ex-post allocation in this case is also efficient. The individuals in unit $j=1$ are subdivided into the fraction $(1-\phi)$ who do not work, but due to the complete markets in unemployment insurance receive a full wage $z_1 f'(\phi \underline{n})$, and the fraction ϕ that work $\underline{n}$ hours receiving the same wage $z_1 f'(\phi \underline{n})$. Thus, within and across units, individuals consume the same amount $c_1 = c_2 = c$. Because of the linearity of the utility function of all $j=1$ in ϕ (risk neutrality), and separability between consumption and labor, we have that

$$v_1{}^1(\phi\underline{n}) = v_1{}^1(0) = constant \tag{6.66}$$

i.e. independent of $\phi\,\underline{n}$ ex-post. Thus, efficiency ex-post holds for all $j=1$, that is

$$[-v_1{}^1(\phi\underline{n})/u_1{}^1(c_1)] = [-v_1{}^1(0)/u_1{}^1(c_1)] = z_1 f'(\phi\underline{n}). \tag{6.67}$$

For all $j=2$, efficiency holds as well, or

$$[-v_1{}^2(n_2)/u_1{}^2(c_2)] = z_2 f'(n_2). \tag{6.68}$$

To sum, for $v_{11}{}^1(n_1)=0$, $v_{11}{}^2(n_2)<0$, separable utility between consumption and labor, and private information, ex-ante efficiency is consistent with ex-post efficiency with lotteries as a revelation mechanism. There will be no differences in consumption within and across units, but there will be ex-post differences in actual labor supply within unit $j=1$ and across units as well.

(iii) Private Information in Individual Preferences with Heterogeneous Types and Ex-Post Inefficiency: Complete Markets

Consider the model in (ii) above with $J=2$, but with a slight modification in the preferences described in (6.61). Let $v_{11}{}^2(n_2) < v_{11}{}^1(n_1) < 0$. Therefore, individuals in unit $j=1$ are uniformly less risk averse than $j=2$ or similarly have a uniformly higher elasticity of substitution. The only difference from (ii) is that now both types are risk averse. Individual preferences are private information and the revelation mechanism is identical: introduce a lottery for $j=1$ to make it unattractive for $j=2$, the more risk averse, while acceptable to $j=1$, the less risk averse.

The important issue here is the difference across units not the specific risk neutrality versus risk aversion per se. For small risk aversion of $j=1$, ex-ante Pareto Efficiency holds in this case as well: for all of $j=1$, ex-ante expression (6.65a) holds and for all of $j=2$ ex-ante expression (6.65b) holds.

However, ex-post allocations may not be the same. The individuals in unit $j=1$ are subdivided into the fraction $(1-\phi)$ who do not work, but due to the complete markets in unemployment insurance receive a full wage $z_1 f'(\phi\,\underline{n})$, and the fraction ϕ that work $\underline{n}$ hours and receive the same wage $z_1 f'(\phi\underline{n})$. Thus, within and across units, individuals consume the same amount $c_1=c_2=c$. But now, both are risk averse implying that

$$v_1{}^1(\phi\underline{n}) \neq v_1{}^1(0) \qquad (6.69)$$

ex-post, since, the marginal rate of substitution is a function of the labor supply when all are risk averse. For the proportion $(1-\phi)$ of individuals in unit $j=1$ who do not work, there will be ex-post inefficiency, or

$$[-v_1{}^1(0)/u_1{}^1(c_1)] \neq z_1 f'(\phi\underline{n}). \qquad (6.70)$$

For the proportion ϕ of individuals in unit $j=1$ who do work, there is ex-post efficiency, or

$$[-v_1{}^1(\phi\underline{n})/u_1{}^1(c_1)] = z_1 f'(\phi\underline{n}). \qquad (6.71)$$

For all of $j=2$, consumption is constant $c_1=c_2=c$, and there is ex-post efficiency as in (6.68). Hence, in this case ex-ante efficiency is consistent with ex-post inefficiency at least for some in the population of $j=1$.

(iv) The Incomplete Markets Case

In cases (ii) and (iii) above, we assumed that there are complete markets for unemployment insurance so that a lottery holder can receive full wage in case of unemployment. This presumes that markets provide full insurance at actuarially fair prices. In this part, we assume that there are no insurance mechanisms available for $j=1$, i.e. there are incomplete markets in unemployment insurance. The individuals of unit $j=1$ are faced with idiosyncratic uninsurable risk. We apply the same revelation mechanism except that the lottery contract specifies that the individual who does not work ex-post will *not* receive a payment. In this case, ex-ante efficiency holds exactly as before, i.e. Pareto efficiency ex-ante (in expectations) holds, but ex-post allocations are inefficient and different even from case (iii) above.

The fraction $(1-\phi)$ of individuals in unit $j=1$ who do not work, will not be able to consume the same amount as the other fraction ϕ ex-post since with incomplete markets they receive nothing in terms of wages. Therefore, the fraction $(1-\phi)$ cannot consume (or consumes just a fixed endowment more realistically) and the fraction ϕ consumes ex-post $c_{1,\phi} = c_1 / \phi > c_1$, where c_1 is the allocation in (ii) or (iii). Thus, in the case of incomplete markets for unemployment insurance there is ex-ante efficiency as before but there is ex-post inefficiency for some of $j=1$. The nature of the inefficiency includes the one discussed in equation (6.69) relating to the marginal disutility of labor, and in addition, it includes inefficiency in the marginal utility of consumption, so that ex-post $u_1^{\,1}(c_1) \neq u_1^{\,1}(0)$ since consumption for all individuals in unit $j=1$ is not going to be identical ex-post.

6.14 Consequences for Asset Allocation

We consider infinitesimal risk as small deviations from a nonstochastic equilibrium induced by z_j, i.e. dz_j, from $z_j = 1$. First, we linearize the first order necessary conditions for an interior equilibrium from a nonstochastic equilibrium for all the problems examined in (i)-(ii)-(iii)-(iv) of section 6.13, to obtain the small deviations dc_j and dn_j as a function of dz_j for all j. Let $\mathbf{dc}$ be a vector column with dimension $(1 \times J)$ with elements $(dc_1, dc_2, \ldots dc_J)$ and similarly $\mathbf{dn}$ and $\mathbf{dz}$ be vector columns with dimension $(1 \times J)$ with elements $(dn_1, dn_2, \ldots dn_J)$ and $(dz_1, dz_2, \ldots dz_J)$. The solution for the linearized system of first order conditions takes the general form

$$\mathbf{dc} = \mathbb{C}\ \mathbf{dz} \tag{6.72a}$$

$$\mathbf{dn} = \mathbb{N}\ \mathbf{dz} \tag{6.72b}$$

where $\mathbb{C}$ and $\mathbb{N}$ are $(J \times J)$ gradient matrices evaluated at the nonstochastic equilibrium with $z_j = 1$. In general, the elements of $\mathbb{C}$ and $\mathbb{N}$ are a function of income and substitution effects through the parameters of preferences and technology. However, if income and substitution effects exactly cancel out, then $\mathbb{N}$ is singular. Since we are interested in the contribution of endogenous labor supply to asset allocation, we rule out the case where income and substitution effects cancel out with $\mathbb{N}$ nonsingular throughout.

Units cannot trade or observe directly the technology z_j. However, there are available organized markets for equity trade where the equity is the profit of each unit given in (6.55b). Using (6.53) and (6.55b), small deviations dz_j induce small changes in equity values (at $z_j = 1$) given by

$$de_j = (1-\alpha)\,[\,f(n_j)\, dz_j + f'(n_j)\, dn_j\,], \quad all\, j. \tag{6.73}$$

In general, letting de be a vector column with dimension $(1 \times J)$ and elements $(de_1, de_2, \ldots de_J)$ we obtain

$$de = \mathbb{E}_1\, dz + \mathbb{E}_2\, d\mathbf{n} \tag{6.74}$$

where $\mathbb{E}_1$ is a $(J \times J)$ diagonal matrix with elements $(1-\alpha) f(n_j)$ and $\mathbb{E}_2$ is a $(J \times J)$ diagonal matrix with elements $(1-\alpha)\, f'(n_j)$. Substituting above for $d\mathbf{n}$ from (6.72b) and solving for dz we obtain

$$dz = [\, \mathbb{E}_1 + \mathbb{E}_2\, \mathbb{N}\,]^{-1}\, de \tag{6.75}$$

for the case where the $(J \times J)$ matrix $[\, \mathbb{E}_1 + \mathbb{E}_2\, \mathbb{N}\,]$ is invertible. Equation (6.75) maps the sources of stochastic deviations into equity values available in the organized market for trade. Thus, even though technology is not observed, the equity market provides an observable variable. Then, consumption deviation $d\mathbf{c}$ in (6.72a) can be mapped into the equity funds as

$$d\mathbf{c} = \{\, \mathbb{C}\ [\, \mathbb{E}_1 + \mathbb{E}_2\, \mathbb{N}\,]^{-1}\,]\,\}\, de \tag{6.76}$$

and a deviation $d\mathbf{c}$ can be optimally supported by holding

$$\mathbb{C}\,[\, \mathbb{E}_1 + \mathbb{E}_2\, \mathbb{N}\,]^{-1}$$

"shares" of equity fund $\boldsymbol{e}$. Denoting elements of the asset allocation matrix $\mathbb{C}\,[\, \mathbb{E}_1 + \mathbb{E}_2\, \mathbb{N}\,]^{-1}$ by c_{ij} for all $i,j=1,2,\ldots J$, the net "foreign" asset position of each unit (region or country) is given by

$$F_i = \Sigma_{j, i \neq j}\, c_{ij} - \Sigma_{j, i \neq j}\, c_{ji} \tag{6.77}$$

which denotes the shares held by type i on equity funds issued by units j minus the shares held by type j on equity funds issued by units i satisfying $\Sigma_i F_i = 0$. Recognizing that the level of the current account of each unit is just y_j minus c_j, we have that

$$CA_j = z_j\, f(n_j) - c_j \qquad all\, j \tag{6.78}$$

and, using (6.72)-(6.76), we obtain deviations of the current account from the nonstochastic equilibrium as a function of the equity funds given by

$$d\mathbb{CA} = \{\, [1/(1-\alpha)]\, \mathbb{I} - \mathbb{C}\,[\, \mathbb{E}_1 + \mathbb{E}_2\, \mathbb{N}\,]^{-1}\,\}\, de \tag{6.79}$$

where $d\mathbb{CA}$ is a vector column with dimension $(1 \times J)$ and elements $(dCA_1, dCA_2, \ldots dCA_J)$, the term $[1/(1-\alpha)]$ is a scalar, and $\mathbb{I}$ is a $(J \times J)$ identity matrix. The intuition for (6.79) is that the product $[1/(1-\alpha)] \times \mathbb{I} \times de$ is the portfolio of tradable equities for each individual of unit j as can be seen from (6.79b), i.e. output in terms of equities. This can be subtracted from consumption in (6.76) yielding the current account directly. We denote the elements of the $(J\times J)$ matrix

$$[1/(1-\alpha)]\,\mathbb{I} - \mathbb{C}\,[\,\mathbb{E}_1 + \mathbb{E}_2\,\mathbb{N}\,]^{-1}$$

by ca_{ij} for all $i,j=1,2,...J$ and each denote the change in asset allocation that replicates the change in the unit's current account, i.e. a measure of gross capital flows across units.

Equations (6.76)-(6.79) are the main relationships that represent the asset allocation across units. For each case (i)-(ii)-(iii)-(iv) of section 6.13 regarding the nature of private information and market completeness, the set of equations can be evaluated and comparisons across the different regimes governing private information and market completeness can be drawn.

We pursue a quantitative approach in drawing the comparisons. A nonstochastic equilibrium is computed with $J=2$, $\omega_j=\pi_j=\frac{1}{2}$ and $z_j=1$ for $j=1,2$. Preferences take the specific form

$$w^1(c_1, n_1) = \log c_1 - \delta\,(n_1^{1+\gamma_1})/(1+\gamma_1) \qquad (6.80a)$$

$$w^2(c_2, n_2) = \log c_2 - \delta\,(n_2^{1+\gamma_2})/(1+\gamma_2) \qquad (6.80b)$$

where $\delta>0$, and $0\le\gamma_1<\gamma_2$ give the curvature of the utility function relative to labor supply. The technology is given by (6.53)-(6.54). The preferences in (6.80a,b) allow us to examine all models in (i)-(ii)-(iii)-(iv) of section 6.13 by varying essentially the parameter γ_1 given plausible choices of α, δ, γ_2 and $\underline{n}$. We use the plausible choices $\alpha=0.6$, $\delta=1.5$, $\gamma_2=0.75$ and $\underline{n}=0.6$.

First, note that under full information and homogeneous types $c_1=c_2=c$, $n_1=n_2=n$, $c_{ij}=c$, $F_i=CA_j=0$ all $i,j=1,2$, $ca_{ij}=ca_{ji}$ for $i=j$ and $ca_{ij}=-ca_{ji}$ for $i\neq j$; a perfectly symmetric equilibrium as in section 6.12. Naturally, introducing an asymmetry in preferences introduces asymmetry in asset allocation and nontrivial net asset position and current account allocations.

Table 6.1 presents results for the alternative regimes. The values in each box represent the matrices defined above. The main result is that inefficiency ex-post with complete markets does not give rise to any substantive change in asset allocation whereas the presence of incomplete markets does, i.e. uninsurable idiosyncratic risk matters. The first row of the table presents the cases in (i) and (ii). The two allocations are the same with the implied value for the lottery given by $\phi=0.517$, as shown theoretically in expression (6.65c).

The asymmetry in preferences induces a very small bias towards holding assets of the risk averse unit, $c_{11}=c_{21}=1.249$, whereas the risk neutral unit runs a current account deficit (in levels) with a positive net asset position. Diversification across the units is almost even. The change in the asset allocation that replicates the change in the unit's current account, ca_{ij}, shows the extent of the gross flows across the units when there are technology disturbances. In order to replicate the change in the current account, both individuals must go long on their own fund and short on the other fund so that the change in the own technology can be fully reflected in the current account. The second row of the table presents case (iii) ex-post. There is very small qualitative and quantitative change relative to (i)-(ii) in terms of the asset allocation.

The current account level is about ½ less than the previous case but the change in the net asset position is trivial. Thus, we conclude that ex-post inefficiency with complete markets for unemployment insurance seems to have very little marginal impact on the asset allocation of individuals.

The third and fourth rows present the model in (iv) ex-post where there are incomplete markets for unemployment insurance. The third row is a direct comparison with the complete markets case in the second row. We obtain a large shift in holdings of the less risk averse unit towards the fund of the more risk averse unit.

The "share" holdings are $c_{11}=0.014$, $c_{12}=3.076$, so that the portfolio of $j=1$ (that works and consumes ex-post) consists of 0.5% in the own equity fund and 99.5% in the other unit equity fund. As a result, the net asset position of the less risk averse unit increases sharply as well as the asset allocation that replicates the change in the current account. The actual level of the current account deficit does not change, but the marginal effects are rather large. The less risk averse individual ($j=1$)

Table 6.1: Simulated Alternative Regimes

	c_{ij}	F_i	CA_j	ca_{ij}
Full Information, Heterogeneity ≡ Private Information, Heterogeneity $\gamma_1 = 0$, $\gamma_2 = 0.75$	1.249 1.250 1.249 1.250	0.27 E-3 -0.27 E-3	-0.144 0.144	1.250 -1.250 -1.249 1.249
Private Information, Heterogeneity, Ex-Post Inefficiency, Complete Markets $\gamma_1 = 0.25$, $\gamma_2 = 0.75$	1.250 1.250 1.250 1.250	-0.20 E-5 0.20 E-5	-0.073 0.073	1.249 -1.250 -1.250 1.249
Private Information, Heterogeneity, Ex-Post Inefficiency, Incomplete Markets $\gamma_1 = 0.25$, $\gamma_2 = 0.75$	0.014 3.076 1.250 1.250	1.826 -1.826	-0.073 0.073	3.486 -3.076 -1.250 1.249
Private Information, Heterogeneity, Ex-Post Inefficiency, Incomplete Markets $\gamma_1 = 0.5$, $\gamma_2 = 0.75$	0.392 2.369 1.249 1.250	1.113 -1.113	-0.030 0.030	2.108 -2.369 -1.249 1.250

Note: Values are c_{11}, c_{12}, c_{21}, c_{22} for c_{ij}, F_1, F_2 for F_i, CA_1, CA_2, for CA_j, and ca_{11}, ca_{12}, ca_{21}, ca_{22} for ca_{ij}.

holds most of its portfolio in the fund of the other unit. This strong diversification result is due to the willingness to trade away the risk of own technology shocks that will have a large effect on the consumption of the individual that actually works; we recall that ex-post $c_{l,\phi} = c_l / \phi > c_l$ in (iv).

The fourth row shows an allocation when γ_l increases from *0.25* to *0.5* with incomplete markets making the two units more alike in terms of risk aversion. The asset allocation moves in the expected direction. The exodus in holdings of the less risk averse own asset to the to the other asset is relatively moderate.

The results show quite sharply that ex-post inefficiencies consistent with ex-ante Pareto Optimal allocations alone do not give rise to substantive changes in asset allocation when there are complete markets for unemployment insurance, i.e. when there is full insurance for idiosyncratic risk. The differences realized ex-post are too small in terms of asset holdings because it is in the interest of all individuals of the less risk averse unit to hold portfolios that guarantee the variations in the lottery probability, regardless whether they work or not ex-post. When there are incomplete markets the story is quite different. Ex-post, the individuals of less risk averse type who do not work will not receive any payment and will be shut off from the capital market. Therefore, the less risk averse individual that works has an incentive to hold an overwhelming fraction of its portfolio on the foreign equity fund thus diversifying away the exposure to the lottery probability.

6.15 **Summary III**

In section 6.11-6.14, we examined the issue of private information, efficiency and market structure in a multi-unit general equilibrium model with endogenous labor supply and its relationship to asset allocation.

In this framework, it can be shown that both the First and Second Fundamental Theorems hold. The main result of our application is that ex-post inefficiency consistent with ex-ante efficiency does not matter for asset allocation when idiosyncratic risks are insurable. But, incomplete markets for unemployment insurance have sharp implications for asset allocation. Uninsurable idiosyncratic risks matter for asset allocation. The major effect is to induce diversification away from the own equity into other units' equity in order to diversify away the exposure to the employment lottery. Therefore, even though ex-post inefficiencies can be consistent with ex-ante efficiency in consumption and labor allocations with private information, the changes in asset allocation in the case of incomplete markets for unemployment insurance go in the direction of more diversification. Thus, not explaining the lack of international portfolio diversification observed in the real world.

Problems

1. Solve using Kuhn-Tucker methods, e.g. Chapter 1,

$$\underset{\{c,n\}}{Max}\ [u(c) - v(n)]$$

$$\text{subject to} \quad c \leq n\,w + r\,k_o$$

$$c \geq 0$$

$$n \in \{0,1\}$$

$$k_o = 1.$$

2. Solve for

$$Max \ \ [f(K,N) - w\,N - r\,K]$$

$$\text{subject to} \quad K,N \geq 0.$$

3. Explain the Pareto Optimal Allocations in section 6.8.

4. What is the role, if any, played by lotteries in your answer to Problem 3 above? Explain.

5. In the model of section 6.13, what is the context and problem of complete versus incomplete markets?

6. The model in sections 6.12-6.13 presents a problem of what type: moral hazard and/or adverse selection? Explain.

Notes on the Literature

A simple presentation of the Robinson Crusoe model under symmetric information noted in section 6.1 is found in Barro (1997), Chapter 2. Also the classical general equilibrium framework and fundamental theorems of welfare economics are examined by Debreu (1959), Mas-Colell *et al* (1995).

The material with asymmetric information in sections 6.7-6.10 follows the seminal contributions of Prescott and Townsend (1984a,b) who showed how to map economies with private information into the classical general equilibrium framework under uncertainty. Helpman and Laffont (1975) is an early contribution in the same general direction. In particular, the basic structure (with the proofs of Pareto optimality not presented in section 6.8) and the adverse selection problem are from Prescott and Townsend (1984a); where they also present the consumption-labor allocation problem with private information and show that, in this case, both the First and Second Fundamental Theorems hold. They also sketch the problems of moral hazard and signaling in the basic general equilibrium framework.

The results of Rogerson (1988), that we presented in the previous sections 6.1-6.6, are based on the Prescott and Townsend (1984a,b) contributions as well. Hansen (1985) has applied the Rogerson (1988) results to quantitative dynamic stochastic general equilibrium models.

An article by Townsend (1987) presents a rich survey of the general equilibrium approach to economics with further examples of applications of lotteries and is highly recommended. The collection of articles in Townsend (1990) presents further important papers in the theory of contracts, principal-agent relationships and transactions costs. The paper by Hornstein and Prescott (1989) presents further examples of lottery allocations in the context of the theory of the firm under general equilibrium, and provide an endogenous explanation for the labor indivisibility constraint of Rogerson (1988) and Hansen (1985) examined in sections 6.1-6.6.

The results in the adverse selection section 6.9 are complementary to the earlier results of Rothschild and Stiglitz (1976) and Wilson (1977).

Sections 6.11-6.15 present a direct application of the methods learned in the previous sections and follow Bianconi (2001). Bianconi (2002) presents an extension to the two-period dynamic case. An example of asset markets that reveal information is found in Berliant and De (1998).

One of the major applications of the result that ex-ante efficiency may be consistent with ex-post inefficiency is in the literature on contracts with renegotiations, e.g. Hart (1995).

The lottery scheme presented in 6.11-6.15 is based on Prescott and Townsend (1984a,b), and other applications of lotteries may be found in Townsend (1987), Hornstein and Prescott (1989), and Besley et al (1994). Prescott (2002) surveys the quantitative general equilibrium literature with non-convexities and lotteries.

We do not consider any potential moral hazard problem relating to the work effort in the presence of full insurance, but the papers by Hansen and Imrohoroglu (1992) and Atkeson and Lucas (1995) present models where the moral hazard problem in unemployment insurance is fully analyzed.

The issue of portfolio choice with endogenous labor supply has been addressed by Bodie et al (1992) in a partial equilibrium framework, and by Jermann (1998) in a general equilibrium framework. Marcet et al (1998) considers the issue of incomplete markets in unemployment insurance. That an apparent labor market failure has a plausible market based explanation is a result also shown by Chari (1983). Here, we apply the same characterization to the problem of asset allocation.

It is well documented that there is a lack of international portfolio diversification inconsistent with simple complete markets portfolio choice models, see e.g. French and Poterba (1991), Leung (1995), Lewis (1996) and Jermann (1998).

The assumptions that capital markets are perfectly integrated across units, labor is assumed immobile but labor income is assumed to be tradable are found in Leung (1995) and Jermann (1998). The infinitesimally small stochastic variation used in sections 6.11-6.15 has been used by Jermann (1998) and the references therein. It takes into account up to second order derivatives of the functions, associated with first order moments of the distributions (means), but ignores third order derivatives associated with second order moments (variances).

The perfectly pooled equilibrium without asymmetric information is found in Lucas (1982) and Leung (1995) among others.

References

Atkeson, Andrew and Robert E. Lucas Jr. (1995) "Efficiency and Equality in a Simple Model of Efficient Unemployment Insurance." *Journal of Economic Theory*, 66, 64-88.

Barro, Robert J. (1997) *Macroeconomics,* 5th Edition. The MIT Press, Cambridge, MA.

Berliant, Marcus and Sanker De (1998) "On the Revelation of Private Information in Stock Market Economies." *Journal of Mathematical Economics*, 30, 241-256.

Besley, Timothy, Stephen Coate, and Glenn C. Loury (1994) "Rotating Savings and Credit Associations, Credit Markets and Economic Efficiency." *The Review of Economic Studies* , 61, 701-719.

Bianconi, Marcelo (2001) "Heterogeneity, Efficiency, and Asset Allocation with Endogenous Labor Supply: The Static Case." *Manchester School*, 69, 253-268.

Bianconi, Marcelo (2002) "Heterogeneity, Adverse Selection and Equilibrium Valuation with Endogenous Labor Supply." Working paper, Department of Economics, Tufts University, July.

Bodie, Zvi, Robert C. Merton, and William F. Samuelson (1992) "Labor Supply Flexibility and Portfolio Choice in a Life Cycle Model." *Journal of Economic Dynamics and Control*, 16, 427-449.

Chari, V.V. (1983) "Involuntary Unemployment and Implicit Contracts." *Quarterly Journal of Economics*, 98 (Supplement), 107-122.

Debreu, Gerard (1959) *The Theory of Value.* Cowles Foundation for Economic Research, Yale University, New Haven, CT.

French, Kenneth and James Poterba (1991) "International Diversification and International Equity Markets." *American Economic Review*, 81, 222-226.

Hansen, Gary D. (1985) "Indivisible Labor and the Business Cycle." *Journal of Monetary Economics,* 16, 309-327.

Hansen, Gary D. and Asye Imrohoroglu (1992) "The Role of Unemployment Insurance in an Economy with Liquidity Constraints and Moral Hazard." *Journal of Political Economy*, 100, 118-142.

Hart, Oliver (1995) *Firms, Contracts, and Financial Structure*. Claredon Lectures in Economics, Claredon Press, Oxford, UK.

Helpman, Elhanan and Jean J. Laffont (1975) "On Moral Hazard in General Equilibrium Theory." *Journal of Economic Theory,* 10, 8-23.

Hornstein, Andreas and Edward C. Prescott (1989) "The Firm and the Plant in General Equilibrium." Federal Research Bank of Minneapolis, Research Paper No. 126, November.

Jermann, Urban J. (1998) "International Portfolio Diversification and Endogenous Labor Supply Choice." Working Paper, University of Pennsylvania, Wharton School, June.

Leung, Charles K. Y. (1995) "Does Non-traded Input Necessarily Deepen the International Non-diversification Puzzle I?: The One-good Case." *Economics Letters*, 49, 281-285.

Lewis, Karen (1996) "What Can Explain the Apparent Lack of International Consumption Risk Sharing?" *Journal of Political Economy*, 104, 267-297.

Lucas, Robert E. Jr. (1982) "Interest Rates and Currency Prices in a Two-Country World." *Journal of Monetary Economics*, 10, 336-360.

Marcet, Albert, Francisco Obiols-Homs, and Phillipe Weil (1998) "Incomplete Markets, Labor Supply and Capital Accumulation." Working paper, ECARE, Brussels, August.

Mas-Colell, Andreu, Michael Whinston and Jerry Green (1995) *Microeconomic Theory.* Oxford University Press, Cambridge, MA.

Prescott, Edward C. (2002) "Non-Convexities in Quantitative General Equilibrium Studies of the Business Cycles." Federal Reserve Bank of Minneapolis; Research Department Staff Report 312, October.

Prescott, Edward C. and Robert Townsend (1984a) "Pareto Optima and Competitive Equilibria with Adverse Selection and Moral Hazard." *Econometrica*, 52, 21-45.

Prescott, Edward C. and Robert Townsend (1984b) "General Competitive Analysis in an Economy with Private Information." *International Economic Review*, 25, 1-20.

Rogerson, Richard (1988) "Indivisible Labor, Lotteries and Equilibrium." *Journal of Monetary Economics*, 21, 3-16.

Rothschild, Michael and Joseph Stiglitz (1976) "Equilibrium in a Competitive Insurance Market." *Quarterly Journal of Economics,* 90, 629-649.

Townsend, Robert (1987) "Arrow-Debreu Programs as Microfoundations of Macroeconomics." Truman F. Bewley, Editor, *Advances in Economic Theory, Fifth World Congress*. Cambridge University Press, Cambridge, UK.

Townsend, Robert (1990) *Financial Structure and Economic Organization.* Basil Blackwell, New York, NY.

Wilson, Charles (1977) "A Model of Insurance Markets with Incomplete Information." *Journal of Economic Theory,* 16, 167-207.

7. Dynamics I: Discrete Time

7.1 Time and Markets

This chapter presents an introduction to uncertainty methods applied to dynamic situations when time is discrete, or $t=1,2,...$ First, in section 7.2, we present some simple concepts relating to markets for transactions in a dynamic setting. We start with forward and futures contracts and a discussion of prices in those markets. We then present some simple examples of swaps of streams of income with the associated discounting properties, and discuss hedging in the context of the prefect hedge, the imperfect hedge and hedging with expected utility.

The remainder of the chapter is dedicated to the theoretical general equilibrium approach to asset pricing under uncertainty with symmetric information and complete markets, and asymmetric information. Sections 7.4-7.12 present dynamic general equilibrium models of resource allocation and asset pricing under complete markets. We pay special attention to the specific assumptions of the underlying stochastic processes in the multi-period context, and examine the problem of excess returns in some detail. In sections 7.13-7.15 we extend the dynamic general equilibrium model to include money, and examine the Fisher equation under conditions of risk. Next, in sections 7.16-7.17 we apply the dynamic general equilibrium framework to discuss the financial problem of the firm presenting a version of the Modigliani-Miller theorem. Sections 7.18-7.21 extend the framework to the case of asymmetric information where we show a contract mechanism that mitigates adverse selection in a recursive dynamic context. Finally, sections 7.22-7.23 examine the explicit distinction between intertemporal substitution and risk aversion and its effect on asset returns.

7.2 Introduction to Financial Contracts

In general, a derivative security is a security whose payoff is explicitly tied to the value of some other variable or security. An example is a forward contract that works as follows. At present time t, a buyer agrees to purchase from a seller an amount of commodities (or currencies or securities) at a predetermined price to be delivered at a prearranged date in the future. At the future date, $t+\Delta t$, the contract is fulfilled and the transaction takes place. For example:

present, t	*future, t+Δt*
• ⟶	⟶ •
↓	↓
buyer agrees to purchase (go long) from seller (go short) X pounds of commodity Y at price P per pound to be delivered at $t+\Delta t$ *from t; or* $t+\Delta t > t$	*agrees to buy: contract is fulfilled at predetermined terms.*

In this case, P is usually called the forward price.

A distinction between forward and futures contracts can be easily drawn. In forward contracts, two parties agree to exchange some item for a prearranged price at some prearranged date in the future. In this case, usually no cash is required at the present date by either party, but this is a more customized and more rigid agreement between the two parties involved. In the case of futures contracts, it is exactly a forward contract that is standardized and traded in an organized exchange, thus less rigid and more open to arrangements standard in the market place. Hence, futures contracts are more flexible. They can be terminated or closed-out before a specified delivery date. In both cases, the party who agrees to buy in the future is said to take a long position whereas the party who agrees to sell takes a short position. An open market for immediate delivery is called a spot market and the immediate price is the spot price.

(i) Forward Prices and Spot Prices

Let $F_{t,T}$ be the forward price at the present date (say $t>0$) to deliver a widget at some future date T, or $T>t>0$. Let S_t be the current spot prices at $t>0$ of the widget in question. More generally, we have the following scheme:

present, t	*future, $t+\Delta t=T$*
spot price at t: (market for immediate delivery)	*spot price at T*
S_t	S_T
forward price at t for T,	
$F_{t,T}$	

First, we can examine the relationship of prices at the present time t, or the spot price S_t and the forward price, $F_{t,T}$ in the left-hand side column of the general scheme above. Under certain conditions, for example zero costs of storage and short selling of commodity available, the relationship between the two prices is given by

$$F_{t,T} = S_t / d(t,T) \tag{7.1}$$

where $d(t,T)$ is some discount factor between t and T, i.e. between the present and the future dates, or

$$d(t,T) \equiv 1 / (1+R)^{T-t} \tag{7.2}$$

for a constant $R \geq 0$, the relevant interest rate used for discrete compounding. Expressions (7.1) and (7.2) are due to a simple arbitrage argument: premiums or discounts should be eliminated by absence of excess supply and demand. Thus, the forward price at t is the spot price at t compounded by some interest rate between t and T. In case of continuous compounding, we would have:

$$F_{t,T} = S_t / exp\,(-R(T-t)) \quad \Leftrightarrow \quad d(t,T) \equiv exp(-R(T-t)) \tag{7.3}$$

where *exp* is the exponential operator. Usually the interest rate, $R \geq 0$, used in these calculations is the so-called Repo Rate: a rate associated with short term agreements to buy and sell a security. In practice, the repo rate is usually a Treasury bill interest rate plus a small premium.

Note first that we can include carrying costs. These costs may be positive, say the cost of storing a widget; or negative for the income received by holding a bond. Carrying costs have an effect on the relationship between the forward and spot price. We call the (capitalized) carrying cost in the interval t to T, $c(t,T)$. For example, in the case of gold the carrying cost is strictly positive, $c>0$. By an arbitrage argument, the forward price cannot differ from the capitalized spot price plus the cost of carrying the widget, or

$$F_{t,T} = [S_t / d(t,T)] + c(t,T). \tag{7.4}$$

However, in some agricultural commodities markets, there may be periods of short supply and we can observe that $F_{t,T} < [S_t / d(t,T)] + c(t,T)$ where the difference captures a convenience yield representing the benefit of holding the commodity.

Second, in the general scheme presented above, we can examine the relationship between the forward price at t, $F_{t,T}$, and the spot price at T, S_T. The expected value of the spot price at time T, S_T, based upon the information set as of time t is denoted $E[S_T|I_t]$, or conditional on information at time t, where I is the information set. As noted above, $F_{t,T}$ is the forward price at time t. That is:

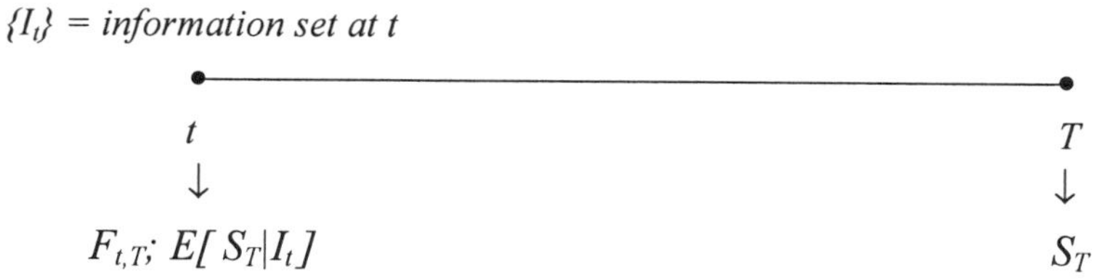

$\{I_t\}$ = *information set at t*

t	T
↓	↓
$F_{t,T}$; $E[S_T|I_t]$	S_T

A natural question that arises is whether $F_{t,T}$ is related (or equal) to the expected value of the future spot price, $E[S_T|I_t]$, or whether the forward price is a good predictor of the actual future price, at least on average. If yes, we have the so-called Efficient Market hypothesis; recall Chapter 4. However, empirically, the likely answer is that there is a discrepancy between the two due to the presence of risk premiums or discounts and other statistically unexplained components, that is

$$F_{t,T} - E[S_T \mid I_t] = \textit{risk premium or discount} + \textit{unexplained component}.$$

Third, we can examine the current value of forward contracts signed in past periods. The current value of a forward contract at $t\text{-}1 < t < T$ follows the scheme:

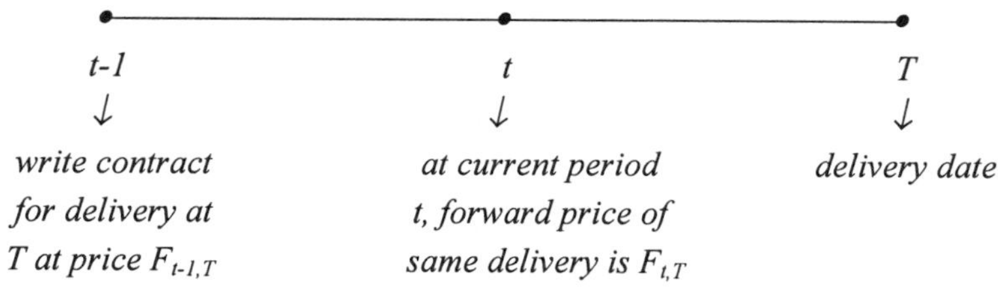

$t\text{-}1$	t	T
↓	↓	↓
write contract for delivery at T at price $F_{t\text{-}1,T}$	*at current period t, forward price of same delivery is* $F_{t,T}$	*delivery date*

What is the current value, denoted $f_t^{\,t\text{-}1,T}$, of the initial contract at time t? Towards an answer, we construct a simple theoretical portfolio at the current date t:

(a) Buy (go long) one unit at t with delivery price $F_{t,T}$ at T;

(b) Sell (go short) one unit at t of initial contract $F_{t\text{-}1,T}$ for delivery at T.

The time t cash value of this operation is denoted $f_t^{\,t\text{-}1,T}$. The final, at time T, cash flow is $F_{t\text{-}1,T} - F_{t,T}$. Thus, the present discounted value (pdv_t) of this operation, at time t, is:

$$pdv_t = f_t^{\,t\text{-}1,T} + (F_{t\text{-}1,T} - F_{t,T})\, d(t,T)$$

which must be zero by an arbitrage argument, i.e. $pdv_t = 0$. Thus, the current (time t) value is the discounted value of the difference between the forward prices, $(F_{t,T} - F_{t\text{-}1,T})d(t,T)$, or

$$f_t^{\,t\text{-}1,T} = (F_{t,T} - F_{t\text{-}1,T})\, d(t,T).$$

(ii) Swaps

A <u>swap</u> is an agreement between two parties to exchange one cash flow stream for another with no initial cash involved. Swaps are usually done with commodities, currencies, securities, and interest payments streams. For example: a <u>plain vanilla swap</u> refers to the following transaction, in the case of an interest rate swap:

Party A × *Party* B

Agrees to pay a series of installments to party B *equal to an interest rate on a "notional" principal amount.* ⟵⟶ *Agrees to pay a series of installments to party* A *equal to the current treasury bill rate of interest on the "notional" principal.*

↑——————————————————↑

Difference is the net payment

Party B hedges the risk of fluctuations on the treasury bill rate with party A. That is, party B gets a constant stream of payments independent of fluctuations in interest rates. Hence, a swap contract is equivalent to a series of forward contracts, and so can be priced using concepts of forward pricing. The value of a commodity swap using forward concepts is as follows:

(a) Party B hedges against fluctuations;

(b) Party A receives the spot price, S_t , for N units of a commodity while paying a fixed amount per unit denoted X. Suppose the agreement is made for M periods:

0	1	2	3	4	M periods
	↓	↓	↓	↓	↓
A *receives* →	S_1	S_2	S_3	S_4	S_M
and pays →	x	x	x	x	x

with net cash flow stream

→ $\{(S_1 - X),\ (S_2 - X),\ \ldots\ldots\ldots\ (S_M - X)\} \times N$

for party A

What is the current value of this swap at $t=0$? Notice that at $t=0$, the structure of forward prices for $t=1,2,.....M$ is denoted $F_{0,[1,2,...M]}$. By an arbitrage argument, we have

$$S_t = F_{0,[t=1,2,...M]}\ d(0,[t=1,2...M]),\ \ each\ t=1,2,...M \qquad (7.5)$$

or the current value is the discounted forward price yielding the stream:

$$S_1 = F_{0,1}\ d(0,1)$$

$$S_2 = F_{0,2}\ d(0,2)$$

$$\ldots$$

$$S_M = F_{0,M}\ d(0,M)$$

where $d(0,i) \equiv \{1 / [(1+r_0)\times(1+r_1)\times...\times(1+r_M)]\},\ i=0,1,2,...,M$ is the compounded discount factor. Thus, the current value of the net cash flow stream is given by the sum of net cash flow

$$V_0 = \{(F_{0,1}-X)\ d(0,1) + (F_{0,2}-X)\ d(0,2) +...+(F_{0,M}-X)\ d(0,M)\} \times N, \quad (7.6)$$

i.e. the value of a swap is determined by a series of forward prices. Usually the constant payment stream X is chosen so that the current value of the net cash flow is zero, $V_0=0$, or

$$V_0 = \Sigma_{t=1}^{M}\{(F_{0,t}-X)\ d(0,t)\} \times N = 0 \qquad (7.7)$$

by choice of X, and the swap has zero present value.

In particular, the value of an interest rate swap for the plain vanilla swap is as follows. Party A agrees to make a payment of a fixed interest rate r on the "notional" principal N, while receiving a floating rate payment on N, denoted r_i, $i=0,1,2,...,M$. The cash flow stream of received payments is given by

$$[\ (r_0 - r),\ (r_1 - r\ ),...,(r_M - r\ )] \times N,$$

for M periods. Next, the compounded discount factor can be written as

$$d(0,i) \equiv \{1 / [(1+r_0)\times(1+r_1)\times...\times(1+r_{M-1})]\}$$

for $i=0,1,2,...,M-1$, and we compute the discounted value

$$r_i \times d(0,i+1) = \{r_i / [(1+r_0)\times(1+r_1)\times...\times(1+r_M)\}. \qquad (7.8)$$

We add and subtract $d(0,i+1)$ from the discounted value (7.8) to obtain

$$(1+r_i)\times d(0,i+1) - d(0,i+1) =$$

$$\{(1+r_i) / [(1+r_0)\times(1+r_1)\times...\times(1+r_M)\} - d(0,i+1)$$

and note that the first term,

$$(1+r_i)\times d(0,i+1) = \{(1+r_i) / [(1+r_0)\times (1+r_1)\times ... \times (1+r_M)\}$$
$$= \{1 / [(1+r_0)\times (1+r_1)\times ... \times (1+r_{M-1})\}$$
$$= d(0,i)$$

where we cancelled out one term, $(1+ r_i)$. Thus, the discounted value $r_i \times d(0,i+1)$ reduces to the difference of discount factors

$$r_i \times d(0,i+1) = d(0,i) - d(0,i+1). \tag{7.9}$$

Then the sum of the stream r_i discounted by $d(0,i+1)$ across the M periods is given by

$$\Sigma_{i=1}^{M} r_i\, d(0,i+1) = \Sigma_{i=1}^{M} [d(0,i) - d(0,i+1)] \tag{7.10}$$

where the left-hand side is the sum of the discounted values of the interest rate r_i and the right-hand side is the sum of the differences of the discount factors. We use (7.8)-(7.10) to derive the current value of the swap as

$$V_0 = \{\Sigma_{i=1}^{M} [d(0,i) - d(0,i+1)] - \Sigma_{i=1}^{M} r\, d(0,i)\} \times N$$
$$= \{\Sigma_{i=1}^{M} [\, (1 - r)\, d(0,i) - d(0,i+1)]\} \times N. \tag{7.11}$$

In this case, we may choose the constant interest rate r in the swap, that makes the discounted sum in expression (7.12) equal to zero, or choose r so that

$$V_0 = \{\Sigma_{i=1}^{M} [\, (1 - r)\, d(0,i) - d(0,i+1)]\} \times N = 0. \tag{7.12}$$

(iii) Hedging with Futures Markets

The concept of hedging regards the elimination of the risk of a loss by giving up the potential for a gain. On the other hand, the concept of insurance regards the payment of a premium to eliminate the risk of a loss, but simultaneously retaining the potential for a gain. We have seen examples of insurance problems with symmetric and asymmetric information in Chapters 5 and 6. Here, we focus on hedging with future contracts to eliminate risk and simultaneously the potential gain.

Consider the following example of a perfect hedge: Professor X (Economics, University Y in the USA) wins the Nobel prize in Economics for the year Z, for example Z=*1992*. The prize is announced on October 9, by the Royal Swedish Academy of Science worth SEK (Swedish Krona) $8 million to be delivered on December 10, of year Z. Hence, the arrangement is

Announcement *Delivery*

Oct. 9 = t *Dec. 10 = T*
SEK$ 8 million

Prof. X wants to hedge possible fluctuations in the SEK$/US$ dollar exchange rate. How can she engage in such hedging? Suppose market conditions are that the spot exchange rate at the announcement date is S_t = *SEK$ 7.80 per US$ 1.00*, or *7.80* Swedish Kronas per US dollar, and the future prize buys *US$ 1,026* million on the spot market; the market forward rate at t for delivery at T is $F_{t,T}$ = *SEK$ 8.0 per US$ 1.00*. In the futures market for foreign exchange, the Swedish Krona is expected to depreciate relative to the US dollar, or the US dollar is expected to appreciate against the Swedish Krona. Thus, the Krona is selling forward at a discount (since it is expected to loose value in the future!).

The perfect hedge scheme is the following. Sell (go short) *SEK$ 8* million at the forward rate $F_{t,T}$ = *SEK 8/US$ 1.00* for delivery at T, receiving *US$ 1* million at T, regardless of the change in the spot rate between t and T: $S_T - S_t$. Henceforth, the perfect hedge consists on taking an equal and opposite (short) position in the futures market giving the perfect hedge, i.e. eliminating all risk involved in possible exchange rate fluctuations. We expand on this idea further.

Suppose there is no perfect hedge available due to some lack of matching in the marketplace. Without a perfect hedge instrument available, a minimum variance hedge can be constructed using other hedging instruments that are correlated with the obligation, a so-called triangular operation. Let the general time frame be

Present date = t *Future date = T.*

At the present time t, an individual wants to hedge the value x given by

$$x_T = s_T w$$

for a future date T, where s_T is the price at T of the quantity w. We must find an instrument that is correlated with $x_T = s_T w$ thus allowing some form of triangular arbitrage. Let $h \lesseqgtr 0$ denote the position (amount) taken at time t in the futures market, and let F_T be the resulting uncertain net payoff at $t=T$ from hedging with some alternative instrument at t, and exchanging at $t=T$.

For example, in Professor X's case, there may not exist a contract available for US$ and SEK$ for the exact amount and timing. But, there may be an available contract for US$ and Deutsche Marks (DM$). The imperfect hedge would be to go short on DM$ and, at the announcement date $t=T$, exchange the SEK$ for DM$ at the spot price $s_T^{s,D}$ under the assumption that the SEK$ and DM$ exchange rates move closely with the US$ currency, i.e. they are strongly positively correlated. Suppose F_T is the netted "price" of this transaction at the future date T, i.e. it includes the US$/DM$ futures from t to T and the exchange at T. Thus, F_T is a random variable as of present time t.

Denote the cash flow at the future date T of this imperfect hedge, y_T, which is given by

$$\begin{aligned} y_T &= x_T + h F_T \\ &= s_T w + h F_T \end{aligned} \tag{7.13}$$

or the initial hedged value plus the value of the uncertain position. The variance of the cash flow is then

$$\begin{aligned} var(y_T) &= E[\{(s_T - E[s_T])\, w + h\,(F_T - E[F_T])\}^2] \\ &= var(s_T)\, w^2 + 2\, cov(s_T, F_T)\, h\, w + var(F_T) h^2. \end{aligned} \tag{7.14}$$

The minimum variance hedge can be appropriately computed by choosing the position h in the futures, at the present date t, that minimizes the variance of the cash flow in (7.14). This is a simple unconstrained problem with first order necessary condition

$$\partial var(y_T)/\partial h = 2\, cov(s_T, F_T)\, w + 2\, h\, var(F_T) = 0 \tag{7.15}$$

yielding a minimum variance solution for the position h:

$$h^* = - cov\ (s_T, F_T)\ w\ /\ var\ (F_T) \qquad (7.16a)$$

where the sufficient second order condition for a minimum is satisfied by the condition

$$\partial^2 var(y_T)/\partial h^2 = 2\ var\ (F_T) > 0.$$

Hence, the minimum variance hedge has the opposite sign of the $cov(s_T, F_T)$ and is inversely related to the $var\ (F_T)$. Recall that the covariance term is just

$$cov\ (s_T, F_T) = \sigma_{s,F} = [var(s_T)\ var\ (F_T)]^{1/2}\ \rho_{s,F}$$

where $\rho_{s,F} = \sigma_{s,F}\ /\ [var(s_T) var\ (F_T)]^{1/2}$ is the coefficient of correlation between s and F. The minimum variance hedge can be written as

$$h^* = - \rho_{s,F}\ w\ /\ [var(s_T)\ var\ (F_T)]^{-1/2} \qquad (7.16b)$$

so that a positive correlation between the price and the net payoff, $\rho_{s,F} > 0$, implies that the hedge should be on the opposite amount (short), or $h^* < 0$ taking a short position at time t. The absence of correlation between the price and the net payoff rules out the possibility of hedging, $\rho_{s,F} = 0 \Rightarrow h^* = 0$. Hence, the more (less) positive the correlation between the price and the net payoff, the more (less) one should hedge on the opposite amount.

It is easy to note that the perfect hedge is just a special case of the imperfect hedge in expressions (7.16a)-(7.16b). If we let

$$F_T = s_T$$

it implies that the correlation between the price and the net payoff is one, or

$$\rho_{s,F} = \sigma_{s,F}\ /\ [var(s_T)\ var\ (F_T)]^{1/2} = 1$$

which, in turn, implies that

$$\sigma_{s,F} = \rho_{s,F}\ [var(s_T)\ var\ (F_T)]^{1/2} = var\ (F_T).$$

Then, substituting into the optimal hedge formula, (7.16a) or (7.16b), we obtain

$$h^* = -w, \qquad with\ \ var\ (y_T) = 0$$

i.e. the perfect hedge of equal and opposite sign position. Therefore, the minimum variance hedge reduces to the perfect hedge if the net value F_T is perfectly correlated to the spot price at the future date T, s_T.

We may compute the optimal hedge under an expected utility representation of preferences as well. Suppose utility is quadratic depending only upon the mean and variance of the cash flow, e.g. section 3.6, and of the form

$$u = E[y_T] - \varphi \, var(y_T) \qquad (7.17)$$

for a real number $\varphi > 0$ measuring the extent of the disutility of variance. Thus, expected utility is increasing and linear in the mean and decreasing and linear in the variance of the cash flow. Maximizing expected utility (7.17) by choice of h is the solution to the simple unconstrained problem

$$\underset{\{h\}}{Max} \{ E[y_T] - \varphi \, var(y_T) \} \qquad (7.18a)$$

or similarly using (7.13)-(7.14)

$$\underset{\{h\}}{Max} \{ (w\, E[s_T] + h\, E[F_T]) - \varphi \, [var(s_T)\, w^2 + 2\, cov(s_T, F_T)\, h\, w + var(F_T) h^2] \} \qquad (7.18b)$$

which gives first order necessary condition

$$E[F_T] - \varphi \, [2\, cov(s_T, F_T)\, w + 2\, var(F_T) h] = 0$$

and a solution for the minimum variance hedge, h^{**} as

$$h^{**} = \{ E[F_T] / 2\, \varphi \, var(F_T) \} - [cov(s_T, F_T)\, w / var(F_T)] \qquad (7.19)$$

where the sufficient second order condition for a maximum is satisfied by the condition

$$- 2\, \varphi \, var(F_T) < 0.$$

In the case of mean-variance utility, the optimal hedge in formula (7.19) has two components. The first is the expected benefit obtained by entering in the futures market to hedge, i.e. hedging is some form of investment and the expected gain on this investment is part of the portfolio. This is given by $E[F_T]/2\varphi \; var(F_T)$, or increasing in the expected payoff and decreasing in the variance of the payoff weighted by the extent of the disutility of the variance, φ. The second term is exactly the minimum variance hedge h^* obtained in (7.16a,b). It is simple to understand expression (7.19) since we are adding a mean term to the objective function relative to the simple minimum variance hedge of expressions (7.14)-(7.16a,b).

Finally, note that it may be difficult to obtain meaningful estimates of the expected net payoff $E[F_T]$. In practice, we may set $E[F_T]=0$ and use the minimum variance solution directly, since it would be somewhat easier to estimate.

7.3 Summary I

Section 7.2 presented a brief introduction to some practical concepts relating to markets and transactions involving discrete time and risk, focusing on financial contracts. Futures markets and the other dynamic financial contracts discussed here are part of a much larger class of assets called derivatives, i.e. assets that derive their value from the value of other assets. In Chapter 8 we present an analysis and discuss in more detail another important class of derivatives called options. In the remainder of this chapter, we focus on the theoretical underpinnings of markets and transactions involving time and risk in general equilibrium.

7.4 General Equilibrium and Asset Pricing under Uncertainty with Complete Markets

In the next few sections, 7.5-7.12, we examine dynamic general equilibrium theory and asset pricing under uncertainty in the case of complete markets and finite state space. We start with a simple exposition of the qualified equivalence between the world envisioned by Gerard Debreu and the world envisioned by Kenneth Arrow. Then, we examine asset prices and returns in a two-period general equilibrium economy. This sets the stage to the multi-period case, where we discuss the unconditional and the conditional cases as well as the widely used Markov process for the evolution of the states of nature. We also study asset prices in the general equilibrium infinite horizon case envisioned by Robert Lucas. Finally, we discuss in some detail the issue of excess returns in the general equilibrium under uncertainty framework.

7.5 General Equilibrium under Uncertainty: Two Equivalent Approaches

In the framework envisioned by Gerard Debreu, all trade occurs in a single timeless market. We can index a commodity by date, location, description, and state of nature. For example: let there be two dates ($t=0,1$), two states of nature ($z=a,b$), and two commodities described as $x_t(z)=X,Y$. There are six objects to be traded since at date 0 the state is observed with certainty: (i) X at date 0; (ii) Y at date 0; (iii) X at date 1 in state a; (iv) X at date 1 in state b; (v) Y at date 1 in state a; (vi) Y at date 1 in state b.

At date 0, agents meet and trade the commodity vectors x_0, $x_1(a)$, $x_1(b)$ and all trade occurs in a single period with all contingencies taken into account. For expositional easy here, we assume that there is a unit of account so that all quantities are expressed in real units of account. However, this is not necessary in Debreu's framework, but makes the comparison with Arrow more transparent.

An individual has a Von-Neumann-Morgenstern (VNM) state separable expected utility function, e.g. Chapter 3, given by

$$U = \Sigma_{z=a,b}\, \pi(z)\, u(x_0, x_1(z)) \qquad (7.20)$$

where the utility function u is strictly increasing and concave, and the number $\pi(z) \in [0,1]$ is the nonnegative probability that state z occurs with $\Sigma_{z=a,b}\, \pi(z)=1$. Individuals face a date-0 budget constraint given by

$$p_0(\omega_0 - x_0) + \Sigma_{z=a,b}\, p_1(z)\,[\omega_1(z) - x_1(z)] \geq 0 \qquad (7.21)$$

where p_0 is the price vector for period 0, ω_0 is the endowment vector for period 0, x_0 is the commodities vector for period 0, $p_1(z)$ is the date-0 state contingent price vector for period 1, $\omega_1(z)$ is the state contingent endowment vector for period 1, and $x_1(z)$ is the state contingent consumption vector for period 1. The state contingent price, $p_1(z)$ can be interpreted as the price of a claim to one unit of commodities, x_1 in period 1 if state z occurs; hence the price of a state contingent claim. However, note that $p_1(z)$ is not a spot price, it is the date-0 price contingent upon a history up to date 1.

A general equilibrium in this case is a set of prices and quantities that maximize expected utility (7.20) subject to the budget constraint (7.21) at the single date 0, and a resources constraint (supply equals demand), taking into account all possible future contingencies. This framework may be extended appropriately for more dates, locations, descriptions, and states of nature, all compactly concentrated at date 0.

Kenneth Arrow envisioned a framework that is alternative but equivalent, under certain qualifications, to Debreu's world. Arrow's idea is to introduce sequential trade in financial securities, i.e. markets open every and each period sequentially. In the world of Debreu there may be redundant trade: many branches of the Debreu's outcome or event tree may never occur because trade occurs only once (as we shall see shortly). Arrow's insight is to replicate Debreu's allocation by introducing trade in a sequence of financial markets thus eliminating redundant trade. The key point here is that allowing trade at sequential dates reduces the number of contingencies traded.

In this sequential context, let an ordinary security, e.g. Chapter 4, be defined as a column vector m of which the jth element is the number of units of account that a holder receives if state $j \in \boldsymbol{Z}$ occurs. For example: money is a security with payoff vector $m' = [1 \ldots 1 \ldots 1]$, (prime here denotes the vector transpose) or a sure payment of one unit of account next period for every state of nature. Suppose, as before, that there are two dates ($t=0,1$), two states of nature ($z=a,b$), and two commodities described as $x_t(z)=X,Y$. According to Arrow, there are two markets that, under certain conditions, can exactly replicate Debreu's allocation. The utility function is as above $U = \Sigma_{z=a,b}\, \pi(z)\, u(x_0, x_1(z))$.

(i) At date 0, a consumer faces a budget constraint of the form

$$p_0\,(\omega_0 - x_0) - \Sigma_{j=1}{}^{K}\, q^j y^j \geq 0 \qquad (7.22)$$

where the index $j=1,2,3,\ldots K$ is the number of ordinary securities available in the market place, belonging to the set $\boldsymbol{K}$, or $\{j \in \boldsymbol{K}: j=1 \ldots K\}$. The securities are bought and sold at a current market price denoted $q^{j \in \boldsymbol{K}}$. Also $y^{j \in \boldsymbol{K}}$ denotes the number of securities $j \in \boldsymbol{K}$ held by an individual at date 0, so that when $y^j > 0$, it indicates the individual is long on security $j \in \boldsymbol{K}$, and when $y^j < 0$, it indicates the individual is short on security $j \in \boldsymbol{K}$. These assets will be available for consumption at date 1. Hence, at date 0, the individual chooses to carry a portfolio of securities up to date 1, i.e. the optimal choices y^j. These securities can be exchanged for consumption at date 1, after the state of nature is realized.

(ii) At date 1, uncertainty is resolved, the state contingent budget constraint in each state of nature is given by

$$p_1(a)\,[\omega_1(a) - x_1(a)] + \Sigma_{j=1}^{K}\, y^j\, m^j(a) \geq 0 \qquad (7.23)$$

$$p_1(b)\,[\omega_1(b) - x_1(b)] + \Sigma_{j=1}^{K}\, y^j\, m^j(b) \geq 0 \qquad (7.24)$$

where $m^{j\in \mathbf{K}}(z)$ is the payoff, in units of account (for example, dollars) of security j in state z, and $p_1(z)$ are date-1 spot prices for exchange of securities for goods at that date. It should be clear that the prices $p_1(z)$ are not date-0 prices, but instead they are spot prices at date 1 (recall that in Debreu's case above, prices are as of date 0 for future delivery).

In summary, Arrow's problem is sequential in the sense that the date-0 problem is

$$\underset{\{x_0,\, y^j\}}{Max}\ \Sigma_{z=a,b}\ \pi(z)\, u(x_0, x_1(z)) \qquad (7.25a)$$

$$subject\ to \qquad p_0(\omega_0 - x_0) - \Sigma_{j=1}^{K}\, q^j y^j \geq 0$$

and at date 1, when uncertainty is resolved, an individual chooses x_1 that satisfies (7.23)-(7.24), or

$$p_1(z)\,[\omega_1(z) - x_1(z)] + \Sigma_{j=1}^{K}\, y^j\, m^j(z) \geq 0 \quad for\ all\ z=a,b. \qquad (7.25b)$$

How can the sequence designed by Arrow replicate Debreu's equilibrium allocation and prices? The answer lies in one of Arrow's brilliant insights. Suppose there is a finite number of states, Z, $\{z\in \mathbf{Z}: z=1...Z\}$ and a finite number of securities K, $\{j\in \mathbf{K}: j=1...K\}$. We can define a payoff matrix, denoted M, consisting of the payoff of each security in each state of nature as

$$M = \begin{bmatrix} m^1(1)... & m^1(Z) \\ & \\ m^K(1)... & m^K(Z) \end{bmatrix} \qquad (7.26)$$

where $m^j(z)$ is the payoff of security j in state z. Notice that the matrix in (7.26) is similar to the familiar matrix of payoffs D of Chapter 4.

If there are at least as many securities as states of nature, or $Z \leq K$, and the payoff matrix M is nonsingular, the market structure is said to be complete, i.e. there must be enough securities to transfer income across all states of nature. If any of these two properties is violated, the market structure is incomplete, e.g. Chapter 4.

In Arrow's framework, there exists a set with $Z=K$ securities for each state of nature, and a diagonal payoff structure given by the identity matrix, or $M=\mathbf{I}$. Thus, the jth 'security' in Arrow's framework pays one unit of account if state j occurs and zero otherwise. This is the so-called Arrow security as seen in Chapter 4. In this case, with $Z=K$ Arrow securities, this payoff structure is consistent with the properties of complete markets and Arrow's world is a complete markets world. This complete markets property turns out to be fundamental for Debreu and Arrow to be compatible. Why? Because, with complete markets, the sequential choice designed by Arrow can be represented as the single period, timeless choice designed by Debreu through an appropriate consolidated budget constraint.

To see this, consider again the two dates $(t=0,1)$, two states of nature $(z=a,b)$, and two commodities described as $x_t(z)=X,Y$. Assume there are two Arrow securities ($j=1,2$) with current price $q_A{}^{j=1,2}$ and with respective quantities held $y^{j=1,2}$. The date-0 budget constraint is

$$p_0\,(\omega_0 - x_0) \; - \; q_A{}^1 y^1 - q_A{}^2 y^2 \geq 0 \qquad (7.27)$$

and the state contingent date-1 budget constraints are

$$p_1\,(a)\,[\omega_1(a) - x_1\,(a)] \; + \; y^1(a) \geq 0 \qquad (7.28)$$

$$p_1\,(b)\,[\omega_1(b) - x_1\,(b)] \; + \; y^2(b) \geq 0 \qquad (7.29)$$

since the Arrow securities pay off $m^1(a)=1$, $m^1(b)=0$, and $m^2(a)=0$, $m^2(b)=1$. It is easy to see that we can substitute the budget constraints (7.28)-(7.29) into the date-0 budget constraint (7.27) to obtain a consolidated budget constraint given by

$$p_0\,(\omega_0 - x_0) + q_A{}^1\, p_1\,(a)[\omega_1\,(a) - x_1\,(a)] + q_A{}^2\, p_1\,(b)\,[\omega_1\,(b) - x_1\,(b)] \geq 0. \qquad (7.30)$$

Then, we can solve Arrow's problem by maximizing expected utility (7.25a) subject to (7.30) exactly as in Debreu's problem of maximizing (7.20) subject to (7.21). Accordingly, the Debreu contingent claim prices, $p_1(z)$, are exactly the Arrow spot prices of one unit of consumption in date 1, weighted by the date-0 price of the appropriate Arrow security, or $q_A{}^j \times p_I(z)$, or

$$p_I(z) = q_A{}^j \, p_I(z).$$

Thus, the date-0 price for the Arrow security is the date-0 price of Debreu state contingent claim deflated by the spot price of one unit of consumption, in date 1 and state z. This framework extends naturally over many periods and sequentially as trade opens up each period. In the absence of complete markets, there are not enough securities to replicate the Debreu budget constraint, we are unable to consolidate (7.28)-(7.29) into (7.30) and the equivalence of the Debreu and Arrow allocations and prices fail. It is important to note also that with incomplete markets the unit of account simplification becomes a problem whereas with complete markets the numeraire is irrelevant.

7.6 **Pricing Contingent Claims in the Two-Period Economy with Complete Markets**

In this section, we first price states of nature or state contingent claims, and then generalize the specific prices to the price of an ordinary security as a bundle of state contingent claims under complete markets. The material here is a direct extension of Chapter 4 to a multi-period setting.

We consider a case where goods exchange for goods, there is no unit of account, and the contingent claims are pure in the sense of promises of units of goods. We study the two-period case with dates 0 and 1. The endowment in period 0 is non-stochastic, denoted ω_0, and the endowment in period 1 is stochastic. The states of nature in period 1 constitute a finite set $\boldsymbol{Z}$ with elements $z=1,2,\ldots Z$, $\{z \in \boldsymbol{Z}: z=1\ldots Z\}$. Thus, the stochastic date-1 endowment is $\omega_I(z)$. An agent in this economy has preferences of the VNM type with the additional assumption of time separability, hence state and time separable, denoted by

$$\begin{aligned} U &= u(c_0) + \beta \; E\,[u(c_I(z))] \qquad (7.31) \\ &= u(c_0) + \beta \, \Sigma_{z \in Z} \; \pi(z) \; u(c_I(z)) \end{aligned}$$

where E is the unconditional expectation operator, c is consumption of the commodity (or bundle of commodities), u is a strictly increasing and (usually strictly) concave utility function, $\pi(z)\in[0,1]$ is the nonnegative probability that state z occurs with $\Sigma_{z\in Z}\, \pi(z)=1$, and β is a subjective discount factor with $\beta\in[0,1)$. The two resources constraints at dates 0 and 1 are respectively

$$c_0 \leq \omega_0 \tag{7.32}$$

$$c_1(z) \leq \omega_1(z), \quad \forall z. \tag{7.33}$$

The agent's problem is to maximize expected utility (7.31) subject to the budget constraints (7.32)-(7.33) by choice of a state contingent consumption bundles $\{c_0, c_1(z)\}$, given the state contingent endowments $\{\omega_0, \omega_1(z)\}$ and probability measure $\pi(z)$. From the constraints (7.32)-(7.33), an interior solution involves consuming the given endowments every period for each state of nature. However, the optimal prices for the constraints, the Lagrange multipliers, contain relevant information for the pricing of state contingent claims to future consumption. We form the Lagrangean function as

$$\mathcal{L} = u(c_0) + \beta\Sigma_{z\in Z}\, \pi(z)\, u(c_1(z)) + q_0(\omega_0 - c_0) + q_1[\omega_1(z) - c_1(z)] \tag{7.34}$$

where $\{q_0, q_1\}$ are nonnegative Lagrange multipliers attached to each constraint, (7.32)-(7.33).

Choices of state contingent bundles are given by the first order necessary conditions for an interior equilibrium as (the second order condition is satisfied by the concavity of the utility function and the linearity of the constraints)

$$u'(c_0) - q_0 = 0 \tag{7.35a}$$

$$\beta\, \pi(z)\, u'(c_1(z)) - q_1 = 0, \quad \text{each } z\in \mathbf{Z} \tag{7.35b}$$

$$\omega_0 - c_0 = 0 \tag{7.35c}$$

$$\omega_1(z) - c_1(z) = 0, \quad \forall z. \tag{7.35d}$$

where $u'(c)=\partial u(c)/\partial c$ is the marginal utility of consumption, and (7.35c,d) hold for $\{q_0, q_1\}>0$. As mentioned, the optimal solution is to consume the entire endowment for all dates and all states of nature, so that there is no saving or dissaving across periods. However, more importantly, the multipliers q_0 and q_1 yield a measure of how much one would be willing to give up today in order to receive one unit of the commodity in the future in some state of nature z. In other words, the relative price $(q_1/q_0)(z)$ prices a state contingent claim, from above, as

$$q(z) \equiv (q_1/q_0)(z) = \beta\,\pi(z)\,u'(c_1(z))\,/\,u'(c_0), \qquad \text{each } z \in \mathbf{Z}. \quad (7.36a)$$

The amount $\beta\,u'(c_1(z))/u'(c_0) > 0$ is the state contingent marginal rate of substitution between dates 0 and 1, denoted $MRS(z)$, or

$$MRS(z) \equiv \beta\;u'(c_1(z))\,/\,u'(c_0) > 0, \qquad \text{each } z \in \mathbf{Z},$$

also called the pricing kernel. Thus,

$$q(z) = \pi(z)\,MRS(z), \qquad \text{each } z \in \mathbf{Z}. \quad (7.36b)$$

Equations (7.36a,b) yield the date-0 price of an 'imaginary' security that has a payoff of one unit of the commodity in state z and zero otherwise, i.e. the price of a pure state contingent claim. Note that this security is distinct from the Arrow security because it pays one unit of the commodity not one unit of account as in Arrow's case. In this case, it is the value of one unit of the good at date 1 and state z in terms of units forgone of the sure good at date 0. The gross one-period rate of return of this imaginary claim is defined as the inverse of its price, or

$$R(z) = 1/q(z), \qquad \text{each } z \in \mathbf{Z}. \quad (7.36c)$$

Consider now a general ordinary security that pays a stochastic dividend denoted by $d(z)$, where $d(z)$ is a payoff list $d(z) = (d(a), d(b), \ldots, d(z), \ldots)$. In fact, we can think of this as a complex security consisting of a bundle of state contingent claims of the type priced in (7.36b). If the agent holds α units of this security, date-1 consumption can be expressed as

$$c_1(z) = \alpha\, d(z), \quad \forall z.$$

The date-1 endowment can be reinterpreted as denominated in units of this asset, so that the date-1 budget constraint becomes

$$c_1(z) \leq \omega_1\, d(z), \quad \forall z.$$

The agent's problem is to maximize the new version of the expected utility in (7.31) with the adjusted budget constraints, or

$$\underset{\{c_0,\alpha\}}{Max} \quad u(c_0) + \beta \sum_{z \in Z} \pi(z)\, u(\alpha\, d(z)) \quad (7.37)$$

$$\text{subject to} \quad c_0 \leq \omega_0$$

$$\alpha \leq \omega_1$$

by choice of $\{c_0, \alpha\}$ where we have substituted $c_1(z) = \alpha\, d(z), \forall z$ into the date-1 constraint. The Lagrangean function in this case is

$$\mathcal{L} = u(c_0) + \beta\ \Sigma_{z\in Z}\ \pi(z)\ u(\alpha\, d(z)) + q_0(\omega_0 - c_0) + q_1^d(\omega_1 - \alpha) \qquad (7.38)$$

where $\{q_0\ ,\ q_1^d\}$ are again the nonnegative Lagrange multipliers. The choices for $\{c_0, \alpha\}$ are given by the first order necessary conditions for an interior equilibrium as (the second order condition is satisfied by the concavity of the utility function and the linearity of the constraints)

$$u'(c_0) - q_0 = 0 \qquad (7.39a)$$

$$\beta\ \Sigma_{z\in Z}\ \pi(z)\ u'(c_1(z))\ d(z) - q_1^d = 0 \qquad (7.39b)$$

$$\omega_0 - c_0 = 0 \qquad (7.39c)$$

$$\omega_1 - \alpha = 0 \qquad (7.39d)$$

so that

$$q^d \equiv (q_1^d/q_0) = \beta\ \Sigma_{z\in Z}\ \pi(z)\ u'(c_1(z))\ d(z)\ /\ u'(c_0) \qquad (7.40a)$$

yields the price of a more general security that has a payoff of $d(z)$ units of the commodity, or using the definition of the marginal rate of substitution,

$$q^d = E[MRS(z)\ d(z)]. \qquad (7.40b)$$

The one-period gross rate of return on this asset is

$$R(z) = d(z)\ /\ q^d, \quad \text{each } z\in \boldsymbol{Z} \qquad (7.41)$$

i.e. the dividend-price ratio; and the expected gross rate of return is

$$E[R(z)] = \Sigma_{z\in Z}\ \pi(z)\ [\ d(z)\ /\ q^d]. \qquad (7.42)$$

The pricing relationships obtained are consistent with no-arbitrage restrictions, i.e. linear pricing. As studied in Chapter 4, linear pricing implies that the price of the bundled security in (7.40a,b) is equal to the linear sum of the appropriate state contingent claims in (7.36a,b), that is

$$q^d = \Sigma_{z\in Z}\ q(z)\ d(z)$$

or the two portfolios have the same payoff. If we multiply and divide the right-hand side of the last expression by the well defined probability of state z, $\pi(z)>0$, we obtain

$$q^d = \Sigma_{z\in Z}\ \pi(z)\ [q(z)/\pi(z)]\ d(z) = E[\ \{q(z)/\pi(z)\}\ d(z)]$$

which when compared with (7.40d) implies

$$q(z)/\pi(z) = MRS(z) = \beta\ u'(c_1(z))\ /\ u'(c_0), \qquad each\ z\in\boldsymbol{Z}$$

which is exactly (7.36a,b). Of course, this represents the Fundamental Theorem of Risk Bearing for State Contingent Claims familiar from Chapter 4. Because an ordinary security is just a bundle of state contingent claims priced separately, sometimes it is referred as a redundant asset.

A special security of the type discussed above is the risk-free security: its payoff is $d(z)=1,\ \forall\ z,$ with price given by (as in (7.40b))

$$q^{d=1} = \beta\ \Sigma_{z\in Z}\ \pi(z)\ u'(c_1(z))\ /\ u'(c_0) \qquad (7.43a)$$

and gross rate of return (as in (7.41c))

$$R = 1\ /\ q^{\ d=1}. \qquad (7.43b)$$

Consider now an asset allocation problem when there are many available ordinary type securities to invest in the market place. Suppose there are $j=1,2,...K$ assets (ordinary securities), belonging to the set $\boldsymbol{K}$, $\{j\in\boldsymbol{K}: j=1...K\}$, available to reproduce the state contingent endowments with $K\geq Z$ where Z is the number of states of nature, $\{z\in\boldsymbol{Z}: z=1...Z\}$. Let α_j units of date-0 commodity be invested in asset j yielding a gross return of $R_j\ (z)$ units of date-1 commodity in each state z, i.e. the gross rate of return is stochastic. The date-0 saving is distributed in the various assets according to the constraint

$$\omega_0\ \text{-}\ c_0 = \ \Sigma_{j\in K}\ \alpha_j. \qquad (7.44)$$

The date-1 consumption opportunities are given by the gross return on savings plus the endowment, or

$$c_1(z) = \ \Sigma_{j\in K} R_j\,(z)\ \alpha_j + \omega_1(z), \qquad \forall\, z. \qquad (7.45)$$

Hence, the problem faced by an individual is simply to

$$\underset{\{\alpha_j\}}{Max}\ \ \{u(\omega_0\ \text{-}\ \Sigma_{j\in K}\ \alpha_j) + \beta\,\Sigma_{z\in Z}\ \pi(z)\ u(\Sigma_{j\in K} R_j\,(z)\ \alpha_j + \omega_1(z))\} \ (7.46a)$$

with first order necessary condition (the second order condition is satisfied by the concavity of the utility function) given by

$$u'(c_0)\ \text{-}\ \beta\,\Sigma_{z\in Z}\ \pi(z)\ u'(c_1(z))\,R_j\,(z) = 0, \qquad \forall j. \qquad (7.46b)$$

As an aside, note that in the case where the state space is continuous with an infinite number of possible states of nature, instead of finite, a continuous distribution function would allow us to substitute the summation for an integral as

$$u'(c_0) - \beta \int_{z\in Z} \{ u'(c_1(z))\, R_j(z)\}\, f(z)\, dz = 0 \qquad (7.46b')$$

where $f(z)$ is the probability density function of z; or

$$u'(c_0) - \beta \int_{z\in Z} \{ u'(c_1(z))\, R_j(z)\}\, dF(z) = 0 \qquad (7.46b'')$$

where $F(z)$ is the cumulative distribution function of the states and $dF(z)/dz = f(z)$.

In general, the first order necessary condition of this problem yields a formula of the type

$$\beta\ E[\, u'(c_1(z))\, R_j(z)/\, u'(c_0)] \ = 1 = E[MRS(z)\, R_j(z)],\ \ z\in\mathbf{Z},\ \ j\in\mathbf{K} \quad (7.47)$$

which is a fundamental asset pricing formula, i.e. a payoff for \$1 with an average return

$$\beta\, E[u'(c_1(z))R_j(z)/\, u'(c_0)]$$

in units of consumption or dollars.

Our model here is a simplified intertemporal version of the Consumption-Based Capital Asset Pricing Model (CCAPM), thus it delivers an asset pricing formula consistent with that model. It is a so-called consumption-based model because the marginal rate of substitution, or stochastic discount factor, is the ratio of marginal utilities of consumption.

The fundamental asset pricing formula has a natural formulation using the covariance decomposition formula,

$$E[XY] = cov(X,Y) + E[X]\, E[Y],$$

to obtain

$$E[MRS(z)]E[\, R_j(z)] + cov(MRS(z),\, R_j(z)) = 1,\ \ z\in\mathbf{Z},\ \ j\in\mathbf{K} \qquad (7.47a)$$

which implies

$$E[\, R_j(z)] - \{1/\, E[MRS(z)]\} = -cov(MRS(z),\, R_j(z))/\, E[MRS(z)],\ z\in\mathbf{Z},\, j\in\mathbf{K}. \qquad (7.47b)$$

Using the fundamental formula (7.47b), note that the term $\{1/ E[MRS(z)]\}$ represents the gross return of an asset that does not covary, or is uncorrelated with the marginal rate of substitution. Thus, formula (7.47b) indicates that the expected excess return, or expected premium, of asset j over the return of an uncorrelated asset depends upon the sign of the covariance between the jth asset and the marginal rate of substitution.

Equations (7.47a,b) present a very useful way to compute excess returns for different assets. In particular, the degree of concavity of the utility function, or risk aversion, implies that individuals may hold other different assets with different rates of return as well. A risk premium or excess return is easily derived using this framework. Without loss of generality, let two assets indexed as m and n, with $\{m,n\}\in \mathbf{K}$, yield two distinct state contingent gross returns, R_m and R_n respectively. Subtracting formula (7.47) for each asset gives

$$\beta\ E[u'(c_1(z))\,[R_m(z) - R_n(z)] / u'(c_0)] = 0$$

$$= E[MRS(z)\,\{R_m(z) - R_n(z)\}]. \qquad (7.48)$$

Using the covariance decomposition formula and the linearity of the expectation operator, we obtain

$$cov[MRS(z)\,R_m(z)] + E[MRS(z)]\,E[R_m(z)] -$$

$$cov[MRS(z)\,R_n(z)] - E[MRS(z)]\,E[R_n(z)] = 0. \qquad (7.49)$$

Thus, the risk premium is determined by formula (7.49). A risk premium vanishes if

(i) $R_m = R_n$, or the gross returns are identical;

(ii) u' is constant, or the utility is linear and the individual is risk neutral;

(iii) $cov(MRS(z), R_m(z)) = cov(MRS(z), R_n(z)) \Rightarrow E[R_m(z)]=E[R_n(z)]$, i.e. the two assets are perfectly correlated.

In general, the differential in rates of returns across assets will depend on whether an asset pays off mostly in good states (low marginal utility) or in bad states (high marginal utility). Suppose that in the case above, $R_n(z)=R,\ \forall\ z$, i.e. it is the risk-free gross return. Then, formula (7.49) above can be rewritten as

$$E[R_m(z)] = R + \{-cov(MRS(z), [R_m(z) - R]) / E[MRS(z)]\} \qquad (7.50)$$

which simply states that the expected return on the risky asset is the risk-free return plus the risk premium; in fact this is a simple application of arbitrage pricing theory. The risk premium or excess return from (7.50) is then

$$E[R_m(z)] - R = \{- cov(MRS(z), [R_m(z) - R]) / E[MRS(z)]\} \qquad (7.51a)$$

and since $E[MRS(z)]>0$ we have that

$$E[R_m(z)] - R \lesseqgtr 0 \Leftrightarrow - cov(MRS(z), [R_m(z) - R]) \lesseqgtr 0. \qquad (7.52a)$$

In general, for any asset $j=1,2,...K$, the risk premium or excess return over the risk-free return is

$$E[R_j(z)] - R = \{- cov(MRS(z), [R_j(z) - R]) / E[MRS(z)]\} \qquad (7.51b)$$

and

$$E[R_j(z)] - R \lesseqgtr 0 \Leftrightarrow - cov(MRS(z), [R_j(z) - R]) \lesseqgtr 0. \qquad (7.52b)$$

Intuitively, suppose that we observe a high consumption state because the asset payoff is good, i.e. low marginal utility. From the definition of the marginal rate of substitution, we know that it varies inversely with date-1 consumption, or $\partial MRS(z)/\partial c_1(z) = \beta\ u''(c_1(z)) / u'(c_0) < 0$, by the (strict) concavity of the utility function, where $u''(c) = \partial^2 u(c)/\partial c^2$. Hence, a high consumption state is associated with a low marginal rate of substitution. But, if the asset payoff is good, it means that the excess return is positive, or $R_j(z)-R>0$. Logically, for the high (low) consumption state asset we have that $cov(MRS(z), [R_j(z) - R]) < (>) 0$.

Most equities have this characteristic: an equity pays a higher rate of return than a treasury bill, which is a relatively safe asset in practice. Later in this chapter, we study the equity premium more closely and examine how it relates to risk aversion.

Another important issue is full risk sharing, which can be suitably examined in the complete markets two-period case as well. As long as individuals have identical preferences, identical discount factors, and full symmetric information, i.e. rational expectations, formulas (7.36a,b) show that for two individuals named ($i=x,y$) with different endowments, their consumption paths are identical since

$$q(z) = \beta\ \pi(z)\ u'(c_{1x}(z)) / u'(c_{0x}) =$$

$$\beta\ \pi(z)\ u'(c_{1y}(z)) / u'(c_{0y}), \quad each\ z \in \mathbf{Z} \qquad (7.53)$$

or the rate of growth of consumption is identical across individuals. Agents use the complete market structure to perfectly share the risk inherent in their endowment streams.

Numerical Example: Let the utility function be the constant relative risk aversion (*CRRA*) type, familiar from Chapter 4, given by

$$u(c) = [c^{1-\gamma} - 1] / (1-\gamma), \quad \gamma > 0$$

where γ is the coefficient of relative risk aversion, $\gamma = 1$ denotes logarithmic utility. The subjective discount factor is $\beta = 0.90$. Let there be two states, $z=1,2$, with equal probabilities, $\pi(1) = \pi(2) = ½$. The endowments are $\omega_0 = 5$ and $\omega_1(1) = 4$, $\omega_1(2) = 6$ units of the commodity respectively. Consider first the case of logarithmic utility, $\gamma = 1$. The price of the risk-free bond is, using expression (7.43b),

$$q^{d=1} = 0.9 \{(1/2)(1/4) + (1/2)(1/6)\}/(1/5) = 0.9375$$

with respective rate of return,

$$R = 1/q^{d=1} = 1.0667.$$

The price and return of a claim to date-1 endowment is an asset that pays a dividend $d(z) = \omega_1(z)$ at date-1, thus $d(1) = \omega_1(1) = 4$, $d(2) = \omega_1(2) = 6$ and

$$q^{d=\omega} = 0.9\{(1/2)(4)(1/4) + (1/2)(6)(1/6)\}/(1/5) = 4.5000$$

and the state contingent return is $R(z) = d(z)/q^{d=\omega}$, by (7.41), with

$$R(1) = 0.8889$$

$$R(2) = 1.3333,$$

and unconditional mean

$$E[R(z)] = ½(0.8889) + ½(1.3333) = 1.1105.$$

The excess return between the risk-free and risky asset can be computed as

$$E[R(z)] - R = 0.0435.$$

We can do the same computation for a risk aversion parameter of $\gamma = 2$ and obtain the results in Table 7.1.

Table 7.1
Price and Returns with Alternative Risk Aversion Parameter

	$Q^{D=1}$	R	$Q^{D=\omega}$	$R(1)$	$R(2)$	$E[R(Z)]$	$E[R(Z)] - R$
$\gamma=1$	0.9375	1.0667	4.5000	0.8889	1.3333	1.1105	0.0435
$\gamma=2$	1.0156	0.9846	4.6875	0.8533	1.2800	1.0665	0.0819

Table 7.1 clearly shows that as risk aversion increases, the price of both the risk-free and risky assets increase and its return decrease. However, the excess return is increasing in risk aversion. We will see, later in this chapter, that risk aversion is a critical parameter in explaining excess returns.

7.7 Introduction to the Multi-Period Economy

We examine next a multi-period economy of the type envisioned by Debreu. In particular, we make use of the Debreu outcome or event tree (distinct from a decision tree). The economy goes on for many discrete time periods indexed by $t=0,1,2,3...T$ with T possibly going to infinity. There may be many individuals named $i=1,2,3...I$. The state of the economy at any period t consists of its history up to that date, denoted by zh^t, assumed to be observed by all agents, i.e. full information. Hence, $zh^T=\{zh^0, zh^1, zh^2, zh^3,..., zh^T\}$ where at $t=0$, $zh^0=z^0$ is known with certainty by all agents.

Figure 7.1 depicts the outcome or event tree for two possible states of nature, $\{a,b\}$, and four time periods, $\{0,1,2,3\}$. The set of all possible histories to any of the nodes is denoted by $\boldsymbol{Z}^t$, also called a filtration, and hence $\{ zh^t \in \boldsymbol{Z}^t: zh^t=[zh^0, zh^1, zh^2, zh^3,..., zh^t]\}$. As we have seen above, Debreu's insight is to define goods in different states and time periods as different commodities. In the case of Figure 7.1, there are $n=2$ states of nature and $T=4$ dates. The possible outcomes, or distinct commodities in Debreu's sense, are $\Sigma_{t=0}{}^T n^t=15$. Thus, the number of nodes may be large in the Debreu's event tree, and many of them may be redundant in the

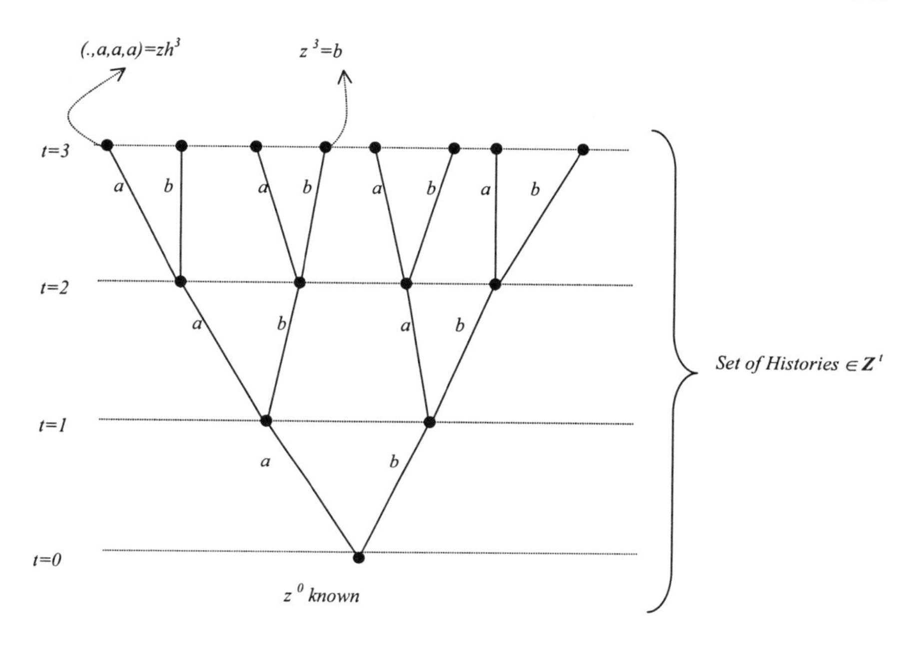

Fig. 7.1 Event Tree

sense of never occurring; recall that one of Arrow's contributions was to eliminate the redundant nodes.

The preferences of the ith individual are of the VNM time and state separable expected utility type denoted by

$$U^i = \Sigma_{t\in T} \; \beta^t \; \Sigma_{zht \in Zt} \; \pi\,(zh^t)\; u^i\,(c(zh^t)) \qquad (7.54)$$

where u is a strictly increasing and (strictly) concave utility function, $\beta\in[0,1)$ is the discount factor, identical across individuals, $\Sigma_{zht \in Zt}$ gives the set of all possible histories of length t, $\pi(zh^t)$ is the positive, unconditional and identical across individuals probability of observing the history zh^t, and $c(zh^t)$ is the consumption bundle when history is zh^t.

Each agent i has a state contingent endowment denoted $\omega^i(zh^t)$, and the aggregate economy-wide endowment is $W(zh^t)$ equal to $\Sigma_i\omega^i(zh^t)$ for each history zh^t.

We assume that $p(zh^t)$, $t \geq 0$, is the date-0, price of the commodity contingent on history zh^t, observed by all agents. Then, the date-0 budget constraint for each agent i may be written as

$$\Sigma_{t \in T} \; \Sigma_{zht \in Zt} \, p(zh^t) \, c^i(zh^t) \leq \Sigma_{t \in T} \; \Sigma_{zht \in Zt} \, p(zh^t) \, \omega^i(zh^t) \qquad (7.55)$$

where the double sums $\Sigma_{t \in T} \Sigma_{zht \in Zt}$ indicates summation over histories of all lengths, i.e. all nodes in the tree. The budget constraint states that the values of the state contingent consumption bundles are no greater than the values of the state contingent endowments for histories of all lengths at all nodes in the tree.

Next, we define a competitive equilibrium for this economy. In Debreu's instantaneous timeless world, a competitive equilibrium is a price system $\{p(zh^t)\}$ and allocations $\{c^i(zh^t)\}$ such that:

(a) Individuals maximize expected utility (7.54) subject to the budget constraint (7.55);

(b) The aggregate resources constraint holds with supply equals demand, or

$$\Sigma_i c^i(zh^t) = \Sigma_i \omega^i(zh^t), \quad \textit{for each history } zh^t.$$

The competitive equilibrium in this case suitably satisfies the First Fundamental Theorem of Welfare Economics, namely it is Pareto Optimal. A social planner may replicate the competitive prices and quantities by solving the equivalent social planning problem, i.e. the Second Fundamental Theorem of Welfare Economics. Thus, the general maximization separates into identical planning problems for each history zh^t:

(i) For the initial non-stochastic state, z^0, the planner's problem is

$$\underset{\{c^i(z^0)\}}{Max} \; \Sigma_i \, \lambda^i \, u^i(c^i(z^0)) \qquad (7.56a)$$

$$\textit{subject to} \qquad \Sigma_i c^i(z^0) \leq \Sigma_i \omega^i(z^0), \quad \forall i$$

by choice of $\{c^i(z^0)\}$, given the welfare weights λ^i, which satisfy property rights and $\Sigma_i \lambda^i = 1$. The Lagrangean function is

$$\mathcal{L}^0 = \Sigma_i \lambda^i \, u^i(c^i(z^0)) + p(z^0) \, [\Sigma_i \omega^i(z^0) - \Sigma_i c^i(z^0)], \qquad (7.56b)$$

where $p(z^0)$ is the nonnegative Lagrange multiplier. The first order necessary conditions for an interior equilibrium (the second order condition is satisfied by the strict concavity of the utility function and the linearity of the constraints) are

$$\lambda^i \; u^{i\,\prime}(c^i(z^0)) - p(z^0) = 0, \; \forall i \tag{7.56c}$$

$$\Sigma_i c^i(z^0) = \Sigma_i \omega^i(z^0), \quad \forall i \text{ and } p(z^0)>0; \tag{7.56d}$$

(ii) For zh^t, $t=0,1,2,3...T$, the planner's problem is

$$\underset{\{c^i(zh^t)\}}{Max} \; \Sigma_i \lambda^i \beta^t \pi(zh^t) \, u^i(c^i(zh^t)) \tag{7.57a}$$

$$\text{subject to} \qquad \Sigma_i c^i(zh^t) \leq \Sigma_i \omega^i(zh^t), \quad \forall i$$

by choice of $\{c^i(zh^t)\}$, given the nonnegative probability $\pi(zh^t)$, and the welfare weights λ^i, which satisfy property rights and $\Sigma_i \lambda^i = 1$. The Lagrangean function is

$$\mathcal{L}^t = \Sigma_i \lambda^i \beta^t \pi(zh^t) \, u^i(c^i(zh^t)) + p(zh^t)\,[\Sigma_i \omega^i(zh^t) - \Sigma_i c^i(zh^t)], \tag{7.57b}$$

where $p(zh^t)$ is the nonnegative Lagrange multiplier. The first order necessary conditions for an interior equilibrium (the second order condition is satisfied by the concavity of the utility function and the linearity of the constraints) are

$$\lambda^i \beta^t \pi(zh^t) \; u^{i\,\prime}(c^i(zh^t)) - p(zh^t) = 0, \; \forall i \tag{7.57c}$$

$$\Sigma_i c^i(zh^t) = \Sigma_i \omega^i(zh^t), \quad \forall i \text{ and } p(zh^t)>0. \tag{7.57d}$$

Thus, for each history zh^t, the solution separates into identical and separate subproblems.

The full risk sharing characteristic of the complete markets dynamic framework can be easily seen in this framework as well, e.g. Chapter 4. Define the pricing $p^*(zh^t) \equiv p(zh^t)/\beta^t \pi(zh^t)$, which is the same for all individuals $i \in \boldsymbol{I}$, i.e. independent of i. Then, from expression (7.57c), we obtain

$$\lambda^i \quad u^{i\,\prime}(c^i(zh^t)) - p^*(zh^t) = 0, \; \forall i \text{ and each } zh^t \tag{7.58a}$$

and logically

$$u^{i\,\prime}(c^i(zh^t))/u^{i\,\prime}(c^i(zh^{t+1})) = p^*(zh^t)/p^*(zh^{t+1}),$$
$$\text{independent of } i \text{ for each } zh^t, \tag{7.58b}$$

i.e. the growth rates of consumption are identical, perfectly correlated, across individuals. Moreover, the growth rate of the shadow prices is also independent of i, thus prices and quantities only change if there are changes in aggregate endowments, $W(zh^t)$, not individual endowments. In effect, individual endowments and consumption profiles can be perfectly pooled together so that risks are efficiently allocated. Under complete markets, full risk sharing obtains.

7.8 Conditional and Transitional Probabilities, Markov Processes, and Conditional Moments

Probabilities for stochastic processes can be expressed in two equivalent ways: unconditional and conditional. Suppose you toss a fair coin twice. The possible equally likely outcomes are: {H,H}, {H,T}, {T,H}, {T,T}. The unconditional probability of the event {H,H} is ¼. However, the probability of the event {H,H} conditional on tossing heads in the first trial is ½, since there are two possibilities out of four. Up to now, we have defined probabilities as unconditional probabilities of hitting a particular node.

For example, in Figure 7.1 at date 3 the unconditional probability of being in a particular node is $1/8$ when each history is equally likely. The conditional probability in the context of Figure 7.1 is the chance of moving up a branch conditional on starting at a given node. In general, given the history up to a particular node, zh^t, the conditional probability of hitting one of the possible states z^{t+1} is written as $\pi(z^{t+1}|zh^t)$, satisfying $\pi(.|.)\geq 0$, or the probability of moving up to z^{t+1} given the history up to zh^t. By definition,

$$\pi(z^{t+1}|zh^t) = \pi(zh^t, z^{t+1}) / \pi(zh^t), \quad \forall \pi(zh^t)\neq 0 \qquad (7.59)$$

where $\pi(zh^t, z^{t+1})$; $\pi(.,.)\geq 0$, is the joint probability of the two events occurring simultaneously, or the probability of the intersection of the event zh^t and z^{t+1}, and $\pi(zh^t)$ is the unconditional probability of the history zh^t as we had before.

An important application of conditional probabilities is a particular stochastic process called the Markov process. By definition, in a Markov stochastic process the next period realization only depends on the realization of the current period, not on all history up to the current period. In other words, all past information is fully incorporated into the current period information set, which is in the current period state. In our notation, a Markov process implies that $zh^t = z^t$. In a Markov process, the conditional probabilities $\pi(z^{t+1}|z^t)$ are called transition probabilities, because they denote the probability that the state will be z^{t+1}, in period $t+1$, given that the current state is z^t.

For example, one widely used Markov process is of the following form: for every $t \geq 1$, there are two possible states, $z=\{a,b\}$, z^0 is either a or b with certainty, and the conditional probabilities are given by

$$\begin{aligned} \pi(z^{t+1}=a \mid z^t) &= (1+\rho)/2 \quad if \quad z^t=a \\ &= (1-\rho)/2 \quad if \quad z^t=b \\ \pi(z^{t+1}=b \mid z^t) &= (1-\rho)/2 \quad if \quad z^t=a \\ &= (1+\rho)/2 \quad if \quad z^t=b \end{aligned} \tag{7.60}$$

where $\rho \in (-1,1)$, is the coefficient of autocorrelation across states of nature. The important property of the process (7.60) is that the conditional probabilities do not depend upon time. In this case, the Markov process is said to be stationary or time homogeneous. In addition, (7.60) implies that the state z^t evolves as a vector difference equation, given by

$$\begin{bmatrix} [(1+\rho)/2]-\lambda & (1-\rho)/2 \\ (1-\rho)/2 & [(1+\rho)/2]-\lambda \end{bmatrix} = 0 \tag{7.61}$$

where the square matrix of coefficients is called the transition or Markov matrix.

A transition (or Markov) matrix has the property that the sum of each element of its rows adds up to one, since probabilities add up to one. Any square matrix with this property is a transition (or Markov) matrix. The transition matrix measures the likelihood of moving from one state to another over time. For example, let the coefficient of autocorrelation across states be $\rho=0.8$. Then, the transition matrix is

$$\begin{bmatrix} 0.9 & 0.1 \\ 0.1 & 0.9 \end{bmatrix} \tag{7.61'}$$

i.e. there is a *90%* chance that if the current state is either a or b, the state in the next period will continue to be a or b respectively, a high degree of persistence of states across time. Alternatively, if $\rho < 0$, the persistence of states across time is very low.

In the dynamic process (7.61), we can examine its long run or convergence properties as in any other dynamic process. The dynamic stability of the process in (7.61) is determined by the eigenvalues of the transition matrix. Denote the eigenvalues by λ, and note that they must satisfy

$$\begin{bmatrix} [(1+\rho)/2]-\lambda & (1-\rho)/2 \\ (1-\rho)/2 & [(1+\rho)/2]-\lambda \end{bmatrix} = 0 \tag{7.62}$$

i.e. the determinant of the square matrix is zero, which reduces to the solution of the quadratic equation,

$$\lambda^2 - \lambda\,(1+\rho) + \rho = 0,$$

yielding the roots $\lambda_1 = 1$ and $\lambda_2 = \rho$. The solution for the pair $z^t(a,b)$ is then

$$\begin{bmatrix} z^t(a) \\ z^t(b) \end{bmatrix} = c_1 \begin{pmatrix} v_1 \\ v_2 \end{pmatrix} \lambda_1^{\,t} + c_2 \begin{pmatrix} v_1' \\ v_2' \end{pmatrix} \lambda_2^{\,t} \tag{7.63}$$

where c_1 and c_2 are constants to be determined, and the column vectors v and v' are the eigenvectors associated with the eigenvalues $\lambda_1 = 1$ and $\lambda_2 = \rho$.

Assuming that $-1 < \rho < 1$, then as $t \to \infty$, $\lambda_2^t = \rho^t \to 0$ and $\lambda_1^t = 1^t \to 1$. Hence, the dynamic system, or Markov chain as it is sometimes called, converges to $c_1\,(v_1 \quad v_2)'$ and the eigenvector associated with the root $\lambda_1 = 1$ is found by computing the pair $\{v_1, v_2\}$ that satisfy

$$\begin{bmatrix} [(1+\rho)/2]-\lambda_1 & (1-\rho)/2 \\ (1-\rho)/2 & [(1+\rho)/2]-\lambda_2 \end{bmatrix} \begin{bmatrix} v_1 \\ v_2 \end{bmatrix} = \begin{bmatrix} 0 \\ 0 \end{bmatrix} \tag{7.64}$$

which yields an eigenvector $(1 \quad 1)'$, or $\{v_1, v_2\} = \{1, 1\}$. Since $c_1\,(v_1 \quad v_2)'$ should be a probability vector whose components sum up to one, the appropriate choice of c_1 is $c_1 = ½$, and the Markov process, or chain, converges to $(½ \ ½)$ as t goes to infinity. This is a useful property of the specific Markov process given by (7.60), which holds for any $-1 < \rho < 1$. In the long run, the probability that the state z^t is a or b is ½, which is equal to the unconditional probability. Figure 7.2 illustrates this property.

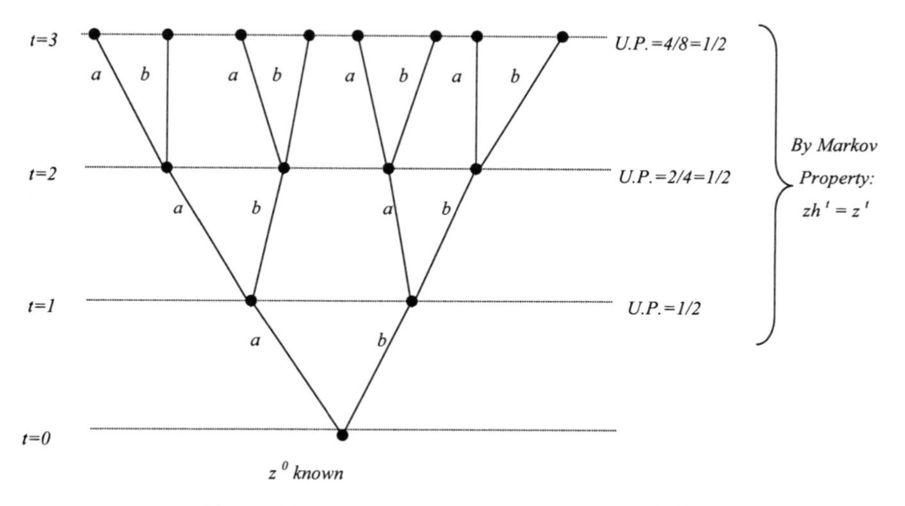

Fig. 7.2 Unconditional Probabilities (U.P.)

The joint probability in this case is given by

$$\pi(zh^t, z^{t+1}) = \pi(z^t, z^{t+1}) = \pi(z^t)\,\pi(z^{t+1} | z^t). \tag{7.65}$$

For example, for $t=1$ we have

$$\pi(z^1=a, z^2=a) = \pi(z^1=a)\,\pi(z^2=a | z^1=a) = ½\,[(1+\rho)/2] \tag{7.66}$$

giving the joint probability that $z^1=a$, $z^2=a$, etc. From expression (7.66), in the case $\rho = 0.8$, the joint probability is $\pi(z^1=a, z^2=a)=0.45$.

In summary, the way conditional probabilities evolve is called a Markov chain, the transition matrix is

$$\Pi = \begin{bmatrix} (1+\rho)/2 & (1-\rho)/2 \\ (1-\rho)/2 & (1+\rho)/2 \end{bmatrix} \tag{7.67}$$

whose sum of elements across rows is one, and each element is a conditional probability described as

$$\pi_{k\ell} = PROB(state\ \ell\ at\ t+1 | state\ k\ at\ t), \quad \{k,\ell\} \in \mathbf{Z}. \tag{7.68}$$

In general, the conditional moments of a random variable follow from above. For a random variable $x(z)$, denote the conditional expectation, up to the time t information set, as E_t, given by

$$E[x(z^{t+1}) | zh^t] = E_t[x^{t+1}] = \Sigma_{zt+1 \in Zt}\, \pi(z^{t+1} | z^t)\, x(z^{t+1}). \tag{7.69}$$

The law of iterated mathematical expectations allows us to compute expectations recursively, or

$$E_t[x^{t+k}] = E_t[E_{t+k-1}[x^{t+k}]] \quad for\ k=2,3,... \tag{7.70}$$

For example, in the Markov case studied above, we let $x(z^t=a)=1$ and $x(z^t=b)=2$. Then, by expression (7.69) we obtain the conditional expectation as

$$E[x(z^{t+1}) | z^t=a] =$$

$$\pi(z^{t+1}=a | z^t=a)\, x(z^{t+1}=a) + \pi(z^{t+1}=b | z^t=a)\, x(z^{t+1}=b)$$

$$= [(1+\rho)/2]\, 1 + [(1-\rho)/2]\, 2$$

$$= (3-\rho)/2$$

and the $E[x(z^{t+1}) | z^t=b]$ may be computed in the same fashion.

7.9 **The Multi-Period Economy Again**

Now consider a dynamic portfolio problem of asset allocation in the sequential multi-period economy envisioned by Arrow. Assume there are many individuals, and individuals are all alike so that one is a representative agent. There are many assets available in the market place indexed by $j=1,2,3...K$, or $j \in \mathbf{K}$ as before. At date t, the endowment of the representative agent is $\omega(zh^t)$, given a history zh^t, up to time t. The total sources of income (in real units of account) of this representative individual at date t is

$$\omega(zh^t) + \Sigma_{j \in \mathbf{K}}\, \alpha_j(zh^{t-1}) R_j(zh^t),$$

where the first term is the endowment and the second term is the income from ordinary securities held from last period to the current period. This income is to be allocated in date-t consumption and holdings of assets, or

$$c(zh^t) + \Sigma_{j\in \mathbf{K}}\, \alpha_j(zh^t).$$

In turn, the representative agent faces a sequence of budget constraints

$$c(zh^t)+\Sigma_{j\in \mathbf{K}}\, \alpha_j\,(zh^t) \leq \omega(zh^t)+\Sigma_{j\in \mathbf{K}}\, \alpha_j\,(zh^{t-1})\, R_j\,(zh^t), \quad \forall\, zh^t \qquad (7.71)$$

i.e. there is one budget constraint for every node in the event tree. Date-0 asset holdings are arbitrarily given as $\{\alpha_j\,(z^0)\}$. The terminal condition is that $\alpha_j\,(zh^{T+1})=0,\ \forall\ zh^t,\ \ \forall\, j$ such that as $T\rightarrow\infty$ the intuition is the same. The representative agent has VNM state and time separable expected utility denoted by

$$U = \Sigma_{t\in T}\ \beta^t\ \Sigma_{zht\,\in Zt}\, \pi\,(zh^t)\ u\,(c(zh^t)) \qquad (7.72)$$

as in (7.54) where $\pi\,(zh^t)$ is the unconditional probability of the history zh^t. The representative agent solves the asset allocation problem

$$\underset{\{c(zh^t),\ \alpha_j\,(zh^t)\}}{Max}\ \Sigma_{t\in T}\ \beta^t\ \Sigma_{zht\,\in Zt}\, \pi\,(zh^t)\ u\,(c(zh^t)) \qquad (7.73)$$

subject to

$$c(zh^t)+\Sigma_{j\in \mathbf{K}}\, \alpha_j\,(zh^t) \leq \omega(zh^t)+\Sigma_{j\in \mathbf{K}}\, \alpha_j\,(zh^{t-1})\, R_j\,(zh^t), \quad \forall\, zh^t$$

given $\{\alpha_j\,(z^0),\, R_j\,(zh^t)\}$ and the terminal condition $\alpha_j\,(zh^T)=0,\ \forall\ zh^t,\ \forall j.$

It is easier to substitute the budget constraint into the objective function and solve the resulting unconstrained sequential problem. This yields first order necessary condition (the second order condition is satisfied by the concavity of the utility function and linearity of the constraint) as

$$-\beta^t\ \pi\,(zh^t)\ u'\,(c(zh^t)) + \beta^{t+1}\ \Sigma_{zht+1\,\in Zt+1}\, \pi\,(zh^t,\, z^{t+1}) \times$$

$$u'\,(c(zh^t,\, z^{t+1}))\ R_j\,(zh^t,\, z^{t+1}) = 0 \qquad \text{for each } j\in \mathbf{K} \qquad (7.74a)$$

where $\pi\,(zh^t,\, z^{t+1})$ is the joint probability of zh^t and z^{t+1} and c and R_j are the state contingent consumption and gross returns. Intuitively, zh^t is a node in the event tree and $\{zh^t,\, z^{t+1}\}$ is any branch stemming from it, given the history zh^t.

We can rewrite (7.74a) as

$$\beta \Sigma_{zht+1 \in Zt+1} [\pi (zh^t, z^{t+1}) / \pi (zh^t)] \{ u' (c(zh^t, z^{t+1})) R_j (zh^t, z^{t+1}) / u' (c(zh^t))\} = 1 \quad \text{for each } j \in \mathbf{K} \qquad (7.74b)$$

where $[\pi (zh^t, z^{t+1}) / \pi (zh^t)]$ is the conditional probability of hitting the node z^{t+1} given history zh^t, $\pi (z^{t+1} | zh^t)$ as in expression (7.59). Thus,

$$\Sigma_{zht+1 \in Zt+1} [\pi (zh^t, z^{t+1}) / \pi (zh^t)]$$

is the conditional expectation operator given information up to date t. In short,

$$\beta E_t [u' (c(zh^t, z^{t+1})) R_j (zh^t, z^{t+1}) / u' (c(zh^t))] = 1 =$$

$$E_t [MRS(zh^t, z^{t+1}) \ R_j (zh^t, z^{t+1})] \quad \text{for each } j \in \mathbf{K} \qquad (7.74c)$$

where $MRS(zh^t, z^{t+1}) \equiv \beta u' (c(zh^t, z^{t+1})) / u'(c(zh^t))$ is the conditional intertemporal marginal rate of substitution between dates t and $t+1$.

All forms of formula (7.74) are almost the same as the fundamental asset pricing formula equivalent to the two-period economy in (7.47), consistent with the Consumption-Based Capital Asset Pricing Model (CCAPM). However, now the expectation is conditional. This is because, given the history, zh^t, the marginal rate of substitution between the two nodes depends on the conditional probability of going from one node to the other.

In addition, if we assume that the states of nature evolve according to the Markov process examined above, then $zh^t = z^t$, and the marginal rate of substitution and gross return do not depend on the entire history, but only on the two successive nodes. In this case, the fundamental asset pricing formula simplifies to

$$E_t [MRS(z^t, z^{t+1}) \ R_j (z^t, z^{t+1})] = 1 \quad \text{for each } j \in \mathbf{K}. \qquad (7.75)$$

The Markovian assumption is quite useful to compute and estimate equilibrium asset prices and returns in dynamic stochastic economies; see e.g. Notes on the Literature.

In particular, Markov processes allow a close relationship between the time dimension and the state dimension. If the state process is Markov with conditional probability of state ℓ next period given state k in the current period, as in expression (7.68), or $\pi_{k \ell} = PROB(state\ \ell\ at\ t+1 | state\ k\ at\ t)$, $\{k, \ell\} \in \mathbf{Z}$, then these transition probabilities are suitably stored in the transition matrix Π given in (7.67). Possibly, the time horizon goes to infinity. The first order necessary condition in state k obeys the asset pricing formula

$$\Sigma_{\ell}\, \pi_{k\ell} MRS(k,\ell)\, R_j\,(k,\ell) = 1, \quad each\ k \qquad (7.76)$$

where $MRS(k,\ell) \equiv \beta u'(c(\ell))/u'(c(k))$ is the state contingent marginal rate of substitution between the current state k in this period and ℓ next period, and $R_j\,(k,\ell)$ is the respective gross return for the asset in those states. This relationship will be very useful below in the study of the Lucas exchange economy.

7.10 Asset Prices in an Infinite Horizon Exchange Economy

Robert Lucas envisioned the economy here. The Lucas economy is one where there is an infinite horizon, a representative agent, and one asset. The representative agent owns the asset. This asset can be thought of as a "tree" in the "backyard." Each period, the "tree" yields a random amount of "fruit." This is the endowment, or dividend indeed received every period. The insight is that with this simple structure one can compute prices and returns for this "tree" and any other asset of any other maturity from a general equilibrium perspective. The economy has a finite state space denoted $z=\{1,2,3...Z\}$. In state z, the random endowment (dividend) is $\omega(z)$. The states of nature evolve according to a Markov process exactly as the one described in (7.60)-(7.64). In Lucas's economy, a competitive equilibrium consists of quantities $\{c,\alpha\}$, consumption and number of shares respectively, and equity prices $\{q^e\}$ so that

(i) The representative agent maximizes expected utility subject to the relevant budget constraints;

(ii) The aggregate resources constraint holds with supply equals demand.

We examine condition (i) first. The representative agent has strictly increasing and strictly concave utility function over current consumption and the VNM state and time separable expected discounted utility in period t is $E_t[\Sigma_{j=0}^{\infty} \beta^j u(c_{t+j})]$, where $\beta \in [0,1)$ is the individual subjective discount factor. The individual sources of income are the value of the current asset holdings plus the dividends, or $\alpha_t\,(q^e_t+\omega_t\,)$, to be spent on consumption or accumulation of assets, $(c_{\,t}+\alpha_{t+1}\ q^e_t\ )$. Thus, the individual date-t budget constraint is

$$c_t + \alpha_{t+1}\, q^e_t \leq \alpha_t\,(q^e_t + \omega_t). \tag{7.77}$$

It is useful to determine the gross return on equities in this economy as

$$R_{t+1} = (\omega_{t+1}/\, q^e_t) + (q^e_{t+1}/\, q^e_t) \tag{7.78}$$

i.e. the dividend-price ratio plus capital gains. Using (7.78), we can rewrite the sequence of budget constraints in (7.77) as

$$A_{t+1} = R_{t+1}\,(A_t - c_t), \quad A_0 \text{ given} \tag{7.79}$$

where $A_t \equiv \alpha_t\,(\omega_t + q^e_t)$ is loosely speaking the individual asset stock. The specific problem for the representative agent can be recast in terms of the state variable A_t, the random gross return R_{t+1}, and control c_t, or

$$\underset{\{c_{t+j},\, A_{t+j+1}\}_{j=0}^{\infty}}{Max}\ E_t\,[\Sigma_{j=0}^{\infty}\ \beta^j\ u\,(c_{t+j})] \tag{7.80}$$

$$\text{subject to} \qquad A_{t+1} = R_{t+1}\,(A_t - c_t), \quad A_0 \text{ given.}$$

where R_{t+1} is only observed after the decision at time t is taken. A solution for (7.80) is naturally obtained by using stochastic discounted dynamic programming techniques, e.g. Chapter 1. Let the function $V(A_t,R_t)$ denote the maximum value when the agent holds assets A_t and the last gross rate of return observed is R_t, so that the state is (A_t,R_t). Let the control be defined as the current saving given by $(A_t - c_t)$. Then, the value function satisfies the Bellman functional equation

$$V(A_t,R_t) = Max_{\{At\,-\,ct\}}\,\{\, u\,(A_t\ - (A_t - c_t)\,) + \ \beta E_t\,[\, V(A_{t+1},R_{t+1})] \mid A_{t+1} = R_{t+1}\,(A_t - c_t)\}. \tag{7.81}$$

The first order necessary condition for the inner maximum is

$$u'\,(c_t)\ \{\partial\,[A_t\ - (A_t - c_t)]/\,\partial\,(A_t - c_t)\} + \ \beta E_t\,[\{\partial\, V(A_{t+1},R_{t+1})/\,\partial A_{t+1}\} \times \{\partial A_{t+1}/\partial\,(A_t - c_t)\}] = 0. \tag{7.82}$$

The Benveniste and Sheinkman formula yields

$$\partial\, V(A_t,R_t)/\,\partial A_t = u'\,(c_t) \tag{7.83}$$

which upon substitution into (7.82) gives the Euler equation as

$$E_t\,[\ \{\beta\, u'\,(c_{t+1})/\, u'\,(c_t)\}\, R_{t+1}] = 1, \tag{7.84}$$

the familiar fundamental asset pricing formula.

Condition (ii) implies that in equilibrium, $c_t(z)=\omega_t(z)$, $\forall t$, $\forall z$, and because there is only one representative agent in the economy, equilibrium in the capital market implies $\alpha_t(z)=1$, $\forall t$, $\forall z$, i.e. the individual holds the total asset. Together with expression (7.84), these conditions deliver the competitive equilibrium for this economy.

In addition, equilibrium satisfies the transversality condition that over the infinite horizon the representative individual does not accumulate infinite debt or credit, or

$$lim_{t\to\infty}\ \beta^{t+1}\ E_t\,[\{\partial V(A_{t+1},R_{t+1})/\,\partial A_{t+1}\}\ A_{t+1}] = 0. \qquad (7.85)$$

Condition (7.85) is guaranteed to hold by the condition on the discount factor $\beta<1$, which we assume to be satisfied (see Chapter 1, and Problem 6 of this chapter).

As before, we can express the asset pricing formula (7.84) in the state dimension in terms of the states of nature, k, ℓ, instead of the time dimension t, i.e.

$$E_t\ [MRS(k,\ell)\ R\ (k,\ell)] = 1 \qquad (7.86)$$

where $MRS\ (k,\ell)\equiv[\beta\, u'\,(c_\ell)/\,u'\,(c_k)]$ is the state contingent marginal rate of substitution between the current state k in this period and ℓ next period, and $R(k,\ell)$ is the respective gross return for the asset in those states. Taking into account the conditional probabilities in the Markov matrix, we write expression (7.86) as

$$\Sigma_\ell\ \pi_{k\ell}\ MRS(k,\ell)\ R\,(k,\ell) = 1 \qquad (7.87)$$

for each current state k.

We can price a variety of assets (in fact any asset) with formula (7.87). The current price, $q_k{}^d$, of a one-period equity at date t and state k, which pays a dividend d_ℓ in state ℓ next period, yields the gross return as $R\,(k,\ell)= d_\ell/q_k{}^d$ which may be substituted into (7.87) to deliver the price

$$q_k{}^d = \Sigma_\ell\ \pi_{k\ell}\ MRS(k,\ell)\ d_\ell,\quad each\ k. \qquad (7.88)$$

Similarly, the risk-free bond yields $d_\ell=1$, $\forall\ \ell$, whose price is

$$q_k{}^{d=1} = \Sigma_\ell\ \pi_{k\ell}\ MRS(k,\ell),\quad each\ k \qquad (7.89a)$$

and its gross return is

$$R_k{}^{d=1} = 1/\,q_k{}^{d=1} = 1/\Sigma_\ell\ \pi_{k\ell}\ MRS(k,\ell)\ ,\ each\ k. \qquad (7.89b)$$

A claim to one unit of the good in all states two periods from the current period is equivalent to a claim to a one-period risk-free bond next period priced at

$$q_k^{d=2} = \Sigma_\ell \, \pi_{k\ell} \, MRS(k,\ell) \, q_k^{d=1}, \quad each \; k \tag{7.90a}$$

with one-period gross rate of return

$$R_k^{d=1} = q_\ell^{\;d=1} / q_k^{\;d=2} = 1 / \Sigma_\ell \, \pi_{k\ell} \, MRS(k,\ell), \quad each \; k \tag{7.90b}$$

i.e. the ratio of return to cost [as in expression (7.89b) above].

Consider now the price of the "tree" itself, the value of all future "fruit" (dividends). Recall R_{t+1} in formula (7.78) and express it in state contingent terms as

$$R(k,\ell) = (\omega_\ell / q^e_k) + (q^e_\ell / q^e_k) \tag{7.91}$$

which upon substitution into (7.87) gives

$$q^e_k = \Sigma_\ell \, \pi_{k\ell} \, MRS(k,\ell) \, (\omega_\ell + q^e_\ell), \quad each \; k \tag{7.92}$$

or the current price of the "tree" as a function of the price next period for each current state k. Equations (7.92) represent a system of k linear equations in k unknowns that may be solved for q^e_k directly. An analogous result obtains in the time dimension as well. Furthermore, recalling the formula (7.84) and the conditional covariance decomposition $E_t[XY]=cov_t(X,Y)+E_t[X]E_t[Y]$, yields

$$E_t[(\omega_{t+1}/q^e_t) + (q^e_{t+1}/q^e_t)] \, E_t[\beta u'(c_{t+1})/u'(c_t)] +$$

$$\beta \, cov_t([u'(c_{t+1})/u'(c_t)], [(\omega_{t+1}/q^e_t)+(q^e_{t+1}/q^e_t)]) = 1. \tag{7.93}$$

where cov_t is the conditional covariance as of time t.

If we assume that consumption does not grow across periods, which can be obtained by assuming that agents are risk neutral, i.e. u is linear in c, it implies that $u'(c_{t+1})/u'(c_t)=1$. In this special case, the conditional covariance is zero, or

$$cov_t(1, [(\omega_{t+1}/q^e_t)+(q^e_{t+1}/q^e_t)]) = 0,$$

which simplifies formula (7.93), and uncovers the Martingale property of asset prices, namely

$$q^e_t = \beta \, E_t[\omega_{t+1} + q^e_{t+1}], \tag{7.94}$$

that is, the current price of equity is a function of the expected future price and dividend.

A general solution for the stochastic difference equation (7.94) shows several important properties of asset prices in general equilibrium. First, denote by L the lag or backshift operator and rewrite equation (7.94) as

$$(1-\beta^{-1}L)\, E_{t-1}\,[q^{e}_{t}] = -\,E_{t-1}\,[\omega_t], \qquad (7.94a)$$

which yields a general solution for $E_{t-1}\,[q^{e}_{t}]$ as

$$E_{t-1}\,[q^{e}_{t}] = -[\,(1-\beta^{-1}L)]^{-1}\; E_{t-1}\,[\omega_t] + g(t), \qquad (7.94b)$$

where $g(t)$ is a parametric function that must satisfy

$$(1-\beta^{-1}L)\, g(t) = 0.$$

It can be shown that the g function is of the form

$$g(t) = b\,(\beta^{-1})^{t}$$

for b, an undetermined parameter ultimately determined by some initial or boundary condition. Thus, a general iterative forward solution for equation (7.94), applying the law of iterated mathematical expectations as in (7.70), is given by

$$q^{e}_{t} = E_t\,[\Sigma_{j=1}^{\infty}\, \beta^{j}\, \omega_{t+j}\,] + b\,(\beta^{-1})^{t}. \qquad (7.94c)$$

It is clear that for $\beta\in[0,1)$ as $t\to\infty$, $(\beta^{-1})^{t}\to\infty$, and to guarantee asymptotic convergence of the equity price q^{e}_{t}, we must choose a particular solution by setting $b=0$, to obtain

$$q^{e}_{t} = E_t\,[\Sigma_{j=1}^{\infty}\, \beta^{j}\;\; \omega_{t+j}\,], \qquad (7.95)$$

so that the price of the "tree" is the present discounted stream of future dividends, consistent with the familiar Efficient Markets hypothesis, e.g. Chapter 4 and Notes on the Literature.

There are three basic remarks regarding the solution in (7.94)-(7.95). First, there is a multiplicity of paths for the equity price q^{e}_{t} in the general solution (7.94c). For each arbitrary b, the price q^{e}_{t} follows a specific time path so that the general solution is consistent with multiple equilibrium paths.

Second, in the general solution (7.94c), the term $b(\beta^{-1})^{t}$ is sometimes called a "bubble" term. This is because as t gets large, $b(\beta^{-1})^{t}$ gets large as well, and the fundamental component of the equity price, given by the term $E_t\,[\Sigma_{j=1}^{\infty}\,\beta^{j}\omega_{t+j}\,]$, becomes negligible relative to the "bubble" component. Hence, for $b\neq 0$, the asset price follows asymptotically a "bubble" path in the sense that it diverges from the fundamental given by the present value of future dividends. The condition for the particular solution, $b=0$, rules out the "bubble" path and ties the price of the asset to its fundamental, in consistency with the transversality condition (7.85).

Third, the backward solution for (7.94) is not consistent with a stable and strictly positive dynamic path for the equity price, so the forward-looking solution is the appropriate one, i.e. expression (7.95).

Numerical Example: Let the utility function be the familiar constant relative risk aversion (*CRRA*) type given by

$$u(c) = [c^{1-\gamma} - 1] / (1-\gamma), \quad \gamma > 0$$

where γ is the coefficient of relative risk aversion ($\gamma = 1$ denotes logarithmic utility) with marginal utility $u'(c) = c^{-\gamma}$. The subjective discount factor is $\beta=0.95$. Let the states evolve as a Markov process with transitional probabilities, $\pi_{11}=\pi_{22}=0.9$, and $\pi_{12}=\pi_{21}=0.1$ as in (7.61'). The state contingent endowments are $\omega_1=0.8$, $\omega_2=1.2$.

Consider first the case of logarithmic utility, $\gamma = 1$. The state contingent marginal rates of substitution are

$$MRS(1,1)=\beta\, u'(\omega_1)/u'(\omega_1)=(0.95)(0.8)/(0.8)=0.95$$

and similarly

$$MRS(1,2)=(0.95)(0.8)/(1.2)=0.63$$

$$MRS(2,1)=(0.95)(1.2)/(0.8)=1.43$$

$$MRS(2,2)=(0.95)(1.2)/(1.2)=0.95.$$

The price of the risk-free bond is, using expressions (7.89a) and (7.89b),

$$q_1^{d=1} = \pi_{11}\, MRS(1,1) + \pi_{12}\, MRS(1,2) = (0.9)(0.95)+(0.1)(0.63)=0.92$$

$$q_2^{d=1} = \pi_{21}\, MRS(2,1) + \pi_{22}\, MRS(2,2) = (0.1)(1.43)+(0.9)(0.95)=0.998$$

with respective rate of return,

$$R_1^{d=1} = 1/q_1^{d=1} = 1.087$$

$$R_2^{d=1} = 1/q_2^{d=1} = 1.002.$$

The price of equities is the solution of the linear system, by (7.93),

$$q_1^{e} = \pi_{11}\, MRS(1,1)\, (q_1^{e}+\omega_1) + \pi_{12}\, MRS(1,2)\, (q_2^{e}+\omega_2)$$

$$q_2^{e} = \pi_{21}\, MRS(2,1)\, (q_1^{e}+\omega_1) + \pi_{22}\, MRS(2,2)\, (q_2^{e}+\omega_2)$$

yielding

$$q_1{}^e = 15.15$$

$$q_2{}^e = 22.80$$

and returns matrix from (7.91), $R(k,\ell) = (\omega_\ell / q^e_k) + (q^e_\ell / q^e_k)$, equal to

$$\begin{vmatrix} 1.053 & 1.584 \\ 0.699 & 1.053 \end{vmatrix}.$$

Hence, the conditional expected returns are

$$E_t\,[R^e{}_1] = \pi_{11}\,R^e{}_{11} + \pi_{12}\,R^e{}_{12} = (0.9)(1.053)+(0.1)(1.584)=1.1061$$

$$E_t\,[R^e{}_2] = \pi_{21}\,R^e{}_{21} + \pi_{22}\,R^e{}_{22} = (0.1)(0.699)+(0.9)(1.053)=1.0176.$$

The state contingent excess returns are

$$E_t\,[R^e{}_1] - R_1{}^{d=1} = 1.1061 - 1.087 = 0.019$$

$$E_t\,[R^e{}_2] - R_2{}^{d=1} = 1.0176 - 1.002 = 0.016$$

which are rather small in magnitude. We turn now to a more detailed analysis of the excess returns.

7.11 **Excess Returns**

Rajnish Mehra and Edward C. Prescott advanced the following inquiries: if one calibrates the Lucas asset pricing model to real world data, can it explain the pattern of asset returns observed in the real world? If not, why? Table 7.2 below presents some sample unconditional moments from the real world for the relevant variables.

Table 7.2
Properties of Consumption Growth and Asset Returns
Annual Data – 1889-1978

	CONSUMPTION GROWTH	*EQUITY RETURN*	*RISK-FREE RETURN*	*EXCESS RETURN*
Mean	*0.0183*	*0.0698*	*0.0080*	*0.0618*
Std Dev	*0.0357*	*0.1654*	*0.0567*	*0.1667*
Autocorrelation	*-0.14*			

Source of data: Mehra and Prescott (1985), Table 1.

The moments in the table can be suitably used to infer properties about the marginal rate of substitution and asset returns. In particular, note that the average equity premium is large, about 6.18% average annual excess return. Can the Lucas model replicate this pattern?

Towards an answer, we first respecify Lucas's model in terms of growth rates. Each period, let the endowment grow at the stochastic gross rate g so that

$$g_{t+1} = \omega_{t+1} / \omega_t \tag{7.96}$$

where g can take two possible values, $g(L)$ and $g(H)$ (for low and high) with transition probabilities given by the Markov chain (7.60)-(7.64). Since, in equilibrium, the agent consumes the entire endowment, consumption also grows at the stochastic gross growth rate $g \in \{g(L), g(H)\}$. The parametric utility function is the constant relative risk aversion (*CRRA*) type given by

$$u(c) = [c^{1-\gamma} - 1] / (1-\gamma), \quad \gamma > 0 \tag{7.97}$$

where γ is the coefficient of relative risk aversion ($\gamma = 1$ denotes logarithmic utility) with marginal utility $u'(c) = c^{-\gamma}$. For this utility function, the marginal rate of substitution across periods is

$$MRS(t,t+1) = \beta (c_{t+1} / c_t)^{-\gamma} > 0 \tag{7.98}$$

and with the convention that the date t state is k and the date $t+1$ state is ℓ, the intertemporal state contingent marginal rate of substitution is

$$MRS(\ell) = \beta g(\ell)^{-\gamma}, \quad \ell = L, H. \tag{7.99}$$

Formula (7.99) shows that the marginal rate of substitution is only a function of the future state ℓ and not of the current state k. This is because, unlike the original Lucas model, in this version it depends only on the growth rate not in the levels. However, the price of the "tree" or the equity price, q^e, depends on both the level of the endowment and the its growth rate. To see this, note that the fundamental asset pricing formula in this case is, from (7.92),

$$\Sigma_\ell \, \pi_{k\ell} \{MRS(\ell) [(\omega_\ell + q^e_\ell) / q^e_k]\} = 1. \tag{7.100}$$

We conjecture that the current price q^e_k is linear in the endowment, or

$$q^e_k(\omega) = q^e_k \, \omega \tag{7.101}$$

hence, in state ℓ next period, we obtain

$$q^e_\ell [g(\ell)\,\omega] = q^e_\ell \, g(\ell)\,\omega, \qquad \ell = L,H \tag{7.102}$$

because the endowment grows at the stochastic rate $g(\ell)$ or $\omega_\ell = g(\ell)\omega$. Then, the asset pricing formula in (7.100) may be written as

$$\Sigma_\ell \, \pi_{k\ell} \, \{MRS(\ell)\,([g(\ell)\,\omega + q^e_\ell \, g(\ell)\,\omega] \,/\, q^e_k \,\omega)\} = 1 \tag{7.103}$$

which upon simplification and use of (7.99) gives the price of the "tree" as

$$q^e_k = \beta \, \Sigma_\ell \, \pi_{k\ell} \, [\, g(\ell)^{1-\gamma} \, (1 + q^e_\ell)], \quad \ell = L,H, \ \ each\ k \tag{7.104}$$

The conjecture is indeed correct because the equilibrium satisfies all budget constraints and the transversality condition.

The equations in (7.104) form a system of k equations for each state k, in k unknowns, q^e_k, which may be solved directly. The associated gross return on the equity is, from (7.91) and (7.101)-(7.102),

$$R(k,\ell) = g(\ell)\,(1 + q^e_\ell) \,/\, q^e_k, \quad \ell = L,H, \ \ each\ k \tag{7.105}$$

The one-period risk-free bond has a price, from (7.100),

$$q_k^{\,d=1} = \Sigma_\ell \, \pi_{k\ell} \, MRS(\ell), \quad each\ k \tag{7.106}$$

and gross rate of return

$$R^{d=1}(k,\ell) = 1/\Sigma_\ell \, \pi_{k\ell} \, MRS(\ell), \quad each\ k \tag{7.107}$$

Equations (7.96)-(7.107) form the basis for a numerical evaluation of the model using the unconditional moments from Table 7.2. To replicate the data, first note that the unconditional probabilities of the two states $\{L,H\}$ from the Markov chain (7.60)-(7.64) are ½ identically. Thus, a choice of state contingent (gross) growth rates from Table 7.2 as *mean ± one standard deviation* yields

$$g(H) = 1.0540, \quad g(L) = 0.9826$$

whose average is 1.0183 and standard deviation is 0.0357 exactly as in the data.

The autocorrelation for the Markov chain from Table 7.2 is $\rho = -0.14$ and thus the transition matrix is

$$\Pi = \begin{bmatrix} 0.43 & 0.57 \\ 0.57 & 0.43 \end{bmatrix}$$

which yields the same autocorrelation of growth rates as in the data. Let preferences be logarithmic and the discount factor be 0.95, or

$$\gamma = 1, \quad \beta = 0.95.$$

We are ready to compute the relevant variables. We start with the price of the one-period risk-free bond in (7.106),

$$q_1^{d=1} = \beta\,[(0.43)(1/1.0540)+(0.57)(1/0.9826)],$$

$$q_2^{d=1} = \beta\,[(0.57)(1/1.0540)+(0.43)(1/0.9826)],$$

for the two states, $k=1,2,$ and since the price in the data is the unconditional mean, the unconditional average price from the model is (independent of $\pi_{k\ell}$)

$$E[q^{d=1}] = ½\,(q_1^{d=1} + q_2^{d=1}) = \beta\,(0.9832).$$

Hence, for $\beta=0.95$, the average risk-free gross return from the model, as in (7.107), is

$$R^{d=1}(k,\ell) = 1/(0.95)(0.9832) = 1.0706,$$

i.e. *7.06%*. From Table 7.2, notice that the average risk-free return is much lower, *0.80%*. The only way to reconcile the theory with the data, in this dimension, is to choose a discount factor that matches the observed risk-free return, or

$$\beta = 1/(1.0080)(0.9832) = 1.0090$$

which gives the risk-free gross return of *1.0080* as in the data. However, this discount factor is greater than one, outside the range of plausible values; see Notes on the Literature. This discrepancy between theory and data regarding the risk-free return is the so-called the Risk-free Rate puzzle.

In order to get some heuristic intuition for this puzzle, consider the non-stochastic version of the asset pricing formula (7.84) with *CRRA* preferences as in (7.97), or

$$\beta\,(c_{t+1}/c_t)^{-\gamma}\,R = 1.$$

If there is no growth in consumption, or $c_{t+1} = c_t$, this reduces to $\beta R=1$, the 'modified golden rule'. However, when $(c_{t+1}/c_t) = g \neq 0$ and $\gamma=1$, this becomes

$$R/g = 1/\beta,$$

and $\beta \in [0,1)$ implies that $R > g$, an inequality not supported by the data. A higher gross growth rate requires a higher discount factor to account for a given risk-free gross return. Even when risk aversion is different than one, or

$$R/(g)^{\gamma} = 1/\beta,$$

if the risk aversion parameter γ is large, we need β large as well for a given risk-free return. In summary, the risk-free rate puzzle regards the mismatch between theory and data in the sense that the theory cannot reconcile the observed low risk-free return with higher positive growth rates and plausible risk aversion and discounting.

Next, we examine the price and return on equities. Notice that for $\gamma = 1$, the price of equities from (7.104) becomes

$$q^e_k = \beta \Sigma_{\ell} \pi_{k\ell} (1 + q^e_{\ell})$$

independent of $g(\ell)$, thus

$$q^e_k = q^e_{\ell}$$

and

$$q^e = \beta/(1-\beta)$$

independent of k. For $\beta=0.95$, the equity price is $q^e=19$.

The gross return on equities is obtained from (7.105) as

$$R(k,\ell) = g(\ell)(1+q^e)/q^e$$

independent of k. Thus, $R(k,\ell)$ is

$$= \begin{bmatrix} 1.1095 & 1.0343 \\ 1.1095 & 1.0343 \end{bmatrix}$$

and the unconditional average gross return is

$$E[R(k,\ell)] = ½\,(1.1095) + ½\,(1.0343) = 1.0720$$

or 7.20%. In fact, the model predicts an excess return between equities and the risk-free bond of the magnitude of

$$7.20\% - 7.06\% = 0.14\%,$$

well short of the *6.18%* excess return observed in the data. This discrepancy between theory and data relating to the excess return of equities is called the Equity Premium puzzle. However, the discrepancy is sensitive to the value of the risk aversion parameter. Table 7.3 presents computations from the model for alternative values of the risk aversion parameter.

Table 7.3
Model Sensitivity Analysis
Unconditional Means (%)

	EQUITY RETURN	RISK-FREE RETURN	EXCESS RETURN
$\gamma = 1$	*7.20*	*7.06*	*0.14*
$\gamma = 5$	*14.21*	*13.22*	*0.98*
$\gamma = 10$	*21.06*	*18.35*	*2.71*

Table 7.3 shows that as the risk aversion parameter increases, the excess return increases as well, as we have seen before. The problem is that too much risk aversion is implausible and that the risk-free return also increases with risk aversion. The model does not adequately predict the equity premium under plausible parameter values.

Lars P. Hansen and Ravi Jagannathan have a more practical take on the issue of the excess return of an asset. In their framework, they consider a balanced portfolio to be a zero net worth portfolio in which the agent goes short on asset j to finance an investment of equal value in another asset i, $\{j,i\} \in \boldsymbol{K}$. The gross return on this portfolio is the return differential $R_i - R_j$ which satisfies expression (7.74c), or

$$E_t\,[\,MRS(zh^t, z^{t+1})\ \{R_i\,(zh^t, z^{t+1}) - R_j\,(zh^t, z^{t+1})\}] = 0. \quad (7.108)$$

We show now that large excess returns imply high variability of the marginal rate of substitution, $MRS(zh^t, z^{t+1})$. To see this, take the unconditional expectation of (7.108) [see (7.48) as well] as

$$E\,[\,MRS(z)\ \{R_i\,(z) - R_j\,(z)\}] = 0. \quad (7.109)$$

which by the covariance decomposition formula implies

$$E[MRS(z)]\ E[R_i\,(z) - R_j\,(z)] + cov(MRS(z), \{R_i\,(z) - R_j\,(z)\}) = 0. \quad (7.110)$$

Note that $cov(X,Y) = \rho\ \sigma_X\ \sigma_Y$ where $\rho \in [-1,1]$ is the coefficient of correlation between X and Y. Thus, (7.110) becomes

$$E[MRS(z)]\, E[R_i(z) - R_j(z)] + \rho\, \sigma_m\, \sigma_{Ri\text{-}Rj} = 0 \qquad (7.111)$$

which yields

$$\rho\, \sigma_m / E[MRS(z)] = -\, E[R_i(z) - R_j(z)] / \sigma_{Ri\text{-}Rj}. \qquad (7.112)$$

Taking absolute values on both sides of (7.112) yields

$$|\rho|\sigma_m / E[MRS(z)] = |\, E[R_i(z) - R_j(z)]| / \sigma_{Ri\text{-}Rj}. \qquad (7.113)$$

because σ_m, $\sigma_{Ri\text{-}Rj}$, and $E[MRS(z)]$ are all strictly positive. Moreover, we have that $|\rho| \leq 1$ because the coefficient of correlation is within the unit circle. Hence, (7.113) implies

$$\sigma_m / E[MRS(z)] \geq |\, E[R_i(z) - R_j(z)]| / \sigma_{Ri\text{-}Rj} \qquad (7.114)$$

where the right-hand side of the inequality, $|\, E[R_i(z) - R_j(z)]| / \sigma_{Ri\text{-}Rj}$, is known as the "Sharpe Ratio" (or "Sharpe Index"). Formula (7.114) provides a useful lower bound on the ratio of the standard deviation of the marginal rate of substitution to its mean. The Sharpe ratio (or index) of the balanced portfolio provides the lower bound for the ratio of the standard deviation of the marginal rate of substitution to its mean. The largest Sharpe ratio is the slope, in absolute value, of the mean-standard deviation frontier in the traditional finance approach, discussed in Chapter 2. From (7.114), it is easy to note that a large excess return implies a large variability of the marginal rate of substitution, σ_m.

Consider the data in Table 7.2. The Sharpe ratio on equities is given by

$$|\, E[R_i(z) - R_j(z)]| / \sigma_{Ri\text{-}Rj} = 0.0618/0.1667 = 0.37$$

which represents the lower bound for $\sigma_m / E[MRS(z)]$. Using the same parameters as in the Mehra and Prescott economy, we obtain

$$E[MRS(z)] = \tfrac{1}{2}\,(0.95)(1.0540) + \tfrac{1}{2}\,(0.95)\,(0.9826) = 0.9674$$

and $\sigma_m = 0.0339$, thus

$$\sigma_m / E[MRS(z)] = 0.0350 < 0.37 = |\, E[R_i(z) - R_j(z)]| / \sigma_{Ri\text{-}Rj}$$

and the obtained ratio of the standard deviation of the marginal rate of substitution to its mean is less than the lower bound determined by the Sharpe ratio (or index). Increasing the parameter for risk aversion to $\gamma = 5$, gives $\sigma_m / E[MRS(z)] = 0.1534 < 0.37$ still substantially less than the lower bound.

The problem is that the marginal rate of substitution in the theory shows too little variability to be compatible with the data, i.e. σ_m is too small. The theory should have characteristics that increase the variability of the marginal rate of substitution. For example, borrowing constraints, variable discount factors, habit formation, heterogeneous agents (to be seen later in this chapter), incomplete markets as seen in Chapter 4, asymmetric information as in the moral hazard case of Chapter 5, are only a few to mention.

7.12 Summary II

In sections 7.4-7.11, we discussed the general equilibrium approach to consumption and investment in a dynamic world with full information. We have shown the asset pricing characteristics of this model. In the next sections, we examine the introduction of money in this framework.

7.13 Stochastic Monetary Theory

Introducing a specific medium of exchange into the models examined above is not an easy task. Money is an asset that plays several roles in the economy, from store of value to record-keeping device, etc. However, money is an asset that is dominated in rate of return by other assets in the economy. Thus, it is not as simple and straightforward to introduce it in simple general equilibrium dynamic models. In this section and the next, we discuss some implications of introducing money and the effect of monetary risk in the Lucas dynamic general equilibrium model of section 7.10.

First, suppose there is an asset called money whose nominal amount held by an individual at time t is denoted m_t. The price level in money terms at time t is denoted p_t. Hence, one unit of money has real purchasing power in terms of the goods and services it can buy of $1/p_t$, and the real value of the nominal money balances held is m_t/p_t. A fictitious monetary authority supplies nominal money balances. Now, suppose we introduce this asset into the Lucas exchange economy of section 7.10. There, we learned that assets are valued according to the stream of dividends (consumption) they provide the holder, e.g. expression (7.95). Money, by definition, pays zero dividends thus its value in the pure exchange economy is zero, or $1/p_t \to 0$; see e.g. Notes on the Literature. The consequence is that for money to be held in this class of models, a specific 'exogenous' reason must be given. Several reasons have been proposed. Here, we examine one of them, the so-called cash-in-advance constraint for transactions.

(i) Cash-in-Advance Constraint

Consider an economy where individuals are identical. They can hold two assets: capital and money. A simple capital accumulation relationship is

$$k_{t+1} = I_t + k_t \ (1-\delta) \tag{7.115}$$

where k_t is the physical capital stock, I_t is gross investment (or the quantity of capital goods acquired in period t for use in period $t+1$), and $\delta \in [0,1]$ is the rate of depreciation of physical capital between t and $t+1$. The individual has access to a technology exhibiting diminishing marginal physical products to the variable physical capital input denoted $f(k_t)$, where $f' > 0$ and $f'' < 0$. This is the usual neoclassical production function, strictly increasing and strictly concave on the variable input in the production of a single good. The budget constraint faced by the individual is given by

$$f(k_t) - c_t - I_t \geq (m_{t+1} - m_t) / p_t \tag{7.116}$$

where c_t is the consumption of the individual. The budget constraint (7.116) indicates that the excess production over real expenditures is held in real money balances. The monetary authority supplies the nominal money balance, and, as we'll see in more detail below, this is the source of risk in this economy.

The cash-in-advance constraint is the 'exogenous' reason why individuals hold money. Suppose that in order to consume and invest individuals must use cash. This is expressed as

$$m_t / p_t \geq c_t + I_t. \tag{7.117}$$

In this framework, we can envision the individual maximizing expected utility subject to the appropriate budget constraints, or

$$\underset{\{c_{t+j},\, I_{t+j},\, m_{t+j+1}\}_{j=0}^{\infty}}{Max}\; E_t\,[\Sigma_{j=0}^{\infty}\; \beta^j\; u(c_{t+j})] \tag{7.118}$$

subject to (7.116)-(7.117) holding with equality, with $(k_0, m_0)>0$ given,

where $\beta \in [0,1)$ is the individual subjective discount factor. The monetary authority supplies money to individuals according to the rule

$$m_{t+1} = \omega_t\, m_t \tag{7.119}$$

where ω_t is an exogenous random variable denoting the gross growth of nominal money balances, i.e. the source of risk in the economy. A realization of ω_t is assumed to be observed by all in the market place, at the beginning of period t, before decisions to consume and invest. In particular, we can consider ω_t as a Markov process with Prob($\omega_t \leq \omega' \mid \omega_{t-1}=\omega)=H(\omega', \omega)$ where there exists a conditional density $h(\omega', \omega)$ for H.

A solution for (7.118) is naturally obtained by using stochastic discounted dynamic programming techniques. Let the function $V(I_{t-1}, m_t)$ denote the maximum value when the agent invested I_{t-1} in physical capital and holds m_t in nominal money balances. The state is given by the pair $\{I_{t\ 1}, m_t\}$ and the control is $\{c_t, I_t, m_{t+1}\}$. Then, the value function satisfies the Bellman functional equation

$$V(I_{t-1}, m_t) = Max_{\{ct,\, It,\, mt+1\}}\, \{\, u(c_t) + \; \beta E_t\,[\, V(I_t, m_{t+1})] \mid$$

$$f(I_{t-1} + k_{t-1}(1-\delta)) - c_t - I_t = (m_{t+1} - m_t)/p_t\, ; \;\; m_t / p_t = c_t + I_t\}. \tag{7.120}$$

where $k_t = I_{t-1} + k_{t-1}(1-\delta)$, and the function f is from (7.115).

The first order necessary conditions for the inner maximum with respect to (c_t, I_t, m_{t+1}), are respectively

$$u'(c_t) - \lambda_t - \gamma_t = 0 \tag{7.121a}$$

$$\beta E_t\,[\{\partial V(I_t, m_{t+1}) / \partial I_t\}] - \lambda_t - \gamma_t = 0 \tag{7.121b}$$

$$\beta E_t\,[\{\partial V(I_t, m_{t+1}) / \partial m_{t+1}\}] - (\lambda_t / p_t) = 0. \tag{7.121c}$$

where $\lambda_t > 0$ and $\gamma_t > 0$ are Lagrange multipliers associated with the constraints (7.116) and (7.117) respectively. The Benveniste and Sheinkman formula yields

$$\partial V(I_{t-1}, m_t)/\partial I_{t-1} = \lambda_t f'(k_t) \tag{7.122a}$$

$$\partial V(I_{t-1}, m_t)/\partial m_t = (\lambda_t/p_t) + (\gamma_t/p_t). \tag{7.122b}$$

A rational expectations equilibrium for this problem is thus

$$\beta E_t[\lambda_{t+1} f'(k_{t+1})] - \lambda_t - \gamma_t = 0 \tag{7.123a}$$

$$\beta E_t[(\lambda_{t+1}/p_{t+1}) + (\gamma_{t+1}/p_{t+1})] - (\lambda_t/p_t) = 0. \tag{7.123b}$$

together with (7.121a), the goods market equilibrium,

$$f(k_t) - c_t - I_t = 0, \tag{7.123c}$$

the cash-in-advance constraint (7.117) holding with equality, and the money market equilibrium (7.119). The solution is completed with the transversality conditions

$$\lim_{t\to\infty} E_t[\beta^{t+1}\{\partial V(I_t, m_{t+1})/\partial I_t\} I_t] = 0$$

$$\lim_{t\to\infty} E_t[\beta^{t+1}\{\partial V(I_t, m_{t+1})/\partial m_{t+1}\} m_{t+1}] = 0.$$

We can learn more about the equilibrium by assuming the special case of *100%* depreciation, $\delta = 1$. In this case, the policy function for the evolution of the capital stock (investment) is, from expressions (7.116), (7.117), (7.119), (7.121a), (7.123a,b), given by the formula

$$u'(f(k_t) - k_{t+1}) =$$

$$\beta^2 E_t[f'(k_{t+1})\{f(k_{t+2})/f(k_{t+1})\}(1/\omega_{t+1})\, u'(f(k_{t+2}) - k_{t+3})]. \tag{7.124}$$

In general, the formula (7.124) yields a stochastic first order difference equation for the evolution of the capital stock. Further specializing by assuming the exogenous stochastic process to be independent and identically distributed (i.i.d.), preferences to be logarithmic, $u(c) = \log c$, and technology to be of the of the Cobb-Douglas form, $f(k) = Ak^\alpha$ for $\alpha \in (0,1)$ and $A > 0$ constant, we obtain a closed form solution for the policy function given by

$$k_{t+1} = \alpha\beta^2 A (k_t)^\alpha (1/\omega_{t+1}). \tag{7.125a}$$

Taking logarithms of both sides of (7.125a), the evolution of the logarithm of the capital stock is given by the stochastic difference equation

$$\log k_{t+1} = \log(\alpha\beta^2 A) + \alpha \log k_t + \log(1/\omega_{t+1}) \tag{7.125b}$$

and monetary surprises have a negative effect on capital accumulation since requiring cash for both consumption and investment involves taxing consumption twice, in the present and in the future. Starting from k_0, the solution for the difference equation (7.125b) is

$$\log k_t = \alpha^t \log k_0 + [(1-\alpha^t)/(1-\alpha)] \log (\alpha\beta^2 A) + \Sigma_{j=0}^{t-1} \alpha^j \log (1/\omega_{t-j}) \qquad (7.126)$$

where we used the geometric series result, $\Sigma_{j=0}^{t-1}\alpha^j=(1-\alpha^t)/(1-\alpha)$, for $\alpha\in(0,1)$. For $\log (1/\omega_t)$ i.i.d. with mean $\mu_{1/\omega}$ and constant variance $\sigma_{1/\omega}^2$, we have

$$E[\log k_t] = \alpha^t \log k_0 + [(1-\alpha^t)/(1-\alpha)][\mu_{1/\omega} + \log (\alpha\beta^2 A)] \qquad (7.127a)$$

$$var(\log k_t) = [(1-\alpha^{2t})/(1-\alpha^2)]\, \sigma_{1/\omega}^2. \qquad (7.127b)$$

Thus, an increase in the variance of the money growth factor, ω, increases the average capital stock, $E[\log k_t]$, because, by Jensen's inequality, $\mu_{1/\omega}$ is increasing in the variance of ω. The intuition behind this effect is that there are only two assets in the model, money and capital. If the volatility of money growth increases, the risk averse agent is going to shift towards capital thus increasing its expected value.

The issue of rate of return dominance can be seen in this model by considering a stationary equilibrium where $\omega=1$ constant. In this case, the rate of return on money is $p_t / p_{t+1}=1$, which is the cost of carrying cash between t and $t+1$. The return on capital, by (7.123a,b), is

$$f'=(1/\beta^2)>1=p_t/p_{t+1}.$$

Money is indeed dominated in rate of return in this framework.

7.14 **Fisher Equation and Risk**

The Fisher equation presents a link between nominal and real interest rates. According to the equation, the real interest factor is the nominal factor discounted by the inflation factor.

Consider the equilibrium conditions (7.121a) and (7.123b) to obtain

$$E_t[\beta u'(c_{t+1})(p_t/p_{t+1})] - \lambda_t = 0. \qquad (7.128)$$

When we divide both sides of (7.128) by $u'(c_t)$, we obtain the link between nominal and real interest rates in the cash-in-advance economy as

$$E_t[\{\beta u'(c_{t+1})/u'(c_t)\}(p_t/p_{t+1})]=\lambda_t/u'(c_t)=[u'(c_t)-\gamma_t]/u'(c_t). \quad (7.129)$$

First, suppose there is no risk in this economy. The sure version of (7.129) is

$$\{\beta u'(c_{t+1})/u'(c_t)\}(p_t/p_{t+1})=\lambda_t/u'(c_t)=[u'(c_t)-\gamma_t]/u'(c_t). \quad (7.129')$$

The term $\lambda_t/u'(c_t)=[u'(c_t)-\gamma_t]/u'(c_t)$ is the marginal utility of nominal wealth priced by the marginal utility of consumption, or the inverse of the opportunity cost of postponing consumption in terms of nominal wealth. This is exactly the inverse of the nominal interest factor. The term on the left-hand side, $\{\beta u'(c_{t+1})/u'(c_t)\}(p_t/p_{t+1})$, is the product of two terms. The first is the marginal rate of substitution in consumption which represents the inverse of the return on capital, or the real interest factor, $\{\beta u'(c_{t+1})/u'(c_t)\}$. The second, (p_t/p_{t+1}), is the inverse of the inflation factor. Thus, equation (7.129') represents the link between the (inverse of the) nominal interest factor equated to the (inverse of the) real interest factor times the inflation factor, the Fisher equation.

Adding risk, we have (7.129), and using the usual alternative covariance formula, we have

$$E_t[\{\beta u'(c_{t+1})/u'(c_t)\}]E_t[(p_t/p_{t+1})]+cov_t(\{\beta u'(c_{t+1})/u'(c_t)\},(p_t/p_{t+1}))$$
$$=\lambda_t/u'(c_t)=[u'(c_t)-\gamma_t]/u'(c_t). \quad (7.130)$$

Under risk, the Fisher equation is adjusted by a term reflecting the inflation risk premium,

$$cov_t(\{\beta u'(c_{t+1})/u'(c_t)\},(p_t/p_{t+1})).$$

If agents are risk neutral and u is linear, or if (p_t/p_{t+1}) is deterministic, the covariance is zero and the traditional Fisher equation in (7.129') holds. However, if

$$cov_t(\{\beta u'(c_{t+1})/u'(c_t)\},(p_t/p_{t+1}))<0,$$

when consumption growth is high (low), inflation is high (low). Alternatively, if

$$cov_t(\{\beta u'(c_{t+1})/u'(c_t)\},(p_t/p_{t+1}))>0,$$

when consumption growth is low (high), inflation is high (low).

7.15 Summary III

In sections 7.13-7.14, we discussed a simple general equilibrium monetary model where the source of risk is the growth of nominal money balances. We use the cash-in-advance constraint to allow money to co-exist with assets that yield a higher rate of return. Including both money and investment as cash goods yields real effects of monetary growth and changes in the volatility of money growth can have effects on the average level of investment as well. We showed that the Fisher equation has an additional inflation risk premium under uncertainty. In the next sections, we discuss the financial problem of firms in a dynamic, stochastic general equilibrium environment.

7.16 The Financial Problem of the Firm in General Equilibrium

In Chapter 4, we discussed the stochastic neoclassical theory of the firm from a production perspective. In this section, we extend the model of the neoclassical theory of the firm to include the financial decisions and dynamics under complete markets and general equilibrium. We abstract from monetary issues; all variables are in real terms. The framework provides a general equilibrium assessment of the Modigliani-Miller theorem, or that the market value of the firm does not depend upon its equity-debt capital structure.

We study a decentralized framework with households and firms. Households engage in consumption of the single commodity produced by firms, make asset allocation decisions by holding firm issued debt or equity, and offer labor. Firms operate a production technology choosing production, investment and labor demand. Firms also make decisions about how to finance new investment with debt or equity issues, paying dividends to shareholders and interest to bondholders. We start with the problem of the representative household.

Households - The labor supply of households is assumed inelastic and normalized to unity. The household solves a dynamic problem to maximize the stream of discounted expected utility subject to the budget constraint. The problem is as follows:

$$\underset{\{c_{t+j},\, z_{t+j+1},\, b_{t+j+1}\}_{j=0}^{\infty}}{Max}\ E_t[\Sigma_{j=0}^{\infty}\ \beta^j\ u(c_{t+j})] \qquad (7.131a)$$

subject to

$$c_t + q_t z_{t+1} + b_{t+1} \leq w_t + R_t b_t + z_t (q_t + d_t), \qquad (7.131b)$$

$\{z_0, b_0\}$ given, and processes (possibly stochastic) for $\{q_t, w_t, R_t, d_t\}$ given,

where $\beta \in [0,1)$ is the individual subjective discount factor, and u is the instantaneous utility, strictly increasing and strictly concave as usual.

The flow budget constraint of the individual in (7.131b) consists of sources (right-hand side) and uses (left-hand side) of income. The sources are w_t the total wages received, $R_t b_t$ the gross interest rate received on one-period firm issued bonds, plus the value of the equity holdings and dividends, $z_t (q_t + d_t)$. The uses consist of consumption, c_t, equity holding $q_t z_{t+1}$, and bond holdings, b_{t+1}. Thus, q_t is the price of equity and z_t is the units of equity held by the individual.

An interior solution for the household problem yields demand functions for consumption, equity and bond holdings. This is naturally obtained by using stochastic discounted dynamic programming techniques. Let the function $V(z_t, b_t)$ denote the maximum value when the agent holds z_t units in equity and b_t in bonds. The state is given by the pair (z_t, b_t) and the control is (z_{t+1}, b_{t+1}). Then, the value function satisfies the Bellman functional equation

$$V(z_t, b_t) = Max_{\{z_{t+1}, b_{t+1}\}} \{ u(w_t + R_t b_t + z_t (q_t + d_t) - q_t z_{t+1} - b_{t+1}) + \beta E_t [V(z_{t+1}, b_{t+1})]\} \qquad (7.132)$$

where we substituted consumption, c_t, from the budget constraint (7.131b) into the utility function u. The first order necessary conditions for the inner maximum with respect to (z_{t+1}, b_{t+1}), are respectively

$$-u'(c_t)\, q_t + \beta E_t[\{\partial V(z_{t+1}, b_{t+1}) / \partial z_{t+1}\}] = 0 \qquad (7.133a)$$

$$-u'(c_t) + \beta E_t[\{\partial V(z_{t+1}, b_{t+1}) / \partial b_{t+1}\}] = 0. \qquad (7.133b)$$

The Benveniste and Scheinkman formula yields

$$\partial V(z_t, b_t)/\partial z_t = u'(c_t)(q_t + d_t) \tag{7.134a}$$

$$\partial V(z_t, b_t)/\partial b_t = u'(c_t) R_t \tag{7.134b}$$

which can be appropriately updated and substituted into the expressions (7.133a,b) to give the stochastic Euler equations:

$$E_t[\{\beta u'(c_{t+1})/u'(c_t)\}\{(q_{t+1} + d_{t+1})/q_t\}] = 1 \tag{7.135a}$$

$$E_t[\{\beta u'(c_{t+1})/u'(c_t)\} R_{t+1}] = 1. \tag{7.135b}$$

The usual asset pricing expressions (7.135a,b) together with the budget constraint (7.131b) give solutions for the demand functions for $\{c_t, z_{t+1}, b_{t+1}\}$ as a function of the given state and prices. The solution is completed by the transversality conditions

$$lim_{t\to\infty} E_t[\beta^{t+1}\{\partial V(z_{t+1}, b_{t+1})/\partial z_{t+1}\} z_{t+1}] = 0 \tag{7.136a}$$

$$lim_{t\to\infty} E_t[\beta^{t+1}\{\partial V(z_{t+1}, b_{t+1})/\partial b_{t+1}\} b_{t+1}] = 0. \tag{7.136b}$$

Firms - Firms operate a production process for a single good, y_t, using physical capital and labor as inputs, subject to a multiplicative stochastic technology disturbance, recall section 4.11. Noting that labor is supplied inelastically and normalized to unity, the production function is

$$y_t = \theta_t f(k_t), \tag{7.137}$$

where k_t is the capital-labor ratio, f is a strictly increasing and strictly concave function, $f' > 0$, $f'' < 0$, and θ_t is an i.i.d. lognormal random variable, e.g. Chapter 8. As usual, the technology is constant returns to scale in all inputs and exhibits diminishing marginal products to each input. Using Euler's theorem, production can be decomposed as

$$y_t = \theta_t f(k_t) = \theta_t f'(k_t) k_t + w_t, \tag{7.138}$$

and the revenue from capital or gross profit is

$$\pi_t = \theta_t f'(k_t) k_t = \theta_t f(k_t) - w_t. \tag{7.139}$$

(i) The Financial Structure of the Firm

The gross profits are assumed to be disbursed in three different categories: paid as dividends to shareholders, $d_t\, z_t$; paid as interest to bondholders, $R_t\; b_t$; or held as retained earnings, re_t. Hence, the disbursement of gross profits is given by the expression

$$\pi_t = d_t\, z_t + R_t\, b_t + re_t\,. \tag{7.140}$$

On the production side, the firm owns the physical capital stock and must decide how much to invest in new capital (demand for capital investment). The investment in physical capital is governed by the expression [as in (7.115)]

$$k_{t+1} = I_t + k_t\; (1\text{-}\delta) \tag{7.141a}$$

where I_t is gross investment (or the quantity of capital goods acquired in period t for use in period $t+1$) and $\delta \in [0,1]$ is the rate of depreciation of physical capital between t and $t+1$. In this case, we also assume the depreciation rate to be *100%*, $\delta=1$, implying that investment is just the capital stock available for use at $t+1$,

$$k_{t+1} = I_t\,. \tag{7.141b}$$

On the financial side, the firm must decide how to finance the new investment, either issuing new equity, $(\, z_{t+1}\; - \; z_t\,)$, new (one-period) bonds, b_{t+1}, or out of retained earnings, re_t, that is

$$k_{t+1} = I_t = q_t\; (z_{t+1} - z_t) + b_{t+1} + re_t\,. \tag{7.142}$$

(ii) Firm's Valuation and Decision-making in General Equilibrium

First, we define the net cash flow out of the firm, denoted N_t, as the cash flow net of investment expenditures or

$$N_t \equiv \pi_t - I_t = \pi_t - k_{t+1}. \tag{7.143}$$

Substituting for gross profits from expression (7.140) we obtain

$$N_t = d_t\, z_t + R_t\, b_t + re_t - k_{t+1}, \tag{7.144}$$

and substituting for retained earnings from (7.142) we obtain

$$N_t = (d_t + q_t)\, z_t + R_t\, b_t - q_t\, z_{t+1} - b_{t+1}. \tag{7.145a}$$

Expression (7.145a) shows the net cash flow as the difference between the disbursements in equity and bonds, $(d_t + q_t)\, z_t + R_t\, b_t$, minus the funds from new equity and bonds, $q_t \, z_{t+1} + b_{t+1}$. Also, from expressions (7.139) and (7.143), the net cash flow can be written as

$$N_t = \theta_t f'(k_t)\, k_t - k_{t+1}. \qquad (7.145b)$$

or gross profits (under constant returns to scale) minus investment.

Now, we are ready to define the market value of the firm. We use two versions of the market value depending upon the position within the time period. The beginning-of-period outstanding market value of the firm, denoted W_t, is the sum of the net cash flow plus the new shares and debt issued, or

$$W_t \equiv N_t + q_t \, z_{t+1} + b_{t+1}, \qquad (7.146a)$$

and the end-of-period outstanding market value is just the value of the firm net or after dividends and debt payments, or

$$W^a_t \equiv q_t \, z_{t+1} + b_{t+1}, \qquad (7.146b)$$

so that the relationship between the two is

$$W_t = N_t + W^a_t. \qquad (7.146c)$$

Under complete markets, there is no discrepancy between shareholders, bondholders and firms about the market valuation, e.g. Chapter 4. The firm should make decisions in order to maximize its beginning-of-period market value given in expression (7.146a,c). However, it appears that the outstanding market value of the firm in (7.146a,c) depends upon the financial decisions of new equity and bond issues, $W^a_t = q_t \, z_{t+1} + b_{t+1}$.

We show next that this appearance is misleading. In general equilibrium, the market value of the firm does not depend upon the firm's financial decision, it only depends upon the firm's production decisions. There is complete separation between the production side and the financial side of the firm's operation (you can imagine a building where the production engineers are in one floor and the financial engineers are in a different floor, so that their decisions are independent). This is a general equilibrium version of the Modigliani-Miller theorem: the outstanding market value of the firm is independent of the debt or equity financial mix.

First, we examine a preview of the general equilibrium. Under general equilibrium, individuals choose optimal demands for $\{c_t, z_{t+1},$

$b_{t+1}\}$ by solving the problem (7.131a,b) given $\{q_t\ ,\ w_t\ ,\ R_t\ ,\ d_t\}$; firms maximize market value; and all resources constraints are satisfied, or supply equals demands in the equity market, bonds market and goods market, or

$$y_t = \theta_t f(k_t) = c_t + I_t \quad all\ t.$$

Thus, the end-of-period outstanding market value of the firm in expression (7.146b) must be consistent with the associated household demands in expressions (7.135a,b). We substitute the optimal demands into the end-of-period market value in the following way: we multiply both sides of the condition (7.135a) by $z_{t+1} \neq 0$, to obtain

$$E_t [\{\beta u'(c_{t+1}) / u'(c_t)\} \{(q_{t+1} + d_{t+1}) z_{t+1}\}] = q_t z_{t+1}; \qquad (7.147a)$$

and similarly both sides of the condition (7.135b) by $b_{t+1} \neq 0$, to obtain

$$E_t [\{\beta u'(c_{t+1}) / u'(c_t)\} R_{t+1} b_{t+1}] = b_{t+1}. \qquad (7.147b)$$

Then, we may substitute expressions (7.147a,b) into (7.146b) appropriately to obtain

$$W^a_t = E_t [\{\beta u'(c_{t+1}) / u'(c_t)\} \{(q_{t+1} + d_{t+1}) z_{t+1} + R_{t+1} b_{t+1}\}]. \ (7.148a)$$

Next, we add and subtract $W^a_{t+1} \equiv q_{t+1} z_{t+2} + b_{t+2}$ to both sides of the last expression, (7.148a), and use the law of iterated mathematical expectations and rearrange, to obtain

$$W^a_t = E_t [\{\beta u'(c_{t+1}) / u'(c_t)\} \{W^a_{t+1} + [q_{t+1} (z_{t+1} - z_{t+2}) - b_{t+2}] + d_{t+1} z_{t+1} + R_{t+1} b_{t+1}\}], \qquad (7.148b)$$

and we substitute the one-period updated version of the net cash flow (7.145a)

$$q_{t+1} (z_{t+1} - z_{t+2}) - b_{t+2} = N_{t+1} - d_{t+1} z_{t+1} - R_{t+1} b_{t+1}$$

into (7.148b) to obtain the end-of-period market value as

$$W^a_t = E_t [\{\beta u'(c_{t+1}) / u'(c_t)\} (W^a_{t+1} + N_{t+1})]. \qquad (7.149)$$

Expression (7.149) is a stochastic difference equation familiar from Chapter 1. A solution is obtained by iterating forward, thus obtaining

$$W^a_t = E_t [(\Sigma_{j=1}^{\infty} \{\beta^j u'(c_{t+j}) / u'(c_t)\} N_{t+j}) + (\lim_{j\to\infty} \{\beta^j u'(c_{t+j}) / u'(c_t)\} W^a_{t+j})]. \qquad (7.150)$$

The end-of-period (net) value of the firm is the sum of two components. The first is the expected present discounted value of all future cash flows, $\Sigma_{j=1}^{\infty}\{\beta^j u'(c_{t+j})/u'(c_t)\}N_{t+j}$, where the discount factor is the stochastic marginal rate of substitution. The second term is a "bubble" term that is guaranteed to approach zero by the individual (and firm as we shall see below) transversality conditions (7.136a,b), or

$$lim_{j\to\infty}\{\beta^j u'(c_{t+j})/u'(c_t)\}W^a_{t+j}=0.$$

We substitute the appropriate solution obtained in (7.150) into (7.146c), and the outstanding market value of the firm is given by

$$W_t = N_t + E_t[\Sigma_{j=1}^{\infty}\{\beta^j u'(c_{t+j})/u'(c_t)\}N_{t+j}]$$
$$= E_t[\Sigma_{j=0}^{\infty}\{\beta^j u'(c_{t+j})/u'(c_t)\}N_{t+j}] \qquad (7.151)$$

or the present discounted value of the expected cash flow. Using expressions (7.139) and (7.143), we can express the cash flow as

$$N_t = \theta_t f(k_t) - w_t - k_{t+1}$$

and substitute into (7.151), to obtain

$$W_t = E_t[\Sigma_{j=0}^{\infty}\{\beta^j u'(c_{t+j})/u'(c_t)\}\{\theta_{t+j}f(k_{t+j}) - w_{t+j} - k_{t+1+j}\}] \qquad (7.152)$$

i.e. the outstanding market value of the firm only depends on the production decisions, or how much to produce and to invest. Expression (7.152) shows that the outstanding market value of the firm does not depend upon the financial decision of the firm in terms of the equity-debt mix issued to finance investment. In the absence of taxes, bankruptcy, asymmetric information, incomplete markets and other forms of distortions, the financial decisions of the firm have no impact upon its market value, the Modigliani-Miller theorem, see Notes on the Literature.

The explicit solution for the value maximization of the firm is as follows. Production managers, separate from financial managers, choose the level of investment to

$$\underset{\{k_{t+j}\}_{j=0}^{\infty}}{Max}\ W_t = E_t[\Sigma_{j=0}^{\infty}\{\beta^j u'(c_{t+j})/u'(c_t)\}\{\theta_{t+j}f(k_{t+j}) - w_{t+j} - k_{t+1+j}\}] \qquad (7.153)$$

where $k_0 > 0$ is given. The first order necessary condition determines a demand for investment where (the sufficient second order condition is satisfied by the strict concavity of the production function)

$$-1 + E_t[\{\beta u'(c_{t+1}) / u'(c_t)\} \theta_{t+1} f'(k_{t+1})] = 0 \qquad (7.154)$$

an expression familiar from the asset pricing models examined in this chapter. Expression (7.154) yields a solution for the demand for investment, which can be appropriately substituted into the aggregate resources constraint to close the general equilibrium.

(iv) The Market Valuation of the Firm in General Equilibrium

One remaining important issue is the extent to which the market valuation of the firm relates to the physical value reflected in the firm's capital stock. We need to make some calculations to show the relationship between market and physical value in this framework.

Substituting expression (7.142) into (7.140), rearranging and using (7.139) yields

$$d_t z_t = \theta_t f'(k_t) k_t - k_{t+1} + q_t (z_{t+1} - z_t) + b_{t+1} - R_t b_t. \qquad (7.155)$$

Expression (7.155) can be updated and inserted into (7.147a) to obtain

$$E_t[\{\beta u'(c_{t+1}) / u'(c_t)\} \{\theta_{t+1} f'(k_{t+1}) k_{t+1} - k_{t+2} + q_{t+1} z_{t+2} + b_{t+2} - R_{t+1} b_{t+1}\}] = q_t z_{t+1}. \qquad (7.156)$$

Next, we multiply condition (7.154) by $k_{t+1} \neq 0$, to obtain

$$E_t[\{\beta u'(c_{t+1}) / u'(c_t)\} \theta_{t+1} f'(k_{t+1}) k_{t+1}] = k_{t+1}. \qquad (7.157)$$

and substitute expressions (7.147b) and (7.157) into (7.156) to obtain the expression

$$E_t[\{\beta u'(c_{t+1})/u'(c_t)\} \{q_{t+1} z_{t+2} + b_{t+2} - k_{t+2}\}] = q_t z_{t+1} + b_{t+1} - k_{t+1}. \qquad (7.158)$$

Expression (7.158) is a stochastic difference equation in the difference between the end-of-period market value, $W^a_t = q_t z_{t+1} + b_{t+1}$, minus investment $I_t = k_{t+1}$. This difference follows a Martingale process, discussed in more detail in Chapter 8, section 8.4. The only stationary solution for this process is

$$W^a_t - I_t = q_t z_{t+1} + b_{t+1} - k_{t+1} = 0 \quad \text{all } t \qquad (7.159a)$$

consistent with the individual transversality conditions (7.136a,b) [see (7.150) as well]. Thus, in general equilibrium we have that

$$W^a_t = I_t \Leftrightarrow q_t z_{t+1} + b_{t+1} = k_{t+1} \qquad all\ t \tag{7.159b}$$

or the outstanding end-of-period market value is equal to the capital stock (investment) available for production, i.e. there is no distinction between the market value and the physical value of the firm.

This is a general equilibrium version of Tobin's Q, where Q is defined as the market value divided by the physical capital value, or

$$Q \equiv (q_t z_{t+1} + b_{t+1})/ k_{t+1} = 1 \qquad all\ t \tag{7.160}$$

The household choices of equity and debt holdings are mapped directly into the choice of the firm's investment. In this case, there is no incentive to buy and sell capital based on $Q \neq 1$. An investment theory based upon variations of Q require some friction absent from this model, see Notes on the Literature.

(v) The Non-Ponzi Game Condition

We shall use the individual's budget constraint (7.131b) to understand a common question that arises in the type of general equilibrium models studied here. Can an individual debt be rolled over forever without ever making a principal payment? A scheme of rolling over debt forever is the so-called Ponzi scheme, see Notes on the Literature. We can easily reinterpret the individual's budget constraint, with equality, as

$$b_{t+1} = R_t b_t + x_t, \qquad b_0\ given \tag{7.161}$$

where $x_t \equiv w_t + z_t (q_t + d_t) - c_t - q_t z_{t+1}$ is the flow of income from wages and equity net of consumption plus new equity holdings. When $b>0$, the individuals hold claims on firms and receive interest payments (for example, the individual holds ATT bonds). When $b<0$, the individual borrows from the firm and make interest payments to the firm (for example, the individual borrows from an ATT credit card). Notice that from the perspective of firms, a parallel definition of x_t is given from expression (7.155) as

$$x_t \equiv d_t z_t - \theta_t f'(k_t) k_t + k_{t+1} - q_t (z_{t+1} - z_t) = k_{t+1} + z_t d_t - q_t (z_{t+1} - z_t) - \pi_t$$

or the flow investment plus dividend payments, $k_{t+1} + z_t d_t$, minus value of new outstanding equity plus profits, $q_t (z_{t+1} - z_t) - \pi_t$. Expression (7.161) is a non-homogeneous and non-autonomous difference equation, whose backward solution, form the initial condition b_0, is given by

$$b_{t+1} = (\Pi_{j=0}{}^{t} R_j)\,[\,b_0 + \Sigma_{n=0}{}^{t}\,(\Pi_{i=0}{}^{n} R_i)^{-1} x_n\,], \qquad b_0 \text{ given.} \qquad (7.162)$$

This solution must satisfy the transversality condition (7.136b), or using (7.134b),

$$\lim_{t\to\infty} E_t\,[\,\beta^{t+1}\,u'(c_{t+1})\,b_{t+1}\,] = 0.$$

Substituting (7.162) into the transversality condition implies that the initial condition must satisfy

$$b_0 + \Sigma_{t=0}{}^{\infty}\,(\Pi_{i=0}{}^{t} R_i)^{-1} x_t = 0 \qquad (7.163)$$

because from the condition (7.134b), the term

$$\beta^{t+1}\,u'(c_{t+1}) R_{t+1} = u'(c_0)(\Pi_{i=1}{}^{t} R_i)^{-1}$$

does not discount b_{t+1}. Substituting (7.163) into (7.162) yields the level of bond holdings consistent with intertemporal solvency as

$$b_{t+1} = (\Pi_{j=0}{}^{t} R_j)\,E_t[\,\Sigma_{n=t+1}{}^{\infty}\,(\Pi_{i=0}{}^{n} R_i)^{-1}\,(-x_n)],\ \text{all } t.\ (7.164)$$

Hence, solution (7.164) avoids a Ponzi scheme: the individual cannot roll over debt forever without making a principal payment at some future date. This level of bond holdings is the current value of the discounted stream of expenditure flows, $-x_t$. In the case of individuals,

$$-x_t \equiv c_t + q_t z_{t+1} - w_t - z_t (q_t + d_t), \quad \text{all } t$$

is the individual expenditure in consumption and new equity net of wage and equity receipts; whereas in the case of firms

$$-x_t \equiv q_t (z_{t+1} - z_t) + \pi_t - k_{t+1} - z_t d_t, \quad \text{all } t$$

is the revenue of new outstanding equity plus profits minus the expenditure in investment plus dividend payments. Of course, for consumers and firms one is the reverse of the other. When $b>0$, the individual is tied to the present value of net expenditures and the firm is tied to the present value of net revenues. When $b<0$, the individual is tied to the present value of net revenues and the firm is tied to the present value of net expenditures. In any case, individuals and firms cannot borrow or lend more than the present value of net revenues or expenditures to satisfy intertemporal solvency.

7.17 **Summary IV**

We examined the capital structure of the firm in general equilibrium under complete markets thus providing a general equilibrium version of the Modigliani-Miller theorem. The rest of this chapter will focus on problems of asymmetric information and the distinction between intertemporal substitution and risk aversion under general equilibrium.

7.18 **Private Information, Stochastic Growth and Asset Prices**

In this part, we extend the dynamic framework to include asymmetric information. We study a simple general equilibrium framework with endogenous growth and potential private information about the productivity of individuals. In particular, growth is driven by a stochastic *"Ak"* type linear technology. The potential private information is about the individual productivity in operating a given stock of capital. The extent of insurance against the idiosyncratic shock is endogenously determined by an efficient long term contract with intermediaries. Thus, we analyze the effects on growth and asset prices of private versus full information of the idiosyncratic shock, in an economy with capital accumulation and aggregate risk.

We consider a one-good model with a large number of individual households. All variables are for each individual unless otherwise noted. Time is discrete and a prime next to a variable denotes its next period value. Figure 7.3, which will be recalled throughout, presents a sketch of the timings and activities in the model. It is important to emphasize that the long term contract characterized in this paper is only contingent on the initial state, but because we derive a separating equilibrium along a stochastic balanced growth path, in this equilibrium, variables evolve sequentially as in Figure 7.3.

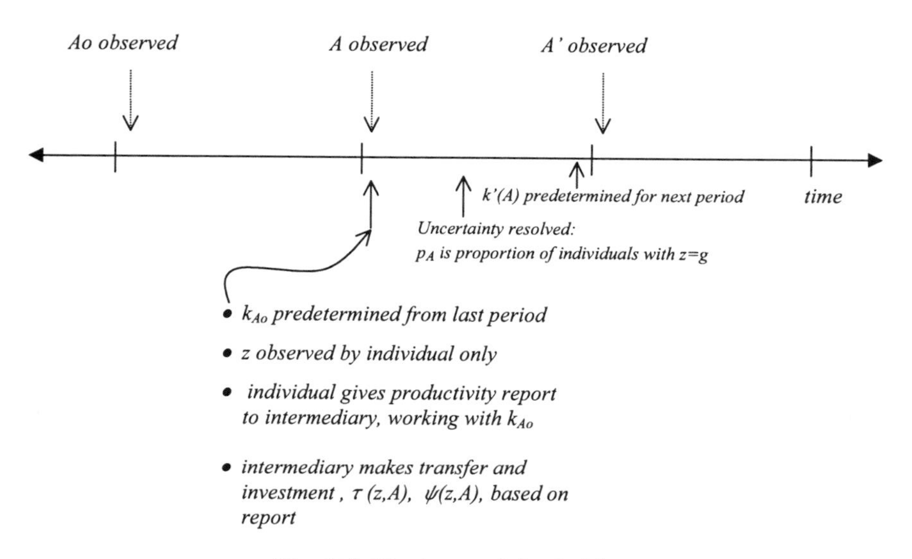

Fig. 7.3 Timing and Activities

We start with the production side. Current output, per quantity of capital, for an individual is denoted by the linear function

$$y(z,A) = z + A \tag{7.165}$$

where $y(z,A)$ is output deflated by the initial level of capital k_{Ao} which is predetermined from last period, as a function of the aggregate state of technology last period A_0, used for production in the current period, e.g. Figure 7.3.

Capital is assumed to fully depreciate every period, i.e. $\delta=1$ in (7.115). A is the current period aggregate state of technology, assumed to be independent and identically distributed, or i.i.d., with probability function

$$A=G \text{ with probability } \pi,$$

$$A=B<G \text{ with probability } 1-\pi$$

where $\pi\in[0,1]$. The unconditional mean of A is assumed to satisfy, $E_A[A]\geq 1$, where E_x is the expectation operator over X. The idiosyncratic component of the technology, z, is assumed to be independently distributed, and may be individual's private information. The probability function of z is

$$z = g \text{ with probability } p_A ,$$

$$z = b < g \text{ with probability } 1 - p_A$$

with $p_A \in [0,1]$. Hence, the probability function of z depends on the aggregate state A and for each A, the probability function may shift. The conditional mean, denoted $\mu(A)$, is

$$E_z[z] = p_A g + (1 - p_A) b \equiv \mu(A) \geq 0.$$

The key issue here is that the effect of aggregate risk on the probability distribution of individual productivity can plausibly make the slope of the function $\mu(A)$ be positive or negative (or zero).

For example, in the good aggregate state, G, the probability of high individual productivity, p_G, may increase when individuals are willing to be more efficient, implying that μ is increasing in A. However, the probability of high individual productivity, p_G, may decrease when individuals are willing to be less efficient, given G, implying alternatively that μ is decreasing in A.

Thus, even though the aggregate risk is systematic, an individual can internalize the aggregate risk, i.e. internalize the externality, so that when p varies negatively with the aggregate state A it allows some hedging of the aggregate risk. However, if p and A are positively correlated, no hedging is possible.

Given the technology (7.165), average aggregate output per individual, deflated by initial capital, is

$$E_{z,A}[z+A] = \pi [\mu(G)+G] + (1-\pi)[\mu(B)+B].$$

Individuals are assumed to be risk averse, with average logarithmic utility

$$v(C) = (1-\beta) \log C \tag{7.166}$$

where $\beta \in [0,1)$ denotes the subjective discount factor assumed to be identical across individuals and C is consumption.

7.19 **Recursive Contracts, General Equilibrium and Asset Prices**

The idiosyncratic component z is the individual's private information so that a revelation mechanism has to be designed, e.g. Chapters 5 and 6.

We proceed by designing a mechanism based on a simple long term principal-agent relationship. There are several risk neutral intermediaries operating competitively and each individual enters a long term relationship with one of the intermediaries.

A typical contract between an individual and an intermediary specifies: (i) A contingent current transfer, per quantity of capital, from the intermediary to the individual denoted by the function $\tau(z,A)$; (ii) Contingent current investment, per quantity of capital, from the intermediary to the individual denoted by the function $\psi(z,A)$ which is identical to the growth factor of the capital stock.

The timing presented in Figure 7.3 shows that all observe the current aggregate state in the beginning of the period. Next, the individual provides a report to the intermediary and the intermediary decides on the appropriate transfer and investment to the individual. Afterwards, uncertainty is resolved and the probability function of the idiosyncratic shock yields the proportion of individuals with respective idiosyncratic components so that the average capital stock is predetermined for next period. Hence, the current contingent expenditure, deflated by the capital stock, for the risk neutral intermediary amounts to

$$\tau(z,A) + \psi(z,A)\,, \quad each\ A\,,\ z. \tag{7.167}$$

Individual current contingent consumption, $c(z,A) \equiv C(z,A,Ao) / k_{Ao}$, then consists of production plus transfers, or

$$c(z,A) = z + A + \tau(z,A)\,, \quad each\ A\,,\ z. \tag{7.168}$$

We let $\mathbb{U}'$ be the current expected discounted lifetime utility entitlement starting from next period onwards, with current full commitment to z, and define a state variable,

$$s' \equiv \mathbb{U}' - \log k'.$$

Considering the lifetime utility entitlement deflated by the capital stock imposes stationarity in the state variable, as it does in all other variables of the balanced growth model, so that $s'=s$. Thus, using the definition of current investment, we have that

$$s'(z',A') = \mathbb{U}'(z,A',A,Ao) - \log k'(z,A,Ao) =$$

$$U'(z,A',A,Ao) - \log \psi(z,A) - \log k_{Ao} = s(z',A') \qquad (7.169)$$

and the linear combination of the lifetime utility entitlement U and $\log k$ is stationary, i.e. s is stationary. The revelation mechanism requires temporary incentive compatibility constraints of the form:

(i) For an individual, currently with $z=b$:

$$\text{if } b + A + \tau(g,B) > 0, \text{ for all } A, \text{ then}$$

$$(1-\beta)\log(b+A+\tau(b,A)) + \beta \log \psi(b,A) + \beta E_{A'}[s'(b,A')] \geq (1-\beta)\times$$

$$\log(b+A+\tau(g,A)) + \beta \log \psi(g,A) + \beta E_{A'}[s'(g,A')], \text{ each } A; \qquad (7.170a)$$

(ii) For an individual with $z=g$:

$$(1-\beta)\log(g+A+\tau(g,A)) + \beta \log \psi(g,A) + \beta E_{A'}[s'(g,A')] \geq (1-\beta)\times$$

$$\log(g+A+\tau(b,A)) + \beta \log \psi(b,A) + \beta E_{A'}[s'(b,A')], \quad \text{each } A. \qquad (7.170b)$$

The constraint in (7.170a) implies that for all A, conditional on the current consumption of individual $z=b$ being strictly positive, when this individual misrepresents, i.e. $b+A+\tau(g,A)>0$, the lifetime utility obtained with truth telling, that is the left-hand side is

$$(1-\beta)\log(b+A+\tau(b,A)) + \beta E_{A'}[U'(b,A',A,Ao)] =$$

$$(1-\beta)\log(b+A+\tau(b,A)) + \beta E_{A'}[s'(b,A')] + \beta \log\psi(b,A)$$

must be no less than the lifetime utility obtained with current misrepresentation and onwards, the right-hand side

$$(1-\beta)\log(b+A+\tau(g,A)) + \beta E_{A'}[s'(g,A')] + \beta \log \psi(g,A).$$

Then, for individual $z=g$, for all A, $g+A+\tau(b,A)>0$ holds by (i), and (7.170b) requires that the lifetime utility obtained with truth telling, i.e. the left-hand side

$$(1-\beta)\log(g+A+\tau(g,A)) + \beta E_{A'}[s'(g,A')] + \beta \log \psi(g,A)$$

must be no less than the lifetime utility obtained with current misrepresentation and onwards, i.e. the right-hand side

$$(1-\beta)\log(g+A+\tau(b,A)) + \beta E_{A'}[s'(b,A')] + \beta \log \psi(b,A).$$

The participation constraint for all individuals is given by

$$s(z,A) \leq E_z[(1-\beta)\log(z+A+\tau(z,A)) + \beta E_{A'}[s'(z',A')] + \beta \log \psi(z,A)],$$

$$\text{each } A \qquad (7.171)$$

which states that for each A, expected (over z) lifetime utility entering the contract can be no less than the initial expected lifetime utility.

In addition, using the law of large numbers, we can average s' across all individuals (z) so that letting the population become large allows the idiosyncratic component of the state variable to vanish in equilibrium, i.e.

$$lim_{i \to \infty} \int_{z'i} s'(z'_i, A') f_{z'i}(z'_i| A')\, dz'_i = s'(A') = s(A'),$$

where $f_{z'i}(z'_i|A')$ is the probability density of the idiosyncratic shocks, conditional on A', across all individuals.

For a characterization of the optimal contract, we use the dual approach: the principal solves an expenditure minimization problem whose solution yields the optimal contract. Hence, using (7.167), the optimal contract is the solution to Bellman's functional equation

$$W(s(A)) = Min\ E_z\,[\, \tau(z,A) + \psi(z,A)\,(\,1 + \rho\, E_{A'}\,[\,W(\,s'(A'))\,]\,)\,]\,, \quad each\ A \qquad (7.172)$$

where the inner minimum is by choice of $\{\tau(z,A),\ \psi(z,A),\ s'(A')\}$ subject to (7.170a,b), (7.171), taking ρ, the discount factor among intermediaries as given, as well as the probability distributions of the shocks.

The solution for the optimal contract must provide the right incentive for each individual to reveal its type truthfully. In principle, deviations from truth telling allow individuals to consume extra hidden output. To avoid this, the optimal contract has the following characteristics: (i) Incentive constraint (7.170a) never binds; (ii) Incentive constraint (7.170b) binds with associated contingent Lagrange multipliers denoted $\lambda(A) \geq 0$; (iii) The participation constraint binds with associated contingent Lagrange multipliers denoted $\eta(A) > 0$. The saddle point for the appropriate Lagrangean function implies first order necessary conditions for $\{\tau(z,A),\ \psi(z,A),\ s'(A'),\ \lambda(A),\ \eta(A)\}$ respectively given by

$$p_A\,(g + A + \tau(g,A)) - (1\text{-}\beta)(\,p_A\,\eta(A) + \lambda(A)) = 0,\ each\ A \qquad (7.173a)$$

$$(1 - p_A) - (1 - \beta)\,(\,[(1\text{-}p_A)\,\eta(A)\,/\,(\,b + A + \tau(b,A))] - [\lambda(A)/(g + A + \tau(b,A))]) = 0\,,\ \ each\ A \qquad (7.173b)$$

$$p_A\psi(g,A)(1 + \rho\, E_{A'}[W(s'(A'))]) - \beta\,(p_A\,\eta(A) + \lambda(A)) = 0,\ \ each\ A \qquad (7.173c)$$

$$(1-p_A)\,\psi(b,A)\,(1+\rho E_{A'}[\,W(s'(A'))\,]) - \beta((1-p_A)\,\eta(A) - \lambda(A)) = 0, \quad \text{each } A \qquad (7.173d)$$

$$p_A\psi(g,A)\,\rho E_{A'}[\partial W(s'(A'))/\partial s'] - \beta(p_A\,\eta(A)+\lambda(A))=0, \quad \text{each } A \qquad (7.173e)$$

$$(1-p_A)\,\psi(b,A)\,\rho E_{A'}[\,\partial W(s'(A'))/\partial s'\,] - \beta((1-p_A)\,\eta(A) - \lambda(A)) = 0, \quad \text{each } A \qquad (7.173f)$$

together with the constraints (7.170b) and (7.171) holding with equality for each A. The set of first order conditions, yield a total of 16 equations in the 16 unknowns $\{\tau(z,A),\ \psi(z,A),\ s'(A'),\ \lambda(A),\ \eta(A)\}$.

The envelope condition (Benveniste-Scheinkman equation) yields:

$$\partial W(s(A))/\partial s = \eta(A), \quad \text{each } A. \qquad (7.174)$$

The Lagrange multiplier $\eta(A)>0$ represents the marginal cost of the initial lifetime utility per unit of capital. The other multiplier $\lambda(A)\geq 0$ plays an important role in the analysis. It represents the marginal cost, in terms of utils, for an individual with $z=g$ to falsely report $z=b$ and receive transfer $\tau(b,A)$.

Hence, $\lambda(A)$ represents the marginal efficiency of the contract. If $\lambda(A)=0$, there is no binding commitment to truth telling and full insurance to the idiosyncratic shock (full risk sharing) is provided by the principal. However, this first best solution does not give any incentive for truth telling when there is private information, so that we observe the usual trade off between risk sharing and incentives. As $\lambda(A)>0$ increases, it gives the value of the contract in terms of the cost of misrepresenting. In particular, the saddle point for the Lagrangean function of (7.172) yields the maximum $\lambda(A)$ that minimizes expenditures. Hence, the larger $\lambda(A)$, the more efficient the contract is in terms of exploring the tradeoff between (partial) risk sharing and incentives.

The optimal contract characterized in (7.173) is a classic separating equilibrium contract; e.g. Chapters 5 and 6. It gives the right incentive for the low productivity individual to reveal truthfully, while making the high productivity individual indifferent. The low productivity individual obtains a small surplus, which induces truth telling, whereas the high productivity individual has no incentive to deviate from truth telling. This is the classic adverse selection problem (Chapters 5 and 6) in a dynamic context.

(i) General Equilibrium and Asset Prices

Perfect competition among intermediaries implies that expenditure will be driven to a minimum or

$$W(s(A)) = Min\ E_z\ [\ \tau(z,A) + \psi(z,A)\ (\ 1 + \rho\, E_{A'}\ [\ W(\ s'(A'))\ ]\ )\ ] = 0\ , \quad each\ A. \qquad (7.175)$$

Any individual with initial capital k, expected lifetime utility U, and marginal product $z+A$, has current consumption, per quantity of capital, given by (7.168), transfer, per quantity of capital, determined by (7.173), investment, per quantity of capital, also determined by (7.173), and output, per quantity of capital, determined by (7.165). Average per capita aggregate quantities can then be computed along the stochastic balanced growth path subject to the economy-wide resources constraint holding for each current aggregate state, that is

$$\int_k E_z\ [\ (z+A)\ k_{Ao} - (z+A+\tau(z,A))\ k_{Ao} - \psi(z,A)\ k_{Ao}\ ]\ f_k\ (\ k|Ao\ )\ dk = 0, \quad each\ A \qquad (7.176)$$

where $f_k(k|Ao)$ is the probability density of the current capital stock, conditional on A_0, across individuals. From expression (7.176), in general equilibrium, ρ has to be such that

$$E_z\ [\tau(z,A) + \psi(z,A)] = 0\ , \quad each\ A \qquad (7.177)$$

or average individual aggregate saving, $E_z[-\tau(z,A)]$, equals average individual investment, $E_z[\psi(z,A)]$. Using (7.175) and (7.177) notice that

$$E_{A'}\ [\ W(\ s'(A'))\ ] = 0 \qquad (7.178)$$

so that it confirms the stationarity of s. The contract is symmetric across all individuals. Using the first order conditions for $\psi(z,A)$ and $s'(A')$, i.e. (7.173c,d,e,f), with (7.174), (7.177) and (7.178) yields

$$\rho = E_{A'}\ [\ 1\ /\ \eta(A')\ ] \qquad (7.179)$$

the risk-free discount factor among intermediaries, which closes the solution for the model.

In fact, ρ is the price of one unit of consumption in every state next period. To see this, note that the intertemporal marginal rate of substitution in the growth framework, here with logarithmic utility, is

$$MRS(z',A') = 1/\psi(z',A'), \quad each\ A',\ z' \tag{7.180}$$

i.e. it only depends on the growth factor not levels. Here, we can explore this property in studying asset prices. From the first order necessary conditions for $\psi(z,A)$, (7.173c,d) using (7.178) yields

$$E_z[\ \psi(z,A)] = \beta\ \eta(A), \quad each\ A \tag{7.181}$$

so that the asset pricing formula for the one period risk-free asset, $\beta E_{z',A'}[MRS(z',A')]$, can be applied to deliver (7.179).

We let $q(z,A)$ denote the price of a claim, among intermediaries, to all future risky dividends from the technology, i.e. the price of equity in this economy. For the logarithmic utility case examined here, it must solve the recursive formula

$$q(z,A) = \beta E_{A',z'}[\ \{1/\psi(z',A')\}\ (\ q(z',A') + y(z',A'))], \quad each\ A,\ z \tag{7.182}$$

which, as in section 7.11, yields a system of $A\times z$ linear equations in the $q(z,A)$ unknowns. We make the assumption that, along the balanced growth path, the price of equity relative to income is constant, or $q(z,A)/y(z,a)=q(z',A')/y(z',A')=q$, constant. Since shocks are i.i.d. and utility logarithmic, the stationary solution is then

$$q(z,A) = q = \beta/(1-\beta), \quad all\ A,\ z \tag{7.183}$$

and the excess return is a function of the discount factor β. Therefore, the price of equity is going to be (7.183) in all economies discussed below, so that we focus on the risk-free asset which relates to the marginal rate of substitution which in turn is only a function of the growth factor.

To sum, the solution is consistent with on going growth of levels, and allocations per quantity of capital and prices, i.e. $\{\tau(z,A),\ \psi(z,A),\ s'(A'),\ \lambda(A),\ \eta(A),\ \rho\}$, all stationary. The optimal contract is offered to all with the right incentives for each type to reveal truthfully, i.e. a separating equilibrium. In particular, the intermediary provides surplus to the low productivity individual to avoid making a larger investment in that individual. On the other hand, the high productivity is indifferent but has no incentive to deviate from truth telling.

7.20 **Growth and Asset Prices with Alternative Arrangements**

We proceed by examining alternative stochastic and informational structures and their impact on asset prices, growth and variability.

(i) Aggregate Risk Only

Consider first the simplest case of no private information in the returns to capital with $z=0$, all z. In particular, there is no discrepancy between aggregate and individual quantities, the typical representative agent framework. As usual, the individual cannot insure against aggregate risk, so that allocations are contingent on the aggregate state. The closed form solution for the model is simple and obtained from the solution of (7.173) with $\lambda(A)=0$, all A, and $z=0$, all z, yielding

$$\psi(A) = \beta A, \quad each\, A \tag{7.184a}$$

$$c(A) = (1 - \beta)\, A, \quad each\, A \tag{7.184b}$$

$$\tau(A) = - \beta\, A, \quad each\, A \tag{7.184c}$$

$$\rho = E_{A'}\,[\, 1 \,/\, A'\,] \tag{7.184d}$$

$$MRS(A') = 1 \,/\, \beta\, A', \quad each\, A'. \tag{7.184e}$$

Thus, we have that

$$\psi(G) > \psi(B)\,, \quad c(G) > c(B)\,, \quad \tau(G) < \tau(B)\,,$$

and ρ (or the *MRS*) depends on the variance of the aggregate disturbance. In equilibrium, consumption, growth (investment), and saving (negative transfers) are larger in the good aggregate state. If the variance of the aggregate shock, A, increases, by Jensen's inequality, ρ increases and the risk-free rate decreases, hence increasing the excess return, i.e. it implies higher variability of the expected marginal rate of substitution. However, aggregate and individual quantities have the same variability as in the representative agent case.

(ii) Idiosyncratic Shocks Only with Full Information (Heterogeneity Only)

Consider the case of no aggregate uncertainty in the returns to capital with $E[A]=A=1$ constant, and let there be no private information of the idiosyncratic shock so that z is fully observed by the intermediary. There is no discrepancy between aggregate and individual quantities as in the representative agent case because the principal, who is risk neutral, bears all the idiosyncratic risk, thus providing full insurance to the risk averse individual. The closed form solution for this economy is obtained from (8) with $\lambda(A)=0$, all A, and $E[A]=A=1$, yielding

$$\psi = \beta\ (\mu + 1) \tag{7.185a}$$

$$c = (1 - \beta)\ (\mu + 1) \tag{7.185b}$$

$$\tau(z)\ = -\beta\ (\ z + 1\ ), \quad each\ z \tag{7.185c}$$

$$\rho = 1 / (\mu + 1) \tag{7.185d}$$

$$MRS = 1 / \beta\ (\mu + 1). \tag{7.185e}$$

where $\mu = p\, g + (1\text{-}p)\ b$. Thus, we have that

$$\psi(g) = \psi(b)\,, \quad c(g) = c(b)\,, \quad \tau(g) < \tau(b)\,,$$

the full insurance (full risk sharing) of idiosyncratic risk solution. In this case, there is a full transfer, $\tau(b) > \tau(g)$, to the low productivity individual to allow equality of consumption and investment across individuals.

(iii) Aggregate and Idiosyncratic Shocks with Full Information

Consider the case of aggregate and idiosyncratic uncertainty, but no private information of the idiosyncratic shock so that z is fully observed by the intermediary. Again, there is no discrepancy between aggregate and individual quantities as in the representative agent case because the principal, who is risk neutral, bears all the risk of the individual uncertainty, thus providing full insurance to the idiosyncratic component of the risk averse individual. However, the individual is not insured against aggregate shocks. The closed form solution for this economy is obtained from (7.173) with $\lambda(A)=0$, all A, yielding

$$\psi(A) = \beta\ (\ \mu\ (A) + A\ ), \quad each\ A \tag{7.186a}$$

$$c(A) = (1 - \beta)\,(\mu(A) + A), \quad each\ A \qquad (7.186b)$$

$$\tau(z,A) = -\beta\,(z + A), \quad each\ A,\ z \qquad (7.186c)$$

$$\rho = E_{A'}[\,1/(\mu(A') + A')\,] \qquad (7.186d)$$

$$MRS(A') = 1/\beta\,(\mu(A') + A'), \quad each\ A'. \qquad (7.186e)$$

Thus, we have that

$$\psi(g,A) = \psi(b,A), \quad c(g,A) = c(b,A), \quad \tau(g,A) < \tau(b,A), \quad each\ A$$

all contingent on the aggregate shock A. Full insurance (full risk sharing) for the idiosyncratic risk is provided by the principal, with full transfer contingent on A.

(iv) Idiosyncratic Shocks Only with Private Information

The three arrangements discussed so far have yielded allocations where the individual quantities are equal to the aggregate per individual quantities due to the provision of full insurance for idiosyncratic shocks.

Consider now a case of no aggregate uncertainty in the returns to capital, or $E[A]=A=1$ constant, with private information of the idiosyncratic shock so that z is not observed by the intermediary. There is discrepancy between aggregate and individual quantities because the principal, who is risk neutral, is not going to bear all the risk of the individual's uncertainty, thus providing only partial insurance to the risk averse individual. The private information requires a revelation mechanism to induce truth telling among heterogeneous individuals. The partial insurance mechanism is endogenously determined by the optimal contract (7.173).

This economy has the empirically appealing property that individual allocations are more variable than aggregate per individual allocations. A closed form solution for this case does not exist. The functional solution obtained from (7.173) with $\lambda>0$ so that the temporary incentive compatibility constraint (7.170b) holds with equality, and $E[A]=A=1$ constant, yields

$$E_z[\psi(z)] = \beta\,\eta \qquad (7.187a)$$

$$E_z[c(z)] = \mu(A) + 1 - \beta\,\eta \qquad (7.187b)$$

$$E_z[\tau(z)] = -\beta\,\eta \qquad (7.187c)$$

$$\rho = \beta E_{z'}[\, 1 / \psi'(z') \,] = 1 / \eta \tag{7.187d}$$

$$MRS = E_{z'}[1 / \psi'(z')\,] = 1 / \beta\, \eta \tag{7.187e}$$

where $\eta > 0$ is the Lagrange multiplier on (7.171) satisfying (7.174). The optimal contract gives

$$\psi(g) > \psi(b), \quad c(g) > c(b), \quad \tau(g) < \tau(b). \tag{7.188}$$

First, the high productivity individual receives a higher investment thus can enjoy higher consumption, and receives a smaller transfer. Hence, we see from (7.187a,b,c)-(7.188) that, a mean-preserving spread of the distribution of the individual shock makes individual quantities more variable than aggregate per individual quantities. Also note that the price of the risk-free asset, ρ (and *MRS*) varies inversely with the marginal cost of lifetime utility, η, and directly with the variance of z, thus improving excess returns. In this case, different individuals have different consumption and investment bundles and the transfer scheme is endogenously partial since the optimal contract provides the right incentive for individuals to reveal their idiosyncratic productivity truthfully. The optimal contract generates a current transfer, in terms of the excess of production over consumption plus investment, from the high productivity to the low productivity, so that

$$b + 1 < c(b) + \psi(b) < c(g) + \psi(g) < g + 1\,, \quad \text{given } A=1 \tag{7.189}$$

Under autarky, each would consume and invest out of its own productivity without net trades and each side of (7.189) would hold with equality; and under full risk sharing the differences would be fully traded so that $c(b)+\psi(b)=c(g)+\psi(g)$.

The mechanism provides the right incentive for the low productivity to reveal truthfully without giving incentive for the high productivity to deviate from truth telling. Hence, the high productivity individual receives higher consumption and investment whereas the low productivity receives lower consumption and investment. In this case, we can show that the marginal efficiency of the contract can be expressed as

$$\lambda = p\,(\,1 - p)\,(\,\psi(g) - \psi(b)\,)\,/\,\beta \tag{7.190}$$

where $p_A = p$, for $A=1$ constant. Thus, the efficiency of the contract increases with the spread of $\psi(z)$, or the variance of z through the term $p(1\text{-}p)$, i.e. the variance of the one trial binomial. In this case, as the variability of z increases, the marginal cost of deviating from truth telling increases and the contract becomes more efficient in partially insuring the increased idiosyncratic risk.

(v) Aggregate and Idiosyncratic Shocks with Private Information

The most general case is the one with aggregate and idiosyncratic uncertainty, and private information of the idiosyncratic shock so that z is not observable by the intermediary. There is discrepancy between aggregate and individual quantities with endogenous partial insurance of the idiosyncratic shock, however, as before, aggregate risk is systematic at the individual level. The closed form solution for this economy does not exist, and the solution from (7.173) with $\lambda(A)>0$, i.e. the temporary incentive compatibility constraint (7.171b) holding with equality, yields

$$E_z[\psi(z,A)] = \beta\eta(A), \quad each\ A \qquad (7.191a)$$

$$E_z[c(z,A)] = \mu(A) + A - \beta\eta(A), \quad each\ A \qquad (7.191b)$$

$$E_z[\tau(z,A)] = -\beta\eta(A), \quad each\ A \qquad (7.191c)$$

$$\rho = \beta E_{z',A'}[1/\psi'(z',A')] = E_{A'}[1/\eta(A')] \qquad (7.191d)$$

$$MRS(A') = E_{z'}[1/\psi'(z',A')] = 1/\beta\eta(A'), \quad each\ A' \qquad (7.191e)$$

where $\eta(A)>0$ is the contingent Lagrange multiplier on (7.171) satisfying (7.174).

First, consider the case where $p_A=p$ for all A. Then, given the probability functions for the aggregate and idiosyncratic shocks, we have that

$$\psi(g,G) > \psi(g,B) \gtreqless \psi(b,G) > \psi(b,B) \qquad (7.192a)$$

$$c(g,G) > c(g,B) \gtreqless c(b,G) > c(b,B) \qquad (7.192b)$$

$$\tau(g,G) < \tau(g,B) \gtreqless \tau(b,G) < \tau(b,B)\,. \qquad (7.192c)$$

Thus, by (7.191)-(7.192), the individual variability is enhanced by the superimposition of the aggregate uncertainty on the idiosyncratic shock relative to the absence of aggregate uncertainty. Again, the optimal contract generates a transfer of current production over consumption plus investment, from the high to the low productivity type, contingent on the aggregate state A, or

$$b + A < c(b,A) + \psi(b,A) < c(g,A) + \psi(g,A) < g + A, \quad each\ A.$$

We can show, using (7.192a), that

$$E_z[\psi(z,G)] - E_z[\psi(z,B)] = \beta(\eta(G) - \eta(B)) > 0, \qquad (7.193)$$

implying that

$$\eta(G) > \eta(B)$$

and the marginal cost of lifetime utility is larger in the good aggregate state relative to the bad aggregate state because there is overall higher consumption and growth in the good aggregate state for all types. However,

$$\lambda(G) - \lambda(B) = (p(1-p)/\beta)(\{\psi(g,G) - \psi(b,G)\} - \{\psi(g,B) - \psi(b,B)\}) \lesseqgtr 0 \qquad (7.194)$$

implies that

$$\lambda(G) \lesseqgtr \lambda(B).$$

Thus, by (7.194), the marginal cost of deviating from truth telling, or the marginal efficiency of the contract may be higher or lower across aggregate states depending on the variability of the idiosyncratic shock across aggregate states. If there is more idiosyncratic variability in the good aggregate state, then $\lambda(G) > \lambda(B)$ and the contract is more efficient in that state, and vice versa. The price of the risk-free asset, ρ (and *MRS*) varies directly with the variance of z and the variance of A. Thus, from the perspective of the excess returns, there is more variability in the *MRS* and thus an improvement in the excess return, relative to the absence of either z or A.

Next, consider the additional effects where p is contingent on A. First, the orderings $\psi(A,g)>\psi(A,b)$, $c(A,g)>c(A,b)$, and $\tau(A,g)<\tau(A,b)$ are preserved for all A. However, depending on how the probability function shifts with changes in A, we can end up with alternative rankings in (7.192). First, examine the case when $p_G > p_B$, or the probability of the high productivity type is larger in the good aggregate state. By expressions (7.192)-(7.194), the individual variability is enhanced by the superimposition of the aggregate uncertainty on the idiosyncratic shock relative to the absence of aggregate uncertainty, or a positive correlation between z and A does not allow for diversification of risk. However, when $p_G<p_B$, the probability of the high productivity type is smaller in the good aggregate state. In this case, an increase in the variance of z, given A, has to take into account the additional effect of A on p_A, which goes in the opposite direction. A negative correlation between z and A allows for some diversification of risk.

Therefore, with aggregate uncertainty, the additional effect dampens the variability of individual quantities enough to make it smaller relative to the case of no aggregate uncertainty. Hence, the main result here is that with aggregate uncertainty, the variability of individual quantities is mitigated. The same is possible for the price of assets in this case. An increase in the variance of z, given A, can decrease the variability in the *MRS* and thus lower the excess return relative to the case of no aggregate uncertainty. In effect, under imperfect risk sharing, a negative correlation between z and A, reduces the variability of individual quantities.

(vi) Comparisons and Simulations

Table 7.4 presents a summary of the results in the alternative arrangements for the expected (over z) growth factor, the discount factor (price of risk-free asset for intermediaries) and the marginal cost of deviations from truth telling or the marginal efficiency of the contract. As seen above, in the case where $p_A=p$ for all A, the expected value of the growth factor with respect to the aggregate shock, $E_{z,A}[\psi\ (z,A)]$, depends on the probability distribution of z and A. But, if p changes with the aggregate shock, then there is the additional channel where the expected growth factor is sensitive to the variability of both z and A. Similarly, for the price of the risk-free asset, it depends on the probability distribution of z and A, and the additional channel if p changes with the aggregate shock. The marginal efficiency of the contract, $\lambda(A)$ depends on the variability of the growth factor, and $p_A(1-p_A\ )$ which is the variability of the one trial binomial for the idiosyncratic shock.

We can show, using the first order conditions (7.173) and the equilibrium condition (7.177) that

Table 7.4
Growth and Prices with Alternative Arrangements Contingent on Aggregate State

	A. AGGREGATE UNCERTAINTY ONLY	*B. IDIOSYNCRATIC SHOCKS ONLY WITH FULL INFORMATION*	*C. AGGREGATE AND IDIOSYNCRATIC SHOCKS WITH FULL INFORMATION*	*D. IDIOSYNCRATIC SHOCKS ONLY WITH PRIVATE INFORMATION*	*E. AGGREGATE AND IDIOSYNCRATIC SHOCKS WITH PRIVATE INFORMATION*
$E_z [\psi(z,A)]$	βA	$\beta \times (E_z[z] + A)$	$\beta \times E_z[z + A]$	$\beta \eta$	$\beta \eta(A)$
ρ	$E_{A'}[A'^{-1}]$	$(E_{z'}[z'] + A')^{-1}$	$E_{A'}[E_{z'}[z' + A']^{-1}]$	η^{-1}	$E_{A'}[\eta(A')^{-1}]$
$\lambda(A)$	0	0	0	$p(1-p) \times (\psi(g) - \psi(b)) \times \beta^{-1}$	$p_A(1-p_A) \times (\psi(g,A) - \psi(b,A)) \times \beta^{-1}$

$$- \lambda(A)(1-\beta)(1- \{(b+A+\tau(b,A))/(g+A+\tau(b,A))\}) = \eta(A) - \mu(A) + A < 0,$$

$$\text{each } A.$$

Hence, we can establish from Table 7.4, columns B. and D. and columns C. and E., that

$$E_z[\psi(z,A)] \mid \textit{full information} > E_z[\psi(z,A)] \mid \textit{private information}, \quad \text{each } A, \tag{7.195a}$$

$$\rho \mid \textit{full information} < \rho \mid \textit{private information} \tag{7.195b}$$

so that, private information reduces the contingent average (over z) growth factor for each aggregate state. Taking into account the additional channel where p depends on A, then

$$E_{z,A}[\psi(z,A)]|\, full\ information > E_{z,A}[\psi(z,A)]|private\ information \quad (7.196a)$$

$$\rho\,|\,full\ information < \rho\,|\,private\ information \quad (7.196b)$$

so that private information reduces the average (over z and A) growth factor and increases the discount factor. However, the key result here is that aggregate uncertainty mitigates the effect of private information, so that

$$E_z[\psi(z,A)]\,|\,full\ information - E_z[\psi(z,A)]\,|\,private\ information \geq$$

$$E_{z,A}[\psi(z,A)]|full\ information - E_{z,A}[\psi(z,A)]|private\ information. \quad (7.197)$$

The gap between expected growth is smaller when there is aggregate uncertainty. Figure 7.4 shows the right-hand side of (7.197) as a function of p_A. Expected growth is in the vertical axis and p_A in the horizontal. The U-shaped thick line represents expected growth with private

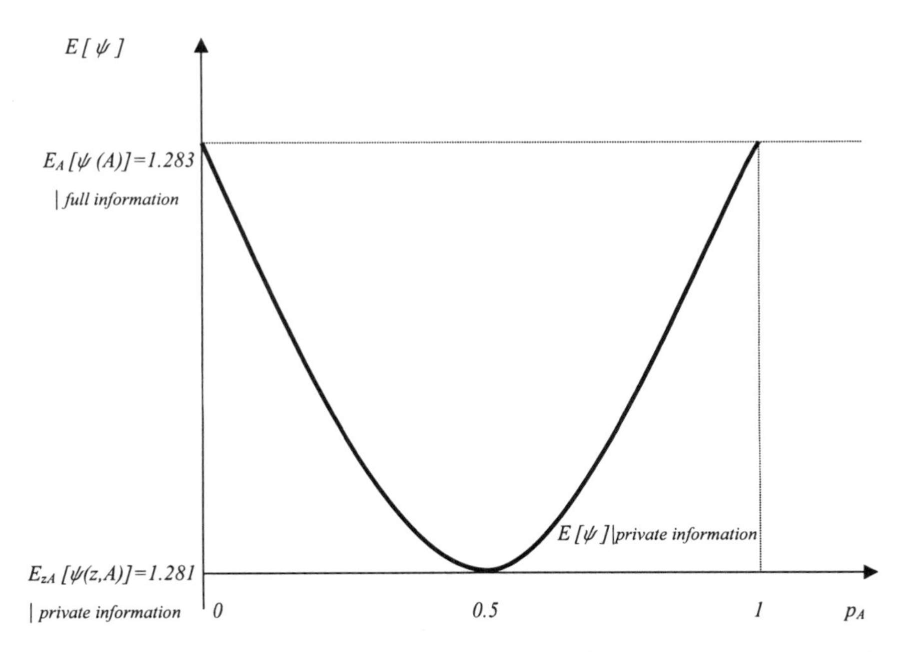

Fig. 7.4 Mitigating Effect of Aggregate Uncertainty on Expected Growth

information, $E_{z,A}[\psi(z,A)]$ | *private information*. At $p_G = p_B = 0.5$, the variance of the idiosyncratic shock is maximum and the difference between private and full information expected growth is the largest. Then, we let the conditional mean of the individual shock depend on the aggregate state A, $\mu(A)$. As $p_G \lesseqgtr 0.5 \lesseqgtr p_B$, the discrepancy between the individual probability across aggregate states widens, the variance of the idiosyncratic shock decreases, and expected growth under private information increases monotonically to the full information value.

The main lessons from Figure 7.4 are: (i) The mitigating effect of aggregate uncertainty is U-shaped in p_A and decreasing in the variance of p_A; (ii) The mitigating effect is never strong enough to reverse the inequality in (7.196a), that private information decreases expected growth. Thus, for either a positive or negative correlation between z and A, the effect of aggregate shocks on p can mitigate, but not reverse, the inefficiency caused by private information on expected growth.

The numerical values in the figure come from Tables 7.5a,b,c, where we present numerical simulations for the case of aggregate and idiosyncratic risk with and without private information. In Table 7.5a, we have the case of maximum variance, $p_G=p_B=0.5$, of the idiosyncratic shock and the discrepancy between expected growth is largest as illustrated in Figure 7.4:

$$E_A[\psi(A)] \mid \textit{full information} = 1.283 >$$

$$E_A[\psi(A)] \mid \textit{private information} = 1.281.$$

In Table 7.5b, we decrease the variance to $p_B=0.2<p_G=0.8$, and the gap between the expected growth decreases to

$$E_A[\psi(A)] \mid \textit{full information} = 1.283 >$$

$$E_A[\psi(A)] \mid \textit{private information} = 1.282.$$

In Table 7.5c, we decrease the variance to $p_B=0.8> p_G=0.2$, and the expected values are symmetric (U-shaped).

One conclusion is that providing public insurance mechanisms for aggregate shocks would be detrimental to expected growth under private information. Insurance to aggregate shocks would counter the mitigating effect of aggregate risk thus leaving agents bearing the negative effect of private information on expected growth.

Comparing expected growth across columns shows that aggregate uncertainty induces more substantive growth effects relative to idiosyncratic uncertainty only (a comparison of the growth factor down each column).

For example, in the case of full information in Table 7.5a, for p_A=*0.5*, comparing columns 1. and 2., we note a change in the growth factor from the bad aggregate state to the good aggregate state of about *18* percentage points, *1.338* minus *1.188*. Examining columns 1. and 2. separately in the case of private information, yields a change in the growth factor across idiosyncratic shocks of approximately *13* percentage points at most, *1.442* minus *1.310*. Therefore, the growth effects due to idiosyncratic shocks only may be "small," but adding aggregate uncertainty has the potential to make the growth effects of private information larger.

Table 7.5a: Simulations:
Aggregate and Idiosyncratic Risk with and without Private Information

$\beta = 0.95$	1. P_A=0.5, G=1.45, G=0.2, B=-0.2, π=0.5	2. P_A=0.5, B=1.25 G=0.2, B=-0.2, π=0.5		
Private Information:				
$c(b,A)$	.0719	.0620	$E_{zA}[c(z,A)]$	.0689
$c(g,A)$	.0759	.0656		
$\lambda(A)$	.0348	.0320		
$\eta(A)$	1.449	1.249		
$\psi(b,A)$	1.310	1.125	$E_{zA}[\psi(z,A)]$	1.281
$\psi(g,A)$	1.442	1.247		
$\tau(b,A)$	-1.178	-.9880	$E_{zA}[\tau(z,A)]$	-1.281
$\tau(g,A)$	-1.574	-1.384		
ρ	.7456	.7456		
Full Information:				
$c(A)$	.0725	.0625	$E_A[c(A)]$	.0675
$\eta(A)$	1.450	1.250		
$\psi(A)$	1.378	1.188	$E_A[\psi(A)]$	1.283
$\tau(b,A)$	-1.178	-.9875	$E_{zA}[\tau(z,A)]$	-1.283
$\tau(g,A)$	-1.578	-1.388		
ρ	.7448	.7448		

Comparing Tables 7.5a,b,c regarding consumption behavior, the variability of individual consumption is slightly larger when $p_G=0.8>p_B=0.2$; and smaller when $p_G=0.2<p_B=0.8$. This confirms that aggregate uncertainty may or may not mitigate the variability of individual quantities. Comparing columns across tables for the case of private information, the variability of individual consumption is larger in the good aggregate state relative to the bad aggregate state.

As a consequence, across all tables $\lambda(G)>\lambda(B)$ and the contract is more efficient in the good aggregate state since there is more variability in that state. In this case, insurance to aggregate shocks can decrease the variability of consumption when $p_G>p_B$ thus making private insurance less efficient.

Table 7.5b: Simulations:
Aggregate and Idiosyncratic Risk with and without Private Information

$\beta = 0.95$	3. $P_B=0.2<P_G=0.8$ $G=1.45$ $G=0.2,$ $B=-0.2,$ $\pi=0.5$	4. $P_B=0.2<P_G=0.8$ $B=1.25$ $G=0.2,$ $B=-0.2,$ $\pi=0.5$		
Private Information:				
$c(b,A)$	.0775	.0563	$E_{zA}[c(z,A)]$	.0684
$c(g,A)$	.0800	.0614		
$\lambda(A)$	.0230	.0196		
$\eta(A)$	1.569	1.129		
$\psi(b,A)$	1.382	1.049	$E_{zA}[\psi(z,A)]$	1.282
$\psi(g,A)$	1.518	1.167		
$\tau(b,A)$	-1.172	-.9937	$E_{zA}[\tau(z,A)]$	-1.282
$\tau(g,A)$	-1.570	-1.389		
ρ	.7615	.7615		
Full Information:				
$c(A)$	.0785	.0565	$E_A[c(A)]$	.0675
$\eta(A)$	1.570	1.130		
$\psi(A)$	1.492	1.074	$E_A[\psi(A)]$	1.283
$\tau(b,A)$	-1.172	-.9935	$E_{zA}[\tau(z,A)]$	-1.283
$\tau(g,A)$	-1.572	-1.394		
ρ	.7609	.7609		

Finally, we look at the row for the discount factor, ρ. First, notice that when $p_G=0.8>p_B=0.2$ (Table 7.5b), the discount factor increases so that the risk-free interest rate decreases improving the excess return. However, when $p_G=0.2<p_B=0.8$ (Table 7.5c), the discount factor decreases so that the risk-free interest rate increases thus reducing the excess return, again confirming that aggregate uncertainty may affect the excess return both ways. In all cases, the inequality in (7.196b) is preserved so that aggregate uncertainty does not reverse the result that private information increases the excess return relative to full information.

7.21 **Summary V**

We argue that aggregate uncertainty is potentially important for the individual decision making process. Idiosyncratic uncertainty alone seems to yield a plausible explanation for the discrepancy in the variability of individual versus aggregate per individual quantities.

Adding aggregate uncertainty provides possible additional channels that can either increase or decrease the variability of individual versus aggregate per individual quantities. The end result is sensitive to the way aggregate uncertainty affects the probability distribution of the idiosyncratic shock, i.e. the sign of the correlation between individual and aggregate risk.

We show that cases where the individual variability may decrease are associated with a probability of high individual productivity being large when the aggregate shock is bad, i.e. the correlation between individual and aggregate risk is negative. We confirm the result that idiosyncratic uncertainty under private information decreases expected growth. We show that aggregate uncertainty can mitigate the effect of idiosyncratic uncertainty but cannot reverse those results. We basically show that effects of private information are sensitive to whether or not aggregate uncertainty is taken fully into account and whether or not aggregate uncertainty affects the probability distribution of idiosyncratic shocks. Thus, aggregate shocks and individual private information may have larger growth effects.

Table 7.5c: Simulations:
Aggregate and Idiosyncratic Risk with and without Private Information

$\beta = 0.95$	5. P_B=0.8>P_G=0.2 G=1.45 G=0.2, B=-0.2, π=0.5	6. P_B=0.8>P_G=0.2 B=1.25 G=0.2, B=-0.2, π=0.5		
Private Information:				
$c(b,A)$	.0663	.0677	E_{zA} [c(z,A)]	.0684
$c(g,A)$	.0718	.0698		
$\lambda(A)$	.0216	.0213		
$\eta(A)$	1.329	1.369		
$\psi(b,A)$	1.237	1.200	E_{zA} [ψ (z,A)]	1.282
$\psi(g,A)$	1.365	1.326		
$\tau(b,A)$	-1.184	-.9823	E_{zA} [τ (z,A)]	-1.282
$\tau(g,A)$	-1.578	-1.380		
ρ	.7414	.7414		
Full Information:				
$c(A)$	.0665	.0685	E_A [c(A)]	.0675
$\eta(A)$	1.330	1.370		
$\psi(A)$	1.264	1.302	E_A [ψ(A)]	1.283
τ (b,A)	-1.184	-.9815	E_{zA} [τ (z,A)]	-1.283
τ (g,A)	-1.584	-1.382		
ρ	.7409	.7409		

Moreover, insurance mechanisms against aggregate shocks would be detrimental to expected growth in the presence of private information. The effects of aggregate and idiosyncratic uncertainty on the risk-free asset price were also examined and they work in the plausible direction of increasing the excess return by decreasing the risk-free return. However, we show that this result is also sensitive to the sign of the correlation between individual and aggregate risk. Further research regarding extensions to the more general isoelastic utility function and issues relating to income distribution is certainly worth pursuing. A more important avenue regards the foundations of the relationship between aggregate shocks and the probability distribution of individual idiosyncratic shocks.

7.22 Risk Aversion, Intertemporal Substitution and Asset Returns

In the discrete time dynamic context of this chapter, we have adopted a discounted welfare function of the VNM expected utility type with time and state separability, as for example in (7.80), or

$$E_t [\Sigma_{j=0}^{\infty} \beta^j \; u(c_{t+j})], \quad \beta \in [0,1)$$

where u is the usual well defined instantaneous utility of consumption c. A widely used functional form used across this book is the power function

$$u(c) = [c^{(1-\gamma)} - 1]/(1-\gamma), \quad \gamma \geq 0 \qquad (7.198)$$

where γ is the coefficient of relative risk aversion *(CRRA)* and $\gamma = 1$ indicates logarithmic utility. However, this specification has the attribute that $1/\gamma$ is exactly the elasticity of intertemporal substitution *(EIS)* in consumption across periods, i.e. *CRRA* is the inverse of the *EIS*. The relationship between those two distinct attributes is undesirable because risk aversion refers to behavior towards timeless gambles, e.g. Chapter 3, and intertemporal substitution refers to behavior towards anticipating or delaying consumption over time.

A plausible characterization of preferences that disentangles *CRRA* from *EIS* must deviate from the traditional VNM time separable framework. In the context of the dynamic asset pricing models presented in this chapter, one characterization that does exactly that is the recursive nonlinear scheme proposed by Larry Epstein and Stanley Zin, and Philippe Weil. Suppose preferences have the general recursive nonlinear form

$$V_t = U(c_t, E_t[V_{t+1}]) \qquad (7.199)$$

where the function U is referred to as an aggregator function. In this case, preferences are not time separable and deviate from the traditional VNM expected utility framework. It is only when the aggregator function U is linear and separable in the expected future value, $E_t[V_{t+1}]$, that (7.199) reduces to the VNM time separable case. For example, if the aggregator function is of the form

$$V_t = u(c_t) + U_2 E_t[V_{t+1}]$$

a simple forward recursion yields

$$V_t = E_t [\Sigma_{j=0}^{\infty} U_2^{\,j} \ u(c_{t+j})],$$

the VNM time separable case, as long as the real number U_2 appropriately discounts.

A common characterization of the nonlinear aggregator function in (7.199) is the isoelastic type

$$\begin{aligned} V_t &= U(c_t, E_t[V_{t+1}]) \\ &= \{(1-\beta)\, c_t^{(1-\rho)} + \beta\,(E_t[V_{t+1}])^{(1-\rho)/(1-\gamma)}\}^{(1-\gamma)/(1-\rho)} \end{aligned} \quad (7.200)$$

where $\gamma \geq 0,\ \gamma \neq 1,$ is the coefficient of relative risk aversion *(CRRA)*, $(1/\rho) \geq 0,\ \rho \neq 1,$ is the elasticity of intertemporal substitution *(EIS)*, and $\beta \in [0,1)$ is the subjective discount factor. For this functional form, the marginal "utilities" are given by

$$\begin{aligned} U_1(c_t, E_t[V_{t+1}]) = V_t (1-\gamma)(1-\beta)\, c_t^{-\rho} / \{(1-\beta)\, c_t^{(1-\rho)} + \\ \beta\,(E_t[V_{t+1}])^{(1-\rho)/(1-\gamma)}\} > 0 \end{aligned} \quad (7.201a)$$

$$\begin{aligned} U_2(c_t, E_t[V_{t+1}]) = V_t \beta\,(E_t[V_{t+1}])^{[(1-\rho)/(1-\gamma)]-1} / \{(1-\beta)\, c_t^{(1-\rho)} + \\ \beta\,(E_t[V_{t+1}])^{(1-\rho)/(1-\gamma)}\} > 0. \end{aligned} \quad (7.201b)$$

When we constrain the parameter space so that

$$\gamma = \rho \Leftrightarrow CRRA = 1/EIS,$$

then the aggregator is linear and separable in the future expected value and we obtain the familiar VNM time separable expected (average) utility

$$V_t = (1-\beta)\, E_t [\Sigma_{j=0}^{\infty} \beta^j \ c_{t+j}^{(1-\gamma)}].$$

When $\gamma \neq \rho \Leftrightarrow CRRA \neq 1/EIS$, we can appropriately distinguish the timeless concept of risk aversion from the dynamic concept of intertemporal substitution. In effect, the more general formulation in (7.199), and the particular parameterization (7.200), allows us to characterize there basic dimensions of preferences: (i) Risk aversion in a timeless gamble context; (ii) Intertemporal substitution in consumption across periods; (iii) Early versus late temporal resolution of risk; where (iii) is intimately related to (i) and (ii). To see this, consider an example of three periods, $t=0,1,2$, where at $t=0$ two lotteries or gambles are proposed to an individual:

Lottery A - with probability *1/2* receive stream of consumption $\{c,c,c\}$; with probability *1/2* receive stream of consumption $\{c,c',c'\}$, $c \neq c'$.
Lottery B - with probability *1* receive $\{c\}$ in period *0;* then gamble in period *1* so that with probability *1/2* receive remaining stream of consumption $\{c,c\}$; with probability *1/2* receive remaining stream of consumption $\{c',c'\}$, $c \neq c'$.

It is easy to note that both lotteries have the same expected value of $2c+c'$. However, lottery A resolves risk earlier at $t=0$ relative to lottery B that resolves risk later at $t=1$. A crucial issue is whether a rational individual is sensitive to the date of risk resolution. The answer is not straightforward because in the utility space, lottery A presents higher expected intertemporal fluctuation in expected utility than lottery B, because lottery B allows one period without risk.

Hence, while early resolution of risk may be beneficial to an individual who is risk averse, it brings about higher expected intertemporal fluctuation in utility, i.e. a tradeoff between timeless risk aversion and intertemporal substitution. In terms of the functional form (7.200), we can elegantly examine the three basic dimensions (i), (ii), and (iii) as follows.

If $\gamma > \rho \Leftrightarrow CRRA > 1/EIS$, risk aversion is greater than the inverse of intertemporal substitution, an individual values more risk aversion than intertemporal substitution in utility, prefers early as opposed to late resolution of risk, and the aggregator function is strictly convex in expected future value, that is

$$U_{22}\left(c_t \, , E_t \left[V_{t+1} \right] \right) = (\gamma - \rho) \, (1-\beta) \, c_t^{\,(1-\rho)} \, U_2 > 0$$

where U_2 is given in (7.201b).

If $\gamma < \rho \Leftrightarrow CRRA < 1/EIS$, risk aversion is smaller than the inverse of intertemporal substitution, an individual values risk aversion less than intertemporal substitution in utility, prefers late as opposed to early resolution of risk, and the aggregator function is strictly concave in expected future value, that is

$$U_{22}\left(c_t \, , E_t \left[V_{t+1} \right] \right) = (\gamma - \rho) \, (1-\beta) \, c_t^{\,(1-\rho)} \, U_2 < 0.$$

If $\gamma = \rho \Leftrightarrow CRRA = 1/EIS$, risk aversion is equal to the inverse of intertemporal substitution and we observe the traditional VNM time separable expected utility case where an individual values risk aversion and intertemporal substitution in utility identically, is indifferent towards the early versus late resolution of risk, and the aggregator function is neither concave nor convex in expected future value, that is

$$U_{22}(c_t, E_t[V_{t+1}]) = (\gamma - \rho)(1-\beta)\, c_t^{(1-\rho)}\, U_2 = 0.$$

We shall put this formulation to work by considering the Lucas model of section 7.10 with the preference structure in (7.199)-(7.200). We will be able to determine the separate effects of risk aversion and intertemporal substitution on asset returns in general. The consumption-saving problem is

$$\underset{\{c_{t+j}, A_{t+j+1}\}_{j=0}^{\infty}}{Max}\; V_t = U(c_t, E_t[V_{t+1}]) = \{(1-\beta)\, c_t^{(1-\rho)} + \beta\, (E_t[V_{t+1}])^{(1-\rho)/(1-\gamma)}\}^{(1-\gamma)/(1-\rho)} \qquad (7.202a)$$

$$\text{subject to} \qquad A_{t+1} = R_{t+1}(A_t - c_t), \qquad (7.202b)$$

$\{A_0\}$ *given and Markov stochastic process for* $\{R_t\}$ *given.*

We will solve this problem in a heuristic manner using the steps of the discounted dynamic programming method used in section 7.10 and Chapter 1. However, you must be aware that the same dynamic programming contraction argument studied in Chapter 1 may not apply generally for the nonlinear preference structure in (7.199)-(7.200), but another contraction argument applies; see Notes on the Literature. As usual, we let the value function $V(A_t, R_t)$ denote the maximum value when the agent holds A_t in wealth and faces gross interest R_t. The state is given by (A_t, R_t) and the control is current saving $(A_t - c_t)$. The value function satisfies the equivalent of Bellman functional equation for preferences (7.199)-(7.200) as

$$V(A_t, R_t) = Max_{\{A_t - c_t\}} \{U(A_t - (A_t - c_t), E_t[V_{t+1}]) \mid A_{t+1} = R_{t+1}(A_t - c_t)\}. \qquad (7.203)$$

The first order necessary conditions for the inner maximum with respect to $(A_t - c_t)$ is

$$- U_1(c_t, E_t[V(A_{t+1}, R_{t+1})]) + U_2(c_t, E_t[V(A_{t+1}, R_{t+1})]) \times E_t[\{\partial V(A_{t+1}, R_{t+1})/\partial A_{t+1}\}\, R_{t+1}] = 0 \qquad (7.204)$$

The equivalent of the Benveniste and Scheinkman formula in this case is

$$\partial V(A_t, R_t)/\partial A_t = U_1(c_t, E_t[V(A_{t+1}, R_{t+1})]) \qquad (7.205)$$

which can be appropriately updated and substituted into the expression (7.204) to give the stochastic Euler equation:

$$E_t[\{U_2(c_t, E_t[V(A_{t+1}, R_{t+1})])U_1(c_{t+1}, E_t[V(A_{t+2}, R_{t+2})])/ \\ U_1(c_t, E_t[V(A_{t+1}, R_{t+1})])\} R_{t+1}] = 1 \qquad (7.206)$$

where the transversality condition (7.85) also applies.

Expression (7.206) is the fundamental asset pricing formula where giving up one unit of consumption today valued at $U_1(c_t, E_t[V(A_{t+1}, R_{t+1})])$, provides on average the next period marginal utility of that unit capitalized by the interest factor and discounted by the marginal utility of the future value, or

$$U_2(c_t, E_t[V(A_{t+1}, R_{t+1})])U_1(c_{t+1}, E_t[V(A_{t+2}, R_{t+2})]) R_{t+1}.$$

The main difference between the formula (7.84) and the more general (7.206) is that the intertemporal marginal rate of substitution in the latter takes the form

$$MRS(.) = U_2(c_t, E_t[V(A_{t+1}, R_{t+1})])U_1(c_{t+1}, E_t[V(A_{t+2}, R_{t+2})])/ \\ U_1(c_t, E_t[V(A_{t+1}, R_{t+1})]) \qquad (7.207)$$

which includes the marginal utility of consumption and the marginal utility of the future value.

The formula (7.206) applies to any traded security in the market place including the risk-free security and the market portfolio. In particular, the problem (7.202a,b) involves holding one risky asset, but can be easily generalized for a portfolio of N assets, say the market portfolio, with gross return

$$R_{M,t+1} = \Sigma_{i=1}^{N} \alpha_{i,t} R_{i,t+1}, \quad \Sigma_{i=1}^{N} \alpha_{i,t} = 1 \qquad (7.208)$$

where the $\alpha_{i,t}$'s are the portfolio weights that can be appropriately chosen separately from the consumption allocation problem. The optimal choices of $\alpha_{i,t}$ satisfy the formula (7.206) with gross return $R_{M,t+1}$.

We proceed by examining the marginal rate of substitution formula (7.207). A solution for the problem involves expressing the MRS as a function of state and control only. First, we substitute (7.201a,b) into (7.207) to obtain

$$MRS(.) = \beta \ (c_{t+1} / c_t)^{-\rho} \ (E_t [\ V_{t+1}] / V_{t+1})^{[(1-\rho)/(1-\gamma)]-1} \quad (7.209)$$

and we note that it requires characterizing the term $E_t[V_{t+1}]/V_{t+1}$ in terms of state and control. Towards that, we apply the guess-and-verify method studied in Chapter 1. From the functional form (7.200), its homogeneity properties and the fact that the accumulation of wealth takes the form (7.202b) and thus includes non-human and human wealth aggregated, we guess that the value function takes the separable form

$$V(A_t, R_t) = \phi (R_t) A_t^{(1-\gamma)} \quad (7.210)$$

where ϕ is a well defined undetermined function of the state. We also propose a consumption function of the form

$$c_t = \mu (R_t) A_t \quad (7.211)$$

where μ is a well defined undetermined function of the state giving the marginal propensity to consume out of current wealth. In general, the marginal propensity to consume is not constant over time, but it may be constant in some special cases. Then, we express the Bellman functional equation using the guess (7.210), the functional form (7.200) and the budget constraint (7.202b) as

$$\phi (R_t) A_t^{(1-\gamma)} = Max_{\{c_t\}} \{(1-\beta) c_t^{(1-\rho)} +$$
$$\beta (E_t [\ \phi (R_{t+1}) R_{t+1}^{(1-\gamma)}] (A_t - c_t)^{(1-\gamma)})^{(1-\rho)/(1-\gamma)}\}^{(1-\gamma)/(1-\rho)} \quad (7.212)$$

with necessary first order condition given by

$$(1-\beta) c_t^{-\rho} - \beta (E_t [\ \phi (R_{t+1}) R_{t+1}^{(1-\gamma)}]^{(1-\rho)/(1-\gamma)} (A_t - c_t)^{-\rho} = 0. \quad (7.213)$$

We can divide (7.213) by $A_t^{-\rho}$ and use (7.211) to obtain the expression

$$(1-\beta) \mu (R_t)^{-\rho} - \beta (E_t [\ \phi (R_{t+1}) R_{t+1}^{(1-\gamma)}]^{(1-\rho)/(1-\gamma)} [1 - \mu (R_t)]^{-\rho} = 0 \quad (7.214a)$$

giving a relationship between the unknown functions ϕ and μ at the optimum. Next, we can insert this last expression (7.214a) into the Bellman equation (7.212) and use (7.211) to obtain

$$\phi (R_t) = \{(1-\beta) c_t^{(1-\rho)} + c_t^{-\rho} (1-\beta) (A_t - c_t)\}^{(1-\gamma)/(1-\rho)} A_t^{-(1-\gamma)} =$$
$$(1-\beta)^{(1-\gamma)/(1-\rho)} \mu (R_t)^{-\rho (1-\gamma)/(1-\rho)} \quad (7.214b)$$

so that (7.214a,b) give two functional equations in the unknown functions ϕ and μ at the optimum.

In principle, we can combine (7.214a,b) to obtain one functional equation in ϕ, the solution of which solves explicitly for the value function (7.210) and the consumption function (7.211).

We proceed here without solving for ϕ explicitly, even though in some special cases a closed form solution exists. The term $E_t\, [\, V_{t+1}\,] \,/\, V_{t+1}$ can be rewritten using the guess (7.210) and the budget constraint (7.202b) as

$$E_t\, [\, V_{t+1}\,] \,/\, V_{t+1} = E_t\, [\, \phi\, (R_{t+1})\, R_{t+1}{}^{(1-\gamma)}] \,/\, \phi\, (R_{t+1})\, R_{t+1}{}^{(1-\gamma)}. \qquad (7.215)$$

When we substitute (7.214a) into (7.214b) and use the budget constraint (7.202b) together with (7.211) to substitute for $1\text{-}\mu\, (R_t)\text{=}A_{t+1}/A_t\, R_{t+1}$, we obtain

$$E_t\, [\, \phi\, (R_{t+1})\, R_{t+1}{}^{(1-\gamma)}] = \phi\, (R_t) \{ \beta^{-1}\, (A_{t+1}/A_t)^{\rho}\, R_{t+1}{}^{-\rho} \}^{(1-\gamma)/(1-\rho)}. \qquad (7.216a)$$

Dividing through by $\phi\, (R_{t+1})\, R_{t+1}{}^{(1-\gamma)}$ yields

$$E_t\, [\, \phi\, (R_{t+1})\, R_{t+1}{}^{(1-\gamma)}] / \phi\, (R_{t+1})\, R_{t+1}{}^{(1-\gamma)} =$$
$$[\phi\, (R_t) / \phi\, (R_{t+1})] \{ \beta^{-1}\, (A_{t+1}/A_t)^{\rho}\, R_{t+1}{}^{-1} \}^{(1-\gamma)/(1-\rho)} \qquad (7.216b)$$

and again substituting (7.214a) into (7.214b), updating one period and dividing yields

$$[\phi\, (R_t) / \phi\, (R_{t+1})] = \{ (c_t / c_{t+1})\, (A_{t+1}/A_t) \}^{-\rho(1-\gamma)/(1-\rho)}$$

which can be inserted into (7.216b) to give

$$E_t\, [\, \phi\, (R_{t+1})\, R_{t+1}{}^{(1-\gamma)}] / \phi\, (R_{t+1})\, R_{t+1}{}^{(1-\gamma)} = \{ \beta^{-1}\, (c_{t+1}/c_t)^{-\rho}\, R_{t+1}{}^{-1} \}^{(1-\gamma)/(1-\rho)}$$
$$(7.217)$$

a function of state and control only. We can then substitute (7.215)-(7.217) into the marginal rate of substitution formula (7.209) to obtain

$$MRS(.) = [\, \beta\, \, (c_{t+1} / c_t)^{-\rho}]^{(1-\gamma)/(1-\rho)}\, (1 / R_{t+1})^{1-[(1-\gamma)/(1-\rho)]} \qquad (7.218)$$

or a geometric weighted average of the marginal rate of substitution in consumption, $[\, \beta\, (c_{t+1} / c_t)^{-\rho}]$, and the market interest discount factor, $(1/R_{t+1})$, where the weights are a function of the degree of risk aversion and intertemporal substitution, $(1 - \gamma) / (1 - \rho)$ and $1 - [(1 - \gamma) / (1 - \rho)]$ respectively. In general, we can express the marginal rate of substitution in terms of the return on the (optimal) market portfolio given in (7.208) and express the stochastic Euler equation in (7.206) as

$$E_t\,[\{[\beta\ (c_{t+1}/c_t)^{-\rho}]^{(1-\gamma)/(1-\rho)}\ (1/R_{M,t+1})^{1-[(1-\gamma)/(1-\rho)]}\}\ R_{i,t+1}] = 1 \quad (7.219)$$

for any asset $i=1,2,...N$, including the risk-free asset; so that for any pair of assets $i\neq j$, including the risk-free asset, we have the excess return as

$$E_t\,[\{[\beta\ (c_{t+1}/c_t)^{-\rho}]^{(1-\gamma)/(1-\rho)}\ (1/R_{M,t+1})^{1-[(1-\gamma)/(1-\rho)]}\}\ (R_{i,t+1} - R_{j,t+1})] = 0. \quad (7.220)$$

When $R_{M,t+1}=R_{i,t+1}=R_{t+1}$ it simplifies to

$$E_t\,[\{\beta\ (c_{t+1}/c_t)^{-\rho}\}^{(1-\gamma)/(1-\rho)}\ R_{t+1}{}^{(1-\gamma)/(1-\rho)}] = 1 \quad (7.221)$$

which also applies to the risk-free asset and any other marketable asset.

The asset pricing formulas (7.219)-(7.221) are revealing. In the special case of VNM time separable expected utility, $\gamma = \rho \Leftrightarrow CRRA = 1/EIS$, we obtain the standard consumption-based capital asset pricing (CCAPM) formula

$$E_t\,[\beta\ (c_{t+1}/c_t)^{-\gamma}\ R_{t+1}] = 1$$

for the power utility (7.298). When $\gamma \neq \rho \Leftrightarrow CRRA \neq 1/EIS$, we can appropriately separate the risk aversion from the intertemporal substitution effects. The general pricing formula in (7.219) takes into account the dynamic consumption discount factor (CCAPM), in the term $[\beta\,(c_{t+1}/c_t)^{-\rho}]^{(1-\gamma)/(1-\rho)}$, and the market discount factor, $(1/R_{M,t+1})^{1-[(1-\gamma)/(1-\rho)]}$ reminiscent of the traditional static capital asset pricing model (SCAPM) or beta model studied in Chapter 2; sometimes called myopic or short term portfolio choice model.

(i) Special Cases of Unit *CRRA* and Unit *EIS*

In general, the characteristics of the solution for the polar cases of unit *CRRA* and unit *EIS* must be examined by computing the appropriate limits of the functional form (7.200).

First we examine the case of unit risk aversion, or $CRRA=\gamma=1$, given an arbitrary value of the $EIS\neq 1$. In this case, it is straightforward to notice that the asset pricing formula in (7.219) reduces to

$$E_t\,[(1/R_{M,t+1})\ R_{i,t+1}] = 1, \quad i=1,2,...N \quad (7.219')$$

independently of the $EIS\neq 1$ and consistent with the SCAPM since the return is driven completely by the covariance of the return of the asset with the return on the market. In order to understand the properties of the consumption function in this case, we take the limit of a monotone affine transformation of the aggregator function (7.200) as

$$lim_{(1-\gamma)\to 0}\, U(c_t, 1+(1-\gamma)(1-\beta)\, E_t[\,V_{t+1}\,])\,/\,(1-\gamma)(1-\beta).$$

Taking logs and using (7.200), we obtain

$$[1/(1-\rho)(1-\beta)]\; lim_{(1-\gamma)\to 0}\, log\,(\,(1-\beta)\, c_t^{\,(1-\rho)} + \beta\,(1+(1-\gamma)(1-\beta)E_t[\,V_{t+1}\,])^{(1-\rho)/(1-\gamma)}\,) \qquad (7.222)$$

which reduces to computing the limit of the term

$$\beta\,(1+(1-\gamma)(1-\beta)E_t[\,V_{t+1}\,])^{(1-\rho)/(1-\gamma)}$$

which can be appropriately written as

$$\beta\, lim_{(1-\rho)/(1-\gamma)\to\infty}\,\{1+[(1-\rho)(1-\beta)/(1-\rho)/(1-\gamma)]E_t[\,V_{t+1}\,])^{(1-\rho)/(1-\gamma)}\,\} = \beta\, exp(\,(1-\rho)(1-\beta)E_t[\,V_{t+1}\,])$$

where *exp* is the exponential operator. Thus, the limit in (7.222) yields

$$V_t|_{\gamma\to 1} = [1/(1-\rho)(1-\beta)]\; log\,(\,(1-\beta)\, c_t^{\,(1-\rho)} + \beta\,(\,exp(\,(1-\rho)(1-\beta)E_t[\,V_{t+1}|_{\gamma\to 1}\,])\,). \qquad (7.223)$$

The guess for the value function in this case is

$$V(A_t, R_t)|_{\gamma\to 1} = \phi\,(R_t)|_{\gamma\to 1} + [(log\, A_t)/(1-\beta)]. \qquad (7.224)$$

The guess in (7.224) can be used together with the budget constraint (7.202b) to compute

$$exp(\,(1-\rho)(1-\beta)E_t[\,V_{t+1}|_{\gamma\to 1}\,]) = (A_t - c_t)^{(1-\rho)}\, exp(\,E_t[\,log\, R_{t+1}^{\,(1-\rho)}] + (1-\rho)(1-\beta)\phi\,(R_{t+1})|_{\gamma\to 1})$$

which can be substituted into the welfare function (7.223) so that Bellman's equation can be written as

$$V_t|_{\gamma\to 1} = Max_{\{c_t\}}\{log(\,(1-\beta)\, c_t^{\,(1-\rho)} + \beta\,(A_t - c_t)^{(1-\rho)}\, exp(\,E_t[\,log\, R_{t+1}^{\,(1-\rho)}] + (1-\rho)(1-\beta)\phi\,(R_{t+1})|_{\gamma\to 1})\,\}. \qquad (7.225)$$

Computing the necessary first order condition and using the consumption function in (7.211) shows the analogous to (7.214a) for the case of unit risk aversion as

$$(1-\beta)\,\mu\,(R_t)^{-\rho}|_{\gamma\to 1} - \beta\,[1-\mu\,(R_t)|_{\gamma\to 1}]^{-\rho}\, exp(\,E_t[\,log\, R_{t+1}^{\,(1-\rho)}] + (1-\rho)(1-\beta)\phi\,(R_{t+1})|_{\gamma\to 1}) = 0 \qquad (7.226)$$

proving that for $CRRA=\gamma=1$, given an arbitrary value of the $EIS\neq 1$, the marginal propensity to consume in the consumption function is not constant, but it is state contingent, i.e. $\mu\,(R_t)|_{\gamma\to 1}$ not constant. The stochastic Euler equation in (7.220) can be computed by substituting (7.226) into the Bellman equation (7.225) as well.

Next, we examine the case of unit elasticity of intertemporal substitution, or $EIS=\rho=1$, given an arbitrary value of the $CRRA\neq 1$. In this case we take logarithms of both sides of (7.200) directly and apply L'Hopital's theorem. Hence,

$$\log V_t = \lim_{(1-\rho)\to 0} (1-\gamma)[(1-\beta)\,c_t^{(1-\rho)} + \beta\,(E_t[\,V_{t+1}\,])^{(1-\rho)/(1-\gamma)}]^{-1} \times$$

$$[(1-\beta)\,c_t^{(1-\rho)} \log c_t + \beta\,(1-\gamma)^{-1}(E_t[\,V_{t+1}\,])^{(1-\rho)/(1-\gamma)} \log (E_t[\,V_{t+1}\,])\,]$$

$$= (1-\gamma)[(1-\beta)\log c_t + \beta\,(1-\gamma)^{-1} \log (E_t[\,V_{t+1}\,])\,], \qquad (7.227)$$

and then the welfare function becomes

$$V_t|_{\rho\to 1} = c_t^{(1-\gamma)(1-\beta)}\,(E_t[\,V_{t+1}|_{\rho\to 1}\,])^{\beta}. \qquad (7.228)$$

In this case the marginal utilities are

$$U_1\,(c_t\,,\,E_t[\,V_{t+1}\,])|_{\rho\to 1} = V_t\,|_{\rho\to 1}(1-\gamma)(1-\beta)\,c_t^{-1} > 0 \qquad (7.229a)$$

$$U_2\,(c_t\,,\,E_t[\,V_{t+1}\,])|_{\rho\to 1} = V_t\,|_{\rho\to 1}\beta\,(E_t[\,V_{t+1}|_{\rho\to 1}])^{-1} > 0. \qquad (7.229b)$$

We can take an identical guess for the value function as in (7.210) and use the budget constraint (7.202b) to write the Bellman equation as [recall (7.212)]

$$\phi\,(R_t)|_{\rho\to 1}\,A_t^{(1-\gamma)} =$$

$$Max_{\{c_t\}}\,\{c_t^{(1-\gamma)(1-\beta)}\,(E_t[\,\phi\,(R_{t+1})|_{\rho\to 1}\,R_{t+1}^{(1-\gamma)}]\,(A_t - c_t)^{(1-\gamma)})^{\beta}\} \qquad (7.230a)$$

with necessary first order condition given by

$$0 = (1-\beta)\,c_t^{(1-\gamma)(1-\beta)-1}\,(A_t - c_t)^{(1-\gamma)\beta} - \beta\,c_t^{(1-\gamma)(1-\beta)}\,(A_t - c_t)^{(1-\gamma)\beta-1}$$

$$= (1-\beta)\,c_t^{-1} - \beta\,(A_t - c_t)^{-1}$$

$$= c_t - (1-\beta)A_t \qquad (7.230b)$$

so that for $EIS=\rho=1$, the marginal propensity to consume out of wealth is constant, i.e.

$$\mu\,(R_t)|_{\rho\to 1} = \mu\,|_{\rho\to 1} = (1-\beta)\ \text{constant}. \qquad (7.230c)$$

It is clear that using (7.230c) with the budget constraint (7.202b) implies that

$$c_{t+1}/c_t = \beta R_{t+1} \qquad (7.231)$$

and consumption growth and the gross rate of return are perfectly correlated.

The asset pricing formula for unit elasticity of intertemporal substitution is obtained as before, that is using (7.229a,b) to compute the marginal rate of substitution as

$$MRS(.)|_{\rho\to 1} = E_t\,[\,V_{t+1}|_{\rho\to 1}\,]^{-1}\, c_{t+1}{}^{(1-\gamma)\,(1-\beta)}\,\beta\, c_{t+1}{}^{-1}\, E_t\,[\,V_{t+1}|_{\rho\to 1}\,]^{\beta}\, c_t$$

$$= \beta\,(c_{t+1}/c_t)^{-1}\, V_{t+1}|_{\rho\to 1}\, E_t\,[\,V_{t+1}|_{\rho\to 1}\,]^{-1} \qquad (7.232a)$$

then using the guess of the value function (7.210) and the consumption function (7.230b,c) to obtain

$$MRS(.)|_{\rho\to 1} = \phi\,(R_{t+1})|_{\rho\to 1}\, R_{t+1}{}^{-\gamma} / E_t\,[\,\phi\,(R_{t+1})|_{\rho\to 1}\, R_{t+1}{}^{(1-\gamma)}]$$

and applying (7.206) and (7.220) in this case implies

$$E_t\,[\phi\,(R_{t+1})|_{\rho\to 1}\, R_{M,t+1}{}^{-\gamma}\,(R_{M,t+1} - R_{i,t+1})] = 0, \quad i=1,2,\dots N$$

so that the marginal rate of substitution simplifies to

$$MRS(.)|_{\rho\to 1} = \phi\,(R_{t+1})|_{\rho\to 1}\, R_{M,t+1}{}^{-\gamma}. \qquad (7.232b)$$

The asset pricing formula for the case of $EIS=\rho=1$ becomes

$$E_t\,[\phi\,(R_{t+1})|_{\rho\to 1}\, R_{M,t+1}{}^{-\gamma} R_{i,t+1}] = E_t\,[\,\phi\,(R_{t+1})|_{\rho\to 1}\, R_{M,t+1}{}^{(1-\gamma)}] = 1,$$

$$i=1,2,\dots N \qquad (7.233)$$

a function of the $CRRA=\gamma\neq 1$, which generally resembles neither the CCAPM nor the SCAPM because of the unknown function $\phi\,(R_{t+1})|_{\rho\to 1}$.

However, given that in this case the consumption-wealth ratio is constant and consumption growth is perfectly correlated with the rate of return, with additional assumptions regarding the probability distribution of the returns some more insights can be gained regarding the compatibility of (7.233) with the CCAPM and the SCAPM as shown below.

Finally, we can examine the case where both the $EIS=\rho=1$ and $CRRA=\gamma=1$. Of course this implies the VNM time separable logarithmic utility so that from (7.218) we have

$$MRS(.)|_{\rho\to 1,\gamma\to 1} = \beta\;(c_{t+1}/c_t)^{-1}$$

and the Euler equation takes the familiar form

$$E_t[\beta\ (c_{t+1}/c_t)^{-1} R_{t+1}] = 1$$

and the consumption function is constant as well.

(ii) Special Cases of Probability Distributions of Returns

First we study the case where all risk is independent and identically distributed (i.i.d.) with constant finite variance. In this case, the Bellman equation (7.212) can be written as

$$\phi(R_t) A_t^{(1-\gamma)} = Max_{\{c_t\}} \{(1-\beta)\ c_t^{(1-\rho)} +$$

$$\beta(E[\phi(R_{t+1}) R_{t+1}^{(1-\gamma)}]\ (A_t - c_t)^{(1-\gamma)})^{(1-\rho)/(1-\gamma)}\}^{(1-\gamma)/(1-\rho)} \qquad (7.212')$$

rendering the consumption choice problem static because

$$E_t[\phi(R_{t+1})R_{t+1}^{(1-\gamma)}]=E[\phi(R_{t+1}) R_{t+1}^{(1-\gamma)}]$$

is constant and uncorrelated with the current state. Consequently, expression (7.214a) becomes

$$(1-\beta)\mu(R_t)^{-\rho} - \beta(E[\phi(R_{t+1})R_{t+1}^{(1-\gamma)}])^{(1-\rho)/(1-\gamma)}[1-\mu(R_t)]^{-\rho} = 0 \qquad (7.214a')$$

and in this case, the marginal propensity to consume out of wealth is constant, i.e.

$$\mu(R_t)|_{i.i.d.} = \mu|_{i.i.d.}\ constant.$$

In turn, by (7.214b) the function ϕ is also constant since

$$\phi(R_t)=\phi|_{i.i.d} = (1-\beta)^{(1-\gamma)/(1-\rho)}\ \mu|_{i.i.d.}^{-\rho(1-\gamma)/(1-\rho)}\ constant. \qquad (7.214b')$$

Substituting (7.214a') into the Bellman equation (7.212') yields a formula for the marginal propensity to consume as

$$\mu|_{i.i.d.} = 1 - \beta^{(1/\rho)}\ (E[R_{t+1}^{(1-\gamma)}])^{(1-\rho)/\rho(1-\gamma)} \qquad (7.234)$$

and using the budget constraint (7.202b) with (7.234), and substituting into the marginal rate of substitution (7.218) yields the formula

$$MRS(.)|_{i.i.d} = R_{t+1}^{-\gamma} / E[R_{t+1}^{(1-\gamma)}].$$

Again, applying (7.206) and (7.220) in this case implies

$$E_t[R_{M,t+1}^{-\gamma}(R_{M,t+1} - R_{i,t+1})] = 0, \qquad i=1,2,...N$$

so that the marginal rate of substitution simplifies to

$$MRS(.)|_{i.i.d} = R_{M,t+1}^{-\gamma}. \qquad (7.235)$$

The asset pricing formula for i.i.d. risk becomes

$$E[R_{M,t+1}^{-\gamma} R_{i,t+1}] = E[R_{M,t+1}^{(1-\gamma)}] = 1, \quad i=1,2,...N \quad (7.236)$$

independently of the *EIS*, which generally resembles the traditional SCAPM. However, given the constant marginal propensity to consume in (7.234), the budget constraint (7.202b) implies that

$$c_{t+1}/c_t = (1-\mu\,|\,_{i.i.d})\,R_{t+1} \quad (7.237)$$

and consumption growth is also i.i.d., exactly proportional to the gross return. The asset pricing formulas (7.236) can be written as

$$E[(c_{t+1}/c_t)^{-\gamma}(1-\mu\,|\,_{i.i.d})^{\gamma} R_{i,t+1}] =$$

$$E[(c_{t+1}/c_t)^{1-\gamma}(1-\mu\,|\,_{i.i.d})^{-(1-\gamma)}] = 1, \quad i=1,2,...N \quad (7.238)$$

thus consistent with the CCAPM. Hence, under i.i.d. risk, both the SCAPM and CCAPM models are compatible and observationally equivalent in effect making the asset pricing formula very much like the traditional VNM time separable case.

Another widely used assumption regarding the probability distribution of random variables in this model is the joint lognormal distribution. We study the lognormal distribution in more detail in Chapter 8, but here it suffices to note that if a random variable X follows a lognormal distribution, then $log\ X \sim N(\chi,\sigma^2)$ and $log\ E_t[X^a] = a\chi + (1/2)\,a^2\sigma^2$ for a real number a. In the general problem (7.202), we can restrict the probability distribution of all returns and consequently of consumption growth to be jointly lognormal. Then, letting $r_{i,t} \approx log\ R_{i,t}$, $r_{M,t} \approx log\ R_{M,t}$, and $\Delta\, c_t = log\ c_{t+1}/c_t$; we assume they are jointly lognormal (and homoskedastic) with mean $\{\underline{r}_i, \underline{r}_M, \underline{\Delta\, c}\}$ and variance-covariance matrix

$$\begin{vmatrix} \sigma_i^2 & \sigma_{iM} & \sigma_{ic} \\ .. & \sigma_M^2 & \sigma_{Mc} \\ .. & .. & \sigma_c^2 \end{vmatrix}$$

where σ_{iM} is the covariance between the return on asset i and the market return, etc. Under the joint lognormal distribution assumption, we can take logarithms of the stochastic Euler equation (7.219), or

$$log\, E_t[\{[\beta\ (c_{t+1}/c_t)^{-\rho}]^{(1-\gamma)/(1-\rho)}\ (1/R_{M,t+1})^{1-[(1-\gamma)/(1-\rho)]}\}\ R_{i,t+1}] = 0,$$

$$i=1,2,...N \quad (7.239)$$

and using the properties of the lognormal distribution, we can rewrite (7.239) as a function of mean, variances and covariances of returns and consumption growth as

$$\{[(1-\gamma)/(1-\beta)]\log\beta - \rho[(1-\gamma)/(1-\rho)]\underline{\Delta c} - (1-[(1-\gamma)/(1-\rho)])\underline{r}_M + \underline{r}_i\} +$$
$$(1/2)\{\rho^2[(1-\gamma)/(1-\rho)]^2\sigma_c^2 + ([(1-\gamma)/(1-\rho)]-1)^2\sigma_M^2 + \sigma_i^2\} +$$
$$\{-\rho[(1-\gamma)/(1-\rho)]\sigma_{ic} + \rho[(1-\gamma)/(1-\rho)](1-[(1-\gamma)/(1-\rho)])\sigma_{Mc} -$$
$$(1-[(1-\gamma)/(1-\rho)])\sigma_{iM}\} = 0. \quad (7.240)$$

In particular, when the ith asset is the risk-free asset, $R_{i,t+1}=R_{f,t+1}$, and $r_{f,t+1} \approx \log R_{f,t+1}$, this expression applies as well:

$$\{[(1-\gamma)/(1-\beta)]\log\beta - \rho[(1-\gamma)/(1-\rho)]\underline{\Delta c} - (1-[(1-\gamma)/(1-\rho)])\underline{r}_M + \underline{r}_f\} +$$
$$(1/2)\{\rho^2[(1-\gamma)/(1-\rho)]^2\sigma_c^2 + ([(1-\gamma)/(1-\rho)]-1)^2\sigma_M^2\} +$$
$$\{\rho[(1-\gamma)/(1-\rho)](1-[(1-\gamma)/(1-\rho)])\sigma_{Mc}\} = 0. \quad (7.241)$$

Subtracting (7.241) from (7.240) we obtain the excess return for any asset $i=1,2,...N$ as

$$\underline{r}_i - \underline{r}_f + (1/2)\sigma_i^2 = \rho[(1-\gamma)/(1-\rho)]\sigma_{ic} + \rho(1-[(1-\gamma)/(1-\rho)])\sigma_{iM} \quad (7.242a)$$

a weighted average of the covariance of the risky asset with consumption growth (CCAPM) and the covariance of the risky asset with the market (SCAPM) as in expression (7.219). Equation (7.242a) also can be written as

$$\log E_t[R_{i,t+1}] - \underline{r}_f = \gamma\sigma_{ic} + [(\gamma-\rho)/(1-\rho)](\sigma_{iM} - \sigma_{ic}) \quad (7.242b)$$

which makes it crystal clear the results obtained in the special cases of unit *CRRA* and unit *EIS*. First, if we have the VNM time separable expected utility framework, $\gamma = \rho \Leftrightarrow CRRA = 1/EIS$, we obtain the standard consumption-based capital asset pricing (CCAPM) formula

$$\log E_t[R_{i,t+1}] - \underline{r}_f = \gamma\sigma_{ic}$$

and the market plays no role. If we have unit risk aversion, or $CRRA=\gamma=1$, given an arbitrary value of the $EIS\neq 1$, (7.242b) reduces to

$$\log E_t[R_{i,t+1}] - \underline{r}_f = \sigma_{iM}$$

independently of the $EIS\neq 1$ and consistent with the SCAPM. In the case of unit elasticity of intertemporal substitution, or $EIS=\rho=1$, given an arbitrary value of the $CRRA\neq 1$, (7.242b) reduces to

$$\sigma_{ic} = \sigma_{iM}$$

and the market return and consumption growth are perfectly correlated; both the CCAPM and the SCAPM hold simultaneously. Of course, when both the $EIS=\rho=1$ and $CRRA=\gamma=1$, it implies the VNM time separable logarithmic utility and

$$\log E_t[R_{i,t+1}] - \underline{r}_f = \sigma_{iM}$$

only the SCAPM holds in this case.

The excess return formula (7.242b) under joint lognormal distribution can be used to derive implications of the effect of risk aversion on the excess return, holding the elasticity of intertemporal substitution constant. We can compute the differential

$$d\{\log E_t[R_{i,t+1}] - \underline{r}_f\}/d\gamma = [1/(1-\rho)]\sigma_{ic} + \{1-[1/(1-\rho)]\}\sigma_{iM} \qquad (7.243)$$

and note that the effect of risk aversion on the excess return is a weighted average of the covariance of the risky asset with consumption and the covariance of the risky asset with the market with respective weights $[1/(1-\rho)]$, $\{1-[1/(1-\rho)]\}$ a function of the given $EIS\neq 1$.

First, the magnitude of the *CRRA* parameter may not be large to generate a large risk premium as long as the covariance with the market is large relative to the covariance with consumption growth. However, consumption growth and the market can be correlated through the budget constraint (7.202b) making this argument less plausible. On the other hand, the risk-free return in (7.241) may not increase much with *CRRA* given $EIS\neq 1$, so that the risk-free rate puzzle is relaxed in this case. It is not clear whether disentangling risk aversion from intertemporal substitution helps resolve the equity premium puzzle; see Notes on the Literature.

7.23 **Summary VI**

In the last section, we presented an extension to dynamic choice with the desirable property that the coefficient of relative risk aversion is separate from the elasticity of intertemporal substitution. The resulting stochastic Euler equation mixes elements of the CCAPM and the SCAPM where the relative weights depend upon the different measures of *CRRA* and

EIS. However, separating *CRRA* from *EIS* may not resolve the traditional equity premium puzzle after all.

Problems

1. Explain what an interest rate swap is.

2. Explain the difference between a perfect hedge and a minimum variance hedge.

3. In section 7.9, what is Arrow's contribution to general equilibrium under uncertainty envisioned by Debreu? Explain.

4. Under what theoretical conditions does a risk premium vanish? Explain.

5. Explain what a Markov process is and its implications for dynamic stochastic general equilibrium models.

6. In Lucas's exchange economy, is the assumption that the discount factor is strictly less than one important? Why or why not? Note: This is a nontrivial question, you may consult Ljungqvist and Sargent (2000), Stokey, Lucas and Prescott (1989), and Chapter 1 for details.

7. What is the difference between the time dimension and the state of nature dimension? Explain.

8. In the model of section 7.22, explain the underlying dynamic contract. Is the problem under consideration a moral hazard, an adverse selection, or both? Explain.

9. For the purposes of asset pricing, is it more plausible to assume unit elasticity of intertemporal substitution, unit relative risk aversion, or both? Explain.

10. In the Lucas exchange economy, explain the role played by a nominal medium of exchange (money).

11. Does the transversality condition play any role in the dynamic stochastic general equilibrium model of the firm in section 7.16?

12. What is the relationship between the model in section 7.16 and Tobin's Q?

Notes on the Literature

Regarding section 7.2, extended material may be found in several low- to medium-level textbooks in finance and investments, for example Bodie et al (1996). Luenberger (1998), Chapter 10 is an excellent reference on forwards, futures, swaps, and minimum variance hedges. Bodie and Merton (2000), Chapter 14 is an accessible treatment of insuring and hedging with futures markets.

Sections 7.4-7.5 follow the seminal contributions of Arrow and Debreu (1954), Debreu (1959) and Arrow (1964, 1965) with discussions available in Farmer (1993), Chapter 8, Ljungqvist and Sargent (2000), Chapter 7, and Backus (1990). Section 7.6 on pricing claims can be found in Sargent (1987), Chapter 3, Ljungqvist and Sargent (2000), Chapter 10, Cochrane (2000), Chapters 1-3, Backus (1990), Duffie (1988, 1996), and it is a prelude to the seminal contribution of Lucas (1978).

In section 7.7, the event tree under complete markets is due to Debreu (1959), Chapter 7. The fundamental theorems of welfare economics are discussed in MasColell et al (1995), Chapter 10; and the risk sharing result in an international finance context can be found in Obstfeld and Rogoff (1996), Chapter 5 and Van Wincoop (1999).

The discussion of section 7.8 can be found in books on stochastic processes, for example Breiman (1986), a more advanced discussion is found in Lucas, Stokey and Prescott (1989), Chapter 8; and also Simon and Blume (1994), Chapter 23. Section 7.9 is based on Arrow (1964, 1965) sequential trade framework. Computational aspects of the Markov assumption are discussed in Cooley (1995), and the empirical implications are found in Campbell, Lo and MacKinlay (1997). Section 7.10 is based on Lucas (1978) seminal exchange economy contribution also found in Stokey, Lucas and Prescott (1989), Sargent (1987), Ljungqvist and Sargent (2000), Cochrane (2000), Duffie (1988, 1996) among others. Krebs (1999) discusses information issues in the exchange economy.

The Efficient Market hypothesis is covered in many articles and books by Eugene Fama, for example Fama (1976, 1991). An excellent survey of the asset pricing literature is in Cochrane and Hansen (1992), Cochrane (2001); and Pliska (1997) presents mathematical finance from a discrete time perspective. Fama (1996) analyzes the potential bias of discounting under uncertainty on expected payoffs. The issue of bubbles is discussed in Tirole (1982) and more recently in Santos and Woodford (1997) and Abreu and Brunnermeier (2000); Garber (1989) is a classic on the historical evidence of bubbles. Blanchard and Weil (1992) discuss dynamic efficiency and bubbles under uncertainty in an overlapping generation framework. The Benveniste and Scheinkman formula is referred from the article by Benveniste and Scheinkman (1979), see Chapter 1.

Section 7.11 on excess returns presents the Mehra and Prescott (1985) equity premium puzzle, the Weil (1989) risk-free rate puzzle, and the interpretation by Hansen and Jaganathan (1991). Abel (1998) provides a unified approach to risk premium in general equilibrium; and Abel (1994) provides directions for the solution of the equity premium puzzle. The issue of the possibility of discount factors exceeding one is discussed in Kocherlakota (1990a). A review of the equity premium literature is found in Kocherlakota (1996), Cochrane (2000) and Campbell(1999). LeRoy (1973) is an early contribution on the effects of risk aversion on asset prices. Barberis et al (1999) present asset pricing analysis with deviations from the Von-Neumann-Morgenstern expected utility framework, focussing on loss aversion as discussed in the appendix of Chapter 3. Bernardo and Judd (2000) present a general analysis of asset prices with informational asymmetries; and Balduzzi and Yao (2000) examine the effects of heterogeneity on asset prices. He and Modest (1995) examine the effects of short sale, borrowing, solvency and trading costs frictions on asset prices. Aiygari (1994) examines the effect of uninsured idiosyncratic risk on aggregate saving. Abel (1990) and Campbell and Cochrane (1999) examine asset pricing with preferences that exhibit habit formation.

Sections 7.13-7.15 on monetary theory is a stochastic version of the Stockman (1981) cash-in-advance paper. The issue of money in the Lucas exchange economy is discussed in Sargent (1987), pages 133-137, and a good discussion of monetary theory in dynamic stochastic general equilibrium models is Lucas (1987). The closed form solution with logarithmic utility and Cobb-Douglas production was first shown in Long and Plosser (1983). The rate of return dominance is discussed in Townsend (1987), and a discussion of monetary volatility is in Bianconi (1992). Ouellete and Paquet (1999) discuss the cost of inflation uncertainty. The Fisher equation under uncertainty is discussed in Sarte (1998). Campbell and Viceira (2000) discuss general asset allocation problems with inflation. Jovanovic and Ueda (1998) present an analysis of asset returns and inflation in an agency framework, and Faig (2000) presents a monetary framework with idiosyncratic and uninsurable returns to capital. Bewley (1986) is a classic contribution that introduces incomplete insurance and liquidity constraints.

The dynamic model of the financial problem of the firm in sections 7.16-7.17 is based on the contribution of Brock and Turnovsky (1981). The discrete time version presented here is also found in Altug and Labadie (1995), Chapter 4. The Modigliani-Miller theorem is in Miller and Modigliani (1958). Deviations from the Modigliani-Miller result in this framework can be attributed to taxes as in Brock and Turnovsky (1981), Osterberg (1989); conflict of interest among shareholders, bondholders managers as in Myers (1977) and Jensen and Meckling (1976); and signaling mechanisms of dividend policy due to asymmetric information as in Ross (1977). Models of investment of firms based on Tobin's Q [due to Tobin (1969)] are found in Hayashi (1982), and Altug and Labadie (1995), Chapter 4 provide a model and further references.

The dynamic model under asymmetric information and adverse selection, in sections 7.18-7.21, is based on Bianconi (2003), and also Khan and Ravikumar (1997), Marcet and Marimon (1992) and Green and Oh (1991). Kim (2003) presents an extension with money. Altug and Labadie (1994), Chapter 7, provide a useful survey of this literature. Spear and Srivastava (1987) present a seminal contribution on dynamic moral hazard, Marcet and Marimon (1998) present a unified theory on dynamic contracts, and Chiappori et al (1994) analyze dynamic credit markets with moral hazard. Laffont and Martimort (2002), Chapters 6 and 7 review dynamic principal-agent models. Barro and Sala-i-Martin (1995) provide a recent survey on economic growth; and Phelan (1994) provides an analysis of contracts with aggregate uncertainty in an overlapping generations framework. Long and Shimomura (1999) provide an analysis of the effect of moral hazard on economic growth also using an overlapping generations model; Bose and Cothren (1996, 1997) analyze asymmetric information in the growth model; and Sussman (1999) analyze the effects of standardized contracts on economic growth. Lehnert et al (1997) examines the effects of liquidity constraints on contractual arrangements.

The well-known failure of the representative agent paradigm to cope with the fact that the variability of individual consumption and income is much larger than the variability of aggregate per individual consumption and income is documented in Deaton (1991, 1992), Pischke (1995). Models of individual heterogeneity, which include discrepancy between individual and aggregate allocations, can be found in Phelan (1994) and in the survey of Rios-Rull (1995). See also the discussion of Carroll (2000), the applications by Kahn (1990), Atkeson and Lucas (1992), Heaton and Lucas (1992), Den Haan (1996, 1997), Kocherlakota (1998), and Krusell and Smith (1998) among others. Goodfriend (1992) argues that individuals are not able to observe the current aggregate state, but only with a time lag. Brock (1982) and Cochrane (1991) are good references for discussions of asset prices in a production framework.

The point that the marginal rate of substitution depends on growth is emphasized in Mehra and Prescott (1985) for an endowment economy, section 7.11, Hansen and Jagannathan (1991) examine the role of the marginal rate of substitution in asset pricing models. The discussion that adding private information in the general equilibrium asset pricing framework may increase the variability of the marginal rate of substitution thus increasing the excess return is also in Heaton and Lucas (1992). The homogeneity properties of the functions in the model are discussed in Alvarez and Stokey (1998).

The mechanism design based on a simple long term principal-agent relationship is by Townsend (1982), and also the recent review of recursive contracts in Ljungqvist and Sargent (2000), Chapter 15 is recommended. For a recent discussion on the issue of the correlation between investment and growth in the *"Ak"* type of model, see McGrattan (1998). For a more general case of isoelastic preferences, the composite state variable would be $U - (1-\sigma) \log k$, where σ is the coefficient of relative risk aversion as shown in Khan and Ravikumar (1997). The paper by Den Haan (1997) discusses related computational and calibration issues in general equilibrium models with heterogeneous agents and aggregate shocks. The recursive contract structure used is by Green (1987), other papers followed in this tradition including Thomas and Worrall (1990), Green and Oh (1991), Marcet and Marimon (1992), and Khan and Ravikumar (1997). For a characterization of the optimal contract, we use the dual approach of Green (1987). The procedure of imposing an aggregate resources constraint to obtain the market interest rate is due to Atkeson and Lucas (1992); they also discuss the possibility of decentralizing efficient allocations using securities trade in the case of private information.

Section 7.20-(i) is the Brock and Mirman (1972) economy with the *"Ak"* technology of Rebelo (1991). Sections 7.20-(ii)-(iii) are due to Khan and Ravikumar (1997) and to a certain extent Marcet and Marimon (1992) and Green and Oh (1991). Taub (1997) presents an analysis of dynamic agency. Benabou and Ok (1998) is an interesting application of risk analysis to social mobility and income distribution.

Sections 7.22-7.23 follow the papers by Epstein and Zin (1989, 1991), Weil (1989, 1990), and in particular Giovannini and Weil (1989), in adopting a framework of preferences proposed by Kreps and Porteus (1978). The review by Epstein (1992) is also useful and competent. Many other authors provide an analysis of asset pricing when risk aversion and intertemporal substitution are detached including Campbell and Viceira (2000), Campbell (2002), Restoy and Weil (1998). Epstein and Zin (1989) present a discussion of the dynamic programming algorithm for this case. Hansen and Singleton (1983) have presented the joint lognormal case and Kocherlakota (1990b) has presented the i.i.d results; both cases are also analyzed in Giovannini and Weil (1989). Tallarini (2000) provides a quantitative assessment of changes in risk aversion with intertemporal substitution held constant with Epstein-Zin-Weil preferences, and study particular cases suggested by Hansen and Sargent (1995), see also Hansen and Sargent (2001) for a discussion.

The manuscript by Campbell and Viceira (2000) is an excellent presentation of general asset allocation and the case of separate risk aversion and intertemporal substitution. Campbell (2002) concludes that separating risk aversion from intertemporal substitution in the homoskedastic case presented here cannot solve the equity premium puzzle in part because consumption and wealth are correlated through the intertemporal budget constraint, see also Restoy and Weil (1998). Vissing-Jorgensen and Attanasio (2002) study the role of stock market participation with separate risk aversion and intertemporal substitution.

References

Abel, Andrew B. (1994) "Exact Solutions for Expected Rates of Return Under Markov Regime Switching: Implications for the Equity Premium Puzzle." *Journal of Money, Credit and Banking,* 26, 345-361.

Abel, Andrew B. (1998) "Risk Premia and Term Premia in General Equilibrium." NBER Working Papers Series No. 6683, August.

Abreu, Dilip and Markus K. Brunnermeier (2000) "Bubbles and Crashes." Working paper, Department of Economics, Princeton University, April.

Aiygari, S. Rao (1994) "Uninsured Idiosyncratic Risk and Aggregate Saving." *Quarterly Journal of Economics,* 659-684.

Altug, Sumru and Pamela Labadie (1994) *Dynamic Choice and Asset Markets.* Academic Press, San Diego, CA.

Alvarez, Fernando and Nancy Stokey (1998) "Dynamic Programming with Homogeneous Functions." *Journal of Economic Theory,* 82, 167-189.

Arrow, Kenneth J. (1964) "The Role of Securities in the Optimal Allocation of Risk Bearing." *Review of Economic Studies,* 31, 91-96.

Arrow, Kenneth J. (1965) *Aspects of the Theory of Risk Bearing*. Yrjo Jahnssonin Saatio, Helsinki.

Arrow, Kenneth J. and Gerard Debreu (1954) "Existence of Equilibrium for a Competitive Economy." *Econometrica*, 22, 265-290.

Atkeson, Andrew and Robert E. Lucas (1992) "On Efficient Distribution with Private Information." *Review of Economic Studies*, 59, 427-453.

Backus, David (1990) *Notes on Dynamic Macroeconomic Theory.* Unpublished class notes, Stern School, New York University.

Balduzzi, Pierluigi and Tong Yao (2000) "Does Heterogeneity Matter for Asset Pricing?" Working paper, Carroll School of Management, Boston College, May.

Barberis, Nicholas, Ming Huang and Tano Santos (1999) "Prospect Theory and Asset Prices." Working paper, University of Chicago Business School, November.

Barro, Robert and Xavier Sala-i-Martin (1995) *Economic Growth.* McGraw Hill Book Co., New York, NY.

Benabou, Roland and Efe A. Ok (1998) "Social Mobility and the Demand for Redistribution: The POUM Hypothesis." NBER Working Papers Series No. 6795, November.

Benveniste, Lawrence M. and Jose A. Scheinkman (1979) "On the Differentiability of the Value Function in Dynamic Models of Economics." *Econometrica*, 47, 727-732.

Bernardo, Antonio and Kenneth L. Judd (2000) "Asset Market Equilibrium with General Tastes, Returns, and Informational Asymmetries." *Journal of Financial Markets,* 3, 17-43.

Bewley, Truman (1986) "Stationary Monetary Equilibrium with a Continuum of Independently Fluctuating Consumers." In Hildenbrad, W. and A. Mas-Collel, Editors: *Contributions to Mathematical Economics in Honor of Gerard Debreu*, 79-102. North-Holland, Amsterdam.

Bianconi, Marcelo (1992) "Monetary Growth Innovations in a Cash-in-advance Asset Pricing Model." *European Economic Review*, 36, 1501-1521.

Bianconi, Marcelo (2003) "Private Information, Growth and Asset Prices with Stochastic Disturbances." *International Review of Economics and Finance*, 12, 1-24.

Blanchard, Olivier J. and Phillipe Weil (1992) "Dynamic Efficiency, the Riskless Rate and Debt Ponzi Games Under Uncertainty." NBER Working Papers Series No. 3992, February.

Bodie, Zvi and Robert C. Merton (2000) *Finance*. Prentice Hall, New York, NY.

Bodie, Zvi, Alex Kane, Alan J. Marcus (1996) *Investments*. Irwin Publishers, Chicago, IL.

Bose, Niloy and Richard Cothren (1996) "Equilibrium Loan Contracts and Endogenous Growth in the Presence of Asymmetric Information." *Journal of Monetary Economics,* 38, 363-376.

Bose, Niloy and Richard Cothren (1997) "Asymmetric Information and Loan Contracts in the Neoclassical Growth Model." *Journal of Money, Credit and Banking,* 29, 423-439.

Breiman, Leo (1986) *Probability and Stochastic Processes*. The Scientific Press, Palo Alto, CA.

Brock, William A. (1982) "Asset Prices in a Production Economy." McCall, J. J. (Ed.) *The Economics of Information and Uncertainty*. The University of Chicago Press for the NBER, IL.

Brock, William A. and Leonard J. Mirman (1972) "Optimal Economic Growth and Uncertainty: The Discounted Case." *Journal of Economic Theory*, 4, 479-513.

Brock, William A. and Stephen J. Turnovsky (1981) "The Analysis of Macroeconomic Policies in Perfect Foresight Equilibrium." *International Economic Review*, 22, 179-209.

Campbell, John Y. (2002) "Consumption-Based Asset Pricing." Working paper, Harvard University, July.

Campbell, John Y. (1999) "Asset Prices, Consumption and the Business Cycle." In Taylor, John and Michael Woodford, Eds. *Handbook of Macroeconomics*, *Vol. 1*. North-Holland, Amsterdam.

Campbell, John Y. and John H. Cochrane (1999) "By Force of Habit: A Consumption-Based Explanation of Aggregate Stock Market Behavior." *Journal of Political Economy*, 107, 205-251.

Campbell, John Y. and Luis M. Viceira (2000) *Strategic Asset Allocation.* Book manuscript, Harvard University, November. (Published by Oxford University Press, 2002).

Campbell, John Y., Andrew W. Lo, A. Craig MacKinlay (1997) *The Econometrics of Financial Markets*. Princeton University Press, Princeton, NJ.

Carroll, Christopher (2000) "Requiem for the Representative Consumer? Aggregate Implications of Microeconomic Consumption Behavior." *American Economic Review Papers and Proceedings,* 90, 110-115.

Chiappori, Pierre-A., Ines Macho, Phillipe Rey, and Bernard Salanie (1994) "Repeated Moral Hazard: The Role of Memory, Commitment, and the Access to Credit Markets." *European Economic Review,* 38, 1527-1553.

Cochrane, John H. (1991) "Production-Based Asset Pricing and the Link Between Stock Returns and Economic Fluctuations." *Journal of Finance,* 46, 209-237.

Cochrane, John H. (2001) *Asset Pricing*. Princeton University Press, Princeton , NJ.

Cochrane, John H. and Lars P. Hansen (1992) "Asset Pricing Explorations for Macroeconomics." *NBER Macroeconomics Annual 1992,* 115-182.

Cooley, Thomas, Ed. (1995) *Frontiers of Business Cycles Research*. Princeton University Press, Princeton, NJ.

Deaton, Angus (1991) "Saving and Liquidity Constraints." *Econometrica*, 59, 1221-1248.

Deaton, Angus (1992) *Understanding Consumption.* Claredon Lectures in Economics, Oxford University Press, Oxford, UK.

Debreu, Gerard (1959) *Theory of Value*. Cowles Foundation, Yale University, New Haven, CT.

Den Haan, Wouter J. (1996) "Heterogeneity, Aggregate Uncertainty and the Short term Interest Rate." *Journal of Business and Economic Statistics*, 14, 399-411.

Den Haan, Wouter J. (1997) "Solving Dynamic Models with Aggregate Shocks and Heterogeneous Agents." Working paper, Department of Economics, UC San Diego, February.

Duffie, Darrell (1988) *Security Markets: Stochastic Models*. Academic Press, Boston, MA.

Duffie, Darrell (1996) *Dynamic Asset Pricing Theory*, Second Edition. Princeton University Press, Princeton, NJ.

Epstein, Larry (1992) "Behavior Under Risk: Recent Developments in Theory and Applications." In Laffont, Jean-J., Ed. *Advances in Economic Theory*, 1-63. Cambridge University Press, Cambridge.

Epstein, Larry and Stanley Zin (1989) "Substitution, Risk Aversion and the Temporal Behavior of Asset Returns: A Theoretical Framework." *Econometrica*, 57, 937-969.

Epstein, Larry and Stanley Zin (1991) "Substitution, Risk Aversion and the Temporal Behavior of Asset Returns: An Empirical Analysis." *Journal of Political Economy*, 99, 263-286.

Faig, Miguel (2000) "Money with Idiosyncratic Uninsurable Returns to Capital." Working paper, Department of Economics, University of Toronto, March. Forthcoming *Journal of Economic Theory.*

Fama, Eugene (1976) *Foundations of Finance*. Basic Books, New York, NY.

Fama, Eugene (1991) "Efficient Capital Markets II." *Journal of Finance*, 46, 1575-1618.

Fama, Eugene (1996) "Discounting Under Uncertainty." *Journal of Business,* 69, 415-428.

Farmer, Roger (1993) *The Macroeconomics of Self-Fulfilling Prophecies*. The MIT Press, Cambridge, MA.

Garber, Peter (1989) "Tulipmania." *Journal of Political Economy,* 97, 535-560.

Giovannini, Alberto and Phillipe Weil (1989) "Risk Aversion and Intertemporal Substitution in the Capital Asset Pricing Model." NBER Working Papers Series No. 2824, January.

Goodfriend, Marvin (1992) "Information-Aggregation Bias." *American Economic Review,* 82, 508-519.

Green, Edward J. (1987) "Lending and the Smoothing of Uninsurable Income." In Prescott, E. and N. Wallace (Eds.), *Contractual Arrangements for Intertemporal Trade*. University of Minnesota Press, MN.

Green, Edward J. and Soo-Nam Oh (1991) "Contracts, Constraints and Consumption." *Review of Economic Studies*, 58, 883-899.

Hansen, Lars P. and Thomas J. Sargent (1995) "Discounted Linear Exponential Gaussian Control." *IEEE Transactions on Automatic Control*, 40, 968-971.

Hansen, Lars P. and Thomas J. Sargent (2001) "Time Inconsistency of Robust Control?" Working paper, University of Chicago and Stanford University, October.

Hansen, Lars P. and Kenneth Singleton (1983) "Stochastic Consumption, Risk Aversion and the Temporal Behavior of Asset Returns." *Journal of Political Economy*, 91, 249-268.

Hansen, Lars P. and Ravi Jagannathan (1991) "Implications of Security Markets Data for Models of Dynamic Economics." *Journal of Political Economy,* 99, 225-262.

Hayashi, Fumio (1982) "Tobin's Marginal q and Average q: A Neoclassical Interpretation." *Econometrica,* 50, 213-224.

He, Hua and David M. Modest (1995) "Market Frictions and Consumption-Based Asset Pricing." *Journal of Political Economy,* 103, 94-117.

Heaton, John and Deborah Lucas (1992) "The Effects of Incomplete Insurance Markets and Trading Costs in a Consumption-Based Asset Pricing Model." *Journal of Economic Dynamics and Control,* 16, 601-620.

Jensen, Michael and William Meckeling (1976) "The Theory of the Firm: Managerial Behavior, Agency Costs and Ownership Structure." *Journal of Financial Economics,* 3, 305-360.

Jovanovic, Boyan and Masako Ueda (1998) "Stock Returns and Inflation in a Principal-Agent Economy." *Journal of Economic Theory,* 82, 223-247.

Kahn, James A. (1990) "Moral Hazard, Imperfect Risk Sharing, and the Behavior of Asset Returns." *Journal of Monetary Economics,* 26, 27-44.

Khan, Aubhik and Bashkar Ravikumar (1997) “Growth and Risk Sharing with Private Information.” Working paper, Department of Economics, University of Virginia, September. (Published, *Journal of Monetary Economics*, 47, 499-521, 2001).

Kim, Young Sik (2003) "Money, Growth and Risk Sharing with Private Information." *Review of Economic Dynamics*, 6, 276-299.

Kocherlakota, Narayana (1990a) "On the 'Discount' Factor in Growth Economies." *Journal of Monetary Economics*, 25, 43-47.

Kocherlakota, Narayana (1990b) "Disentangling the Coefficient of Relative Risk Aversion from the Elasticity of Intertemporal Substitution: An Irrelevance Result." *Journal of Finance* 45, 175-190.

Kocherlakota, Narayana (1996) "The Equity Premium: It's Still a Puzzle." *Journal of Economic Literature,* 34, 42-71.

Kocherlakota, Narayana (1998) “The Effects of Moral Hazard on Asset Prices when Financial Markets are Complete.” *Journal of Monetary Economics*, 41, 39-56.

Krebs, Thomas (1999) "Information and Asset Prices in Complete Markets Exchange Economies." *Economics Letters,* 65, 75-83.

Kreps, David M. and Evan L. Porteus (1978) "Temporal Resolution of Uncertainty and Dynamic Choice Theory." *Econometrica*, 46, 185-200.

Krusell, Per and Anthony Smith (1998) "Income and Wealth Heterogeneity in the Macroeconomy." *Journal of Political Economy*, 106, 867-896.

Laffont, Jean-J. and David Martimort (2002) *The Theory of Incentives: The Principal-Agent Model*. Princeton University Press, Princeton, NJ.

Lehnert, Andreas, Ethan Ligon, and Robert M. Townsend (1997) "Liquidity Constraints and Incentive Contracts." Working paper, Department of Economics, University of Chicago, November.

LeRoy, Stephen (1973) "Risk Aversion and the Martingale Property of Stock Prices." *International Economic Review,* 14, 436-446.

Ljungqvist, Lars and Thomas J. Sargent (2000) *Recursive Macroeconomic Theory*. The MIT Press, Cambridge, MA.

Long, John and Charles Plosser (1983) "Real Business Cycles." *Journal of Political Economy*, 91, 39-69.

Long, Ngo V. and Kazuo Shimomura (1999) "Education, Moral Hazard and Endogenous Growth." *Journal of Economic Dynamics and Control,* 23, 675-698.

Lucas, Robert E. (1978) "Asset Prices in an Exchange Economy." *Econometrica*, 46, 1426-1445.

Lucas, Robert E. (1987) *Models of Business Cycles*. Yrjo Jahansson Lectures Series. Blackwell, London.

Luenberger, David (1998) *Investment Science*. Oxford University Press, Oxford, UK.

Marcet, Albert and Ramon Marimon (1992) “Communication, Commitment and Growth.” *Journal of Economic Theory*, 58, 219-249.

Marcet, Albert and Ramon Marimon (1998) "Recursive Contracts." Working paper, European University Institute, Florence, October.

MasColell, Andreu, Michael Whinston and Jerry Green (1995) *Microeconomic Theory*. Oxford University Press, Oxford, UK.

McGrattan, Ellen (1998) "A Defense of *AK* Growth Models." *Federal Reserve Bank of Minneapolis Quarterly Review,* Fall, 13-27.

Mehra, Rajnish and Edward Prescott (1985) "The Equity Premium: A Puzzle." *Journal of Monetary Economics*, 15, 145-161.

Miller, Merton and Franco Modigliani (1958) "The Cost of Capital, Corporate Finance and the Theory of Investment." *American Economic Review,* 48, 261-297.

Myers, Stuart (1977) "Determinants of Corporate Borrowing." *Journal of Financial Economics,* 5, 147-175.

Obstfeld, Maurice and Kenneth Rogoff (1995) *Foundations of International Macroeconomics*. The MIT Press, Cambridge, MA.

Osterberg, William P. (1989) "Tobin's *q*, Investment and the Optimal Financial Structure." *Journal of Public Economics*, 40, 293-318.

Ouellette, Pierre and Alain Paquet (1999) "Inflation Uncertainty and the Cost of Inflation." *Canadian Journal of Economics,* 32, 195-204.

Phelan, Christopher (1994) "Incentives and Aggregate Shocks." *Review of Economic Studies*, 61, 681-700.

Pischke, Jorn-Steffen (1995) "Individual Income, Incomplete Information and Aggregate Consumption." *Econometrica*, 63, 805-840.

Pliska, Stanley R. (1997) *Introduction to Mathematical Finance.*Basil Blackwell Publishers, New York, NY.

Rebelo, Sergio (1991) "Long Run Policy Analysis and Long Run Growth." *Journal of Political Economy*, 99, 500-521.

Restoy, Fernando and Phillipe Weil (1998) "Approximate Equilibrium Asset Prices." Working paper, ECARE, Brussels, March.

Rios-Rull, Jose V. (1995) "Models with Heterogeneous Agents." In Cooley, Thomas, Ed., *Frontiers of Business Cycle Research.* Princeton University Press, Princeton, N.J.

Ross, Stephen (1977) "The Determinants of Financial Structure: The Incentive-Signaling Approach." *Bell Journal of Economics,* 8, 23-40.

Santos, Manuel and Michael Woodford (1997) "Rational Asset Pricing Bubbles." *Econometrica*, 65, 19-87.

Sargent, Thomas J. (1987) *Dynamic Macroeconomic Theory.* Harvard University Press, Cambridge, MA.

Sarte, Pierre-D. G. (1998) "Fisher's Equation and the Inflation Risk Premium in a Simple Endowment Economy." Federal Reserve Bank of Richmond *Quarterly Review*, 84, 53-72.

Simon, Carl and Lawrence Blume (1994) *Mathematics for Economists.* Norton, New York, NY.

Spear, Stephen E. and Sanjay Srivastava (1987) "On Repeated Moral Hazard with Discounting." *Review of Economic Studies,* 54, 599-617.

Stockman, Alan (1981) "Anticipated Inflation and the Capital Stock in a Cash-in-advance Economy." *Journal of Monetary Economics,* 8, 387-393.

Stokey, Nancy, Robert E. Lucas with Edward Prescott (1989) *Recursive Methods in Economic Dynamics*. Harvard University Press, Cambridge, MA.

Sussman, Oren (1999) "Economic Growth with Standardized Contracts." *European Economic Review,* 43, 1797-1818.

Tallarini, Thomas (2000) "Risk-sensitive Real Business Cycles." *Journal of Monetary Economics,* 45, 507-532.

Taub, Bart (1997) "Dynamic Agency with Feedback." *RAND Journal of Economics,* 28, 515-543.

Thomas, Jonathan and Thomas Worrall (1990) "Income Fluctuations and Asymmetric Information: An Example of a Repeated Principal Agent Problem." *Journal of Economic Theory*, 51, 367-390.

Tirole, Jean (1985) "Asset Bubbles and Overlapping Generations." *Econometrica,* 53, 1499-1528.

Tobin, James (1969) "A General Equilibrium Approach to Monetary Theory." *Journal of Money Credit and Banking,* 1, 15-29.

Townsend, Robert M. (1982) "Optimal Multiperiod Contracts and the Gain from Enduring Relationships under Private Information." *Journal of Political Economy,* 90, 1166-1186.

Townsend, Robert M. (1987) "Asset Return Anomalies in a Monetary Economy." *Journal of Economic Theory*, 41, 219-247.

Van Wincoop, Eric (1999) "How big are Potential Welfare Gains from Risksharing?" *Journal of International Economics*, 47, 109-135.

Vissing-Jorgensen, Annette and Orazio Attanasio (2002) "Stock Market Participation, Intertemporal Substitution and Risk Aversion." Working paper, Northwestern University and University College London; December.

Weil, Phillipe (1990) "Nonexpected Utility in Macroeconomics." *Quarterly Journal of Economics,* 105, 29-42.

Weil, Phillipe (1989) "The Equity Premium Puzzle and the Risk-Free Rate Puzzle." *Journal of Monetary Economics,*24, 401-421.

8. Dynamics II: Continuous Time

8.1 Asset Price Dynamics, Options and the Black-Scholes Model

The next few sessions present dynamic models of random variables that may be interpreted as asset prices. The models are discussed initially in a discrete time setting as a prelude to the continuous time framework that is used in modern finance applications. The class of processes discussed is the random walk type processes. Random walk models fit well asset prices because of the Efficient Market hypothesis seen in Chapters 4 and 7. According to this hypothesis, past history is fully reflected in the current price, thus a Markov process does well; e.g. Chapter 7 for a discussion of Markov processes. Secondly, markets respond instantaneously to new information and this is reflected in the properties of the random component. Third, there are no systematic gains or losses, the no arbitrage condition holds.

Next, we discuss the models of asset price dynamics together with options strategies and the no arbitrage condition to obtain the Black-Scholes formula for the price of a simple European Call option.

Later in this chapter, we apply the continuous time stochastic calculus methodology to discuss equilibrium models of consumption and wealth growth, portfolio choice, asset prices and capital accumulation with constant relative risk aversion and briefly with recursive preferences.

8.2 Discrete Time Random Walks

Consider some variable S_t for a discrete time index $t=0,1,2,3...$ determined by the following stochastic difference equation

$$S_{t+1} = S_t + \epsilon_t, \quad S_0 \text{ given} \tag{8.1}$$

where ϵ_t is an independent and identically distributed or i.i.d. discrete random variable with $E[\epsilon_t]=0$, $var(\epsilon_t)=\sigma^2 \in (0,\infty)$, $Cov(\epsilon_s, \epsilon_t)=0$, $s \neq t$. Given the stochastic structure, this model is sometimes called the discrete state model. A stochastic process for S_t determined by equation (8.1) is called an autoregressive process of order one, and is indeed a simple stochastic difference equation of first order as studied in Chapter 1. It belongs to the general class of Markov processes since all the past history is fully reflected in the current value, S_t. Because the coefficient associated with the term S_t is one, this process is known as a random walk. In fact, given S_0, a backward recursion in (8.1) gives a solution for S_t as

$$S_{t+1} = \epsilon_t + (\epsilon_{t-1} + (\epsilon_{t-2} + (\epsilon_{t-3} + (\epsilon_{t-4} + \ldots)\ldots) + S_0 \tag{8.2}$$

or

$$S_{t+1} = S_0 + \Sigma_{i=1}^{t} \epsilon_i. \tag{8.3}$$

Thus, the behavior of S is governed by the sum of two components: (a) A stochastic term which is the sum of the random variables ϵ_t, $\Sigma_{i=1}^{t} \epsilon_i$, which behaves like a stochastic trend; (b) A deterministic component initially given, S_0. In the random walk process denoted by (8.1), the solution (8.3) indicates that the expected value of S_{t+1} is

$$E[S_{t+1}] = S_0 \tag{8.4a}$$

a constant, and its variance is

$$var(S_{t+1}) = t\,\sigma^2 \tag{8.4b}$$

proportional to the time index t. As $t \rightarrow \infty$, $t\sigma^2 \rightarrow \infty$, and a random walk process is said to be non-stationary in variance, but stationary in mean.

Now, consider a minor modification to the process (8.1) as

$$S_{t+1} = \mu + S_t + \epsilon_t, \quad \{S_0, \mu\} \text{ given} \tag{8.5}$$

for ϵ_t as above and μ constant and given. The stochastic process (8.5) is again an autoregressive process of order one (a simple stochastic difference equation of first order). Because the coefficient associated with the term S_t is one and there is a constant term added to the right-hand side, this process is known as a random walk with drift. The recursive solution for (8.5) is

$$S_{t+1} = \mu + \epsilon_t + (\mu + \epsilon_{t-1} + (\mu + \epsilon_{t-2} + (\mu + \epsilon_{t-3} + (\mu + \epsilon_{t-4} + ...)) + S_0 \quad (8.6)$$

or

$$S_{t+1} = t\,\mu + S_0 + \Sigma_{i=1}^{t}\,\epsilon_i. \quad (8.7)$$

Thus, the behavior of S is now governed by the sum of three components: (a) A stochastic term which is the sum of the random variables ϵ_t, $\Sigma_{i=1}^{t}\epsilon_i$, which behaves like a stochastic trend; (b) A deterministic component initially given, S_0; (c) A deterministic component proportional to the time index, t, given by $t\mu$, which behaves like a deterministic trend. Hence, in the random walk with drift denoted by (8.5), the solution (8.7) indicates that the expected value of S_{t+1} is

$$E[S_{t+1}] = S_0 + t\,\mu \quad (8.8a)$$

not constant, and its variance is

$$var(S_{t+1}) = t\,\sigma^2 \quad (8.8b)$$

both proportional to the time index t. As $t \to \infty$, both the average and the variance of S_{t+1}, approach infinity and a random walk with drift process is non-stationary in both mean and variance.

The stochastic component that drives the processes in (8.1) and/or (8.4) may follow a normal distribution, the so-called continuous state model, or $\epsilon_t \sim N(0, \sigma^2)$ and independent and identically distributed (i.i.d.). In this case, the random variable S_t also follows a normal distribution because it is a linear function of a sum of i.i.d. normally distributed random variables, which is a direct consequence of the Central Limit theorem, see e.g. Notes on the Literature.

8.3 A Multiplicative Model in Discrete Time and a Preview of the Lognormal Random Variable

A model that is very useful for the analysis of asset prices is the multiplicative model. The structure is analogous to the random walk models examined above. Let the time index run as $t=0,1,2,3...N-1$ for N periods. The variable S evolves according to

$$S_{t+1} = S_t \quad u_t, \quad S_0 \text{ given} \tag{8.9a}$$

for a random variable u assumed to be i.i.d. across periods. The term u governs the gross change in S across periods. Assuming $\{S, u\} \in (0, \infty)$, the process in (8.9a) may be transformed by taking natural logarithms of both sides yielding

$$\log S_{t+1} = \log S_t + \log u_t \tag{8.9b}$$

an additive model as in (8.1), i.e. a "random walk" in logarithms. If we define

$$\log u_t \equiv \omega_t \tag{8.10}$$

and let $\omega_t \sim N(\mu_\omega, \sigma_\omega^{\ 2})$, i.e. ω_t follows a normal distribution with mean μ_ω and variance $\sigma_\omega^{\ 2}$; then the recursive solution for $\log S_{t+1}$ is

$$\log S_{t+1} = \omega_t + (\omega_{t-1} + (\omega_{t-2} + (\omega_{t-3} + (\omega_{t-4} + ...)...))) + \log S_0 \tag{8.11}$$

or

$$\log S_{t+1} = \log S_0 + \Sigma_{i=1}^{\ t} \, \omega_i. \tag{8.12}$$

Thus, the behavior of $\log S$ is governed by the sum of two components: (a) A stochastic term which is the sum of the random variables ω_t, $\Sigma_{i=1}^{\ t}\omega_i$, which behaves like a stochastic trend; (b) A deterministic component initially given, $\log S_0$. The expected value of $\log S_{t+1}$ is

$$E[\log S_{t+1}] = S_0 + t\,\mu_\omega \tag{8.13a}$$

not constant, and its variance is

$$var(\log S_{t+1}) = t\,\sigma_\omega^{\ 2} \tag{8.13b}$$

both proportional to the time index t. As $t \to \infty$, both the average and the variance of $\log S_{t+1}$, approach infinity and a random walk process in logarithms is also non-stationary in mean and variance.

Now, note by (8.10) that

$$u_t = \exp(\omega_t)$$

where *exp* is the exponential operator. Since $\omega_t \sim N(\mu_\omega, \sigma_\omega^{\ 2})$, the random variable u_t is said to follow a Lognormal distribution. By (8.12), $\log S_{t+1}$ is governed by a sum of normally distributed random variables; thus it is itself normally distributed with respective mean and variance given by

(8.13a)-(8.13b). Analogously, S_{t+1} follows a lognormal distribution. The probability distribution function, $P_U(u)$, for a discrete lognormal random variable is illustrated in Figure 8.1. The domain of u is the nonnegative real line and the distribution is bell shaped but skewed to the right. The mean of a lognormal random variable is given by

$$E[u_t] = [exp(\mu_\omega + (1/2)\ \sigma_\omega^2)] \tag{8.14a}$$

and its variance is

$$var(u_t) = [exp(2\ \mu_\omega + \ \sigma_\omega^2)]\ [(exp(\sigma_\omega^2\ )) - 1]. \tag{8.14b}$$

The point to notice here is that in (8.14a), the mean of the lognormal depends on the variance of the underlying normal random variable ω_t. The reason is that as the variance of the distribution increases, it spreads the distribution in both directions but the domain of the lognormal is the nonnegative real line. Thus, as the variance changes, the mean must change as well in order to maintain the domain of the distribution. We shall examine this distribution in more detail in section 8.6 below.

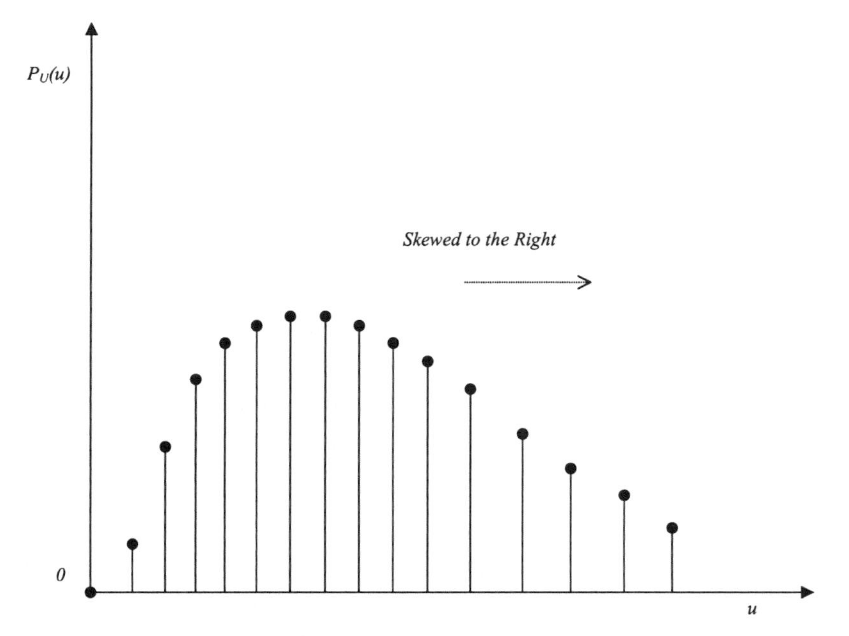

Fig. 8.1 Illustration of Discrete Probability Distribution of Lognormal Random Variable

8.4 Introduction to Random Walk Models of Asset Prices in Continuous Time

Let the time interval be reduced to an infinitesimal amount denoted dt as is usual in ordinary infinitesimal calculus. If the asset price at time t is S, in the infinitesimal interval dt the asset price changes to $S+dS$. For example, in Figure 8.2 a discretized version of this logic is presented in $\{S,t\}$ space.

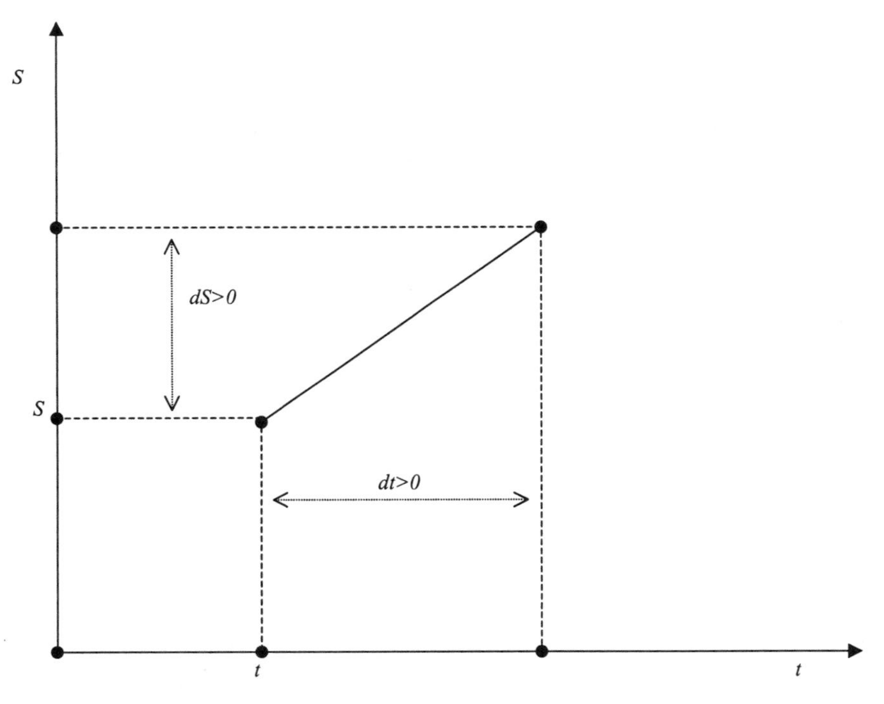

Fig. 8.2 Discretized "Small" Interval

The question we are interested in answering is how to model the corresponding relative or percent infinitesimal change in the asset price, S, denoted dS/S, under uncertainty. Towards an answer, we look for a general class of models based on the decomposition of the process in stochastic and non-stochastic parts analogously to the discrete time random walk models studied in sections 8.2-8.3. A common way to decompose the relative change, dS/S, is:

(a) A non-stochastic component denoted by

$$\mu\, dt$$

where μ measures the average rate of growth of the asset price, or the average of dS/S, in the "small" interval dt. The variable μ is a drift term: if μ is constant and nontrivial, the process exhibits a constant drift; if μ is not constant and nontrivial the process exhibits a variable drift; and if μ is trivial the process has no drift.

For example, in pure finance applications, this non-stochastic component is related to the opportunity cost of investing in a risk-free asset available in the market place; in macroeconomic applications, this term is related to the mean growth of the economy;

(b) A stochastic component denoted by

$$\sigma\, dZ$$

where σ is the standard deviation of the rate of growth of the asset price, or the standard deviation of dS/S, and dZ is a process of increments in Z which are independent and identically distributed (i.i.d.) random variables with mean zero and distribution to be discussed below.

Hence, the sum of the two components gives a formula for dS/S as

$$dS/S = \mu\, dt + \sigma\, dZ, \tag{8.15}$$

a mathematical representation of the rate of change of S in the infinitesimal interval dt governed by a non-stochastic component, $\mu\ dt$ and a stochastic component, $\sigma\ dZ$. Equation (8.15) is a stochastic differential equation as studied in Chapter 1. This equation satisfies the Markov property since the increments in S only depend on variables as of time t. All randomness comes from the stochastic component, the term σ dZ.

Take, for example, the special case where $\sigma = 0$. Then, equation (8.15) reduces to an ordinary non-stochastic differential equation

$$dS/S = \mu\, dt$$

which can be integrated indefinitely (for constant μ) as

$$\int (1/S)\, dS = \mu \int dt$$

giving

$$log\, S = \mu\, t + k$$

for an arbitrary constant of integration k. At $t=0$, the initial condition gives $k=log\, S_0$ and the solution is

$$log\, (S/S_0) = \mu\, t$$

or

$$S = S_0 \quad exp(\mu\, t)$$

an exponential non-stochastic process for S.

In the general case where $\sigma \neq 0$, the stochastic component, $\sigma\, dZ$, is driven by the increments on Z. The random variable Z follows a standard Wiener process (Brownian Motion or diffusion process). The increments are usually written as

$$dZ = \varepsilon_t\, (\, dt\,)^{1/2} \tag{8.16}$$

proportional to ε_t where $\varepsilon_t \sim N(0,1)$, i.e. ε_t follows a standard normal distribution with mean 0 and variance 1; and is i.i.d. across periods; and proportional to the square root of dt, $(dt)^{1/2}$. Thus,

$$E[dZ] = 0 \tag{8.17a}$$

$$var\, (dZ) = dt \tag{8.17b}$$

and it is assumed that initially $Z_0=0$ with probability one. In (8.17b), the variance of the increment dZ is proportional to dt, not $(dt)^2$, because in (8.16) dZ is proportional to $(\, dt\,)^{1/2}$; this is an important characteristic that will be recalled commonly.

A standard Wiener process denoted by equation (8.16) can be suitably derived as the limit of a discrete time random walk process of the type we examined above in sections 8.2-8.3. First, we envision a time line and pick a finite time interval denoted T (or $t=0$ to $t=T$). Then, we divide this finite time interval in k units [for $k=0,1,2,3,\ldots,T$] of equal length denoted Δt, so that

$$\Delta t = T/k \;\Rightarrow\; T = k \times \Delta t.$$

For example, if $k=5$ and $\Delta t=2$, the finite time interval is $T=10$. Now, we can position the random variable Z at any point on the time line, denoted $Z(t_k)$, for an arbitrary k, and an arbitrary Δt. In particular, the discrete interval from t_k to t_{k+1} has appropriate length $\Delta t = t_{k+1} - t_k$, since the difference is one arbitrary unit of length, Δt. More generally, for $k,j=0,1,2,\ldots,T$, when $k > j$,

$$t_k - t_j = \Delta t \times (k - j).$$

Then, a random walk for Z in the interval t_{k+1}, t_k, for $k=0,1,2,3,\ldots,T$ is given by

$$Z(t_{k+1}) = Z(t_k) + \varepsilon(t_k)(\Delta t)^{1/2} \tag{8.18}$$

where $\Delta t = t_{k+1} - t_k$, $\varepsilon(t_k) \sim N(0,1)$, and $Z(t_0)=0$ with probability one. The solution for $Z(t_k)$ is thus

$$Z(t_k) = \Sigma_{i=0}^{k-1}\, \varepsilon(t_i)(\Delta t)^{1/2}. \tag{8.19}$$

Now, let there be two fixed time points denoted $\{k,j=0,1,2,\ldots T: k>j\}$ and compute the difference

$$Z(t_k) - Z(t_j) = \Sigma_{i=j}^{k-1}\, \varepsilon(t_i)(\Delta t)^{1/2} \tag{8.20}$$

where the equality comes from the fact that all terms in the interval $0<i<j$ cancel out. This quantity, $Z(t_k) - Z(t_j)$, is distributed normally because it is the sum of standard normals. Hence, the mean is

$$E[Z(t_k) - Z(t_j)] = 0$$

and the computation of the variance involves a few steps:

$$var[Z(t_k) - Z(t_j)] = E[\Sigma_{i=j}^{k-1}\, \varepsilon(t_i)(\Delta t)^{1/2}]^2$$

$$= \Sigma_{i=j}^{k-1}\, E[\varepsilon(t_i)^2]\,[(\Delta t)^{1/2}]^2, \text{ since } \varepsilon(t_i) \text{ is i.i.d.}$$

$$= \Sigma_{i=j}^{k-1}\, (\Delta t), \text{ since } E[\varepsilon(t_i)^2]=1$$

$$= (k-j)(\Delta t), \text{ proportional to unit time interval } \Delta t$$

$$= t_k - t_j \quad , \text{ since } t_k - t_j = \Delta t \times (k-j)$$

i.e. the variance of the difference $Z(t_k) - Z(t_j)$ is proportional to the time interval, $t_k - t_j$. The standard Wiener process (Brownian motion or diffusion process) for Z can be obtained as the appropriate limit of the process (8.18) as $\Delta t \rightarrow 0$, and from the mean and variances above we obtain $E[dZ]=0$, $var(dZ)=dt$ as in (8.17a,b).

It is very important to note that a standard Wiener process (Brownian motion or diffusion process) is not differentiable with respect to time everywhere, i.e. dZ/dt does not exist in the normal sense. A process of this type, dZ/dt, is referred in the engineering literature as white noise.

The result that standard Wiener processes are not differentiable with respect to time, can be seen by considering, from (8.18),

$$\Delta Z/\Delta t = (1/2)\varepsilon(t_k)(\Delta t)^{-1/2},$$

so that, taking the limit as $\Delta t \to 0$, the derivative is not well defined.

A stochastic process given by (8.15) is a more generalized Wiener process (Brownian motion or diffusion process) described by

$$d\,x(t) = a\,dt + b\,dZ \tag{8.21}$$

for constant real numbers $\{a,b\}$ and the standard Wiener process (Brownian motion or diffusion process) dZ. An important characteristic of this class of generalized processes is that they have an analytic solution. This is obtained by integrating both sides of (8.21), for constants $\{a,b\}$, from 0 to t as

$$\int_0^t d\,x(t) = a\int_0^t dt + b\int_0^t dZ \tag{8.22a}$$

obtaining , recall $Z(0)=0$ with probability one,

$$x(t) = x(0) + a\,t + b\,Z(t). \tag{8.22b}$$

In the specific case of (8.15), $a \equiv \mu S$, and $b \equiv \sigma S$; $\{\mu, \sigma\}$ constant. An even more generalized Wiener process (or Brownian motion), called an Ito process, is described by

$$d\,x(t) = a(x,t)\,dt + b(x,t)\,dZ \tag{8.23}$$

where a and b are now functions of $\{x,t\}$. A general analytic solution for this process cannot be obtained, but Ito's Lemma (to be discussed in the next section) is a very useful tool in understanding these processes.

Before we move on, we call your attention to another important property of the process Z, the standard Wiener process. We recall expression (8.18) and compute the conditional expectation at time t_k for t_{k+1} and obtain

$$E_{tk}[Z(t_{k+1})] = Z(t_k)$$

which says that the expected value of the random variable Z conditional on the information available at time t_k is just the current (observed) realization of the random variable at time t_k. More generally, computing the conditional expectation at time t_k of Z in (8.20) and applying the law of iterated mathematical expectations yields

$$E_{tk}[Z(t_k)] = Z(t_j)\,, \quad all\ k>j$$

so that the prediction of Z is the current (observed) realization at time t_k.

Roughly speaking, a dynamic process that satisfies this property is called a Martingale process, and we have just shown that a standard Wiener process (Brownian motion or diffusion process) is a Martingale process. Martingale processes are an important tool in modern finance theory. They imply that future variations are unpredictable given the current information available. For example, imagine that Z is the price of a stock. Note that Z is not stationary. In effect, do you know why Z is not stationary? (It is because the variance of Z is proportional to t itself; however, the variance of dZ is proportional to Δt, the time interval, and not t itself, thus constant!). In any event, when Z is thought of as the price of stock, the martingale property implies that no arbitrage opportunities are available. If the expectations of the future price were above (below) the current price, market participants would rush to buy (sell) it now and sell (buy) later at a profit pushing the current price up (down) up until the martingale property holds. This is in fact the simplest version of the Efficient Market hypothesis, studied in Chapters 4 and 7.

8.5 A Multiplicative Model of Asset Prices in Continuous Time

The discrete time multiplicative model in (8.9)-(8.14) has its continuous time stochastic counterpart given by

$$d \log S(t) = \nu\, dt + \sigma\, dZ \qquad (8.24)$$

where ν and $\sigma \geq 0$ are constants, and dZ is a standard Wiener process. In (8.24), $\nu\, dt$ is the mean value of $d\log S(t)$ and gives the non-stochastic growth component proportional to dt. The term $\sigma\, dZ$ is the stochastic component so that the standard deviation of $d\log S(t)$ is $\sigma\ (dt)^{1/2}$, proportional to the square root of the length of the period.

The stochastic process in equation (8.24) is a generalized Wiener process as in (8.21) because $\{\nu, \sigma\}$ are constants. Its analytic solution is, from (8.22a,b),

$$\log S(t) = \log S(0) + \nu\, t + \sigma\, Z(t). \qquad (8.25)$$

for $\log S(0)$ given. Hence,

$$E[\log S(t)] = E[\log s(0)] + \nu t \tag{8.26a}$$

and

$$var(\log S(t)) = \sigma^2 t \tag{8.26b}$$

both proportional to t, and the process is non-stationary in mean and variance. Because the mean of $\log S(t)$ grows linearly with t, a process as in (8.24) is called a geometric Brownian motion, i.e. it has a constant drift term. In addition, because Z is normally distributed, $Z(t) \sim N(0,t)$, then

$$\log S(t) \sim N\{\log S(0) + \nu t, \sigma^2 t\}$$

is also normally distributed. As a result, $S(t)$ follows a lognormal distribution. The important point to notice is that $S(t) = \exp(\log S(t))$ and by direct substitution

$$S(t) = S(0) \exp(\nu t + \sigma Z(t)). \tag{8.27}$$

However, by (8.26a), one may conjecture that $E[S(t)]$ is equal to $S(0) \exp(\nu t)$. This conjecture is incorrect because Wiener processes do not follow the rules of ordinary calculus. In this case,

$$E[S(t)] = S(0) \{\exp([\nu + (1/2) \sigma^2]t)\} \tag{8.28a}$$

and

$$var[S(t)] = S(0)^2 \{\exp(2\nu t + \sigma^2 t)\} [\exp(\sigma^2 t) - 1]. \tag{8.28b}$$

The reason for this discrepancy is that, unlike ordinary calculus, under stochastic calculus

$$d \log S(t) \neq dS(t)/S(t). \tag{8.29}$$

If $\log S(t)$ follows the process in (8.24), the appropriate process for $S(t)$ is

$$d S(t)/S(t) = [\nu + (1/2) \sigma^2] dt + \sigma dZ(t) \tag{8.30a}$$

i.e. the mean is corrected by $(1/2)\sigma^2$. As a consequence, in (8.15) and (8.24) the means are related by

$$\mu \equiv \nu + (1/2) \sigma^2. \tag{8.30b}$$

The correction from $\log S(t)$ to $S(t)$ discussed above is a special case of a more general transformation equation defined by Ito's Lemma. In this particular case, note that the discrepancy is due to Jensen's inequality: $\log S(t)$ is a concave function of $S(t)$. We discuss Ito's Lemma in the next two sections.

As a simple numerical example of a Geometric Brownian motion, consider that the current price of a share is $S(t)=\$1.00$, the drift parameter is $\mu=1$, and the standard deviation is $\sigma=0.2$, or 20%. Let the time interval be one day as a proportion of about 250 business days in a year, so that $\Delta t\equiv 1/250$. Thus, in the process (8.30a), it implies that the stochastic component $dZ(t)$ is normally distributed with mean zero and variance $\Delta t=1/250$. Suppose we pick one random draw from the $N(0,1/250)$ distribution and obtain $dZ(t)=0.08352$. Then, by direct computation of formula (8.30a) we obtain

$$dS(t)=[1\times 1\times(1/250)] + (1\times 0.2\times 0.08352) = \$0.02,$$

a two cents increment. The new stock price at $t+\Delta t$ is

$$S(t) + dS(t) = 1 + 0.02 = \$1.02$$

and a series obtained in this fashion yields a time path for the stock price driven by the deterministic and stochastic components as in (8.30a).

8.6 Introduction to Ito's Lemma and the Lognormal Distribution Again

In this section, we follow a heuristic approach to Ito's Lemma. Ito's Lemma is one of the most important results about the manipulation of random variables in a continuous time stochastic setting. Loosely speaking, Ito's Lemma is for functions of random variables what Taylor's theorem is for functions of deterministic variables. To see this, let f be a twice continuously differentiable function of S only, $f(S)$. Assuming for the moment that S is non-stochastic, we can vary S by a small amount, dS, and find the corresponding small variation in $f(S)$ by applying Taylor's theorem. Noting that derivatives are written since f is a function of S only, we obtain

$$df(S) = [df(S)/dS]\, dS + (1/2)\, [d^2f(S)/dS^2]\, dS^2 + o(dS^2). \quad (8.31)$$

where $o(dS^2)$ indicates terms of order higher than dS^2.

Now, let S be stochastic where, for example, dS is determined by the process (8.15). Since dS is a random number, we may square it to obtain

$$dS^2 = (\mu S\,dt + \sigma S\,dZ)^2$$
$$= \mu^2 S^2 dt^2 + 2 S^2 \mu \sigma\, dt\, dZ + \sigma^2 S^2 dZ^2 \qquad (8.32)$$

and note that because $dZ=\varepsilon_t (dt)^{1/2}$, only the last term in (8.32) is of order of magnitude one in dt. The other two terms are of order of magnitude greater than one, and can be ignored for infinitesimally small dt. Thus, we may write (8.32) as

$$dS^2 = \sigma^2 S^2 dZ^2 + o(dt) = \sigma^2 S^2 dZ^2 \qquad (8.33)$$

where $o(dt)$ indicates terms of higher order than dt, with $o(dt) \approx 0$. The final step to obtain a version of Ito's Lemma regards the term dZ^2 in (8.33). The assumption invoked is that, for sufficiently small dt, we may write

$$dZ^2 \to dt \quad as \quad dt \to 0 \qquad (8.34)$$

or for sufficiently small dt, the variance of dZ becomes small and dZ^2 becomes non-stochastic equal to dt, i.e. recall that $E[dZ^2]=var(dZ^2)=dt$. Using expression (8.34), (8.33) becomes

$$dS^2 = \sigma^2 S^2 dt \qquad (8.35)$$

and substituting dS from (8.15) and dS^2 from (8.35) into the Taylor's expansion (8.31), ignoring higher order terms or $o(dS^2) \approx 0$, we obtain

$$df(S) = \{[df(S)/dS]\, \mu S + (1/2)\, [d^2 f(S)/dS^2]\, \sigma^2 S^2\}\, dt + [df(S)/dS]\, \sigma S\, dZ \qquad (8.36)$$

a simplified version of Ito's Lemma. This formula represents a stochastic differential equation for the function $f(S)$ when we compute the small change in the function of the random variable, $df(S)$, in response to a small change in the random variable itself, dS. The first term is the mean of $df(S)$ proportional to dt, and the second term is the stochastic component driven by the standard Wiener process dZ.

A useful example to consider is the natural logarithmic function, $f(S)=log\, S$. In this case,

$$df(S)/dS = 1/S, \quad and \quad d^2 f(S)/dS^2 = -1/S^2.$$

Substituting into Ito's formula (8.36) yields the generalized Wiener process (Brownian motion or diffusion process)

$$d\, log\, S = [\mu - (1/2)\, \sigma^2]\, dt + \sigma\, dZ \qquad (8.37)$$

a stochastic differential equation in *log S*, as we studied in equations (8.15), (8.24), and (8.30a,b). The increment *d log S* is normally distributed with mean

$$E[d \log S] = [\mu - (1/2)\, \sigma^2]\, dt \tag{8.38a}$$

and variance

$$var(d \log S) = \sigma^2\, dt. \tag{8.38b}$$

Also, equation (8.37) has an analytic solution [recall (8.22a,b)] given by

$$\log S(t) = \log S(0) + [\mu - (1/2)\, \sigma^2]\, t + \sigma\, Z(t) \tag{8.39}$$

given *log S(0);* and *log S(t)* is normally distributed with mean and variance as in (8.26a,b) using (8.30), yielding

$$\log S(t) \sim N\{\log S(0) + [\mu - (1/2)\, \sigma^2]\, t,\ \sigma^2 t\}. \tag{8.40}$$

Then, the probability density function (p.d.f.) of $f(S)=\log S$ is given by

$$p.d.f.(\log S) =$$

$$[1/(2\pi\sigma^2 t)^{1/2}]\ \{exp(-[(\log [S/S(0)]-[\mu -(1/2)\sigma^2]\, t)^2/2\sigma^2\, t])\} \tag{8.41}$$

where $\pi=3.141592...$ is a physical constant, for the domain $-\infty < \log S < \infty$. To fully understand the lognormal distribution, we have to find the p.d.f. of *S*. In order to find the p.d.f. of *S*, we use a result from statistics that relates the cumulative distribution function of *log S* to the cumulative distribution function of *S*. Recall that the cumulative distribution function of a continuous random variable *Y*, for the given domain $-\infty<Y<\infty$, is

$$Prob(Y \leq y) = F(y) = \int_{-\infty}^{y} p.d.f.\ (x)\, d\, x \tag{8.42}$$

for *x* an index of integration. In turn,

$$p.d.f.(y) = dF(y)/dy \tag{8.43}$$

i.e. the p.d.f. is the derivative of the cumulative distribution function, and expression (8.42) can be equivalently written as

$$Prob(Y \leq y) = F(y) = \int_{-\infty}^{y} d\, F(x). \tag{8.42'}$$

Figure 8.3 illustrates the p.d.f. and the cumulative distribution function respectively for a continuous lognormal random variable.

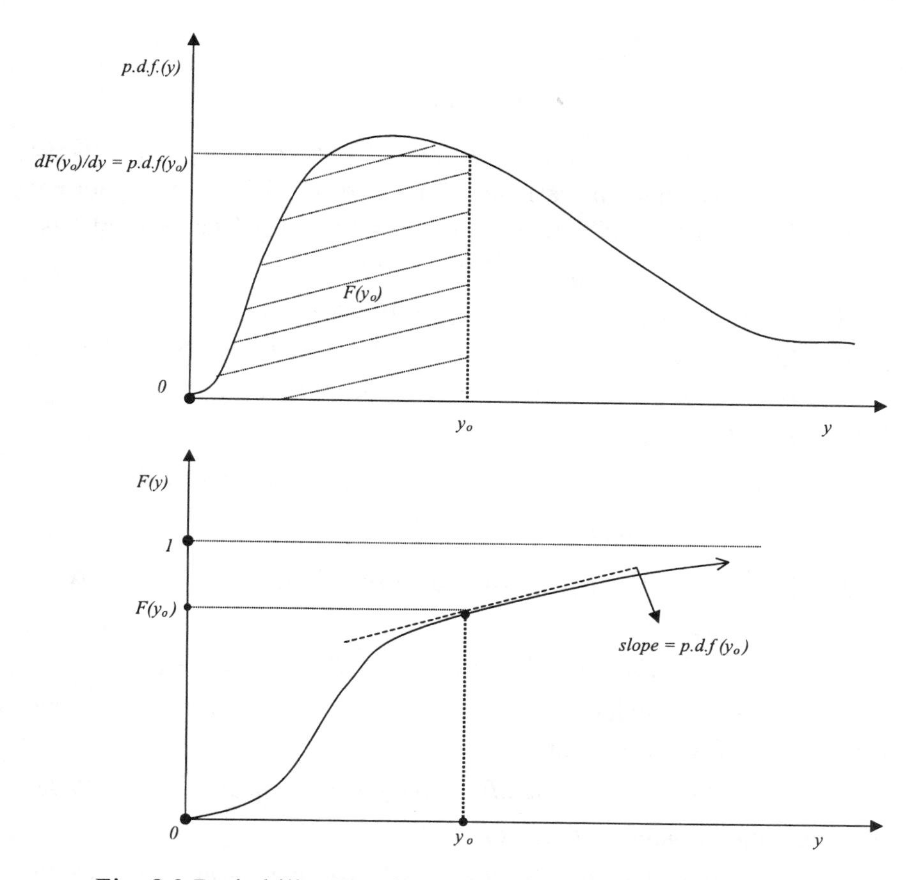

Fig. 8.3 Probability Density Function (p.d.f.) and Cumulative Distribution Function for Continuous Lognormal Random Variable

The result we use to relate the cumulative distribution function of $\log S$ to the cumulative distribution function of S is

$$Prob(\log S \leq \log s) = Prob(S \leq s) \qquad (8.44)$$

or the cumulative distribution function of $\log S$, $F(\log S)$, is equal to the cumulative distribution function of S, $F(S)$, given that the logarithm is a monotone transformation. Hence, by (8.43)-(8.44) the p.d.f. of S is

$$p.d.f.(S) = dF(\log S)/dS = p.d.f.(\log S)/S \qquad (8.45)$$

or

$$p.d.f.(S) =$$

$$[1/S\ (2\pi\sigma^2 t)^{1/2}]\ \{exp\ (-[(log\ [S/S(0)] - [\mu - (1/2)\ \sigma^2]\ t)^2/2\sigma^2\ t])\} \quad (8.46)$$

for $0<S<\infty$. A random variable S is said to follow a lognormal distribution if it has a p.d.f. given by expression (8.46). The mean of S is

$$E[S] = \int_0^\infty S\ p.d.f.(S)\ d\,S. \quad (8.47)$$

Evaluating this integral involves the following steps:
(i) Let $log\ S = x + log\ S(0) + [\mu - (1/2)\ \sigma^2]t$ so that $dlog\ S = dx$ for the domain $-\infty < log\ S < \infty$; then we can rewrite (8.47) as

$$E[S] = \int_{-\infty}^{\infty} [1/S\ (2\pi\sigma^2 t)^{1/2}] \times$$

$$\{exp\ (-x^2/2\sigma^2 t)\}\{\ exp\ (x + [log\ S(0)] + [\mu - (1/2)\ \sigma^2]\ t)\}\ d\,x$$

$$= \{exp\ \ (log\ S(0) + \mu t - (1/2)\sigma^2 t +$$

$$[(1/2)\sigma^4\, t^2/\sigma^2 t])\}\ \int_{-\infty}^{\infty} [1/(2\pi\sigma^2 t)^{1/2}]\ \{exp\ (-[x - \sigma^2 t]^2/2\sigma^2 t)\}\ dx$$

where the integral

$$\int_{-\infty}^{\infty} [1/(2\pi\sigma^2 t)^{\ 1/2}]\ \{exp\ (-[x - \sigma^2 t]^2/2\sigma^2 t)\}\ dx = 1$$

since it is the area under a normal random variable with mean $\sigma^2 t$ and variance$\sigma^2 t$. Then, we obtain

$$E[S] = \{exp\ (log\ S(0) + \mu\, t)\} = S(0)\ exp\ \mu\, t; \quad (8.48)$$

(ii) Applying the same methodology,

$$var(S) = S(0)^2\ \{exp(\ 2\ \mu\, t)\}\ [(exp(\ \sigma^2 t)) - 1] \quad (8.49)$$

[recall expressions (8.28a,b) and (8.30a,b)].

In summary, for dynamic models of asset prices driven by standard Wiener processes, $Z(t)$, and using (8.30b), Table 8.1 presents the relevant moments.

Table 8.1
Moments for Asset Price with Normal and Lognormal Distribution

$dS(t)/S(t) = \mu\, dt + \sigma\, dZ(t)$ $\Rightarrow S$ *is lognormal with mean* $S(0) \exp(\mu t)$ *and variance* $S(0)^2 \{\exp(2\mu t)\} [(\exp(\sigma^2 t)) - 1]$	$dS(t)/S(t) = [\nu + (1/2)\sigma^2]\, dt + \sigma\, dZ(t)$ $\Rightarrow S$ *is lognormal with mean* $S(0) \{\exp([\nu + (1/2)\sigma^2]t)\}$ *and variance* $S(0)^2 \{\exp((2\nu + \sigma^2)t)\}[(\exp(\sigma^2 t)) - 1]$
$d \log S(t) = [\mu - (1/2)\sigma^2]\, dt + \sigma\, dZ(t)$ $\Rightarrow \log S(t) \sim N(\log S(0)$ $+ [\mu - (1/2)\sigma^2]\, t, \sigma^2 t)$	$d \log S(t) = \nu\, dt + \sigma\, dZ(t)$ $\Rightarrow \log S(t) \sim N(\nu t, \sigma^2 t)$.

8.7 Ito's Formula: The General Case

The two stochastic equations for $S(t)$ and $\log S(t)$ in the summary Table 8.1 are different and the difference is from the correction term in the mean. This is because the random variables have order $(dt)^{1/2}$ and their squares produce first order effects, dt, rather than second order effects, dt^2. Hence, for example, if X follows a lognormal distribution, then $\log X \sim N(\mu_x, \sigma_x^2)$ and $\log E[X] = \mu_x + (1/2)\sigma_x^2$. As seen above, Ito's Lemma gives a formula for the transformation. A more general formula is obtained from a stochastic process for x, as in (8.23) repeated here for convenience, given by a more generalized Wiener process (Brownian motion or diffusion process), called an Ito process,

$$d\,x(t) = a(x,t)\, dt + b(x,t)\, dZ$$

where Z is a standard Wiener process (Brownian motion or diffusion process) as before. Let the process $y(t)$ be defined as

$$y(t) = G(x,t) \tag{8.50}$$

where the function G is twice continuously differentiable. Expanding (8.50) according to Taylor's theorem yields

$$dy(t) = [\partial G(.)/\partial x]\, [a\, dt + b\, dZ] + (1/2)\, [\partial^2 G(.)/\partial x^2]\, [a\, dt + b\, dZ]^2 + [\partial G(.)/\partial t]\, dt + o(dt, dZ^2) \tag{8.51}$$

where $o(dt, dZ^2)$ are terms of order higher than $\{dt, dZ^2\}$. Applying the same method as in (8.32)-(8.35), we obtain a more general formula for Ito's Lemma as

$$dy(t) = \{a\,[\partial G(.)/\partial x] + (1/2)\,b^2\,[\partial^2 G(.)/\partial x^2] + [\partial G(.)/\partial t]\}\,dt + b\,[\partial G(.)/\partial x]\,dZ \qquad (8.52)$$

for $\{a,b\}$ constants. As before for example, if the process for S is (8.15) and $G(S)=\log S$, then

$$a = \mu S, \quad b = \sigma S$$

and

$$dG(.)/dS = 1/S, \quad d^2G(.)/dS^2 = -1/S^2, \quad \text{and} \quad dG(.)/dt=0.$$

Hence, applying Ito's Lemma yields

$$d \log S(t) = [\mu - (1/2)\sigma^2]dt + \sigma\, dZ(t)$$

as seen in expression (8.37) as well.

8.8 Asset Price Dynamics and Risk

We have derived and examined, in some detail, two basic stochastic processes in continuous time. One for S and another one for some function of S by applying Ito's Lemma, as, for example, in the cases summarized in Table 8.1. Both processes were driven by the same stochastic component, dZ. There is a useful way to use this information to essentially hedge the risk inherent in an asset price. Suppose we construct a portfolio consisting of the asset whose price is S, and some derivative product of this asset, mathematically represented by the function $f(S)$. Let the portfolio consist of a long position in $f(S)$ and a short position of *DELTA* units of the underlying asset S, for example

$$\pi = f(S) - DELTA\ S \qquad (8.53)$$

where π is the portfolio value. For a constant short position *DELTA*, an infinitesimal change in the value of this portfolio is given by the formula

$$d\pi = df(S) - DELTA\ dS. \qquad (8.54)$$

We may suitably substitute (8.15) and (8.36) into (8.54) to obtain

$$d\pi = \{\mu S\,([df(S)/dS] - DELTA) + (1/2)\,[d^2 f(S)/dS^2]\,\sigma^2 S^2\}\,dt + \sigma S\,([df(S)/dS] - DELTA)\,dZ \tag{8.55}$$

which is an incremental process in π driven by a non-stochastic component,

$$\{\mu S\,([df(S)/dS] - DELTA) + (1/2)[d^2 f(S)/dS^2]\sigma^2 S^2\}\,dt$$

and a stochastic component

$$\sigma S\,([df(S)/dS] - DELTA)\,dZ.$$

Thus, by choosing a short position on S given by

$$DELTA = df(S)/dS, \tag{8.56}$$

the incremental process becomes non-stochastic, or

$$d\pi = (1/2)\,[d^2 f(S)/dS^2]\,\sigma^2 S^2\,dt. \tag{8.57}$$

The lesson here is that by exploring the correlation between S and $f(S)$, through $DELTA = df(S)/dS$, one is able to construct a portfolio that completely eliminates the risk inherent in the underlying asset. We shall see next that this result is crucial for the discussion of option pricing and the Black-Scholes model.

8.9 **Options**

In this section, we discuss the models of asset price dynamics together with options strategies and arbitrage opportunities to obtain the Black-Scholes formula for the price of a simple European Call option. The main assumption is the no arbitrage restriction that all existing profit opportunities vanish instantaneously. An options market has the usual structure that at the current time of contract, say t, an individual is willing to commit to buy or sell an asset at a prearranged price at some date $T > t$ in the future.

We denote by $V(S,t)$ the value of an option at time of contract t to an underlying asset whose price is S. A Call option gives its holder the right to purchase the asset, and we denote its value by $V(S,t)=C(S,t)$. A Put option gives its holder the right to sell the asset, denote its value by $V(S,t)=P(S,t)$.

The common distinction regarding the maturity structure of options is the following. A European option allows its holder to exercise the option only at expiration date, for example if T is the expiration date, only at $T > t$. An American option allows its holder to exercise the option at any time $t \leq T$, before and including the expiration date. In examining option strategies, we define the following parameters:

(i) σ is the standard deviation of the growth rate of the price of the underlying asset;

(ii) B is the agreed upon exercise price of the option at $T > t$, or $T \geq t$ depending on the type of option negotiated;

(iii) $r > 0$ is the risk-free interest rate of a risk-free asset available in the market place.

We first analyze strategies at time of expiry, $T > t$. Consider a European call option at date of expiry, $T > t$. If $S > B$, the best strategy for the holder is to exercise the call option paying B for an asset worth $S > B$, thus making a profit of $S-B$. However, if $S < B$, the best strategy for the holder is not to exercise the call option because it would incur a loss of B minus S, hence the option expires with no value. This logic gives the value of a European call option, at expiry $T > t$, as

$$C(S,T) = MAX\,(S-B,\ 0) \tag{8.58}$$

the so-called the intrinsic value of the option. Figure 8.4a shows how the intrinsic value $C(S,T)$ varies with the price of the underlying asset at $T > t$, given the exercise price B. To the left of $S=B$, the value is zero (no profit, no value), whereas to the right of $S=B$ the value is increasing in S (more profit, value increases).

Similarly, a European put option has intrinsic value

$$P(S,T) = MAX\,(B-S,\ 0) \tag{8.59}$$

and Figure 8.4b shows that to the left of $S=B$ the value is increasing in S (more profit, value increases), whereas to the right of $S=B$ the value is zero (no profit, no value).

Next, we examine the value of the option at time of contract, so that we have to include the risk between time of contract and time of expiry.

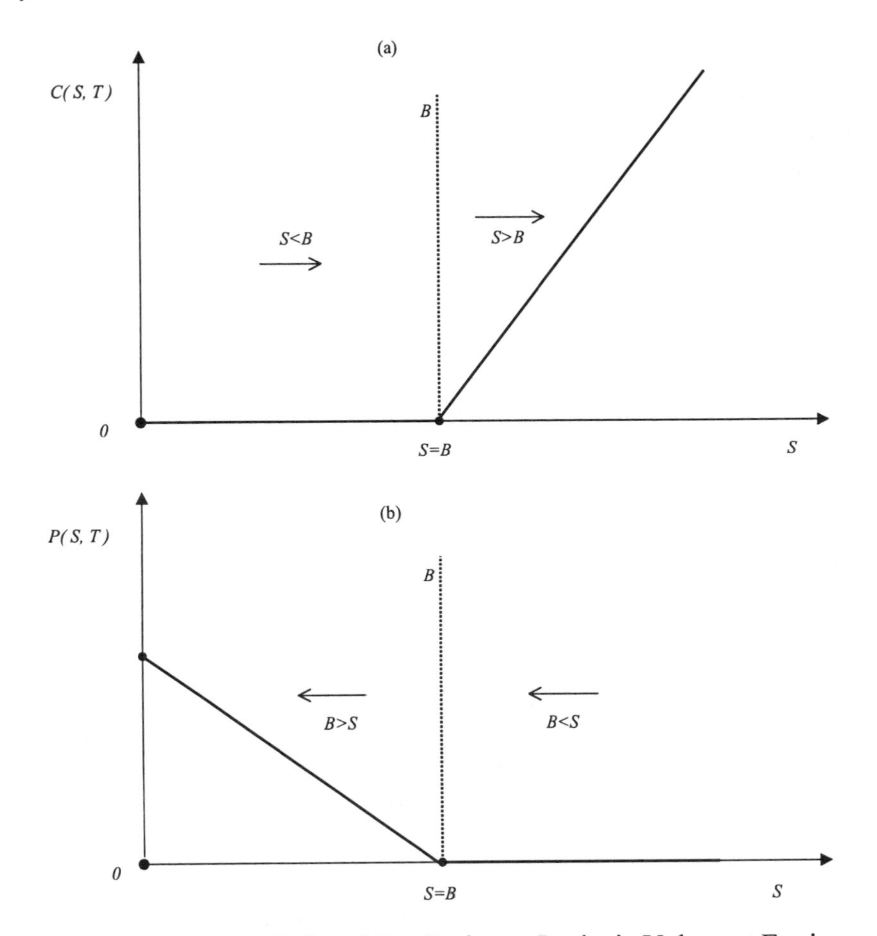

Fig. 8.4 European Call and Put Options - Intrinsic Values at Expiry

8.10 The Black-Scholes Partial Differential Equation

Let the asset price follow a geometric Brownian motion (as studied before) denoted by

$$dS/S = \mu\, dt + \sigma\, dZ, \tag{8.60}$$

where μ is the average of the growth rate of the asset price, σ is its standard deviation, and Z is a standard Wiener process. The underlying assumptions are as follows. The risk-free interest rate $r>0$ is taken as given and non-stochastic. There are no transaction costs and no dividends paid by the asset in question. Trade is continuous, short selling is allowed, and assets are perfectly divisible.

Consider a European option with value $V(S,t)$ on the underlying asset whose price is S. The stochastic differential equation followed by the value $V(S,t)$ is easily obtained by applying Ito's Lemma, or

$$dV(.) = \{ \mu S [\partial V(.)/\partial S] + (1/2) \sigma^2 S^2 [\partial^2 V(.)/\partial S^2] + [\partial V(.)/\partial t]\} dt + \sigma S [\partial V(.)/\partial S] dZ \qquad (8.61)$$

giving the generalized Ito process followed by $V(S,t)$.

The underlying asset and the option may be used to construct a portfolio that hedges the risk inherent in the asset price. As seen in section 8.8, go one unit long on the option and *DELTA* units short on the underlying asset obtaining a portfolio worth

$$\pi = V - DELTA\ S \qquad (8.62)$$

with respective change (for constant *DELTA*) given by

$$d\pi = dV - DELTA\ dS. \qquad (8.63)$$

Substituting (8.60) and (8.61) into (8.63), gives the stochastic differential equation for the value of the portfolio π as

$$d\pi = \{\mu S ([\partial V(.)/\partial S] - DELTA) + (1/2) \sigma^2 S^2 [\partial^2 V(.)/\partial S^2] + [\partial V(.)/\partial t]\} dt + \sigma S ([dV(.)/dS] - DELTA) dZ, \qquad (8.64)$$

and the random component of this process is eliminated by setting

$$DELTA = \partial V(.)/\partial S \qquad (8.65)$$

at the start of the small interval dt. The resulting non-stochastic process for the value of the portfolio is thus given by

$$d\pi = \{(1/2) \sigma^2 S^2 [\partial^2 V(.)/\partial S^2] + [\partial V(.)/\partial t]\} dt. \qquad (8.66)$$

Now, consider the next best alternative for the value of the portfolio, π. The amount π can be invested in a risk-free asset yielding a non-stochastic rate of return

$$d\pi = r\, \pi\, dt \qquad (8.67)$$

over the interval dt, i.e. the value π grows at the risk-free rate $r>0$. By the no arbitrage assumption (and zero transactions costs), an individual must be indifferent between the two strategies, and (8.66) and (8.67) must be equated, or

$$r\,\pi\,dt = \{(1/2)\,\sigma^2 S^2\,[\partial^2 V(.)/\partial S^2] + [\partial V(.)/\partial t]\}\,dt. \qquad (8.68)$$

Substituting (8.62) and (8.65) into (8.68), we obtain the formula

$$(1/2)\sigma^2 S^2[\partial^2 V(.)/\partial S^2]+[\partial V(.)/\partial t]+r\,S\,[\partial V(.)/\partial S] - rV(.)=0, \qquad (8.69)$$

known as the Black-Scholes partial differential equation in the function $V(.)$. A solution for this equation gives a formula for the pricing of European options, the so-called Black-Scholes formula.

Before we examine the Black-Scholes formula, we make three remarks about the Black-Scholes stochastic partial differential equation. First, the equation is the sum of two general components:

$$(1/2)\sigma^2 S^2[\partial^2 V(.)/\partial S^2]+[\partial V(.)/\partial t]$$

is the return on the hedged portfolio given in (8.66); and

$$r\,V(.) - r\,S\,[\partial V(.)/\partial S]$$

is the net return on the 'bank' risk-free deposit.

In the case of European options, those two terms are equated, but for American options this need not be the case. Second, the equation is not a function of the growth rate of the stock price, μ. Hence, the value of the option does not depend on μ. A consequence of this remark is that different individuals may have different estimates of μ, but may indeed value the option on the underlying asset identically. Third, $DELTA=\partial V(.)/\partial S$ is a measure of association between the value of the option and the value of the underlying stock, a variable of utmost importance both in theory and practice, see Notes on the Literature.

8.11 The Black-Scholes Formula for a European Call Option

The partial differential equation for V in (8.69) is a backward parabolic equation. As is, there are many possible solutions for this equation. In order to obtain a unique solution, we must impose boundary conditions. The equation has two differentials in S thus requiring two boundary conditions on S. It has one differential in t and it is a backward equation, thus requiring one boundary endpoint condition in t, so that the equation can be solved backwards.

The boundary conditions for a European call option, whose value is $V(S,t)=C(S,t)$, are given by:

$$(i)\ As\ S \rightarrow 0,\ dS \rightarrow 0,\ and\ C(0,t)=0;$$

$$(ii)\ As\ S \rightarrow \infty,\ C(S,t)\rightarrow S\ and\ C(S,t)=S;$$

$$(iii)\ As\ t\rightarrow T,\ C(S,T)\rightarrow MAX\ (S\text{-}B,\ 0)\ and\ C(S,T)=MAX\ (S\text{-}B,\ 0).$$

The first two conditions are the initial and terminal conditions of the value of the option as a function of the price of the underlying asset. In particular, condition (ii) indicates that as the price of the asset becomes large, the value of the call converges to the price itself, since for large S the exercise price becomes less important. The last condition is the terminal condition at $T > t$. At this terminal date, the value of the call is the non-stochastic intrinsic value.

Given conditions (i)-(iii), the exact solution for (8.69) in the case of the European call option is

$$C(S,t) = S\, N(d_1) - B\, N(d_2)\ [exp(-r\ (T - t))] \qquad (8.70)$$

where

$$d_1 \equiv \{log(S/B)+[r+(1/2)\ \sigma^2](T\text{-}t)\}\ /\ \sigma\ (T\text{-}t)^{\ 1/2};$$

$$d_2 \equiv \{log(S/B)+[r\text{-}(1/2)\ \sigma^2](T\text{-}t)\}\ /\ \sigma\ (T\text{-}t)^{\ 1/2} = d_1 - \sigma\ (T\text{-}t)^{\ 1/2};$$

and

$$N(x) = \int_{-\infty}^{x} [1/(2\pi)^{1/2}]\ [exp\ (-(1/2)\ y^2)]\ dy,$$

is the cumulative distribution function of a standard normal random variable, for y the index of integration.

Equation (8.70) is the Black-Scholes formula for the valuation of a European call option. Consider the exact solution at two extreme points: (a) At expiration, $t = T$, $d_1 = d_2 = +\infty$ *(if* $S > B$*)*, or $d_1 = d_2 = -\infty$ *(if* $S < B$*)*. The cumulative distribution function at those extremes is $N(+\infty)=1$, $N(-\infty)=0$, and hence

$$C(S,T)= MAX\ (S\text{-}B,\ 0)$$

the intrinsic value discussed above;

(b) At $T\rightarrow\infty$, $d_1 = \infty$, $N(+\infty) = 1$, and $[exp\ (-r\ (T - t))] = 0$, and hence

$$C\ (S,\infty) = S,$$

a perpetual call on the asset.

A numerical example by pencil and paper is instructive to illustrate the workings of the Black-Scholes formula. Consider a simplified normalized timeframe so that the current period, at time of contract, is $t=0$; and the expiration is in 3 months with $T=3/12$. A 3-month $(T=3/12>t=0)$ European call option with exercise price $B=\$94.00$ at expiry, on a stock with current price $S(t)=\$100.00$, standard deviation of price change of 20%, $\sigma=0.2$, and risk-free interest rate of 2%, $r=0.02$, has a current price at contract ($t=0$) of $\$7.93$. How? In pencil and paper, with the values given we compute the values $d_1=0.71875$ and $d_2=0.61875$. We can use a numerical polynomial approximation for the cumulative distribution function of a standard normal (within about six decimals) given by

$$N(x) = \begin{cases} 1 - N'(x)\,[a_1 k + a_2 k^2 + a_3 k^3 + a_4 k^4 + a_5 k^5], & \text{for } x \geq 0; \\ 1 - N(-x), & \text{for } x < 0, \end{cases}$$

where

$$N'(x) = [1/(2\pi)^{1/2}][\exp(-(1/2)\,x^2)];$$

$$k=1/(1+0.2316\,x);$$

and

$$a_1=0.3194;\ a_2=-0.3565;\ a_3=1.7815;\ a_4=-1.8212;\ a_5=1.3303.$$

Thus, using the approximation we obtain $N(0.71875)=0.7639$ and $N(0.61875)=0.7319$ for the probabilities under the standard normal. The value at contract ($t=0$) of this call option is then

$$C(\$100,0) = (100.00 \times 0.7639) - (94.00 \times 0.7319 \times 0.9950) = \$7.93.$$

It is important to note that this is the price at time of contract $t=0$, not at expiry $T>t$. At expiry, when S varies we obtain alternative intrinsic values as in Figures 8.4 with linear portions. However, if we vary S in the Black-Scholes formula (8.70), the relationship is nonlinear because of the risk carried between the time of contract and the time of expiry. Notice also that in the case the stock is risk-free itself, $\sigma=0$, then $d_1 = d_2$

$= +\infty$ for $S>B$, and $N(+\infty)=1$, thus the value of the option at contract ($t=0$) is

$$C(S,t) = S - B\,[\exp(-r\,(T - t))] = \$6.47$$

that is the time value of money with the adjustment for the delay between contract and expiry at *2%* risk-free interest, i.e. it is not the intrinsic value *S-B=$6.00.*

Also, from equations (8.65) and (8.70), the *DELTA* of the call option, or the measure of association between the value of the option and the value of the underlying stock, is

$$DELTA = \partial C(.)/\partial S = N(d_1)$$

which may be used to implement strategies with the call option.

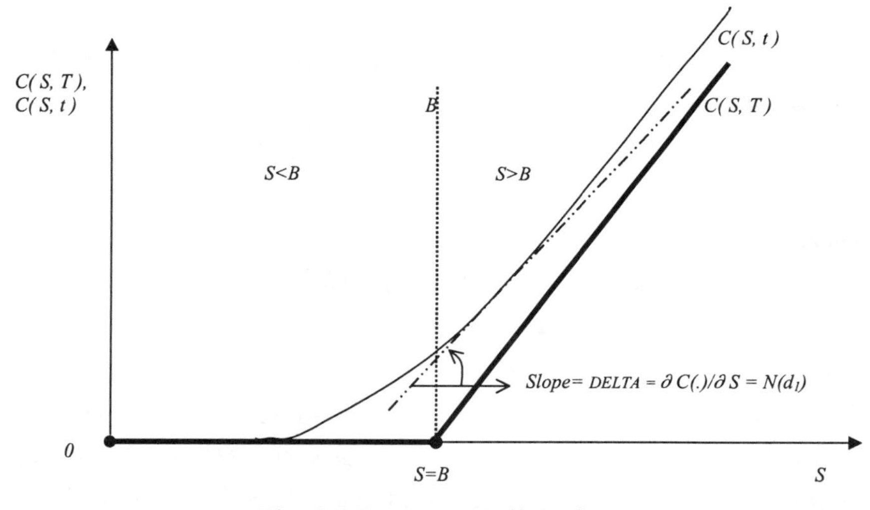

Fig. 8.5 European Call Option

Figure 8.5 illustrates the points made above for the European call option case. The intrinsic value is the linear relationship in the two ranges, *C(S,T)*. The call option, according to the Black-Scholes formula, is the nonlinear relationship that takes into account the risk between the time of contract and expiry, *C(S,t)*. The slope of the call option, evaluated at the point at the tangent line to *C(S,t)*, is the *DELTA* of the call option, which equals $N(d_1)$. Alternatively, in the case of an European put option, we have $DELTA = \partial P(.)/\partial S = N(d_1) - 1$.

8.12 Summary I

In sections 8.1-8.11, we have covered basic material relating to the modeling of stochastic processes in continuous time. One important application of this methodology is in modern finance theory, the so-called Black-Scholes analysis discussed here. In the remainder of this chapter, we apply the methodology to study dynamic stochastic general equilibrium models with focus on economic growth and asset pricing.

8.13 Introduction to Equilibrium Stochastic Models

We start examining simple models of wealth accumulation and portfolio choice when returns to a risky asset are driven by Wiener processes (Brownian motion or diffusion processes). We obtain solutions to the allocation problem by using a continuous time version of Bellman's dynamic programming techniques, the same techniques used for the discrete time case in Chapter 7 and presented in Chapter 1. We then extend the models to include discussions of the role of risk aversion and capital accumulation.

The models lend itself naturally to the problem of costs of economic fluctuations, and we examine this issue in some detail as well. We start with the case of logarithmic utility without capital accumulation, and extend to the case of constant relative risk aversion (*CRRA*) utility. Finally, we introduce capital accumulation and briefly discuss an extension that separates risk aversion and intertemporal substitution.

8.14 Consumption Growth and Portfolio Choice with Logarithmic Utility

We start presenting the basic macroeconomic structure of a simple consumption growth and portfolio choice model without capital accumulation. Identical individuals populate the economy. There are two assets available in the market place:

(i) a is a risky asset with return per unit of time

$$dR_a = r_a dt + dZ \qquad (8.71)$$

where $r_a > 0$ is the mean return per unit of time and dZ is a normally distributed temporally independent random component with mean zero, $E[dZ]=0$, constant variance per unit of time, $E[(dZ)^2]=\sigma_z^2 dt$, and $Z(0)=0$ with probability one. Formula (8.71) is the familiar geometric Brownian motion studied in sections 8.4-8.5;

(ii) b is a risk-free asset with return per unit of time

$$dR = r\ dt \qquad (8.72)$$

where $r>0$ is the non-stochastic return.

Real wealth for a typical individual is the sum of the two assets

$$w = a + b \qquad (8.73a)$$

and we may define the shares in an individual portfolio as

$$n \equiv a/w, \quad 1 - n \equiv b/w \qquad (8.73b)$$

where no restrictions on n are imposed momentarily.

(a) Consumer Problem

The typical consumer chooses non-stochastic consumption to wealth ratio and portfolio shares to maximize expected utility subject to appropriate budget constraints and given initial conditions and prices. In this first simple case, we assume that utility is logarithmic

$$u(c) = \log c \qquad (8.74)$$

which implies a coefficient of relative risk aversion of unity equal to the inverse of the elasticity of intertemporal substitution. The individual flow budget constraint is written as

$$da + c\, dt + db = b\, dR + a\, dR_a \qquad (8.75a)$$

where income from holding the two assets, $b\ dR + a\ dR_a$, is spent in either consumption, $c\ dt$, or investment in the two available assets, $da+db=dw$. Using expressions (8.73a,b), (8.75a) yields the evolution of individual wealth as

$$dw = w\,[\,n\ dR_a + (1-n)\ dR\,] - c\ dt \qquad (8.75b)$$

which upon substitution of (8.71)-(8.72) yields the stochastic differential equation

$$dw/w = [\,n\ r_a + (1\text{-}n)\ r - (c\,/\,w\,)]\ dt + n\ dZ \qquad (8.75c)$$

where $E[n\ dZ]=0$ and $var(n\ dZ)= n^2\sigma_z^{\ 2}dt$.

Each individual solves a stochastic control problem to

$$Max\ E_o \int_0^{\infty} u(c\ )\ exp(-\beta\, t)\ dt \qquad (8.76)$$

subject to the accumulation equation (8.75c), by choice of consumption to wealth ratio, $(c\,/w)$, and the share of the risky asset in the portfolio, n, given prices, $\{r_a\,,r\}$, and the initial wealth, $w(0)$, where utility is in (8.74), and $\beta > 0$ is the individual subjective rate of time preference common across all individuals.

Along the balanced growth path from the appropriate initial conditions, we have that both the portfolio shares and the marginal propensity to consume out of wealth, $n>0$ and $(c\ /w\ )>0$, must be constant. Hence, individual wealth and consumption grow at an average constant rate denoted ψ, that is

$$E[dw\,/w\,]= E[dc\,/c\,]= \psi\ dt. \qquad (8.77)$$

From the accumulation equation (8.75c) we have that wealth evolves according to

$$dw\,/w= \psi\ dt + n\ dZ, \qquad (8.78a)$$

a geometric Brownian motion, where $\psi\ dt$ is the deterministic component and $n\ dZ$ is the stochastic component. Hence, the mean growth and variance of growth are respectively

$$\psi \equiv n\ r_a + (1\text{-}n)\ r\ - (c\,/w\,) \qquad (8.78b)$$

$$E[(dw/w)^2]/dt= \sigma_{\psi}^{\ 2} = n^2\sigma_z^2 \qquad (8.78c)$$

A solution for the stochastic control problem (8.76) is obtained as follows, e.g. Chapter 1. A value function for the problem with logarithmic utility is given by

$$V(w,t) = \underset{\{c/w,\, n\}}{Max} \quad E_o \int_0^\infty [\log w + \log (c/w)] \exp(-\beta t)\, dt. \qquad (8.79)$$

Given the exponential discount, we can write the value function in the time separable form

$$V(w,t) = \exp(-\beta t)\, X(w) \qquad (8.80)$$

where X is a twice continuously differentiable function to be determined below. Let the differential generator of the value function be defined as

$$L(V) \equiv (\partial V/\partial t) + \{w\, [n\, r_a + (1-n)\, r - (c\,/\,w\,)]\}\, (\partial V/\partial w) +$$
$$(1/2)\, n^2 \sigma_z^2\, w^2\, (\partial^2 V/\partial w^2). \qquad (8.81)$$

Individual optimization consists of choices of $\{c/w,n\}$, that maximize the Lagrangean function

$$\mathcal{L} = \exp(-\beta t)\, [\log w + \log (c/w)] + L(\, V\, (w\,,t)) \qquad (8.82)$$

with necessary first order conditions for $\{c/w\,,\, n\}$ given by

$$\exp(-\beta t)\, (c/w)^{-1} - w\, (\partial V/\partial w) = (c/w\,)^{-1} - \, w\, \, (d\, X\,/\, d\, w) = 0 \qquad (8.83a)$$

$$(r_a - r)\, w\, (\partial V/\partial w) + [cov(ndZ\,,dZ\,)/dt]\, w^2\, (\partial^2 V/\partial w^2) =$$
$$(r_a - r)\, w\, (d\, X\,/\partial w) + (n\, \sigma_z^2\,)\, w^2\, (d^2 X\,/dw^2) \;=\; 0 \qquad (8.83b)$$

where $cov(ndZ\,,dZ\,)/dt = n\, \sigma_z^2$.

The value function must satisfy the Bellman functional equation:

$$\underset{\{c/w,\, n\}}{Max} \{\, \exp(-\beta t)\, [\log w_i + \log (c\,/\,w\,)\,] + L(\, V\, (w\,,t))\} = 0. \qquad (8.84)$$

Substituting for L from (8.81), where $\{c/w,n\}$ are the optimized values satisfying the first order necessary conditions (8.83a,b), into the Bellman equation and canceling $\exp(-\beta t)$ yields

$$\log w + \log (c/w) - \beta X + \{w\, [\, n\, r_a + (1-n)\, r - (c\,/\,w\,)]\}\, (dX/d\, w\,) +$$
$$(1/2)\, n^2\, \sigma_z^2\, w^2 (\, d^2 X/d\, w^2\,) = 0. \qquad (8.85)$$

This is an ordinary differential equation in the function X. A solution for (8.85) can be found by a suitable guess for the function $X(w\,)$, i.e. a guess-and-verify method. Given the logarithmic form of utility, we propose a guess of the form

$$X(w\,) = \kappa_0 \; + \kappa_1 \log w \qquad (8.86)$$

where (κ_0, κ_1) are constants to be appropriately determined. For the guess given by (8.86), we have

$$dX/dw = \kappa_1 w^{-1} \tag{8.87a}$$

$$d^2X/dw^2 = -\kappa_1 w^{-2}. \tag{8.87b}$$

First, we substitute the guess into the first order conditions (8.83a,b) for $\{c/w, n\}$ to obtain

$$\kappa_1 = (c/w)^{-1} \Rightarrow \quad log\, \kappa_1 = -log\,(c/w) \tag{8.88a}$$

$$n = (r_a - r)/\sigma_z^2 \tag{8.88b}$$

where $\{c/w, n\}$ are the optimal values, and $n > 0$ implies that $(r_a - r) > 0$ or the risky asset yields a positive premium. Note that for this simple case of logarithmic utility, the solution for the share of the risky asset is obtained directly by (8.88b). The next step is to substitute the condition (8.88a), using (8.78b), into the Bellman functional equation to obtain

$$log\, w - log\, \kappa_1 - \rho\,(\kappa_0 + \kappa_1\, log\, w) + \kappa_1\, [\psi - (1/2)\, n^2\, \sigma_z^2\,] = 0. \tag{8.89}$$

This is an equation in $log\, w$ and constants, given the optimal values of $\{c/w, n\}$. Thus, (8.89) is satisfied for a recursive choice of constants given by

$$\kappa_1 = 1/\beta \tag{8.90a}$$

$$\kappa_0 = [\,\beta\, log\, \beta + \psi - (1/2)\, n^2\, \sigma_z^2\,]\,/\,\beta^2. \tag{8.90b}$$

The solution for the stochastic control problem (8.76), along a stochastic balanced growth path with $dc/c = dw/w$ stationary and constant share $n > 0$, is then

$$c/w = \beta \tag{8.91a}$$

$$n = (r_a - r)/\sigma_z^2 \tag{8.91b}$$

with mean growth given by

$$\psi = [(r_a - r)/\sigma_z^2\,]\, r_a + (1 - [(r_a - r)/\sigma_z^2])\, r - \beta, \tag{8.91c}$$

and $(r_a - r) > 0$, where the variance of the growth rate per unit of time is given by (8.78c). Expressions (8.91a,b) yield solutions for c/w, and n as a function of the exogenous stochastic processes and prices.

We can verify whether the guess (8.86) is indeed correct by checking the transversality condition

$$lim_{t \to \infty}\, E_0\, [\, exp(-\beta\, t)\, X(w)] = 0. \tag{8.92}$$

Given the form of the value function in (8.86), we have to solve the stochastic differential equation in (8.78a) in order to check the transversality condition. Towards a solution for (8.78a), we integrate both sides in the range $(0,t)$ to obtain

$$\int_0^t dw/w = \psi \int_0^t dt + n \int_0^t dZ = \psi\, t + n\, Z(t) \tag{8.93}$$

where $Z(0)=0$ with probability one. In order to evaluate the left-hand side, $\int_0^t dw/w$, we use Ito's Lemma for the function $g(w)=\log w$. In this case, by Ito's Lemma

$$\begin{aligned} d \log w &= (\partial \log w / \partial w)\, dw + (1/2)\, (\partial^2 \log w / \partial w^2)\, (dw)^2 \\ &= (dw/w) - (1/2)\, n^2 \sigma_z^2\, dt \end{aligned} \tag{8.94}$$

where we used the fact that by the Brownian motion dw/w in (8.78a), $(dw)^2 = n^2 w^2 \sigma_z^2 dt$. Hence, from (8.94),

$$dw/w = d \log w + (1/2)\, n^2 \sigma_z^2 dt \tag{8.95}$$

which can be integrated over the range $(0,t)$ to obtain

$$\begin{aligned} \int_0^t dw/w &= \int_0^t d\log w + (1/2)\, n^2 \sigma_z^2 \int_0^t dt \\ &= \log w(t) - \log w(0) + (1/2)\, n^2 \sigma_z^2\, t \end{aligned} \tag{8.96}$$

where $w(t)$ is written as a function of time. Equating (8.96) to (8.93), we obtain a solution

$$\log w(t) = \log w(0) + [\psi - (1/2)\, n^2 \sigma_z^2\,]\, t + n\, Z(t). \tag{8.97}$$

Finally, to substitute into the transversality condition, we compute the expected value as

$$E_0 [\log w(t)] = \log w(0) + [\psi - (1/2)\, n^2 \sigma_z^2\,]\, t \tag{8.98}$$

since $E_0\ [Z(t)]=Z(0)=0$. Substituting into the transversality condition, reduces to

$$\lim_{t\to\infty} \{\exp(-\beta t)\, [\psi - (1/2)\, n^2 \sigma_z^2\,]\, t\} = 0 \;\Leftrightarrow\; \beta > 0. \tag{8.99}$$

As long as $\beta > 0$, the transversality condition is satisfied because the expression in the limit decreases exponentially and increases linearly with t so that the limit converges to zero. We confirm that the solution for the proposed value function is correct. In effect, the condition $\beta > 0$, implies that $c/w>0$ as well.

As we mentioned, this framework lends itself naturally for the study of the welfare costs of risk. Since here diffusion processes drive risk, variances have an effect on means thus providing an additional channel for risk to impact on welfare. We can compute cumulative welfare and welfare changes in this simple framework. The welfare criteria in (8.76) with logarithmic utility is:

$$\Omega \equiv E_o \int_0^\infty [\log w(t) + \log \beta] \exp(-\beta t)\, dt. \qquad (8.100)$$

Substituting the solution for $\log w(t)$ from (8.97) and rearranging we obtain

$$\Omega \equiv \log \beta \int_0^\infty \exp(-\beta t)\, dt + E_o \int_0^\infty \{ \log w(0) + [\psi - (1/2)\, n^2 \sigma_z^2]\, t + n\, Z(t)\} \exp(-\beta t) dt.$$

$$= (1/\beta) \log \beta + E_o \int_0^\infty \{\log w(0) + [\psi - (1/2)\, n^2 \sigma_z^2]\, t + n\, Z(t)\} \exp(-\beta t)\; dt.$$

$$= (1/\beta) \log \beta + \log w(0) \int_0^\infty \exp(-\beta t)\, dt + E_o \int_0^\infty \{[\psi - (1/2)\, n^2 \sigma_z^2]\, t + n\, Z(t)\} \exp(-\beta t) dt.$$

$$= (1/\beta)[\log \beta + \log w(0)] + [\psi - (1/2)\, n^2 \sigma_z^2] \int_0^\infty t \exp(-\beta t)\, dt$$

$$= (1/\beta)[\log \beta + \log w(0)] + [\psi - (1/2)\, n^2 \sigma_z^2]\, (1/\beta)^2 \qquad (8.101)$$

where we used the assumption that $\beta > 0$, the fact that $E_o[Z(t)] = Z(0) = 0$, and integration by parts implies that

$$\int_0^\infty t \exp(-\beta t) dt = \int_0^\infty d[t(-1/\beta)\exp(-\beta t)] - \int_0^\infty (-1/\beta)\exp(-\beta t) dt = (1/\beta)^2.$$

The formula for cumulated welfare in (8.101) can be used to measure welfare changes in an equivalent variation manner. We use (8.101) to find the change in $w(0)$ so that $d\Omega = 0$. In effect, we answer the question of how much would an individual be willing to give up in initial wealth to keep cumulative welfare unchanged in the presence of shocks. First, we totally differentiate (8.101) to obtain

$$d\Omega = (1/\beta\, w(0))\, dw(0) + (1/\beta)^2\, [\, d\psi - n\, \sigma_z^2\, dn - (1/2)\, n^2\, d\sigma_z^2] \qquad (8.102)$$

which may be appropriately solved for $dw(0)$ so that $d\Omega = 0$. After using (8.78b) for $d\psi$, the result is

$$dw(0)|_{d\Omega=0} = [w(0)/\beta]\, [(1/2)\, n^2\, d\sigma_z^2 + d(c/w)] \qquad (8.103)$$

which may be appropriately evaluated for changes in the exogenous moment σ_z^2. We use the model to gain insight on the role of shocks on mean growth, welfare, and consumption.

(b) Qualitative Comparative Statics

Qualitative comparative statics of changes in the variance of the return on the risky asset, $d\sigma_z^2$, are obtained from (8.91a,b), (8.78c), and (8.103). The results are

$$d(c/w)\,/\,d\sigma_z^2 = 0 \tag{8.104a}$$

$$dn\,/\,d\sigma_z^2 = -\,n\,/\,\sigma_z^2 < 0 \tag{8.104b}$$

$$d\psi/\,d\sigma_z^2 = -\,n^2 < 0 \tag{8.104c}$$

$$d\,\sigma_\psi^2/\,d\sigma_z^2 = n^2 + \sigma_z^2\,2\,n\,(dn\,/\,d\sigma_z^2) = -\,n^2 < 0 \tag{8.104d}$$

$$d\,w\,(0)|_{d\Omega=0}\,/\,d\sigma_z^2 = [w(0)/\beta]\,(1/2)\,n^2 > 0 \tag{8.104e}$$

In (8.104a), we obtain the usual result that the marginal propensity to consume out of wealth is the rate of time preference so that an increase in the variance of the return of the risky asset has no effect. This is a consequence of the logarithmic utility assumption. From (8.104b), the share of the risky asset in the portfolio is decreasing on the variance of the risky asset. This is because the riskier the prospect the more the individual will hold the risk-free asset in the portfolio. In turn, the effect on the mean growth of consumption and wealth in (8.104c) is also negative. The variance of the growth rate is decreasing in the variance of the return because the portfolio effect in (8.104b) dominates the direct variance effect of σ_z^2. The welfare effect is in (8.104e). As mentioned, this is an equivalent variation measure. It gives the necessary change in initial wealth required to offset the change in risk to maintain lifetime utility constant. The sign is positive indicating that in order to maintain constant utility, an increase in risk requires the agent to receive additional initial wealth at the rate $(1/2\beta)\,n^2$. It indicates the cost of an additional unit of fluctuation (variance) in terms of initial wealth to maintain constant utility. The cost is increasing in n, the share of the risky asset, and decreasing in β, the subjective rate of time preference.

8.15 Consumption Growth and Portfolio Choice with *CRRA* Utility

The basic asset structure of equations (8.71),(8.72),(8.73a,b) above is maintained in this section. The consumer problem is changed because we assume utility to take the constant relative risk aversion (*CRRA*) form

$$u(c) = (1/\gamma)\, c^{\gamma} \tag{8.105}$$

where $\gamma < 1$ yields a coefficient of relative risk aversion of $(1-\gamma)$, for $\gamma=0$ denoting the logarithmic utility case in (8.74); and $(1-\gamma)^{-1}$ is the elasticity of intertemporal substitution. The budget constraint in (8.75) remains the same as well as the problem (8.76) and growth characterization (8.77)-(8.78a,b). The solution for the stochastic control problem is where the differences start to show. A value function for (8.76) takes the same form as (8.80) and the differential generator of the value function is the same as (8.81). Individual optimization consists of choices of $\{c/w, n\}$, that maximizes the Lagrangean function

$$\mathcal{L} = \exp(-\beta t)\, (w^{\gamma}/\gamma)\, [(c/w)]^{\gamma} + L(V(w,t)). \tag{8.106}$$

In the case of *CRRA* utility, the necessary first order conditions for $\{c/w,n\}$ are given by

$$w^{\gamma} (c/w)^{\gamma-1} = w\, dX/dw \tag{8.107a}$$

$$(r_a - r)\;\; w\, dX/dw + [cov(ndZ, dZ)/dt]\, w^2 d^2X/dw^2 = 0. \tag{8.107b}$$

where $cov(ndZ,dZ)/dt = n\sigma_z^2$.

The value function must satisfy the Bellman functional equation:

$$\underset{\{c/w,n\}}{Max} \{ \exp(-\beta t)\, (w^{\gamma}/\gamma)\, (c/w)^{\gamma} + L(V(w,t))\} = 0. \tag{8.108}$$

Substituting for L into the Bellman equation and canceling $\exp(-\beta\, t)$ yields the ordinary differential equation in the function X as

$$(w^{\gamma}/\gamma)\, (c/w)^{\gamma} - \beta X + \{w\, [n\, r_a + (1-n) r - c/w\,]\}\, (dX/dw) + (1/2)\, n^2 \sigma_z^2\, w^2\, (d^2X/dw^2) = 0. \tag{8.109}$$

A solution for (8.109) can be found by a suitable guess for the function $X(w)$, but with *CRRA* utility the guess takes a different the form compared to (8.86). In this case, we postulate

$$X(w) = \kappa\, w^{\gamma} \tag{8.110a}$$

where κ is a constant to be determined. For a guess given by (8.110a), we have

$$dX/dw = \kappa\gamma\, w^{\gamma-1} \qquad (8.110b)$$

$$(d^2X/dw^2) = \kappa\gamma\,(\gamma-1)\, w^{\gamma-2}. \qquad (8.110c)$$

Substituting into the first order condition for $\{c/w, n\}$, (107a,b) we obtain

$$\kappa = (c/w)^{\gamma-1}/\gamma \quad \Leftrightarrow \quad (c/w) = (c/w)^{\gamma}/\gamma\kappa. \qquad (8.111)$$

$$n = (r_a - r)/(1-\gamma)\,\sigma_z^2. \qquad (8.112a)$$

A solution for the share of the risky asset is obtained directly from (8.112a), for $n>0$ with $(r_a-r)>0$. Substituting (8.111) into the Bellman equation (8.109) we obtain a solution for (c/w) as

$$c/w = (1-\gamma)^{-1}\{\beta - \gamma[n\, r_a + (1-n)\, r] - (1/2)\gamma(\gamma-1)\, n^2\sigma_z^2\} \qquad (8.112b)$$

and the solution for the mean growth rate is

$$\psi = n\, r_a + (1-n)\, r - (c/w). \qquad (8.112c)$$

In this case, the mean growth and variance of growth are the same as presented in (8.77), (8.78a,b,c). The solution for consumption and portfolio shares, in this case, depends on the parameter $\gamma<1$, where $(1-\gamma)$ is the coefficient of relative risk aversion (*CRRA*). It is easy to note from (8.112a,b) that in the logarithmic utility case, $\gamma=0$, we have the results obtained in section 8.14. The *CRRA*, $(1-\gamma)$, is usually taken to be above one, implying that the parameter $\gamma<-1$; however other cases cannot be ruled out. In (8.112a), we note that risk aversion has the same qualitative effect as the variance of the return on the risky asset.

Consumption in expression (8.112b) is more complicated. The marginal propensity to consume out of wealth has two additional terms. The first term is a portfolio effect, given by the expression

$$-\gamma[n\, r_a + (1-n)\, r]/(1-\gamma) > 0,$$

for $\gamma<0$. The positive effect on consumption out the portfolio effect is associated with the benefit of investing, given risk aversion. The second is a precautionary motive effect, $(1/2)\gamma\, n^2\sigma_z^2<0$, for $\gamma<0$. This is negative because risk aversion interacts with the variance of the return on the risky asset making the individual consume less to buffer the volatility in asset markets. In the case of logarithmic utility, those two effects cancel out. In the *CRRA* case, they do not cancel out, and a better understanding of the effect on consumption can be obtained when risk aversion and intertemporal substitution are separated, e.g. Chapter 7 and section 8.17 below.

In this case, for a solution along a stochastic balanced growth path with $dc/c=dw/w$ stationary, constant share $n>0$, and using (8.110a), the transversality condition takes the form

$$lim_{t\to\infty} E_0 [\, exp(-\beta t)\, w_i^{\gamma}] = 0. \qquad (8.113)$$

A solution for the evolution of wealth is similar to (8.93)-(8.97), where here we rewrite (8.97) as

$$w(t) = w(0)\, exp([\, \psi - (1/2)\, n^2\sigma_z^2\,]\, t + n\, Z(t)\,) \qquad (8.114)$$

where $Z(0)=0$ with probability one. Upon substitution of (8.114) into (8.113) yields

$$[w(0)^{\gamma}]\, lim_{\, t\to\infty} E_0[exp(\gamma\, [\psi -(1/2)\, n^2\sigma_z^2]t - \beta t + \gamma n\, Z(t))] = 0. \qquad (8.115)$$

The expectation of the exponential operator is in this case

$$E_o\, [exp(\gamma\, [\, \psi - (1/2)\, n^2\sigma_z^2\,]\, t - \beta t + \gamma n\, Z(t))] =$$
$$exp(\, [\gamma\psi - \beta + (1/2)\gamma\, (\gamma -1)\, n^2\, \sigma_z^2\,]t) \qquad (8.116)$$

so that the limit converges to zero as long as

$$[\gamma\psi - \beta + (1/2)\gamma\, (\gamma -1)\, n^2\, \sigma_z^2\,] < 0 \qquad (8.117a)$$

which by (8.112b) implies

$$c/w > 0. \qquad (8.117b)$$

Hence, as long as $c/w > 0$, the guess is correct and the value function (8.110a) is the appropriate choice that satisfies the transversality condition.

We can compute cumulative welfare and welfare changes in this case as well. The welfare criteria in (8.76) using the functional form (8.105) is:

$$\Omega_1 \equiv (1/\gamma)(c/w)^{\gamma}\, E_o \int_0^{\infty} w\,(t)^{\gamma}\, exp(-\beta t)\, dt. \qquad (8.118)$$

Using the expressions (8.114)-(8.116) above, and defining

$$H \equiv [\gamma\psi - \beta + (1/2)\gamma\, (\gamma -1)\, n^2\sigma_z^2],$$

the integral in (8.118) becomes:

$$w\,(0)^{\gamma} \int_0^{\infty} exp(H\, t)\, dt = -\, w\,(0)^{\gamma}/\, H, \; for\;\; H<0,$$
$$where\;\;\; H<0 \Leftrightarrow (c/w)>0$$

is the condition that satisfies the transversality condition as in (8.117a,b). Thus, the solution for cumulative welfare is

$$\Omega_1 = (1/\gamma)\,(c/w)^{\gamma}\,w(0)^{\gamma} / \{\beta - \gamma[\psi + (1/2)(\gamma - 1)\,n^2\sigma_z^2]\}. \qquad (8.119)$$

The measure of welfare changes is the equivalent variation measure. As before, we use (8.119) to find the change in $w\,(0)$ so that $d\Omega_1 = 0$. Totally differentiating (8.119), we obtain

$$d\Omega_1 = ([\,(c/w)^{\gamma-1}\,d(c/w)\,w(0)^{\gamma} + (c/w)^{\gamma}\,dw(0)\,w(0)^{\gamma-1}\,]\,\{\beta - \gamma[\psi + (1/2)(\gamma - 1)\,n^2\sigma_z^2]\} + [\gamma\,d\psi + \gamma(1/2)(\gamma - 1)\,n^2\,d\sigma_z^2 + \gamma(\gamma - 1)\,n\,\sigma_z^2\,dn] \times \{(1/\gamma)\,(c/w)^{\gamma}\,w(0)^{\gamma}\}) / \{\beta - \gamma[\psi + (1/2)(\gamma - 1)n^2\sigma_z^2]\}^2 \qquad (8.120)$$

which may be appropriately solved for $dw\,(0)$ so that $d\Omega_1 = 0$. The result is

$$dw\,(0)|_{d\Omega 1=0} = -\,\{\Omega_1\,\gamma\,[\,d\psi + n(\gamma - 1)\,(\,(1/2)\,n\,d\sigma_z^2 + \sigma_z^2\,dn\,)] - (c/w)^{\gamma-1}\,d(c/w)\} / (c/w)^{\gamma}\,w(0)^{\gamma-1} \qquad (8.121)$$

which may be appropriately evaluated for changes in the exogenous moment σ_z^2, and risk aversion $(1-\gamma)$. In this case, we may use the model to gain insight on the role of shocks and relative risk aversion on mean growth, portfolio choice, consumption, and welfare.

(a) Qualitative Comparative Statics

Qualitative comparative statics of changes in the variance of the return on the risky asset, $d\sigma_z^2$, are obtained from (8.112a,b). The results are

$$\partial(c/w) / \partial\sigma_z^2 = \gamma\,\{(1/2)\,n^2 - [(r_a - r)(1-n)/(1-\gamma)\sigma_z^2]\} \lesseqgtr 0 \qquad (8.122a)$$

$$\partial n / \partial\sigma_z^2 = -\,n / \sigma_z^2 < 0 \qquad (8.122b)$$

$$\partial\,\sigma_\psi^2 / \partial\sigma_z^2 = -\,n^2 < 0 \qquad (8.122c)$$

In (8.122a), we observe that an increase in the variance of the return on the risky asset has two opposing effects on the consumption to wealth ratio. An increase in σ_z^2 activates the precautionary saving motive, and saving (consumption) is pushed upwards (downwards) by the term $\gamma\,(1/2)\,n^2 < 0$, for the case $\gamma < 0$. On the other hand, an increase in σ_z^2, indicates opportunities for higher portfolio gains pushing consumption upwards by the term $-\gamma\,[(r_a - r)(1-n)/(1-\gamma)\sigma_z^2] > 0$, for $\gamma < 0$. In (8.122b), the result is unambiguous, higher σ_z^2 is associated with lower investment in the risky asset, and in (8.122c) the effect on the variance of the growth rate is identical to (8.104d).

We can repeat the same exercise for changes in the level of risk aversion. Qualitative comparative statics of changes in risk aversion, $\partial(1-\gamma)$, are obtained from (8.112a,b) as

$$\partial(c/w)/\partial(1-\gamma) = (1-\gamma)^{-1}\{\psi + [2n\gamma(r_a - r)/(1-\gamma)] + n^2((1-\gamma-(1/2))\sigma_z^2\} \lesseqgtr 0 \tag{8.123a}$$

$$\partial n/\partial(1-\gamma) = -n/(1-\gamma) < 0 \tag{8.123b}$$

$$\partial \sigma_\psi^2/\partial(1-\gamma) = -2n^2\sigma_z^2/(1-\gamma) < 0 \tag{8.123c}$$

In (8.123a), we again observe two opposing effects on the consumption to wealth ratio. An increase in $(1-\gamma)$ activates the precautionary motive, and saving (consumption) moves upwards (downwards) by the term $(1-\gamma)^{-1}[2n\gamma(r_a - r)/(1-\gamma)]<0$, for $\gamma<0$. On the other hand, an increase in $(1-\gamma)$ indicates the need for higher portfolio gains thus pulling consumption upward. This effect can be seen in the term $(1-\gamma)^{-1}\{\psi+n^2((1-\gamma-(1/2))\sigma_z^2\}>0$, for $\gamma<0$. A better understanding of the risk aversion effect on the marginal propensity to consume requires separation of risk aversion and intertemporal substitution as mentioned above. In (8.123b), the result is unambiguous, higher risk aversion is associated with lower investment in the risky asset. In (8.123c), an increase in risk aversion decreases the variance of the growth rate through the portfolio channel in (8.123b).

In both cases, the effect on the mean growth rate is obtained as

$$\partial\psi_i/\partial j = (r_a - r)(\partial n/\partial j) - [\partial(c/w)/\partial j], \quad j \equiv \{(1-\gamma), \sigma_z^2\}. \tag{8.124}$$

The welfare effect can be appropriately obtained from (8.121), and left as an exercise for the reader.

8.16 **Capital Accumulation and Asset Returns**

We extend the framework in sections 8.14-8.15 by introducing physical capital accumulation. As before, all agents are identical and there is only one good, but now the good may be consumed or invested in physical capital, the variable factor input. A typical individual has access to an instantaneous stochastic technology yielding a flow of output per unit of time dq given by

$$dq = A\, k\, (\, dt + dy\,) \tag{8.125}$$

for a constant $A>0$. The variable input is k, denoting the individual instantaneous stock of capital. The stochastic component of the technology is as follows: dy is a normally distributed temporally independent economy-wide technology shock common to all individuals with mean zero, $E[dy]=0$, and constant variance per unit of time, $E[(dy)^2]=\sigma_y^2 dt$. Equation (8.125) denotes a simple continuous time version of the stochastic *"Ak"* technology studied in Chapter 7, sections 7.18-7.21.

The equilibrium return is obtained accordingly from (8.125). The private return to capital is denoted

$$dR_k = (r_k - \delta)dt + dz_k \tag{8.126}$$

where $\delta\, dt$ is the non-stochastic rate of depreciation of capital per unit of time, $\delta \in [0,1]$ and

$$r_k = A \tag{8.127a}$$

$$dz_k = A\;dy \tag{8.127b}$$

Note again that $r_k - \delta$ is the mean net return to capital, and the return to capital, using (8.127a,b) follows the geometric Brownian motion

$$dR_k = (A - \delta)dt + A\, dy. \tag{8.128}$$

As for private wealth, individuals hold their own capital stock, k, and trade in risk-free bonds, denoted b, yielding the non-stochastic return, $r\, dt$, per unit of time. Hence, real wealth for a typical individual is

$$w = k + b \tag{8.129a}$$

and we may define the shares in an individual portfolio as

$$n_k \equiv k/w, \quad 1 - n_k \equiv b/w \tag{8.129b}$$

where no restrictions on n_k are imposed momentarily.

The consumer problem is the usual. The typical consumer chooses non-stochastic consumption to wealth ratio and portfolio shares to maximize expected utility subject to appropriate budget constraints and given initial conditions and prices. Individual instantaneous utility takes the *CRRA* form (8.105) repeated here

$$u(c) = (1/\gamma)\, c^{\gamma} \qquad (8.130)$$

where $\gamma < 1$ yields a coefficient of relative risk aversion of $(1-\gamma)$, for $\gamma=0$ denoting the logarithmic utility case; with elasticity of intertemporal substitution is $(1-\gamma)^{-1}$. The individual flow budget constraint is written as

$$dk + c\,dt + db = b\,r\,dt + k\,dR_k \qquad (8.131a)$$

where income from capital and bond holdings is spent in either consumption or investment in capital and bonds. Using expressions (8.126), (8.127a,b), (8.129a,b) yields the evolution of individual wealth as

$$dw = \{ w\,[n_k (A - \delta) + (1- n_k)\, r - (c/w)]\}\, dt + w\, n_k A\, dy \qquad (8.131b)$$

where $n_k A\, dy$ is the stochastic component of individual wealth.

Each individual solves a stochastic control problem as in (8.76), repeated here

$$Max\ E_o \int_0^{\infty} u(c)\, exp(-\beta t)\, dt$$

subject to the accumulation equation (8.131b), by choice of consumption to wealth ratio, (c/w), and the share of capital in the portfolio, n_k, given $r>0$ and initial individual wealth, where utility is in (8.130), and $\beta >0$ is the individual subjective rate of time preference common across all individuals.

A value function for (8.76) takes the form

$$V(w,t) = exp(-\beta t)\, X(w) \qquad (8.132)$$

where X is a twice continuously differentiable function. Let the differential generator of the value function be defined as

$$L(V) \equiv (\partial V/\partial t) + \{w\,[n_k (A - \delta) + (1- n_k)\, r - (c/w)]\}\,(\partial V/\partial w) + (1/2)\, A^2 n_k^2 \sigma_y^2 w^2\, (\partial^2 V/\partial w^2). \qquad (8.133)$$

Individual optimization consists of choices of $\{c/w, n_k\}$, that maximize the Lagrangean function

$$\mathcal{L} = exp(-\beta t)\,(w^{\gamma}/\gamma)\,[(c/w)]^{\gamma} + L(V(w,t)) \qquad (8.134)$$

with necessary first order conditions for $\{c/w, n_k\}$ given by

$$w^{\gamma}(c/w)^{\gamma-1} = w\,(dX/dw) \qquad (8.135a)$$

$$[(A-\delta)-r]\ w\,(dX/dw) + [cov(A\,n_k\,dy, A\,dy)/dt]\,w^2\,(d^2X/dw^2) = 0. \qquad (8.135b)$$

The value function must satisfy the Bellman functional equation:

$$\underset{\{c/w,\ n_k\}}{Max}\ \{\exp(-\beta t)\,(w^{\gamma}/\gamma)\,(c/w)^{\gamma} + L(V(w,t))\} = 0. \qquad (8.136)$$

Substituting for L into the Bellman equation and canceling $exp(-\beta\ t)$ yields

$$(w^{\gamma}/\gamma)\,(c/w)^{\gamma} - \beta X + \{w\,[n_k(A-\delta) + (1-n_k)\,r - (c/w)]\}\,(dX/dw) + (1/2)\,A^2\,n_k^2\sigma_y^2\,w^2\,(d^2X/dw^2) = 0. \qquad (8.137)$$

This is an ordinary differential equation in the function X. A solution for (8.137) can be found by a suitable guess for the function $X(w)$. We propose a guess of the form

$$X(w) = \kappa\, w^{\gamma} \qquad (8.138)$$

where κ is a constant to be determined. For a guess given by (8.138), we have

$$dX/dw = \kappa\gamma\ w^{\gamma-1} \qquad (8.138a)$$

$$(d^2X/dw^2) = \kappa\gamma\ (\gamma-1)\,w^{\gamma-2}. \qquad (8.138b)$$

Substituting into the first order condition for (c/w), (8.135a) we obtain

$$\kappa = (c/w)^{\gamma-1}/\gamma \qquad (8.138c)$$

or $(c/w) = (c/w)^{\gamma}/\gamma\kappa$. Substituting the latter into Bellman equation we obtain a solution for (c/w) as

$$c/w = (1-\gamma)^{-1}\{\beta - \gamma\,[n_k(A-\delta) + (1-n_k)\,r\,] - (1/2)\gamma(\gamma-1)\,A^2\,n_k^2\sigma_y^2\} \qquad (8.139a)$$

and from the first order condition (8.135b),

$$(A - r) = \delta + (1-\gamma)\,[\,cov(A\,n_k\,dy, A\,dy)]/dt. \qquad (8.139b)$$

The solutions above imply that:

$$n_k = [(A-\delta) - r\,]/\,(1-\gamma)\,A^2\sigma_y^2; \qquad (8.139c)$$

and the solution for the mean growth rate is

$$\psi = n_k (A - \delta) + (1-n_k)\, r - (c / w) =$$

$$E[dc/c]/dt = E[dw/w]/dt = E[dk/k]/dt \qquad (8.140a)$$

where the variance of the growth rate per unit of time is

$$E[(dw/w)^2]/dt = \sigma_\psi^{\ 2} = A^2\, n_k^{\ 2} \sigma_y^{\ 2} \qquad (8.140b)$$

Hence, equations (8.139a,b,c), (8.140a,b) yield equilibrium solutions for (c/w), n_k, mean growth ψ, and variance of growth $\sigma_\psi^{\ 2}$, given the exogenous moment, σ_y^2, the risk-free rate, $r>0$, and the *CRRA*, $(1-\gamma)$.

As in sections 8.14-8.15, for a solution along a stochastic balanced growth path with $dc/c=dw/w=dk/k$ stationary, and constant share $n_k>0$, the transversality condition takes the form

$$lim_{t\to\infty} E_0 [\, exp(-\beta t)\, w^{\gamma}] = 0. \qquad (8.141)$$

A solution for the evolution of wealth is similar to (8.93)-(8.97), where here we rewrite (8.97) as

$$w(t) = w(0)\, exp(\, [\, \psi - (1/2)\, A^2 n_k^{\ 2} \sigma_y^{\ 2}\,]\, t + \; A\, n_k\, y(t)\,) \qquad (8.142)$$

where $y(0)=0$ with probability one. Upon substitution of (8.142) into (8.141) yields

$$[w(0)^{\gamma}] lim_{t\to\infty} E_0\, [exp(\, \gamma\, [\, \psi - (1/2)\, A^2 n_k^{\ 2} \sigma_y^{\ 2}\,]\, t - \beta t + \gamma A\, n_k\, y(t)\,)] = 0. \qquad (8.143)$$

The expectation of the exponential operator is in this case

$$E_o\, [exp(\gamma\, [\, \psi - (1/2)\, A^2\, n_k^{\ 2} \sigma_y^{\ 2}\,]\, t - \beta t + \gamma A\, n_k\, y(t))] =$$

$$exp(\, [\gamma \psi - \beta + (1/2) \gamma (\gamma - 1)\, A^2 n_k^{\ 2} \sigma_y^{\ 2}\,]t) \qquad (8.144)$$

so that the limit converges to zero as long as

$$[\gamma \psi - \beta + (1/2) \gamma (\gamma - 1)\, A^2\, n_k^{\ 2}\, \sigma_y^{\ 2}\,] < 0 \qquad (8.145a)$$

which by (8.139a) implies

$$c/w > 0. \qquad (8.145b)$$

Hence, as long as $c/w > 0$, the guess is correct and the value function (8.132) is the appropriate choice that satisfies the transversality condition.

We can compute cumulative welfare and welfare changes in this case as well. The welfare criteria in (8.76) using the functional form (8.130) is:

$$\Omega_2 \equiv (1/\gamma)(c/w)^{\gamma} E_o \int_0^{\infty} w(t)^{\gamma} \exp(-\beta t)\, dt. \qquad (8.146)$$

Using the expressions (8.142)-(8.144) above, and defining $H_1 \equiv [\gamma\psi - \beta + (1/2)\gamma(\gamma-1) A^2 n_k^2 \sigma_y^2]$, the integral in (8.146) becomes:

$$w(0)^{\gamma} \int_0^{\infty} \exp(H_1 t)\, dt = -\, w(0)^{\gamma} / H_1, \; for \; H_1 < 0,$$

$$where \quad H_1 < 0 \Leftrightarrow (c/w) > 0$$

is the condition that satisfies the transversality condition as in (8.145a,b). Thus, the solution for cumulative welfare is

$$\Omega_2 = (1/\gamma)\,(c/w)^{\gamma}\, w(0)^{\gamma} / \{\beta - \gamma[\psi + (1/2)(\gamma-1) A^2 n_k^2 \sigma_y^2]\}. \qquad (8.147)$$

The measure of welfare changes is the equivalent variation measure. As before, we use (8.147) to find the change in $w(0)$ so that $d\Omega_2=0$. Totally differentiating (8.147), we obtain

$$d\Omega_2 = ([\,(c/w)^{\gamma-1}\, d(c/w)\, w(0)^{\gamma} + (c/w)^{\gamma}\, dw(0)\, w(0)^{\gamma-1}]\, \{\beta - \gamma[\psi + (1/2)(\gamma-1) A^2 n_k^2 \sigma_y^2]\} + [\gamma\, d\psi + \gamma(1/2)(\gamma-1)\, n_k^2 A^2 d\sigma_y^2 + \gamma(\gamma-1) n_k \times A^2 \sigma_y^2\, dn_k]\{(1/\gamma)(c/w)^{\gamma}\, w(0)^{\gamma}\}) / \{\beta - \gamma[\psi + (1/2)(\gamma-1) A^2 n_k^2 \sigma_y^2]\}^2 \qquad (8.148)$$

which may be appropriately solved for $dw(0)$ so that $d\Omega_2=0$. The result is

$$dw(0)|_{d\Omega 2=0} = -\, \{\Omega_2\, \gamma[\, d\psi + n_k(\gamma-1) A^2 ((1/2)\, n_k\, d\sigma_y^2 + \sigma_y^2\, dn_k)] - (c/w)^{\gamma-1} d(c/w)\} / (c/w)^{\gamma}\, w(0)^{\gamma-1} \qquad (8.149)$$

which may be appropriately evaluated for changes in the exogenous moment σ_y^2, or changes in risk aversion $(1-\gamma)$. In this case, we may use the model to gain insight on the role of shocks and relative risk aversion on mean growth, variance of growth, portfolio choice, consumption, and welfare.

Comparative statics and welfare analysis may be performed in a similar fashion as in sections 8.14-8.15. We make one remark about the comparative statics in this case. It is possible that an increase in the variance of the technology shock increases the mean growth of the economy with capital accumulation. This is because additional risk requires a higher rate of return on investment for the risk averse individual to be willing to bear risky capital. The higher rate of return increases the amount of investment increasing mean growth; see Notes on the Literature. We leave the computation of the comparative statics as an exercise to the reader.

Here, we extend the analysis to examine the issue of pricing assets and Zero-level pricing.

(a) Pricing Assets

All financial claims can be priced in this framework and are measurable with respect to the productivity shocks, dy. Suppose we let the price of a financial claim (stock), $S(t)$, evolve according to the usual Geometric Brownian motion, as in (8.15),

$$dS(t)/S(t) = r_S dt + dz_S \tag{8.150}$$

where r_S is the mean return and dz_S is a normally distributed temporally independent disturbance with mean zero, $E[dz_S]=0$, and constant variance per unit of time, $E[(dz_S)^2]=\sigma_S{}^2 dt$. Then, the price of the claim is consistent with the usual marginal rate of substitution (stochastic discount factor) pricing. The stochastic discount factor is defined over consumption growth and, for the utility function in (8.130), the asset pricing formula for an arbitrary time interval Δt takes the form

$$E\,[\,\{c(t+\Delta t)/c(t)\}^{-(1-\gamma)}\exp(-\beta\Delta t)\,\{S(t+\Delta t)/S(t)\}] = 1 + o(\Delta t) \tag{8.151}$$

where $o(\Delta t)$ indicates terms of order higher than Δt. The first term $\{c(t+\Delta t)/c(t)\}^{-(1-\gamma)} \times \exp(-\beta\Delta t)$ is the marginal rate of substitution between $(t,t+\Delta t)$, and the term $\{S(t+\Delta t)/S(t)\}$ is the capital gain between $(t,t+\Delta t)$ for carrying the asset. In order to determine the price of the claim, we expand the left-hand side of (8.151) according to Ito's Lemma to obtain

$$\begin{aligned} E\,[\,\{c(t+\Delta t)/c(t)\}^{-(1-\gamma)}\exp(-\beta\Delta t)\,\{S(t+\Delta t)/S(t)\}] &\approx 1 - \beta\Delta t + E[\Delta S(t)/S(t)] - \\ (1-\gamma)\,E[\Delta c(t)/c(t)] + (1/2)\,(1-\gamma)\,(2-\gamma)\,E[\{\Delta c(t)/c(t)\}^2] &- \\ (1-\gamma)\,E[\{\Delta c(t)/c(t)\}\,\{\Delta S(t)/S(t)\}] + o\,(\Delta t). & \end{aligned} \tag{8.152}$$

We can compute the expected values in the expansion using (8.140a,b), (8.150) to obtain

$$E[\Delta S(t)/S(t)] = r_S\Delta t$$

$$E[\Delta c(t)/c(t)] = \psi\Delta t$$

$$E[\{\Delta c(t)/c(t)\}^2] = A^2 n_k{}^2\sigma_y{}^2\Delta t$$

$$E[\{\Delta c(t)/c(t)\}\ \{\Delta S(t)/S(t)\}] = cov(A\ n_k dy,\ dz_S)\Delta t = A\ n_k \sigma_{yS}\ \Delta t$$

where $\sigma_{yS} \equiv cov(dy,\ dz_S)$ is the covariance between the productivity shock and the asset price. Substituting the expressions above into the expansion (8.152), collecting terms in Δt, and ignoring higher order terms, or $o(\Delta t) \approx 0$, yields

$$E\,[\,\{c(t+\Delta t)/c(t)\}^{-(1-\gamma)} \exp(-\beta\Delta t)\ \{S(t+\Delta t)/S(t)\}] \approx 1+\Delta t\ \{-\beta + r_S - (1-\gamma)\psi + (1/2)\ (1-\gamma)\ (2-\gamma)\ A^2 n_k^{\ 2}\sigma_y^2 - (1-\gamma)\ A\ n_k \sigma_{yS}\} \qquad (8.153)$$

We evaluate the following limit using (8.151) and ignoring higher order terms, $o(\Delta t) \approx 0$, to obtain

$$lim_{\Delta t \to 0}\ E\,[\,\{c(t+\Delta t)/c(t)\}^{-(1-\gamma)} \exp(-\beta\Delta t)\ \{S(t+\Delta t)/S(t)\}] \approx \ lim_{\Delta t \to 0}\,[1 + \Delta t\ \{-\beta + r_S - (1-\gamma)\ \psi + (1/2)\ (1-\gamma)\ (2-\gamma)\ A^2 n_k^{\ 2}\sigma_y^2 - (1-\gamma)\ A\ n_k \sigma_{yS}\}] \Rightarrow$$

$$-\beta + r_S - (1-\gamma)\psi + (1/2)\ (1-\gamma)\ (2-\gamma)\ A^2 n_k^{\ 2}\sigma_y^2 - (1-\gamma)\ A\ n_k \sigma_{yS} = 0. \qquad (8.154)$$

Then, the equilibrium mean return on the stock is obtained by solving (8.154) as

$$r_S = \beta - (1-\gamma)\ \psi - (1/2)\ (1-\gamma)\ (2-\gamma)\ A^2 n_k^{\ 2}\sigma_y^2 + (1-\gamma)\ A\ n_k \sigma_{yS} \qquad (8.155)$$

a function of the underlying parameters in the model, in particular of the covariance between the productivity shock and the disturbance in the stock price. The risk-free returns and excess returns may be computed as well, and we apply the Zero-level pricing method below.

(b) Zero-Level Pricing

The framework examined so far lends itself to the concept of pricing assets optimally. In particular, the shadow price of the asset is the one that makes the holder indifferent between holding it or not, the so-called Zero-level pricing. In the context of our model with production, the optimal "price" of the risk-free asset is obtained by evaluating (8.139c) at $n_k=1$, obtaining the implicit return on an asset that is uncorrelated with any other in the economy as

$$r\,|_{nk=1} = (A-\delta) - (1-\gamma)A^2\sigma_y^2. \qquad (8.156)$$

This yields a risk premium over the return on capital $r_k=A-\delta$, using (8.139b), as

$$(r_k - \delta) - r = (1-\gamma)A^2\sigma_y^2 = (1-\gamma)\,[\,cov(A\,dy,\,A\,dy)]/dt \qquad (8.157)$$

proportional to the covariance of the productivity shock and the wealth shock that are perfectly correlated, i.e. the risk in wealth is measurable only with respect to the productivity shock.

In turn, evaluating the equilibrium variables at $n_k=1$, yields a formula for the mean return on stock in (8.155) as

$$r_S\,|_{nk=1} = [(A-\delta) - (1/2)A^2\sigma_y^2] + (1-\gamma)\,A\,\sigma_{yS} \qquad (8.158)$$

and the risk premium for stock is, by (8.157)-(8.158)

$$r_S - r = (1-\gamma)\,A\,\sigma_{yS} = (1-\gamma)cov(A\,dy,\,dz_S)/dt \qquad (8.159)$$

proportional to the covariance of the productivity shock and the stochastic component of the stock price. Combining (8.157) and (8.159) we obtain a simple arbitrage pricing representation of assets as

$$\begin{aligned} r_S - r &= [cov(A\,dy,\,dz_S)/cov(A\,dy,\,A\,dy)]\,[(r_k - \delta) - r] \\ &= [cov(A\,dy,\,dz_S)/var(dw)]\,[(r_k - \delta) - r] \\ &= \beta\,[(r_k - \delta) - r] \qquad (8.160) \end{aligned}$$

where $\beta \equiv cov(Ady,dz_S)/var(dw)$, is the stock beta defined as the covariance of the productivity and the stock price divided by the variance of the wealth measured as the variance of the productivity shock. The simple asset pricing formula obtains, and at the optimal pricing scheme $n_k=1$, the interpretation is the usual that the premium on the stock is related to the premium on the market portfolio though the stock beta, see e.g. the SCAPM model in Chapter 2.

8.17 **Risk Aversion and Intertemporal Substitution**

In Chapter 8, we extended the basic *CRRA* preferences to the non-time-additive case that can distinguish between risk aversion and intertemporal substitution. Here, we extend the framework of section 8.16 with risky physical capital accumulation, to include nonlinear recursive preferences that can appropriately distinguish risk aversion from intertemporal substitution in a continuous time stochastic setting.

The approach we take is to consider a stochastic continuous time extension of the preference structure of section 7.22. As before, all agents are identical, there is only one good, and the good may be consumed or invested in physical capital, the variable factor input. The stochastic technology is given in (8.125) and return in (8.126). There is trade in risk-free bonds, denoted b, yielding the non-stochastic return, $r\, dt$, per unit of time, and wealth is in expressions (8.129a,b). In this case, we write the individual flow budget constraint from (8.213a,b) as the evolution of individual wealth

$$w(t+\Delta t) - w(t) =$$

$$\{ w(t)\, [n_k\, (A - \delta) + (1 - n_k)\, r - (c(t) / w\, (t)\,)]\}\, \Delta t + w(t)\, n_k\, A\, \Delta y \qquad (8.161)$$

for an arbitrary (small) time interval Δt, where $n_k\, A\, \Delta y$ is the i.i.d. stochastic component of individual wealth with $E[\Delta y]=0$, $var(\Delta y)=\sigma_y^2 \Delta t$. In this case, we adopt a formulation of welfare based on a close continuous time extension of the discrete time case studied in Chapter 7, formula (7.200). This formulation is also related to (8.76) and (1.110) in Chapter 1. This is the case where we deviate from the traditional VNM time-separable expected utility framework. In the alternative non-time-separable case, welfare is defined as

$$V(w(t)) = lim_{\Delta t \to 0}\, U\, (\, c(t)\, ,\, E_t\, [\, V(w(t+\Delta t)]\,)$$

$$= lim_{\Delta t \to 0}\, \{\, c(t)^{\rho}\, \Delta t + exp(-\beta \Delta t)\, (E_t\, [\, V(w(t+\Delta t))^{\gamma}\,])^{\rho / \gamma} \}^{(1/\rho)} \qquad (8.162)$$

where U is the aggregator function, $(1-\gamma) \geq 0$, $\gamma \neq 0$, is the coefficient of relative risk aversion *(CRRA)*, $1/(1-\rho) \geq 0$, $\rho \neq 0$, is the elasticity of intertemporal substitution *(EIS)*, and $\beta > 0$ is the subjective rate of time preference. For the time-separable VNM expected utility case we have the restriction in the parameter space $\gamma = \rho \Leftrightarrow CRRA = 1/EIS$, and we obtain the familiar VNM time-separable expected utility. When $\gamma \neq \rho \Leftrightarrow CRRA \neq 1/EIS$, we can appropriately distinguish the timeless concept of risk aversion from the dynamic concept of intertemporal substitution, e.g. section 7.22.

The problem facing an individual is to maximize expression (8.162) subject to the evolution of wealth (8.161), with $w(t)$ given, by choice of consumption, $c(t)$, and the share of the risky capital, n_k. In this case, the equivalent of Bellman's equation for this problem can be written as

$$V(w(t)) = lim_{\Delta t \to 0} \underset{\{c(t), n_k\}}{Max} \{c(t)^{\rho} \Delta t + exp(-\beta \Delta t)\, (E_t [\, V(w(t+\Delta t))^{\gamma}])^{\rho/\gamma} \}^{(1/\rho)} \tag{8.163}$$

subject to (8.161).

Dividing both sides by $V(w(t))$, we obtain

$$1 = lim_{\Delta t \to 0} \underset{\{c(t), n_k\}}{Max} \{[c(t)/V(w(t)]^{\rho} \Delta t + exp(-\beta \Delta t) \times (E_t [\{V(w(t+\Delta t))/V(w(t))\}^{\gamma}])^{\rho/\gamma} \}^{(1/\rho)}. \tag{8.164}$$

subject to (8.161).

In this case, given the form of the evolution of the state in (8.161) and the isoelastic function (8.162) we apply the guess-and-verify method with a guess for the value function

$$V(w(t)) = \kappa\, w(t) \tag{8.165a}$$

for κ a constant undetermined coefficient. We also guess that the consumption function is linear in wealth as

$$c(t) = \mu\, w(t) \tag{8.165b}$$

for μ a constant undetermined coefficient giving the marginal propensity to consume out of wealth. Substituting the guesses (8.175a,b) and the constraint (8.161) into Bellman's equation (8.164) yields the problem

$$1 = lim_{\Delta t \to 0} \underset{\{\mu, n_k\}}{Max} \{(\mu/\kappa)^{\rho} \Delta t + exp(-\beta \Delta t)\, (E_t [\{1 + [n_k (A - \delta) + (1 - n_k) r - \mu]\} \times \Delta t + n_k A\, \Delta y\}^{\gamma}])^{\rho/\gamma} \}^{(1/\rho)}. \tag{8.166}$$

In order to solve this problem, we take successive expansions of the term in brackets using Ito's Lemma. The solution could also be obtained using the usual methods used in this chapter and Chapter 1, see Notes on the Literature. We start with the expectation term in (8.166) as

$$E_t [\{1 + [n_k (A - \delta) + (1 - n_k)\, r - \mu]\} \Delta t + n_k A\, \Delta y\}^{\gamma}] \approx 1 + \gamma [n_k (A - \delta) + (1 - n_k)\, r - \mu + (1/2)(\gamma - 1)\, A^2 n_k^{\,2} \sigma_y^2]\, \Delta t + o(\Delta t) \tag{8.167}$$

where $o(\Delta t)$ indicates terms of order higher than Δt. Substituting (8.167) into (8.166) and ignoring higher order terms, or $o(\Delta t) \approx 0$, yields

$$1 = lim_{\Delta t \to 0} \underset{\{\mu, n_k\}}{Max} \{(\mu/\kappa)^{\rho} \Delta t + exp(-\beta \Delta t)\, (1 + \gamma [n_k (A - \delta) + (1 - n_k) r - \mu + (1/2)(\gamma - 1)\, A^2 n_k^{\,2} \sigma_y^2]\, \Delta t\,)^{\rho/\gamma} \}^{(1/\rho)}. \tag{8.168}$$

Next, we expand the term multiplying the exponential as

$$exp(-\beta\Delta t)\ (1+\gamma[n_k(A-\delta)+(1-n_k)r-\mu+(1/2)(\gamma-1)A^2 n_k{}^2\sigma_y^2]\Delta t)^{\rho/\gamma}$$
$$\approx\{-\beta+\rho[n_k(A-\delta)+(1-n_k)r-\mu+(1/2)(\gamma-1)A^2 n_k{}^2\sigma_y^2]\}\ \Delta t+o(\Delta t) \qquad (8.169)$$

and substitute into (8.168) to obtain, ignoring $o(\Delta t)(\approx 0)$,

$$1=\lim_{\Delta t\to 0}\underset{\{\mu,n_k\}}{Max}\ \{(\mu/\kappa)^{\rho}\Delta t+(-\beta+\rho[n_k(A-\delta)+(1-n_k)r-\mu+(1/2)(\gamma-1)A^2 n_k{}^2\sigma_y^2]\Delta t)\}^{(1/\rho)}. \qquad (8.170)$$

Then, we expand the term in brackets as

$$\{[(\mu/\kappa)^{\rho}-\beta+\rho[n_k(A-\delta)+(1-n_k)r-\mu+(1/2)(\gamma-1)A^2 n_k{}^2\sigma_y^2]]\Delta t\}^{(1/\rho)}$$
$$\approx 1+(1/\rho)[(\mu/\kappa)^{\rho}-\beta+\rho[n_k(A-\delta)+(1-n_k)r-\mu+(1/2)(\gamma-1)A^2 n_k{}^2\sigma_y^2]]\Delta t + o(\Delta t) \qquad (8.171)$$

and substitute into (8.170) ignoring $o(\Delta t)(\approx 0)$, to obtain

$$1=\lim_{\Delta t\to 0}\underset{\{\mu,n_k\}}{Max}\ \{1+(1/\rho)[(\mu/\kappa)^{\rho}-\beta+\rho[n_k(A-\delta)+(1-n_k)r-\mu+(1/2)(\gamma-1)A^2 n_k{}^2\sigma_y^2]]\Delta t\}. \qquad (8.172)$$

We then take the limit and obtain an operational version of Bellman's equation as

$$0=\underset{\{\mu,n_k\}}{Max}\{(\mu/\kappa)^{\rho}-\beta+\rho[n_k(A-\delta)+(1-n_k)r-\mu+(1/2)(\gamma-1)A^2 n_k{}^2\sigma_y^2]\}. \qquad (8.173)$$

It is straightforward to compute the necessary first order conditions for consumption/wealth ratio and the share of capital, $\{\mu,n_k\}$, as

$$\mu=\kappa^{\rho/(\rho-1)} \qquad (8.174a)$$

$$n_k=[(A-\delta)-r]/(1-\gamma)A^2\sigma_y^2. \qquad (8.174b)$$

The first result to notice is that the share of risky capital depends only upon the *CRRA* parameter $(1-\gamma)$ and not upon the *EIS* related to ρ. More importantly, expression (8.174b) is identical to (8.139c) for the case of time-separable VNM expected utility, or *CRRA=1/EIS*. The reason for this observationally equivalent result is that all underlying risk is i.i.d., and as shown in section 7.22-(ii) for the discrete time case, under i.i.d. risk the two cases are perfectly compatible, independently of the relationship between *CRRA* and *EIS*. In particular, under zero-level pricing, $n_k=1$, the risk-free interest rate is

$$r\,|_{nk=1} = (A-\delta) - (1-\gamma)A^2\sigma_y^{\,2}$$

as in (8.156) yielding a risk premium over the return on capital $r_k = A-\delta$, as

$$(r_k - \delta) - r = (1-\gamma)A^2\sigma_y^{\,2}$$

exactly as in (8.157). Hence, under i.i.d. risk, separating intertemporal substitution from risk aversion is irrelevant for the risk-free return and the premium over risky capital.

It remains for us to determine the pair of constants $\{\mu,\kappa\}$. This can be done by substituting the optimality condition (8.174a,b) into Bellman's equation (8.173) to obtain

$$\kappa = ([1/(1-\rho)]\ \{\beta - \rho\,[n_k\,(A-\delta) + (1-n_k)r + (1/2)(\gamma-1)A^2\,n_k^{\,2}\sigma_y^{\,2}\,]\}\,)^{(\rho-1)/\rho} \qquad (8.175a)$$

and thus from (8.174a)

$$c(t)/w(t) = \mu = [1/(1-\rho)]\ \{\beta - \rho\,[n_k\,(A-\delta) + (1-n_k)r + (1/2)(\gamma-1)A^2\,n_k^{\,2}\sigma_y^{\,2}\,]\}. \qquad (8.175b)$$

Hence, as in (8.140a,b), the solution for the mean growth rate is

$$\psi = n_k\,(A-\delta) + (1-n_k)\,r - \mu = E[dc/c]/dt = E[dw/w]/dt = E[dk/k]/dt \qquad (8.176a)$$

where the variance of the growth rate per unit of time is

$$E[(dw/w)^2]/dt = \sigma_\psi^{\,2} = A^2\,n_k^2\sigma_y^{\,2}. \qquad (8.176b)$$

In the case of time-separable VNM expected utility, $\gamma = \rho \Leftrightarrow CRRA = 1/EIS$, and we obtain the consumption to wealth ratio from (8.175b) identical to (8.139a), or

$$c/w = [1/(1-\gamma)]\{\beta - \gamma\,[n_k\,(A-\delta) + (1-n_k)\,r + (1/2)\,(\gamma-1)\,A^2\,n_k^2\sigma_y^{\,2}]\}.$$

When $\gamma \neq \rho \Leftrightarrow CRRA \neq 1/EIS$, the effect of risk aversion on the marginal propensity to consume, given the *EIS*, is

$$\partial\mu\,/\,\partial(1-\gamma) = (1/2)\ [\rho/(1-\rho)]\ n_k^{\,2} \lesseqgtr 0 \quad \Leftrightarrow \quad \rho = 1 - (1/EIS) \lesseqgtr 0 \qquad (8.177)$$

and for low (high) *EIS*, it implies $\rho < 0\ (>0)$, and $\partial\mu\,/\,\partial(1-\gamma) < (>) 0$. When $EIS = 1$, then $\rho = 0$ and risk aversion does not impact upon the marginal propensity to consume out of wealth, $\partial\mu\,/\,\partial\,(1-\gamma) = 0$.

The case presented here is restricted to the probability distribution of risk to be i.i.d. and the differences with the cases where $CRRA = 1/EIS$ in terms of asset returns are not dramatic. Welfare analysis can also be performed in the same fashion as in the previous sections, see Notes on the Literature.

8.18 Summary II

In sections 8.13-8.17, we studied the solution of simple dynamic equilibrium stochastic models in continuous time where risk is governed by Wiener processes. This class of models provides a useful framework for the analysis of the effects of risk on consumption, growth, portfolio choice and welfare. We briefly extended to the case where risk aversion and intertemporal substitution are separate showing that the effect on asset returns under the particular i.i.d. distributional assumptions are negligible.

Problems

1. Simulation with Lognormal price distribution: Let the natural logarithm of the price of a stock, *log S,* follow the geometric stochastic process

$$d \log S = v\, dt + \sigma\, dX$$

where dX is a standard Wiener Process (Brownian Motion), $dX = \epsilon(t)(dt)^{1/2}$, and $\epsilon(t) \sim N(0,1)$ is a standard Normal random variable. A discretized form for the geometric process above is:

$$\log S(t_{k+1}) - \log S(t_k) = v\,\Delta t + \sigma\, \epsilon(t)\, (\Delta t)^{1/2}, \quad k=0,1,2...$$

with a particular price for each t_{k+1} given by

$$S(t_{k+1}) = S(t_k)\, \exp[\, v\,\Delta t + \sigma\, \epsilon(t)\, (\Delta t)^{1/2}], \quad k=0,1,2...$$

given $S(t_0)$. If historically, $\log S(t_{k+1}) / S(t_k) \sim N(v, \sigma^2)$, use the following parameters in your simulation:

$\Delta t = 1/250$, *i.e. let the time interval be a day in 250 yearly business days;*

$$S(t_0) = \$35;$$

$$v = 0.15;$$

$$\sigma^2 = (0.30)^2.$$

Let $k=0,1,2...250$, then draw *250* observations from the Random number generator in Excel for the random variable $\epsilon(t)$, and construct a series for $S(t_{k+1})$ from $S(t_0)$. Plot the price of the stock over time and compute and plot a histogram (use the histogram command in Excel) for the distribution of the price. Explain your findings.

2. Use the Black and Scholes (BS) formula to evaluate the price of an European Call Option on a stock with current price *\$25*, exercise price *\$25*, risk-free interest rate *6%*, standard deviation of the stock *0.30*. The option has expiration date of *6* months ($T=6/12=0.5$, from $t=0$). Construct a graph of the value you found, $C(S,t)$, as a function of a grid of S above and below the current price (for example: $S=15,17.5,20,22.5,25,27,5,30,32.5,35$). Explain your finding. [Note: Use

the NormSDist(.) in Excel to find the normal cumulative distribution function values in the Black-Scholes formula].

3. Compute the comparative statics on welfare using the framework of section 8.15.

4. Compute the comparative statics for the model in section 8.16.

Notes on the Literature

The material in section 8.2 is based on simple discrete time series models. Analysis of stochastic difference equations of this type may be found in Sargent (1976), Hamilton (1994), and Ljungqvist and Sargent (2000); and random walks in discrete time are also discussed in Hamilton (1994) and Ermini (2003).

The Central Limit theorem may be found in any mathematical statistics book, for example Hogg and Craig (1978). In section 8.3, the lognormal distribution is found in Aitchison and Brown (1957), Hogg and Craig (1978), Dixit and Pindyck (1994), Chapters 1 and 2, Benninga (1997), and Luenberger (1998), Chapter 11.

Section 8.4 covers Wiener processes. Study of continuous time stochastic processes goes back at least to Einstein (1956). Simple and accessible expositions on Wiener processes may be found in Luenberger (1998), Dixit and Pindyck (1994), and Wilmott, Howison and Dewynne (1995). A more advanced treatment is in Karatzas and Shevre (1991), and also Karatzas and Shreve (1998) and Musiela and Rutkowski (1998).

Sections 8.6 and 8.7 cover Ito's Lemma which is available in Luenberger (1998), Dixit and Pindyck (1994), Wilmott, Howison and Dewynne (1995), Karatzas and Shevre (1991), and Malliaris and Brock (1982) with solutions for stochastic differential equations found in Oksendal (1985). The simple application to asset prices and elimination of risk is found in Wilmott, Howison and Dewynne (1995) and Nefcti (1996).

Options are covered in various finance textbooks. At the introductory level there is Bodie et al (1996) and Bodie and Merton (2000). A medium level and very clear presentation is in Luenberger (1998) and also in Wilmott, Howison and Dewynne (1995), Hull (1993), Nefcti (1996) and Jarrow and Turnbull (2000). More advanced presentations include Duffie (1988, 1996) and Ingersoll (1987).

Sections 8.7 and 8.11 are about the Black-Scholes formula due to Black and Scholes (1973) and Merton (1973). Applications and references on the Black-Scholes analysis and options are surveyed in

Merton (1998). A useful and simple computational approach with Excel is provided in Benninga (1997), and Shaw (1998) provides a computational approach using Mathematica.

Sections 8.13-8.16 present the equilibrium approach pioneered by Merton (1969, 1975), also in the collection Merton (1990). The methodology for the solution of the general equilibrium model based on the stochastic control problem follows Turnovsky (2000a), with material also available in Dixit and Pindyck (1994), see Chapter 1.

The welfare analysis with equivalent variation in a macroeconomics context is due to Lucas (1987), and in a continuos time setting follows Turnovsky (2000b). Alvarez and Jermann (2000) propose an alternative method using asset prices.

The growth framework with capital accumulation is in Merton (1975) with many examples in Malliaris and Brock (1982). Becker and Zilcha (1997) is a general framework of Ramsey equilibrium under uncertainty; and Back and Pliska (1987) is an interesting application regarding information issues. Chow (1997) presents the material using Lagrange Multiplier techniques. An application to international economics is in Obstfeld (1994), Devereux and Saito (1997), and Turnovsky (1997); and to monetary and fiscal policy in Grinols and Turnovsky (1998), Turnovsky (2000b), Rebelo and Xie (1999), and Stulz (1986). Basak (1999) presents an equilibrium model with human capital, Basak (2000) examines asset pricing with frictions, Kyle and Xiong (2000) is a recent application to financial contagion, Wang (1993) is an application of the methodology under asymmetric information, and Turnovsky and Bianconi (2001) is an application with uninsurable idiosyncratic risk.

In particular, results showing that volatility may increase mean growth are shown in Obstfeld (1994) and Turnovsky (2000a,b). Results relating to the consumption function are original to Sandmo (1970), and the precautionary saving motive derived in the consumption function is discussed in Kimball (1990), Hubbard, Skinner and Zeldes (1994), and Carroll and Kimball (1996).

The zero-level pricing method is discussed in Luenberger (1998), asset pricing in Saito (1998), and the beta representation in this framework is found in Turnovsky (2000a). The manuscript by Campbell and Viceira (2000), Chapter 5, present several applications of stochastic calculus to financial markets analysis, and Duffie (2002) is an excellent survey. Ait-Sahalia (2000) presents methodology to uncover continuous time stochastic models from discrete data; and Lo (2000) provides a useful survey of problems in finance. Dumas (1989) presents an application to international financial markets, and Epstein and Miao (2003) extend to the case of Knightian uncertainty.

The final section 8.17 is based on Svensson (1989), which is a direct continuous time extension of Epstein and Zin (1989). There are several important applications of this form of separation between risk aversion and intertemporal substitution and other methods of analysis as well. In the vein of Svensson's (1988) contribution, Obstfeld (1984) focuses on growth and welfare effects and Giuliano and Turnovsky (2003) is an application in open economy macroeconomics. Duffie and Epstein (1992a,b) propose a different approach that is used in Campbell and Viceira (2000), and Fisher and Gilles (1999).

References

Aitchison, John and James A. C. Brown (1957) *The Lognormal Distribution, with Special Reference to its Uses in Economics.* Cambridge University Press, Cambridge, UK.

Ait-Sahalia, Yacine (2000) "Telling from Discrete Time Data Whether the Underlying Continuous-Time Models is a Diffusion." Working paper, Department of Economics, Princeton University.

Alvarez, Fernando and Urban J. Jermann (2000) "Using Asset Prices to Measure the Cost of Business Cycles." Working paper, University of Chicago and University of Pennsylvania, May.

Back, Kerry and Stanley Pliska (1987) "The Shadow Price of Information in Continuous Time Decision Problems." *Stochastics,* 22, 151-186.

Basak, Suleyman (2000) "A Model of Dynamic Equilibrium Asset Pricing with Heterogeneous Beliefs and Extraneous Risk." *Journal of Economic Dynamics and Control,* 24, 63-95.

Basak, Suleyman (1999) "On the Fluctuations in Consumption and Market Returns in the Presence of Labor and Human Capital: An Equilibrium Analysis." *Journal of Economic Dynamics and Control,* 23, 1029-1064.

Becker, Robert and Itzak Zilcha (1997) "Stationary Ramsey Equilibria Under Uncertainty." *Journal of Economic Theory,* 75, 122-140.

Benninga, Simon (1997) *Financial Modeling,* Second Edition. The MIT Press, Cambridge, MA.

Black, Fisher and Miron Scholes (1973) "The Pricing of Options and Corporate Liability." *Journal of Political Economy*, 81, 637-654.

Bodie, Zvi and Robert C. Merton (2000) *Finance*. Prentice Hall, New York, NY.

Bodie, Zvi, Alex Kane, Alan J. Marcus (1996) *Investments*. Irwin Publishers, Chicago, IL.

Campbell, John Y. and Luis M. Viceira (2000) *Strategic Asset Allocation.* Book Manuscript, Harvard University, November. (Published by Oxford University Press, 2002).

Carroll, Christopher and Miles Kimball (1996) "On the Concavity of the Consumption Function." *Econometrica*, 64, 981-992.

Chow, Gregory (1997) *Dynamic Economics: Optimization by the Lagrangian Method.* Oxford University Press, Oxford, UK.

Devereux, Michael and Makoto Saito (1997) "Growth and Risk Sharing with Incomplete International Asset Markets." *Journal of International Economics,* 42, 453-481.

Dixit, Avinash K and Robert S. Pindyck (1994) *Investment Under Uncertainty.* Princeton University Press, Princeton, NJ.

Duffie, Darrell (1988) *Security Markets: Stochastic Models.* Academic Press, Boston, MA.

Duffie, Darrell (1996) *Dynamic Asset Pricing Theory,* Second Edition. Princeton University Press, Princeton, NJ.

Duffie, Darrell (2002) "Intertemporal Asset Pricing Theory." Working paper, Graduate School of Business, Stanford University, July.

Duffie, Darrell and Larry Epstein (1992a) "Asset Pricing with Stochastic Differential Utility." *Review of Financial Studies,* 5, 411-436.

Duffie, Darrell and Larry Epstein (1992b) "Stochastic Differential Utility." *Econometrica,* 60, 353-394.

Dumas, Bernard (1989) "Two-Person Dynamic Equilibrium in the Capital Market." *Review of Financial Studies,* 2, 157-188.

Einstein, Albert (1956) *Investigation on the Theory of the Brownian Movement.* Dover Publications, New York, NY.

Epstein, Larry and Jianjun Miao (2003) "A Two-Person Dynamic Equilibrium under Ambiguity." *Journal of Economic Dynamics and Control,* 27, 1253-1288.

Epstein, Larry and Stanley Zin (1989) "Substitution, Risk Aversion and the Temporal Behavior of Consumption and Asset Returns: A Theoretical Framework." *Econometrica,* 57, 937-970.

Ermini, Luigi (2003) *Empirical Macroeconomic Modeling.* Oxford University Press, New York, NY (Forthcoming).

Fisher, Mark and Christian Gilles (1999) "Consumption and Asset Prices with Homothetic Recursive Preferences." Working paper, Federal Reserve Bank of Atlanta.

Giuliano, Paola and Stephen J. Turnovsky (2003) "Intertemporal Substitution, Risk Aversion and Economic Performance in Stochastically Growing Open Economy." Working paper, Department of Economics, University of Washington. Forthcoming, *Journal of International Money and Finance.*

Grinols, Earl and Stephen J. Turnovsky (1998) "Risk, Optimal Government Finance and Monetary Policies in a Growing Economy." *Economica,* 65, 401-428.

Hamilton, James (1994) *Time Series Analysis.* Princeton University Press, Princeton, NJ.

Hogg, Robert V. and Allen T. Craig (1978) *Introduction to Mathematical Statistics.* Macmillan Publishing Co., New York, NY.

Hubbard, R. Glenn, Jonathan Skinner and Stephen Zeldes (1994) "The Importance of Precautionary Motives in Explaining Individual and Aggregate Savings." *Carnegie-Rochester Conference Series on Public Policy,* 40, 59-126.

Hull, John C. (1993) *Options, Futures and Other Derivative Securities.* Prentice Hall, Englewood Cliffs, NJ.

Ingersoll, Jonathan E. (1987) *Theory of Financial Decision Making*. Rowman & Littlefield, Savage, MD.

Jarrow, Robert and Stephen Turnbull (2000) *Derivative Securities,* Second Edition. SouthWestern Publishing Co., New York, NY.

Karatzas, Ioannis and Stephen Shevre (1991) *Brownian Motion and Stochastic Calculus.* Springer-Verlag, New York, NY.

Karatzas, Ioannis and Stephen Shreve (1998) *Methods of Mathematical Finance.* Springer-Verlag, New York, NY.

Kimball, Miles (1990) "Precautionary Saving in the Small and in the Large." *Econometrica*, 58, 53-73.

Kyle, Albert S. and Wei Xiong (2000) "Contagion as a Wealth Effect." Working paper, Fuqua Business School, Duke University.

Ljungqvist, Lars and Thomas J. Sargent (2000) *Recursive Macroeconomic Theory.* The MIT Press, Cambridge, MA.

Lo, Andrew W. (2000) "Finance: A Selective Survey." Working paper, MIT Sloan School, January.

Lucas, Robert E. (1987) *Models of Business Cycles.* Yrjo Jahansson Lectures Series, Blackwell, London.

Luenberger, David (1998) *Investment Science.* Oxford University Press, Oxford, UK.

Malliaris, A. G. and William A. Brock (1982) *Stochastic Methods in Economics and Finance.* North-Holland, Amsterdam.

Merton, Robert C. (1969) "Lifetime Portfolio Selection Under Uncertainty: The Continuous Time Case." *Review of Economics and Statistics*, 51, 247-257.

Merton, Robert C. (1973) "Theory of Rational Option Pricing." *Bell Journal of Economics and Management Science*, 4, 141-183.

Merton, Robert C. (1975) "An Asymptotic Theory of Growth Under Uncertainty." *Review of Economic Studies*, 42, 375-393.

Merton, Robert C. (1990) *Continuous Time Finance.* Basil Blackwell. New York, NY.

Merton, Robert C. (1998) "Applications of Option-Pricing Theory: Twenty-Five Years Later." *American Economic Review,* 88, 323-349.

Musiela, Marek and Marek Rutkowsky (1998) *Martingale Methods in Financial Modelling,* Springer-Verlag, New York, NY.

Neftci, Salih N. (1996) *Introduction to the Mathematics of Financial Derivatives.* Academic Press, San Diego, CA.

Obstfeld, Maurice (1994) "Risk-Taking, Global Diversification, and Growth." *American Economic Review*, 84, 1310-1329.

Oksendal, B. (1985) *Stochastic Differential Equations*. Springer-Verlag, New York, NY.

Rebelo, Sergio and Danyang Xie (1999) "On the Optimality of Interest Rate Smoothing." *Journal of Monetary Economics,* 43, 263-282.

Saito, Makoto (1998) "A Simple Model of Incomplete Insurance: The Case of Permanent Shocks." *Journal of Economic Dynamics and Control*, 22, 763-778.

Sandmo, Agnar (1970) "The Effects of Uncertainty on Savings Decisions." *Review of Economic Studies*, 37, 353-360.

Sargent, Thomas J. (1976) *Macroeconomic Theory.* Academic Press, San Diego, CA.

Shaw, William (1998) *Modelling Financial Derivatives with Mathematica.* Cambridge University Press, Cambridge, UK.

Stulz, Rene (1986) "Asset Pricing and Expected Inflation." *Journal of Finance,* 41, 209-223.

Svensson, Lars E. O. (1989) "Portfolio Choice with Non-Expected Utility in Continuous Time." *Economics Letters*, 30, 313, 317.

Turnovsky, Stephen J. (1997) *International Macroeconomic Dynamics.* The MIT Press, Cambridge, MA.

Turnovsky, Stephen J. (2000a) *Methods of Macroeconomic Dynamics*, Second Edition. The MIT Press, Cambridge, MA.

Turnovsky, Stephen J. (2000b) "Government Policy in a Stochastic Growth Model with Elastic Labor Supply." *Journal of Public Economic Theory*, 2, 389-433.

Turnovsky, Stephen J. and Marcelo Bianconi (2001) "The Welfare Gains from Stabilization in a Stochastically Growing Economy with Idiosyncratic Shocks and Flexible Labor Supply." Working paper, University of Washington and Tufts University, December.

Wang, Jing (1993) "A Model of Intertemporal Asset Prices Under Asymmetric Information." *Review of Economic Studies,* 60, 249-282.

Wilmott, Paul, Sam Howison and Jeff Dewynne (1995) *The Mathematics of Financial Derivatives*. Cambridge University Press, Cambridge, UK.

Index

Abel, A. B. 440, 445
Abreu, D. 440, 445
actuarially fair price, 218, 265
adverse selection, 212, 216, 217, 227, 233, 234, 237, 238, 241, 243, 244, 263, 279, 280, 281, 294, 300, 307, 308, 315, 326, 327, 328, 332, 404, 437, 441
affine transformation, 429
agency, 212, 213, 214, 280
aggregate resources constraint, 284
aggregator function, 421, 422, 423, 424, 429
Aitchison, J. 507, 510
Ait-Sahalia, Y. 509, 510
Aiygari, S. R. 440, 445
Aizenman, J. 166, 167
Akerlof, G. 280, 282
Allen, F. 208, 210
Allen, R. G. 61, 64, 65
Altug, S. 441, 445
Alvarez, F. 442, 445, 508, 510
American option, 472
Araujo, A. 281, 282
arbitrage opportunity, 98, 100
Arrow security, 348, 349
Arrow, K. J. 114, 117, 132, 134, 136, 176, 208, 210, 344, 345, 346, 347, 348, 349, 351, 359, 366, 437, 439, 445
asset pricing, 172, 173, 176, 188, 332, 344, 354, 368, 369, 371, 375, 376, 377, 378, 382, 390, 395, 406, 421, 425, 428, 431, 432, 433, 434, 437, 440, 442, 443
Asset Returns and Moral Hazard, 270
asymmetric information under general equilibrium, 284
asymmetric information, 212, 216, 217, 219, 220, 223, 242, 244, 269, 274, 278, 281, 284, 285, 293, 294, 301, 308, 327, 329
Atkeson, A. 328, 330, 442, 443, 445
Attanasio, O. 444, 451
autocorrelation across states of nature, 363
autoregressive process, 453

Back, K. 508, 510
backshift or lag operator, 373
Backus, D. 439, 445
Balduzzi, P. 440, 445
Barberis, N. 440, 445
Barro, R. J. 166, 167, 327, 330, 441, 445
Basak, S. 508, 510
Battigalli, P. 280, 282
Bayes Theorem, 12, 257
Bayesian approach, 154
Becker, R. 508, 510
Bellman, R. 64, 62
Bellman's equation, 35, 37, 39, 53, 370, 384, 389, 424, 482
Benabou, R. 443, 446
Benninga, S. 104, 105, 507, 508, 510
Benveniste and Scheinkman formula, 40, 44, 62, 370, 384

Benveniste, L. 40, 44, 46, 62, 64, 370, 384, 390, 404, 425, 440, 446
Berliant, M. 328, 330
Bernardo, A. 440, 446
Berndt, E. 104, 105
Bernstein, P. 104, 105, 134, 136, 166, 167
Besley, T. 328, 330
beta, 94, 95, 499, 508
Bewley, T. 441, 446
Bianconi, M. 328, 330, 441, 446, 508, 513
bilateral relationship, 213
Binmore, K. 61, 64
binomial distribution, 272
Black, F. 104, 105, 452, 471, 473, 475, 476, 477, 478, 479, 505, 507, 510
Black-Scholes formula, 452, 471, 475, 476, 477, 478, 507
Blackwell's sufficient conditions, 37, 38, 39
Blanchard, O. J. 280, 282, 440, 446
Blume, L. 61, 62, 66, 439, 450
Bodie, Z. 328, 330, 439, 446, 507, 510
Bose, N. 442, 446
Breiman, L. 439, 446
Brock, W. A. 63, 65, 441, 442, 443, 446, 507, 508, 512
Brown, J. A. C. 507, 510
Brownian Motion, 48, 50, 52, 459, 505, 512
Brunnermeier, M. 440, 445
bubble, 373, 394
Bugrov, Y. S. 61, 64
Bultel, D. 134, 136

Call option, 471
Camerer, C. 166, 167
Campbell, J. Y. 104, 105, 134, 136, 439, 440, 441, 443, 446, 447, 508, 509, 510
capital gains, 370
capital market line, 92, 93, 94, 95, 96, 143, 144, 145, 147, 149, 151, 152
Carroll, C. 442, 445, 447, 508, 510
carrying costs, 335
Cartesian product, 18, 119
cash-in-advance constraint, 383, 385, 388
Central Limit Theorem, 153, 454
Certainty Equivalent, 99, 103, 112
certainty line, 120
Chamberlain, G. 104, 105
Chari, V. V. 328, 330
Chiang, A. 61, 64
Chiappori, P. 441, 447
Chow, G. 62, 63, 64, 508, 511
Chung, K. L. 61, 64
classical general equilibrium, 284, 294, 296, 297, 308, 327
classical statistical inference, 256
Cochrane, J. H. 104, 105, 208, 210, 439, 440, 442, 447
coefficient of relative risk aversion, 189, 193, 357, 374, 376, 452, 479, 487
comparative statics, 123, 134, 145, 149, 151, 486, 490, 491, 496, 506
complete markets, 183, 184, 186, 187, 188, 193, 195, 196, 198, 199, 203, 204, 207, 314, 317, 332, 344, 348, 349, 356, 361, 362, 388, 392, 398, 439
complex security, 351
conditional probability, 11, 12, 362, 366, 368
cone of diversification, 74, 75, 78
constrained Pareto optimal, 216
consumption bundles, 107, 117
consumption-based capital asset pricing (CCAPM) formula, 169, 185, 188
Consumption-Based Capital Asset Pricing Model, 173, 188, 199, 270, 354, 368
continuity (or Archimedean) axiom, 109
continuous random variables, 15
continuous state model, 454
contraction mapping theorem, 37
convenience yield, 335

Cooley, T. 439, 447, 450
costs of storage, 334
Cothren, R. 442, 446
Craig, A. T. 61, 65, 507, 511
credit rationing, 238, 239
Crooke, P. 61, 65
Cumulative Distribution Function, 13, 354, 466, 467, 476, 477, 506
cumulative welfare, 485, 489, 490, 495, 496
current account, 320, 321, 322

De la Fuente, A. 61, 64
De Meza, D. 280, 282
De, S. 328, 330
Deaton, A. 442, 447
Debreu, G. 134, 136, 208, 210, 327, 330, 331, 344, 345, 346, 347, 348, 349, 358, 360, 437, 439, 445, 446, 447
DELTA, 470, 471, 474, 475, 478
Den Haan, W. 442, 443, 447
depreciation of physical capital, 383, 391
derivative security, 333
Devereux, M. 508, 511
Dewynne, J. 507, 513
Diamond, P. 208, 209, 210
differential generator of the value function, 50, 482, 487, 493
discrete random variable, 12, 13, 15, 29
dividend-price ratio, 352, 370
Dixit, A. 61, 62, 63, 64, 104, 105, 134, 136, 280, 282, 507, 508, 511
Dow, J. 166, 167
Duffie, D. 208, 209, 210, 439, 447, 507, 509, 511
Dumas, B. 509, 511
dynamic general equilibrium, 332, 344, 382
Dynamic Programming, 33, 62, 64, 370, 384, 389, 424, 443, 479

Early versus late temporal resolution of risk, 422
Edgeworth complements, 147
efficient frontier, 76, 78, 80, 82, 86, 88, 89, 92, 102
Efficient Market hypothesis, 173, 208, 335, 440, 452
effort, 213, 246, 247, 248, 249, 250, 251, 252, 253, 254, 255, 256, 257, 258, 259, 260, 261, 262, 263, 264, 265, 266, 267, 268, 271, 272, 273, 274, 275, 277
Eichberger, J. 134, 136, 280, 282
eigenvalues, 364
eigenvectors, 364
Einstein, A. 507, 511
Ekeland, I. 62, 64
elasticity of intertemporal substitution, 421, 422, 430, 431, 434, 435, 437, 480, 487, 493, 500
elasticity of labor supply, 293
elementary (see also Arrow) security, 176
Enders, W. 62, 64
Envelope theorem, 26, 27, 81, 173, 174
Epstein, L. 134, 136, 166, 167, 421, 443, 447, 509, 511
Equity Premium puzzle, 380
equivalent portfolio, 177, 178, 203
equivalent variation, 485, 486, 490, 496, 508
Ermini, L. 507, 511
Ethier, W. 166, 167
Euler equation, 40, 41, 44, 45, 46, 47, 51, 52, 53, 54, 55, 370, 425, 427, 430, 432, 433, 435
Euler's theorem, 390
European Call option, 452, 471
event tree, 346, 358, 367, 439
excess return, 86, 92, 94, 95, 98, 332, 344, 355, 358, 375, 380, 410, 412, 440
expected utility theorem, 109, 119
exponential discount function, 34
exponential operator, 8, 58
exponential probability distribution, 269

exponential utility index, 130, 138, 140, 153
externality, 400

Faig, M. 441, 448
Fama, E. 104, 105, 208, 210, 440, 448
Farmer, R. 439, 448
Ferguson, B. 62, 63, 64
finite state space, 122, 126
First Fundamental Theorem of Welfare Economics, 194, 285, 298, 307, 360
First Order Approach, 246, 263, 265, 273
Fischer, S. 280, 282
Fisher equation, 332, 386, 387, 388, 441
Fisher, M. 509, 510, 511
forward contract, 333, 336
forward price, 333, 334, 335, 336, 338
Freixas, X. 280, 282
French, K. 328, 330
Friedman, J. 281, 282
Full Risk Sharing, 193, 195, 213, 260, 356, 362, 404, 408, 409, 410
Fundamental Theorem of Calculus, 8
Fundamental Theorem of Risk Bearing for Securities Markets, 186
Fundamental Theorem of Risk Bearing for State Contingent Claims, 186, 187, 199, 353
Fundamental Theorem of Risk Bearing, 219, 279
futures contracts, 332, 333

Gale, D. 208, 210
Gandolfo, G. 62, 64
Garber, P. 440, 448
Geanakopolos, J. 208, 210
Gilles, C. 509, 511
Giovannini, A. 443, 448
Giuliano, P. 509, 511
global minimum-variance point, 81
Glosten, L. R. 208, 210
Godfrey, L. 281, 282
Gollier, C. 62, 65, 134, 136
Goodfriend, M. 442, 448
Green, E. J. 441, 443, 448, 449
Green, J. 134, 137
Gresham's Law, 238
Grinols, E. 209, 210, 508, 511
gross capital flows, 321
guess-and-verify method, 482, 501
Gut, A. 61, 65

Hamilton, J. 507, 511
Hammond, P. 61, 62, 66
Hansen, G. D. 327, 328, 330
Hansen, L. P. 380, 440, 442, 443, 447, 448
Harper, I. 134, 136, 280, 282
Harris, M. 62, 65, 280, 282
Hart, O. 208, 210, 280, 281, 282, 283, 328, 330
Hayashi, F. 441, 448
He, H. 440
Heaton, J. 442, 448
hedging, 339
Helpman, E. 327, 331
Henderson, J. M. 134, 136
Hessian, 21, 22, 23, 24, 57
heterogeneity, 311, 312, 316
hidden knowledge, 215
Hirshleifer, J. 166, 167, 208, 210
Hogg, R. V. 61, 65, 507, 511
Holmstrom, B. 281, 283
Hornstein, A. 327, 328, 331
Howison, S. 507, 513
Huang, C.-F. 104, 105, 135, 136
Huang, C. J. 61, 65
Hubbard, R. G. 508, 511

idiosyncratic or non-systematic risk, 95, 128, 129, 271, 274, 275, 276, 277, 301, 322, 324, 408, 409, 411, 416, 440
idiosyncratic shock, 398, 401, 404, 408, 409, 411, 412, 413, 416, 419
imperfect hedge, 332, 341, 342
imperfect risk sharing, 413
Imrohoroglu, A. 328, 330

incentive compatibility constraints, 229, 230, 248, 249, 252, 253, 260, 262, 263, 268, 269, 294, 295, 296, 298, 303, 314, 315
incentive mechanism, 219
incentives, 213, 227, 234, 244, 246, 248, 252, 254, 255, 263, 265, 270, 274, 279, 280
income and substitution effects, 319
incomplete contracts, 214, 215, 280
incomplete markets, 185, 197, 198, 199, 204, 205, 207, 208, 318, 331, 349, 382, 394
indifference set, 142, 220
individual rationality constraints, 224, 229
inferior good, 125, 127, 128, 131
infinite horizon, 344, 369, 371
infinitesimal risk, 319
inflation, 386, 387, 388, 441
Ingersoll, J. 104, 105, 507, 512
insurance, 176, 177, 184, 187, 217, 218, 219, 246, 250, 254, 258, 263, 264, 265, 266, 274, 277, 279, 280, 300, 308, 330, 331, 339, 398, 404, 408, 409, 411, 416, 418, 420, 441
Integration, 6, 9, 10
intermediate capital good, 157, 158, 159
international portfolio diversification, 325, 328
intertemporal substitution and risk aversion, 332, 398
intertemporal substitution, 117, 134
Intriligator, M. 61, 65
intrinsic value, 472, 476, 478
Ito's Lemma, 48, 50, 51, 53, 461, 463, 464, 465, 469, 470, 474

Jacobian, 28
Jagannathan, R. 104, 105, 380, 442, 448
Jarrow, R. 507, 512
Jensen, M. 104, 105, 386, 407, 441, 448
Jensen's inequality, 17, 111, 198, 291, 386, 407, 463
Jermann, U. 328, 331, 508, 510
Jewitt, I. 281, 283
joint lognormal distribution, 433, 435
Jones, L. 281, 283
Jovanovic, B. 441, 448
Judd, K. L. 440, 446

Kahn, J. A. 281, 283
Kahneman, D. 166, 167
Kamien, M. 63, 65
Karatzas, I. 63, 65, 507, 512
Khan, A. 443, 449
Kim, Y. S. 441, 449
Kimball, M. 134, 136, 508, 510, 512
Klein, M. 61, 65
Knightian uncertainty, 154, 157, 159, 161, 162, 164, 166, 509
Kocherlakota, N. 440, 442, 443, 449
Krebs, T. 439, 449
Kreps, D. M. 134, 136, 443, 449
Krusell, P. 442, 449
Kuhn-Tucker Theory, 24, 26, 229, 303, 326
Kyle, A. 508, 512

Labadie, P. 441, 445
labor market, 308, 310, 328
labor supply indivisibility, 285, 296
Laffont, J. J. 280, 281, 282, 283, 327, 331, 441, 447, 449
lag or backshift operator, 33
Lagrange multiplier, 22, 25, 27
law of iterated mathematical expectations, 366, 373, 393, 461
law of large numbers, 403
Lehnert, A. 442, 449
lemons problem, 241
Leonard, D. 61, 65
LeRoy, S. 208, 211, 440, 449
Leung, C. 328, 329, 331
Lewis, K. 328, 331
L'Hopital's theorem, 430
Liebniz's formula, 9, 238

Likelihood Ratios, 256
linear independent, 18, 19
linear pricing, 169, 352
linear sharing rule, 261, 273, 274
Lintner, J. 104, 105
Litzenberger, R. 104, 105, 135, 136
Ljungqvist, L. 62, 65, 437, 439, 442, 449, 507, 512
Lo, A. 439, 447, 509, 512
Lognormal distribution, 455, 456, 463, 466, 468, 469, 507
long position, 67
long term principal-agent relationship, 401, 442
Long, C. 441, 442, 449, 450
Long, N.-V. 62, 65
loss aversion, 440
lottery ticket, 287, 289, 301, 315
lottery, 108, 109, 110, 111, 112, 113, 115, 118, 119, 132, 284, 289, 331
Lucas exchange economy, 369, 383, 437, 440
Lucas, R. E. 62, 66, 328, 329, 330, 331, 344, 369, 375, 376, 382, 383, 424, 437, 439, 440, 442, 443, 445, 448, 449, 450, 508, 512
Luenberger, D. 104, 106, 208, 211, 281, 283, 439, 449, 507, 508, 512

Machina, M. 134, 137, 166, 167
Macho-Stadler, I. 280, 281, 283
MacKinlay, A. C. 439, 447
Maggi, G. 280, 282
Magill, M. 208, 211
Malliaris, A. G. 63, 65, 507, 508, 512
Manuelli, R. 281, 283
Marcet, A. 328, 331, 441, 443, 449
marginal cost pricing, 202
marginal rate of substitution, 120, 142, 150, 152, 174, 175, 176, 186, 189, 199, 204, 220, 312, 313, 316, 318, 351, 352, 354, 355, 356, 368, 369, 371, 376, 380, 381, 382, 387, 394, 405, 406, 407, 425, 427, 431, 432, 442
Marimon, R. 441, 443, 449
market portfolio, 91, 92, 95
market value of the firm, 388, 392, 393, 394
Markov matrix, 363, 371
Markov process, 29, 42, 344, 363, 365, 368, 369, 374, 384, 437, 452
Markowitz formulation, 80
Markowitz, H. 80, 104, 106
Martimort, D. 280, 281, 283, 441, 449
Martingale process, 395, 462
Martingale property, 372
Mas-Colell, A. 62, 65, 134, 137, 327, 331, 439, 449
mathematical expectation, 13, 16
matrix of payoffs, 179, 181, 182, 183, 184, 185, 193, 347
maximum likelihood estimate, 256
McGrattan, E. 104, 105, 442, 450
mean-preserving spread, 410
mean-standard deviation space, 70, 72, 73, 74, 78, 86, 93, 144
mean-variance analysis, 67, 82, 104, 126, 128, 130, 142
mean-variance utility, 343
Meckling, W. 441
Mehra, R. 375, 381, 440, 442, 450
Mendenhall, W. 61, 65
Merton, R. C. 63, 65, 104, 106, 439, 446, 450, 507, 508, 510, 512
metric space, 35
Miao, J. 509, 511
Milgrom, P. 208, 210
Miller, M. H. 169, 208, 211, 332, 388, 392, 394, 398, 441, 450
minimum variance hedge, 340, 341, 342, 343, 437
minimum-variance set, 78, 80, 81, 82, 83, 86, 88, 89, 90, 93, 94, 102
Modest, D. 440, 448
Modigliani, F. 169, 208, 211, 332, 388, 392, 394, 398, 441, 450
Modigliani-Miller theorem, 166, 169, 208, 332, 388, 392, 394, 398, 441
monetary risk, 382

Monotone Likelihood Ratio Condition, 259
moral hazard, 212, 216, 217, 244, 245, 246, 250, 254, 255, 256, 259, 261, 265, 266, 269, 270, 274, 275, 276, 277, 278, 280, 281
Moreira, H. 281, 282
Morgenstern, O. 108, 118, 131, 132, 134, 137, 138, 153, 164, 166
Mossin, J. 104, 106
multi-period economy, 358, 366
multiple equilibrium paths, 373
Musiela, M. 63, 65, 507, 512
Mussa, M. 280, 283
Mutual Fund theorem, 148, 166
Myers, S. 441, 450
Myerson, R. 104, 106

natural logarithms, 6
Nefcti, S. 63, 65, 507
neoclassical theory of the firm, 169, 202, 212, 214, 215, 279, 388
Newman, A. 166, 167
Nikolsky, S. M. 61, 64
no-arbitrage theorem, 99
noise to signal ratio, 258, 259, 269
non-additive probabilities, 108, 154, 155, 159
non-convexities, 284, 293, 294, 328
non-satiation, 109, 110, 114, 126, 132
normal distribution, 16, 48, 138, 141
normal good, 124, 125
Novshek, W. 62, 65

Obstfeld, M. 208, 211, 439, 450, 508, 509, 512
Oh, S.-N. 441
Ok, E. 443, 446
Oksendal, B. 63, 65, 507, 512
One-Fund theorem, 84, 87, 88
optimal contract, 219, 228, 230, 231, 232, 233, 243, 246, 248, 249, 250, 251, 253, 255, 259, 260, 262, 263, 265, 269, 403, 404, 406, 409, 410, 411, 443
ordinary security, 170, 174, 176, 177, 178, 182, 186, 207, 346, 353, 367
Osterberg, W. 441, 450
Ouellete, P. 441

Paquet, A. T. 441, 450
Pareto Optimal, 194, 196, 202, 206, 215, 219, 279, 285, 292, 298, 299, 300, 302, 303, 304, 305, 306, 307, 308, 309, 324, 326, 360
partial insurance, 409
partial risk sharing, 246, 256, 259, 260, 261, 263, 266, 274, 275
Perez-Castrillo, J. D. 280, 281, 283
perfect hedge, 339, 340, 342, 437
Phelan, C. 441, 442, 450
Pindyck, R. 62, 63, 64, 134, 137, 507, 508, 511
Pischke, J.-S. 442, 450
Pithagoras theorem, 18
plain vanilla swap, 336, 338
Pliska, S. 440, 450
Plosser, C. 441, 449
Ponzi scheme, 396, 397
Porteus, E. 443, 449
portfolio weights, 69, 80, 91
Poterba, J. 328, 330
Pratt, J. 61, 66, 114, 117, 132, 134, 137
precautionary saving motive, 116, 134, 277, 490, 508
Prendergast, C. 280, 283
Prescott, E. C. 284, 294, 327, 328, 331, 375, 381, 437, 439, 440, 442, 448, 450
price discrimination, 219, 223, 226, 241, 280
pricing kernel, 351
principal-agent relationship, 216, 217, 219, 223, 234, 245
private information, 215, 216, 217, 218, 220, 221, 223, 226, 227, 232, 233, 234, 236, 241, 242, 244, 245, 246, 247, 249, 251, 252, 255, 263, 267, 271, 277, 294, 295, 296, 297, 298,

300, 301, 302, 308, 309, 310, 311, 314, 316, 317, 321, 324, 325, 327
probability density function, 15, 57, 60, 138, 139, 237, 239, 267, 268, 269, 354, 466
production efficiency, 204
property rights, 212, 215, 360, 361
prudence, 116, 125, 127, 134, 139, 190, 199
Put option, 472

quadratic form, 69, 80
quadratic utility, 107, 125, 126, 127, 128, 132, 135
quality, 213, 220, 222, 224, 225, 227, 231, 232, 241, 242, 243
Quandt, R. 134, 136
Quinzii, M. 208, 211

Rabin, M. 166, 167
random walk, 452, 453, 454, 455, 458, 459, 460
rank of a matrix, 184
rate of return dominance, 386, 441
Ravikumar, B. 441, 443, 449
Raviv, A. 280, 282
Rebelo, S. 443, 450, 508, 512
redundant asset, 353
redundant nodes, 359
Repo Rate, 334
Restoy, F. 443, 444, 450
retained earnings, 391
revelation mechanism, 243, 401, 402, 409
Rigotti, L. 166, 168
Riley, J. 166, 167, 208, 210, 280, 283
Rios-Rull, J. V. 208, 211, 442, 450
risk aversion and intertemporal substitution, 479, 489, 491, 499, 504, 509
risk aversion, 109, 110, 111, 112, 114, 115, 116, 117, 122, 123, 124, 125, 126, 127, 129, 130, 132, 134, 135, 139, 142, 145, 153, 154, 161, 164, 355, 356, 357, 358, 374, 376, 379, 380, 381, 421, 422, 423, 424, 427, 428, 429, 434, 435, 437, 440, 443, 444
risk loving, 114
risk neutral, 114, 142, 155, 157, 160
risk premium, 113, 114, 115, 128, 129, 276, 277, 278
risk-free asset, 84, 85, 86, 88, 91, 92, 93, 97, 99, 122, 123, 127, 139, 144, 149, 151, 152, 153, 172, 174, 175, 179, 188, 192, 193, 275, 353, 425, 458, 472, 474, 480, 486, 498
Risk-free Rate puzzle, 378
risk-neutral probabilities, 188
Robinson Crusoe, 284, 327
Rochet, J. C. 280, 282
Rogerson, R. 285, 327, 331
Rogerson, W. 281, 283
Rogoff, K. 208, 211, 439, 450
Romer, P. 166, 168
Rosen, S. 280, 283
Ross, S. 62, 66, 104, 106, 441, 450
Rothschild, M. 104, 105, 208, 210, 328, 331
Rubinfeld, D. 134, 137
Rutkowski, M. 63, 65, 507

saddle point, 22, 24, 403, 404
Saito, M. 508, 511, 512
Sala-i-Martin, X. 166, 167, 441, 445
Salanie, B. 280, 281, 283
sample space, 10, 11, 12, 13
Sandmo, A. 508, 512
Santos, M. 440, 445, 450
Sargent, T. J. 62, 65, 66, 166, 168, 437, 439, 440, 442, 443, 448, 449, 450, 507, 512
Sarte, P. 441, 450
Savage approach, 120
Savage, L. 134, 137
Scheinkman, J. A. 40, 44, 46, 62, 64, 208, 211, 390, 404, 425, 440, 446
Schmeidler, D. 166, 168
Scholes, M. 452, 471, 473, 475, 476, 477, 478, 479, 505, 507, 510

Schwartz, N. 63, 65
Second Fundamental Theorem of Welfare Economics, 194, 285, 294, 298, 299, 300, 302, 308, 309, 360
separating equilibrium, 233, 398, 404, 406
Sharpe Ratio, 381
Sharpe, W. 104, 106
Shaw, W. 508, 513
Shevre, S. 63, 65, 507, 512
Shimomura, K. 442, 449
short position, 67
short sales, 68, 69, 75, 76, 78, 80, 81, 85, 86, 92, 93
short selling, 334
signaling, 212, 241, 280, 308, 327
Silberberg, E. 62, 66
Simon, C. 61, 62, 66, 439, 450
Singleton, K. 443, 448
Skiadas, C. 166, 168
Skinner, J. 508, 511
Smith, A. 442, 449
spanning sets, 182
Spanning, 18
Spear, S. 441, 450
Spence, M. 280, 283
spot market, 333, 340
spot price, 333, 334, 335, 337, 341, 342, 345, 349
Srivastava, S. 441, 450
Stackelberg game, 216
Standard Debt Contract, 234
state contingent claim, 118, 129, 134, 176, 180, 181, 345, 349, 351
state contingent payoff, 176, 187
State Dependent Utility, 118, 119
state prices, 177, 178, 182, 183, 184, 185, 186, 188, 189, 190, 193, 195, 196, 197, 201, 203, 204, 205
state space, 176, 184, 186, 203, 204
states of nature, 118, 119, 120, 127, 138, 154, 166, 169, 170, 177, 178, 179, 182, 183, 184, 185, 186, 187, 188, 191, 196, 197, 199, 200, 201, 202, 206, 207, 214, 217, 251, 261, 263, 267
Static Capital Asset Pricing Model, (SCAM) 92, 93, 102
static optimization, 6, 20, 61
statistically independent, 11, 17
Stiglitz, J. 280, 283, 328, 331
stochastic difference equation, 29, 373, 385, 393, 395, 453
stochastic differential equation, 48, 65, 458, 465, 466, 474, 481, 484
stochastic dynamic optimization, 43, 49, 54
stochastic process, 332, 362, 439, 453, 459, 461, 462, 469, 474, 505
stochastic Ramsey model, 45, 293
stochastic technology, 203
stochastic total factor productivity, 205
stochastic trend, 453, 454, 455
Stockman, A. 440, 450
Stokey, N. 62, 66, 437, 439, 442, 445, 450
Stulz, R. 508, 513
Subgame Perfect Nash Equilibrium, 245
substitution (or independence) axiom, 109
substitution effect, 149, 151
Sudaram, R. K. 62, 66
sunk cost, 157, 158, 159, 160
supervisor-labor, 246, 251, 263, 266
Sussman, O. 442, 451
Svensson, L. E. O, 509, 513
swap, 336, 337, 338, 339, 437
Sydseater, K. 61, 62, 66
systems of financial markets, 169, 177, 181, 208

Takayama, A. 61, 62, 66
Tallarini, T. 443, 451
Tanis, E. A. 61, 65
Taylor's theorem, 464, 469
temporary incentive compatibility constraints, 402
terminal condition, 367

Thomas, J. 443, 447, 448, 449, 450, 451
Tirole, J. 281, 282, 283, 440, 451
Tobin, J. 104, 106, 166, 168,, 396, 438, 441, 448, 450, 451
Tobin's *Q*, 396, 438, 441
total factor productivity, 310
Townsend, R. M. 284, 294, 327, 328, 331, 441, 442, 449, 451
tradeoff between risk sharing and incentive, 214
transaction costs, 212, 214, 215
transition probabilities, 363, 368, 376
transitivity axiom, 109
transversality condition, 41, 47, 51, 53, 55, 56, 62, 371, 374, 377, 397, 425, 438, 483, 484, 489, 490, 495, 496
Turnbull, S. 507, 512
Turnovsky, S. J. 63, 66, 441, 446, 508, 509, 511, 513
Tversky, A. 166, 167
Two-Fund theorem, 82, 88, 93, 143

Ueda, M. 441, 448
uncertainty aversion, 154, 155, 156, 159, 160, 161, 162, 164
Unconditional Probabilities, 365
unemployment insurance, 285, 309, 315, 316, 317, 318, 322, 324, 325, 328

Van Wincoop, E. 439, 451
Varian, H. 61, 63, 66, 134, 137
vector difference equation, 363
Viceira, L. 104, 105, 134, 136, 441, 443, 447, 508, 509, 510
Vissing-Jorgensen, A. 444, 451
VNM expected utility, 108, 109, 126, 140, 141, 142, 147, 153
Von Neumann, J. 134, 137

Wang, J. 508, 513
Wang, T. 166, 167
Warner, J. 208, 211
wealth effect, 150, 151
Webb, D. 280, 282
Weber, M. 166, 167
Weil, P. 421, 440, 443, 444, 446, 448, 450, 451
Weiss, A. 280, 283
welfare weights, 194, 298, 305, 312, 360, 361
Werlang, S. R. C. 166, 167
Whinston, M. 134, 137
white noise, 460
Wiener process, 459, 460, 461, 462, 465, 469, 473, 505
Wilmott, P. 507, 513
Wilson, C. 328, 331
Woodford, M. 440, 446, 450
Worrall, T. 443, 451
worst case" scenario, 156

Xie, D. 508, 512
Xiong, W. 508, 512

Yao, T. 440, 445
Young's theorem, 21

Zeldes, S. 508, 511
Zero-level pricing, 270, 281, 497
Zilcha, I. 508, 510
Zin, S. 421, 443, 447, 509, 511